PROGRESS IN COLLOID & POLYMER SCIENCE

Editors: F. Kremer (Leipzig) and G. Lagaly (Kiel)

Volume 97 (1994)

Trends in Colloid and Interface Science VIII

Guest Editors:

R. H. Ottewill (Bristol) and
A. R. Rennie (Cambridge)

SPRINGER-VERLAG BERLIN HEIDELBERG GMBH

Die Deutsche Bibliothek – CIP-Einheitsaufnahme

Trends in colloid and interface science.

Früher begrenztes Werk in verschiedenen Ausg.
8 (1994)
(Progress in colloid & polymer science ; Vol. 97)
ISBN 978-3-662-16036-7   ISBN 978-3-7985-1673-1 (eBook)
DOI 10.1007/978-3-7985-1673-1
NE: GT

ISSN 0340-255 X

© 1994 by Springer-Verlag Berlin Heidelberg
Originally published by Dr. Dietrich Steinkopff Verlag
GmbH & Co. KG, Darmstadt in 1994
Softcover reprint of the hardcover 1st edition 1994

Chemistry editor: Dr. Maria Magdalene Nabbe; English editor: James C. Willis; Production: Holger Frey, Bärbel Flauaus.

Type-Setting: Macmillan Ltd., Bangalore, India

The VII Conference of the European Colloid and Interface Society (ECIS) was held at the School of Chemistry of the University of Bristol, England, from the 12th–16th September 1993. The Scientific Sessions were opened by the President, Professor Dominic Langevin and Plenary Lectures on the main themes of the Conference were given by H. Wennerström (Lund), H. N. W. Lekkerkerker (Utrecht), H. Hoffmann (Bayreuth), P. Botherel (Pessac), F. Candau (Strasbourg) and M. N. Jones (Manchester). In all, 35 papers were presented orally and 102 posters were displayed during the 4 days of the meeting. Lively discussions took place during the sessions and around the posters. Our thanks go to all of those who contributed to the cordial scientific atmosphere of the meeting. 159 people attended the meeting from 20 different countries including, from outside Europe, participants from Canada, Russia and Taiwan.

The members of the Scientific Committee for the meeting were: F. Candau, M. Corti, H. F. Eicke, H. Hoffmann, K. Holmberg, P. Laggner, A. R. Rennie and C. Solans, with R. H. Ottewill acting as Chairman and Th. F. Tadros as Co-Chairman.

Generous donations, which helped to finance the meeting were made by Academic Press Limited, Brookhaven Instruments, Camtel Services, Malvern Instruments Limited, J. Wiley and Sons Limited, University of Bristol and Zeneca PLC. Out warmest thanks to these organisations, as well as to a number of people who helped with the day to day organisation of the meeting; Paul Bartlett, Julia Cutler, Grahame Johnson, Phil Taylor and especially to Mrs Jean Proctor, who acted as Secretary for the Conference, before, during and after the meeting.

This volume contains a selection of the papers and posters presented at the meeting sub-divided into the six principle sessions: Applications of the Principles of Colloid Science, Suspensions, Surfactants, Emulsions and Rheology, Microemulsions and Bio-Colloids.

R. H. Ottewill (Bristol)
A. R. Rennie (Cambridge)

CONTENTS

Contents

Contents

Contents

Progr Colloid Polym Sci (1994) 97:1–5
© Steinkopff-Verlag 1994

B.A. Noskov
D.O. Grigoriev

Capillary wave propagation on solutions of surfactants: a new method for kinetic studies

Received: 16 September 1993
Accepted: 25 March 1994

B.A. Noskov (✉), · D.O. Grigoriev
Faculty of Chemistry
St. Petersburg State University
Universitetskiy prospekt 2
198904 St. Petersburg-Stariy Petergof,
Russia

Abstract It is well-known that adsorption kinetics influences the damping coefficient of capillary waves. However, only a few kinetic studies of adsorption based on this effect have been published in the literature. We show that the method of low-frequency capillary waves has certain advantages as compared with other methods traditionally used for this purpose. It can be applied also to the investigation of the micellization kinetics.

The damping coefficient and the wavelength of ripples on the surface of aqueous solutions of sodium dodecylsulfate (SDS) and dodecylpyridinium bromide (DPB) have been measured as a function of surfactant concentration. For both systems the adsorption at the air-water interface is essentially diffusion controlled. Above the Critical micelle concentration (CMC) a slight increase of the damping for solutions of SDS is connected with a corresponding change of the shear viscosity of the bulk phase. For DPB solutions the results can be explained only if the micellization kinetics are taken into account. The estimated values of the relaxation time for the slow stage of the micellization process are in agreement with the results obtained in the course of investigations of bulk phases.

Key words Capillary waves – air-water interface – adsorption kinetics – ionic surfactants – micellization kinetics

Introduction

In recent years experimental methods, based on the measurements of capillary wave characteristics, have become widespread in the physical chemistry of surface phenomena. It is well-known that the data on the damping of surface waves can be used in the studies of the structure of insoluble monolayers [1,2]. Although it was shown in the classical works that the method of low-frequency capillary wave permits evidence to be obtained concerning the adsorption mechanism of surfactants at the gas-liquid interface [3,4], only a few attempts have been made to apply this method to kinetic studies [5–9].

It is noteworthy that kinetic studies of the adsorption of surfactants at the liquid-gas interface have a long history [10]. The interest in this question diminished sometimes, but then revived again under the influence of new experimental evidence or applied problems.

There are two main difficulties hindering kinetic surface studies. Firstly, there is a very high sensitivity of the results to the presence of minor surface-active impurities. Secondly, there is a relatively high adsorption rate of conventional surfactants which forces measurements to be made at time intervals essentially less than a second. Experimental methods used in the case of small adsorption times (0.001 s–0.1 s) (the oscillating jet method, the maximum bubble pressure method, and some others) are usually connected with a strong external perturbation of the system under study. This leads to the appearance of non-linear hydrodynamic phenomena. The account of these makes the task more complicated [11].

In the case of the capillary wave method these external perturbations can be easily minimized. In this work, for example, the ratio of the amplitude to the wavelength is about 0.001. Then the system can be described by linear hydrodynamic equations and the boundary problem corresponding to the experimental conditions allows an exact solution to be obtained.

The main purposes of this work were to show the further applicability of the capillary wave method to kinetic studies at the water-air interface and to determine the adsorption mechanism of sodium dodecylsulfate (SDS) and dodecylpyridinium bromide (DPB), which were chosen as examples of anionic and cationic surfactants.

The results presented below, apparently can be considered also as a first experimental confirmation of the influence of micellization kinetics on the damping of the surface transverse waves. This means that the capillary wave method can be successfully applied for the investigation of various processes, not only in the surface layer, but in the bulk phase close to the interface as well.

Theory

The main result of an analytical investigation of linear surface waves propagating along a flat interface is a dispersion equation, connecting the wave characteristics (the wavelength λ, the damping coefficient α, the angular frequency ω) with the properties of the bulk phase (the density ρ, the shear viscosity μ) and the surface properties (the static surface tension σ and the complex dynamic surface elasticity $\tilde{\varepsilon}$) [4, 8]. For conventional soluble surfactants the surface shear viscosity is small and the dynamic surface elasticity is reduced to the dilational dynamic surface elasticity ε. This quantity can be calculated with the help of non-equilibrium thermodynamics [12]. In particular, if the concentration in the bulk phase can be considered as homogeneous, the quantity ε takes the following form [12]:

$$\varepsilon \equiv \frac{\delta\sigma}{\delta\ln S} = \left(\frac{\partial\sigma}{\partial\ln S}\right)_{\xi} - \sum_{i=1}^{N} \frac{\left(\frac{\partial\sigma}{\partial\ln S}\right)_{\xi} - \left(\frac{\partial\sigma}{\partial\ln S}\right)_{A_i}\xi_j}{1 + i\omega\tau_i},$$

(1)

where N is the number of normal relaxation processes in the surface layer, S is the area of a surface element, τ_i is the isothermal relaxation time of a normal reaction (process) i,

ξ_i is the chemical variable, A_i is the corresponding reaction affinity. The low index ξ_i means that the derivative corresponds to a non-equilibrium process but is taken at the equilibrium values of the thermodynamic variables, in particular at $\xi_i = 0$. The low index A_i indicates equilibrium conditions for a normal process i.

For solutions of surfactants the main relaxation process is adsorption (or desorption). In this case the condition of homogeneous concentration corresponds to pure adsorption kinetics (the largest value of the adsorption barrier). Then $(\partial\sigma/\partial\ln S)_A = 0$, $(\partial\sigma/\partial\ln S)_{\xi} = -\partial\sigma/\partial\ln\Gamma$, where Γ is the adsorption.

In a more general case it is necessary to take into account the diffusion of the surfactant to and from the surface and the expression for the dynamic surface elasticity becomes more complicated. In the case of an arbitrary number of surfactants this quantity takes a relatively simple form only at small rates of the surface coverage:

$$\varepsilon \equiv \left(\frac{\partial\sigma}{\partial\ln S}\right)_{\xi,C} - \sum_{i=1}^{N'} \frac{\left(\frac{\partial\sigma}{\partial\ln\Gamma_i}\right)_{\Gamma_j}}{1 + i\omega\tau_i + (1+i)\sqrt{\frac{\omega}{2D_i}}\left(\frac{\partial\Gamma_i}{\partial c_i}\right)_{A_i}},$$

(2)

where $\tau_i = 1/\alpha_i$ is the relaxation adsorption time, α is the kinetic coefficient of the desorption process, c_i is the subsurface concentration and N' is the number of surfactants in the system.

However, in the important particular case of a single surfactant it is possible to obtain an expression for ε which is justified for an arbitrary surface coverage and concentration, [12]:

$$\varepsilon \equiv \frac{\partial\sigma}{\partial\ln\Gamma} + \frac{\frac{\partial\sigma}{\partial\ln\Gamma}}{1 + i\omega\tau_i + (1+i)\sqrt{\frac{\omega}{2D}}\frac{\partial\Gamma}{\partial c}},$$

(3)

where the relaxation time is determined already by a more general relation $\tau = [\alpha + \beta c/\Gamma_\infty]^{-1}$, β is the kinetic coefficient of the adsorption, Γ_∞ is the maximum value of the adsorption.

Micellization makes the system even more complicated. However, progress in the kinetic theory of this phenomenon [13] allows an analytical investigation of this case also [14]. When the rate of the real adsorption process is high ($\beta \gg \omega D_1$) and the frequency is close to the inverse relaxation time of the slow stage of the micellization process τ_2^{-1}, the following relation can be obtained [14]:

$$\varepsilon = -\frac{\partial\sigma}{\partial\ln\Gamma}\left(1 - \frac{iD_1 t}{\omega}\left(\frac{\partial\Gamma}{\partial c}\right)^{-1}\right)^{-1},$$

(4)

where D_1 is the diffusion coefficient of monomers, and t^{-1} is the width of the diffusional boundary layer which contains a term connected with the micellization:

$$t^2 = k^2 - D_1^{-1}(i\omega - \tau_2^{-1}) \tag{5}$$

k is the complex wave number.

Note that if $\omega \gg \tau_2^{-1}$, one can neglect the influence of the micelles and this quantity is determined only by the diffusion of monomers.

Experimental

The damping coefficient and the length of the capillary waves were measured by means of an electromechanical technique, using a capacity probe (Fig. 1) [7]. The surface waves were created by a mechanical generator 1, made of thin capillary tubes. The generator oscillated perpendicularly to the liquid surface under the action of an electrodynamic vibrator 2, fed from a low-frequency electric generator 3. The surface of the liquid and a stainless steel pate 4 formed a dynamic air condenser. The alternate electric current flowing through the probe was proportional to the surface potentials of the liquid and of the metal, as well as to the vibration amplitude of the liquid surface. The current was amplified by an electrometer unit 5 and a selective unit 6 and was applied to an oscillograph 7. Another signal to the oscillograph was applied from the electric generator 3. The circuit included also a grounded platinum electrode 8 immersed in a rectangular silica trough 9. The mechanical generator moved relative to the wave probe and successive measurements of the amplitude and the phase of the electric signal allowed to determine the damping coefficient and the wavelength. The temperature was $20 \mp 0.5\,°C$.

The surface pressure of solutions of the anionic surfactant was measured by means of the Wilhelmy plate technique.

The viscosity of the solutions was determined by means of a capillary viscosimeter.

Both DPB and SDS were obtained from Reachim (Moscow). DPB was recrystallized four times from a mixture of ethanol and ethyl acetate. The purification of SDS required two stages. At first the powder of SDS was extracted by hexane in a Soxhlet extractor for about 6 h to remove the remains of high chain alcohols. Then the substance was recrystallized four times from a mixture of benzene and ethanol.

Fresh twice-distilled water was used to prepare the aqueous solutions for the experiments. An all-Pyrex still and alkaline permanganate were employed in the second stage of distillation.

In spite of the preliminary purification of the surfactant, the surface properties of the aqueous solutions of SDS still varied slowly with time, presumably because of the hydrolysis. Therefore, sweeping barriers and a capillary pipette attached to a vacuum pump were used to clean the liquid surface every 30 min during the course of the experiments [7]. During this time, the variations of the capillary wave parameters were within the accuracy of the measurements.

Results and discussion

Figure 2 shows the measured damping coefficient for several solutions of SDS as a function of the logarithm of the concentration for a frequency of 200 Hz. Corresponding results for solutions of DPB are represented in Fig. 3. It is

Fig. 2 The damping coefficient as a function of concentration for SDS solutions at a frequency of 200 Hz. The curves are calculated according to the dispersion equation and to Eqs. (3)–(7):$\tau = 0$ (dashed line), $\tau = 10^{-2}$ s (solid line). The arrow indicates the CMC value

Fig. 1 Schematic representation of the capillary wave generation and detection system

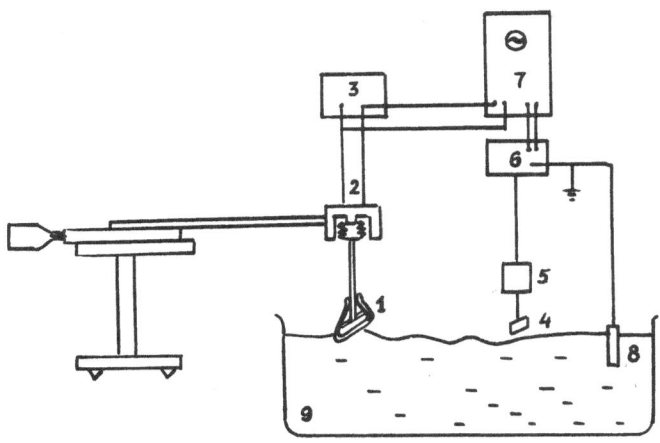

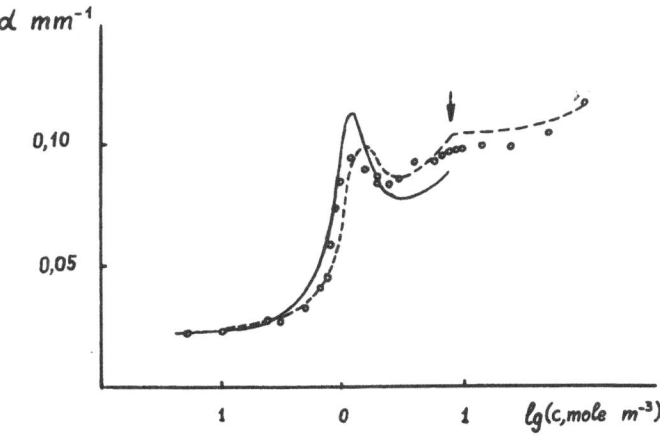

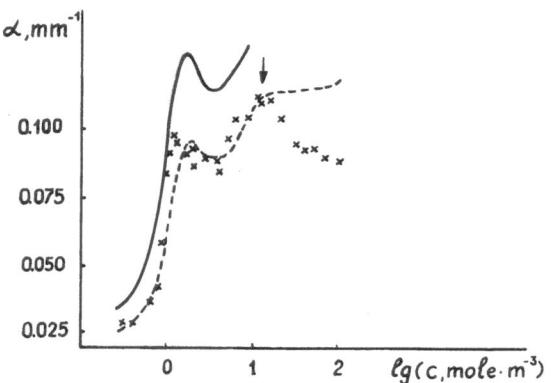

Fig. 3 The damping coefficient as a function of concentration for DPB solutions at a frequency of 200 Hz. The curves are calculated according to the dispersion equation and to Eqs. (3)–(7): $\tau = 0$ (dashed line), $\tau = 10^{-2}$ s (solid line). The arrow indicates the CMC value

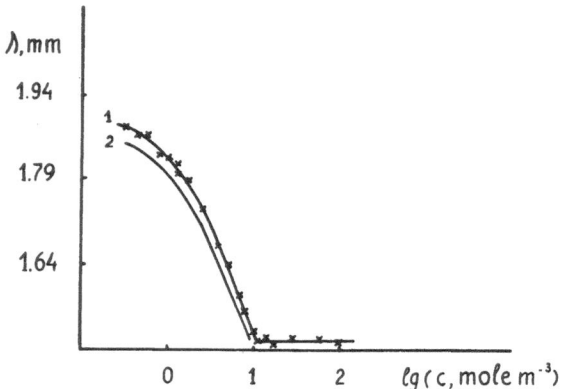

Fig. 4 The wavelength as a function of concentration for DPB solutions at 180 Hz. Curve 1 is a guide to the eye. Curve 2 is calculated according to the dispersion equation and to Eqs. (3)–(7) at $\tau = 0$

noteworthy that a relatively smooth dependence for these surfactants were obtained compared with for example the results for solutions of dodecylammonium chloride [9].

Above the CMC the directions of changes in the presented experimental curves (Figs. 2, 3) become opposite. Apparently, this can be connected with a difference between some properties of micelles for the investigated substances. However, a more careful consideration of this question has to precede an investigation of the adsorption kinetics in non-micellar solutions.

For this purpose, the damping coefficient was calculated using the dispersion equation of ripples [8, 9]. The values of the real and imaginary parts of the dynamic surface elasticity were calculated by Eq. (3). The derivatives $\partial\Gamma/\partial c$ and $\partial\sigma/\partial\ln\Gamma$ corresponding to the equilibrium state were calculated by means of analytical differentiation of the following equalities

$$\sigma = \sigma + nRT\Gamma_\infty\left[\ln\left(1 - \frac{\Gamma}{\Gamma_\infty}\right) + a\left(\frac{\Gamma}{\Gamma_\infty}\right)^2\right] \quad (6)$$

$$bc = \frac{\Gamma_\infty}{\Gamma_\infty - \Gamma}\exp(-2a\Gamma/\Gamma_\infty) \quad (7)$$

where R is the gas constant, T is the temperature, σ_0 is the surface tension of the solvent, $n = 2$ for an ionic surfactant if it is a 1:1 electrolyte. For solutions of SDS the parameters a, b, Γ_∞ were determined from the static surface tension isotherm and have the values 0.10; 0.26 m$^3 \cdot$ mole^{-1}; $3.8 \cdot 10^{-6}$ mole $\cdot$ m^{-2}, respectively.

For solutions of DPB the values of these parameters were determined from the data on the variation of wavelength. In Fig. 4 both the experimental curves and the results of calculations by the dispersion equation and relations (3), (6), (7) are represented for the frequency

180 Hz. The main features of the calculated dependencies and their agreement with the measured values do not change for other frequencies. Because the diffusion coefficient of a surfactant molecule is not sensitive to the nature of the polar group and the calculated wave characteristics are not very sensitive to the magnitude of this quantity, the value for SDS was used in all calculations ($D = 0.6 \cdot 10^{-9}$ m$^2 \cdot$ s^{-1}) [9]. The roots of the dispersion equation were determined by the Newton–Raphson technique. The best agreement was obtained for the following values of the parameters: $a = 1.0$, $b = 0.27$ m$^3 \cdot$ mole^{-1}, $\Gamma_\infty = 3.3 \cdot 10^{-6}$ mole $\cdot$ m^{-2}. One can see a small difference (about 0.7%) between the experimental and theoretical values of the wavelength on average (Fig. 4). This systematic deviation is usual for the applied measurement method [7, 9].

The method of calculations described above is strictly applicable only for the case of dilute solutions, where the mean activity coefficient equals unity. However, the calculations based on the Debye–Huckel theory justify this assumption, at least for the concentration range corresponding to the local maximum of the damping [8, 9].

The obtained values of the parameters a, b, Γ_∞ were used in further calculations of the damping coefficient. The broken lines in Figs. 2 and 3 represent the theoretical results corresponding to $\tau = 0$. The agreement between the calculated and experimental damping coefficients at concentrations below the CMC is reasonably good if one takes into account all the assumptions used in the calculations. The deviation of the adsorption relaxation time from zero leads to a more considerable difference exceeding the scatter of the experimental data, especially in the region of the local maximum of the α vs log c plot (Figs. 2, 3). Therefore, the adsorption process in the investigated systems is essentially diffusion controlled.

It is noteworthy that the difference between the two theoretical curves corresponding to the diffusion controlled kinetics and to the pure adsorption kinetics (Figs. 2, 3; solid lines) is more visible for solutions of DPB. This may be connected with lower surface activity for this substance. Really, when the surface activity increases, the contribution of the adsorption (desorption) to the relaxation of surface stresses diminishes (at a given frequency of perturbations) and the rheology of the adsorption film approaches a pure elastic behavior. Thus, for example, the influence of adsorption on the damping of capillary waves for solutions of cetyltrimethylammonium chloride is so slight that the conclusion about the adsorption mechanism is difficult to achieve [9]. From this point of view it is interesting to apply the method of low frequency capillary waves to surfactants with a lower surface activity. Then, because of higher sensitivity to the adsorption kinetics, one can expect a manifestation of more subtle effects.

The obtained results justify the application of Eq. (4) to concentrations exceeding the CMC. In this case, however, one has to take into account the alterations of the shear viscosity of the bulk phase. The broken lines in Figs. 2 and 3 at concentrations above the CMC represent the results of calculations obtained using the experimental values of the viscosity. The dynamic surface elasticity was determined assuming that $\tau_2 \Rightarrow \infty$, that is the size and the number of micelles do not change for a time comparable with the capillary wave period. This assumption is justified for SDS solutions (Fig. 2) where the theoretical and experimental curves agree reasonably well above the CMC also. It means that $\tau_2 \gg 0.001$ s.

For solutions of DPB at $c >$ CMC, one can see a deviation of the experimental data from the theoretical curve (Fig. 3). In this case the relaxation time τ_2 turns out to be comparable with the wave period. The micelle–monomer exchange contributes significantly to the transport of monomers to the interface during the adsorption process and from the interface in the case of desorption. The application of Eq. (4) leads to an estimation: $\tau_2 \sim 0.001$ s.

These values of the relaxation time for the slow stage of the micellization process τ_2 agree with more exact results obtained during a study of the bulk phases [15]: close to the CMC $\tau_2 = 6 \cdot 10^{-4}$ s for DPB and $\tau_2 = 0.55$ s for SDS. However, our estimates have been obtained on the basis of the measurements of surface properties only and by means of a more simple method.

In summary, as mentioned above, the method of low-frequency capillary waves proves to be convenient for the kinetic studies of adsorption and micellization in the case of conventional surfactants. Apparently, the accuracy of the results will be higher for surfactants with lower surface activity (containing 8 or 10 carbon atoms). For surfactants with higher surface activity the application of surface waves with lower frequency (the longitudinal surface waves, for example) is preferable.

Acknowledgement. This work was done with financial support from the Russian Foundation of Fundamental Research (project nr. 93-03-5478).

References

1. Earnshaw JC, Winch PJ (1990) J Phys: Condens Matter 2:8499–8516
2. Miyano K, Tamada K (1993) Langmuir 9:508–514
3. Davies JT, Vose RW (1965) Proc Roy Soc (London) A286:218–232
4. Lucassen J, Hansen RS (1967) J Colloid Interface Sci 23:319–328
5. Sasaki M, Yasunaga T, Tatsumoto T (1977) Bull Chem Soc Japan 50:858–861
6. Stenvot C, Langevin D (1988) Langmuir 4:1179–1183
7. Noskov BA, Vasiliev AA (1988) Kolloidn Zh 50:909–918
8. Noskov BA, Anikieva OA, Makarova NV (1990) Kolloidn Zh 52:1091–1100
9. Noskov BA (1993) Colloids Surfaces A: Physicochem Eng Aspects 71:99–104
10. Milner SR (1907) Phil Mag (Ser 6) 13:96–111
11. Noskov BA, Shchinova MA (1989) Kolloidn Zh 51:69–77
12. Noskov BA (1982) Kolloidn Zh 44:492–498
13. Anniansson GEA (1983) In: Wyn-Jones E, Gormally J (eds) Aggregation Processes in Solutions. Elsevier, Amsterdam, pp 70–93
14. Noskov BA (1989) Izv Akad Nauk SSSR, Mech zhydkoosty i gaza N2:105–114 (in Russian)
15. Hoffmann H, Nagel R, Platz G, Ulbricht W (1976) Colloid Polymer Sci 254:812–834

Progr Colloid Polym Sci (1994) 97:6–8
© Steinkopff-Verlag 1994

P. Dynarowicz

Interaction between molecules in adsorbed films at the air/water interface

Received: 16 September 1993
Accepted: 25 March 1994

Dr. P. Dynarowicz
Jagiellonian University
Faculty of Chemistry
Ingardena 3
30-060 Krakow
Poland

Abstract A regular solution theory for a system of molecules of different molecular sizes is used to derive parameters which can be related to the interaction between molecules of surface active compounds and water molecules in adsorbed films at the free water surface.

Key words Adsorbed films – parameters of interaction – the water/air interface – regular solution theory

Introduction

Today, many research activities are focused on the interaction between molecules in the surface films formed at the aqueous solution/air interface. The compounds which are of particular interest are surface active agents which adsorb from the bulk water to the interface and form a monolayer. The adsorbed monolayer can be formed by one kind of solute molecule only (here, we have a two-component system of surfactant molecules and water) or by a mixture of compounds of different surface activity. The mixed surfactants systems are very popular in applications as they show unique properties which are not expected for individual surfactants. Such nonideal behavior, due to molecular interaction between film-forming molecules, are of great theoretical and practical importance. The binary surfactants systems are most popular for scientific investigations. The mixed monolayer can be modeled either as a two-dimensional gas or a two-dimensional solution [1]. The two-dimensional solution approach is much closer to the surface region picture than the two-dimensional (2D) gas concept, since the 2D solution model accounts for the solvent as a third component. The equations describing the parameters of interaction in the mixed adsorbed films have been initially derived for mixed micelles [2–6] and then adapted for mixed surfac-

tants systems [7–12]. In this treatment, however, it has been assumed that the interface does not contain water molecules. This assumption limits the applicability of these equations to high surfactant concentrations in the region called "monolayer coverage". In this paper a statistical thermodynamics approach of regular solutions has been applied to derive parameters of interaction between solute molecules and water in the adsorbed films formed by a single surface active compound. This treatment can be easily extended to mixed monolayers of binary surfactants and then enables calculation of the parameters of interaction to be made not only between particular surface active agents, but also between surfactants and water molecules.

Theory

Let us first consider a two-component system, containing N_i molecules of the ith kind, $i = 1, 2$ (1 corresponds to a solute and 2 to solvent molecules), $v_i = V/N_i$ denotes their molecular volume. Let us assume that each molecule has "c" neighbouring molecules. In a random distribution, there is $(N_1 v_1/V)$ molecules of the first kind, and $(N_2 v_2/V)$ molecules of the second kind. The energy of interaction of one molecule of type 1 with its c_1 neighbours, which is also its effective energy of interaction with the whole system, is given by $c_1\phi_{11}N_1v_1/V + c_1\phi_{12}N_2v_2/V$, wherein V is the total volume of the system ($V = N_1v_1 + N_2v_2$). For the second kind of molecules we have: $c_2\phi_{12}N_1v_1/V + c_2\phi_{22}N_2v_2/V$. Here, ϕ_{ij} is the energy of interaction of a pair of molecules of type 1 (ϕ_{11}), type 2 (ϕ_{22}) or 1 and 2 (ϕ_{12}), respectively. As there are N_1 molecules of the first kind, their total energy of interaction with the system (U_1) can be expressed by:

$$U_1 = c_1\phi_{11}\frac{N_1^2 v_1}{V} + c_1\phi_{12}\frac{N_1 N_2 v_2}{V}. \tag{1}$$

Similarly, the energy of interaction of molecules "2" with the system (U_2) is given by:

$$U_2 = c_2\phi_{22}\frac{N_2^2 v_2}{V} + c_2\phi_{12}\frac{N_1 N_2 v_1}{V}. \tag{2}$$

Thus, the total potential energy in a two-component system, U, containing two different kinds of molecules can expressed by:

$$U = \frac{1}{2}(N_1 U_1 + N_2 U_2) = \frac{1}{2V}(c_1\phi_{11}N_1^2 v_1$$
$$+ c_1\phi_{12}N_1 N_2 v_2 + c_2\phi_{12}N_1 N_2 v_1$$
$$+ c_2\phi_{22}N_2^2 v_2) \tag{3}$$

(the factor 1/2 in the above equation is used so as not to count the interaction twice). After rearrangement it becomes:

$$U = \frac{1}{2V}[c_1\phi_{11}N_1(N_1 v_1 + N_2 v_2) + c_2\phi_{22}N_2(N_1 v_1$$
$$+ N_2 v_2) + N_1 N_2\{\phi_{12}(c_1 v_2 + c_2 v_1) - c_1\phi_{11}v_2$$
$$- c_2\phi_{22}v_1\}]. \tag{4}$$

For convenience let us define:

$$\Phi_{12} = \phi_{12}(c_1 v_2 + c_2 v_1) - c_1\phi_{11}v_2 - c_2\phi_{22}v_1 \tag{5}$$

Thus:

$$U = N_1(\tfrac{1}{2}c_1\phi_{11}) + N_2(\tfrac{1}{2}c_2\phi_{22}) + (N_1 N_2/2V)\Phi_{12}. \tag{6}$$

The partition function of this system, Z, has the form [13]:

$$Z = \left(\frac{q_1 Ve}{N}\right)^{N_1}\left(\frac{q_2 Ve}{N}\right)^{N_2} e^{-U/kT}, \tag{7}$$

where q_i terms include the molecular partition function for internal motions ($q_i = (2\pi m_i kT)^{3/2}/h^3$), and e is the base of the natural logarithms. Thus,

$$-kT\ln Z = -kT[N_1\ln(q_1 e) + N_1\ln V - N_1\ln N_1$$
$$+ N_2\ln(q_2 e) + N_2\ln V - N_2\ln N_2] + U \tag{8}$$

The chemical potential of the component 1, $\mu_{1(T,p,N_1,N_2)}$, can be calculated from the following equation:

$$\mu_1 = -kT\frac{\partial \ln Z}{\partial N_1} + p\frac{\partial V}{\partial N_1} = -kT\left[\ln(q_1 e)\right.$$
$$+ \ln V - \ln N_1 - 1 + (N_1 + N_2)\frac{v_1}{V}\right]$$
$$+ \frac{\partial U}{\partial N_1} + pv_1 = -kT\ln q_1 - kT\ln V$$
$$+ kT\ln N_1 - kT(N_1 + N_2)\frac{v_1}{V} + \frac{1}{2}c_1\phi_{11} +$$
$$+ \frac{1}{2}\frac{N_2^2}{V^2}\Phi_{12}v_2 + pv_1. \tag{9}$$

To calculate the standard chemical potential in the pure condensed state, μ_{01}, let us put $N_2 = 0$ in the above equation. Hence, we have:

$$\mu_{01} = \tfrac{1}{2}c_1\phi_{11} - kT\ln(q_1 v_1) - kT + p_0 v_1. \tag{10}$$

It is possible to express the chemical potential, μ_1, as:

$$\mu_1 = \mu_{01} + kT(\ln v_1) + kT\ln\left(\frac{V}{N_1}\right)$$
$$+ kT - kT(N_1 + N_2)\frac{v_1}{V}$$
$$+ \frac{1}{2}\Phi_{12}\frac{N_2^2}{V^2}v_2 + (p - p_0)v_1. \tag{11}$$

Under isobaric conditions, the last term (on the right) in the above equation disappears. Remembering that $V = N_1 v_1 + N_2 v_2$, and introducing $n_i = N_i/V$ (n_i/N_{Av} is the molar concentration), we obtain:

$$\mu_1 = \mu_{01} + kT\ln(n_1 v_1) + kT n_2(v_2 - v_1) + \tfrac{1}{2}\Phi_{12}n_2^2 v_2. \tag{12}$$

The above equation, which is derived for a bulk solution, can be transferred to the surface phase considering the interfacial region as a 2D space and including a surface force field term, $(\sigma - \sigma_{01})s_1$, where σ is surface tension of a solution, and σ_{01} is the surface tension of pure component 1:

$$\mu_1^{surf} = \mu_{01}^{surf} + kT\ln(n_1^{surf}s_1) + kT n_2^{surf}(s_2 - s_1)$$
$$+ \tfrac{1}{2}\Phi_{12}^{surf}(n_2^{surf})^2 s_2 - (\sigma - \sigma_{01})s_1. \tag{13}$$

In equilibrium $\mu_1 = \mu_1^{surf}$; in particular for pure component 1 $\mu_{01} = \mu_{01}^{surf}$. Thus $\mu_1 - \mu_{01} = \mu_1^{surf} - \mu_{01}^{surf}$. Hence,

$$kT\ln(n_1 v_1) + kT n_2(v_2 - v_1) + \tfrac{1}{2}\Phi_{12}n_2^2 v_2$$
$$= kT\ln(n_1^{surf}s_1) + kT n_2^{surf}(s_1 - s_1)$$
$$+ \tfrac{1}{2}\Phi_{12}^{surf}(n_2^{surf})^2 s_2 - (\sigma - \sigma_{01})s_1. \tag{14}$$

Let us assume that the number of neighbouring molecules surrounding molecule "i" (c_i) is proportional to the surface of "i". Considering variables v_i and s_i it is possible to write down that $c_i \sim v_i^{2/3}$ and $c_i^{surf} \sim s_i^{1/2}$. Hence, $c_1 = c_2(v_1/v_2)^{2/3}$ and $c_1^{surf} = c_2^{surf}(s_1/s_2)^{1/2}$.

This implies the following:

$$\Phi_{12} = c_2(v_1)^{2/3}[\phi_{12}(v_2^{1/3} + v_1^{1/3}) - v_2^{1/3}\phi_{11} - v_1^{1/3}\phi_{22}],$$

$$\tag{15}$$

and

$$\Phi_{12}^{surf} = c_2^{surf}(s_1)^{1/2}[\phi_{12}(s_2^{1/2} + s_1^{1/2}) - s_2^{1/2}\phi_{11} - s_1^{1/2}\phi_{22}].$$

$$\tag{16}$$

Based on the spherical model in the most dense face-centered packing, the molecular surface area, s, can be calculated from the expression $s = [M_w/(N_{Av}\rho)]^{2/3}$ [14], where N_{Av} is the Avogadro number, M_w is the molecular weight, and ρ is the density. Since the molecular volume $v = M_w(\rho N_{Av})$, therefore $v^{1/3} = [M_w/(N_{Av}\rho)]^{1/3}$. Similarly, molecular surface area $s = [M_w/(N_{Av}\rho)]^{2/3}$, and thus $s^{1/2} = [M_w/(N_{Av}\rho)]^{1/3}$. Hence, $s^{1/2} = v^{1/3}$. Applying the simplest model of the close-packed spherical molecules in both phases, we find that $c_i^{surf} = (6/12)c_i$. Taking all the above into consideration, the relation between Φ_{12} in the bulk phase and at the surface is seen to be:

$$\Phi_{12}^{surf}/\Phi_{12} = \tfrac{1}{2}(s_1)^{1/2}(v_1)^{2/3} \tag{17}$$

From the experimental data of chances in surface tension as a function of concentration, applying the Gibbs adsorption equation, it is possible to calculate both the number of molecules at the surface, n_i^{surf}, and to write the overdetermined system of equations of type (14) for both components. Solving them numerically, the parameter Φ_{12} can be obtained. Knowing the value of Φ_{12} in the bulk phase and basing on the relation expressed in Eq. (17), the parameter Φ_{12}^{surf}, connected with molecular interaction at the interface, can be determined,

According to Eq. (6), the expression for the increase in potential energy, ΔU, when a solution containing N_1 molecules of the first kind and N_2 molecules of the second kind is formed from components in their pure state, can be obtained from the equation:

$$\Delta U = U - U^0 = N_1(\tfrac{1}{2}c_1\phi_{11}) + N_2(\tfrac{1}{2}c_2\phi_{22})$$
$$+ (N_1N_2/2V)\Phi_{12} - [N_1(\tfrac{1}{2}c_1\phi_{11}) + N_2(\tfrac{1}{2}c_2\phi_{22})]$$
$$= (N_1N_2/2V)\Phi_{12} . \tag{18}$$

Since:

$$\frac{\Delta U}{\Delta N_1} = \frac{N_2}{2V}\Phi_{12} , \tag{19}$$

therefore, for diluted solutions ($N_2 \gg N_1$), the term ($N_2/2V$) can be considered as constant. As seen, ΔU (ΔH under isobaric conditions) is proportional to the parameter Φ_{12}. Thus, it is possible to verify the theory of obtaining the enthalpy of mixing from the values of Φ_{12} and to compare the results with experimental calorimetric data.

In the case of mixed adsorbed films containing two different surface active molecules, the treatment of two components presented above can be extended to that of three components. The analogous approach leads to an equation of type (14) which has the form (see ref. [15] for details):

$$kT\ln(n_1v_1) + kTn_2(v_2 - v_1) + kTn_3(v_3 - v_1)$$
$$+ \tfrac{1}{2}\Phi_{12}n_2^2v_2 + \tfrac{1}{2}\Phi_{13}n_3^2v_3 + \tfrac{1}{2}n_2n_3(\Phi_{12}v_3$$
$$+ \Phi_{13}v_2 + \Phi_{23}v_1) = kT\ln(n_1^{surf}s_1) + kTn_2^{surf}(s_2 - s_1)$$
$$+ kTn_3^{surf}(s_3 - s_1) + \tfrac{1}{2}\Phi_{12}^{surf}(n_2^{surf})^2s_2 + \tfrac{1}{2}\Phi_{13}^{surf}(n_3^{surf})^2s_3$$
$$+ \tfrac{1}{2}n_2^{surf}n_3^{surf}(\Phi_{12}^{surf}s_3 + \Phi_{13}^{surf}s_2 - \Phi_{23}^{surf}s_1)$$
$$- (\sigma - \sigma_{01})s_1 . \tag{20}$$

The values of parameters Φ_{ij} and Φ_{ij}^{surf}, obtained using simple solution theory, can be regarded as tentative only. Nevertheless, the obtained results allow one to draw some conclusions about the interactions which exist between molecules in solutions.

References

1. Lucassen-Reynders EH (1976) Progress Surf Membr Sci 10:253–351
2. Lange H (1953) Kolloid Z 131:96–103
3. Shinoda K (1954) J Phys Chem 58:541–544
4. Shinoda K, Hutchinson E (1962) J Phys Chem 66:577–582
5. Rubingh DN (1979) In: Mittal KL (ed) Solution Chemistry of Surfactants. Plenum Press, New York, pp 337–354
6. Holland PM (1986) Adv Colloid Interface Sci 26:111–121
7. Holland PM (1986) In: Scamehorn JF (ed) Phenomena in Mixed Surfactants Systems. ACS Symposium Series 311, American Chemical Society, Washington DC, pp 102–115
8. Ingram BT (1980) Colloid Polym Sci 258:191–193
9. Rosen MJ, Hua XY (1982) J Colloid Interface Sci 86:164–172
10. Hua XY, Rosen MJ (1982) J Colloid Interface Sci 87:469–477
11. Rosen MJ (1986) In: Scamehorn JF (ed) Phenomena in Mixed Surfactants Systems. ACS Symposium Series 311, American Chemical Society, Washington DC, pp. 144–162
12. Rosen MJ (1991) Langmuir 7:885–888
13. Moelwyn-Hughes EA (1961) Physical Chemistry. Pergamon Press, Oxford, p. 825
14. Young DM, Crowell AD (1962) Physical Adsorption of Gases. Butterworth, London, p. 226.
15. Dynarowicz P (1993) J Colloid Interface Sci 159:119–123

Progr Colloid Polym Sci (1994) 97:9–11
© Steinkopff-Verlag 1994

S. May

Position of the neutral surface in charged monolayers

Received: 16 September 1993
Accepted: 25 March 1994

Abstract Curvature elasticity of monolayers can be described in terms of elastic moduli. These moduli depend on the position of the neutral surface (surface of inextension during pure bending). Based on a combination of a mean field theory of chain packing and a head group contribution for amphiphilic molecules, the position of the neutral surface is determined. The electrostatic part of charged molecules is treated within the framework of the electric double layer. The neutral surface lies nearly in the middle of the region in which a nonzero pressure is acting. Its position is mainly determined by the mechanical pressure and not by the electrostatic contribution. Although the main part of the bending rigidity modulus comes from the chain region, the electrostatic contribution can be several $k_B T$.

Key words Neutral surface – bending rigidity – curvature elasticity – charged monolayers

S. May
Friedrich-Schiller-Universität Jena
Institut für Biochemie und Biophysik
Philosophenweg 12
07743 Jena, FRG

Introduction

In recent years the question of the influence of electrostatic surface charges on the bending rigidity has been theoretically investigated. A critical parameter in treating a charged monolayer within the Gouy–Chapman theory is the position of the neutral surface. In the present paper, a simple model is used to compare the chain and head group contributions to the bending rigidity modulus dependent on the position of the dividing surface within the monolayer. The electrostatic contribution in the case of surface charges is also taken into account. To approximate energy changes caused by deformations, the expansion of the free energy F up to quadratic order in terms of area changes $(a - a_0)$ and the main curvatures c_1 and c_2, or their linear combinations $c_+ = c_1 + c_2$ and $c_- = c_1 - c_2$,

is used

$$F(c_+, c_-, a) = \frac{1}{2} k \, a_0 (c_+ - c_0)^2 + \frac{1}{4} \bar{k} \, a_0 (c_+^2 - c_-^2)$$
$$+ \gamma(a - a_0) + \frac{1}{2} \lambda \frac{(a - a_0)^2}{a_0}$$
$$+ \tau c_+ (a - a_0), \tag{1}$$

where k is the bending rigidity modulus, $\bar{k}$ the modulus of Gaussian curvature, c_0 the spontaneous curvature, λ the stretching elastic modulus, γ the lateral tension, and τ the modulus of mixed deformation [1]. This formulation can be applied relative to each dividing surface, parallel to the boundary between water and the head groups. If the modulus of mixed deformation vanishes, the deformation of the monolayer is described in terms of the neutral

Fig. 1 Position of the
coordinate system within the
monolayer. The coordinate
origin lies in the dividing
surface with its x-axis
perpendicular to it. l_{hg} and l_c are
the widths of the polar and
chain region, respectively. The
distance between the dividing
surface and the polar region is
denoted by δ

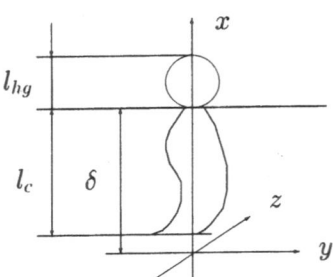

surface, i.e., bending and stretching are decoupled. Description of the internal structure of the monolayer is done using a coordinate system shown in Fig. 1. The coordinate origin lies in the dividing surface with its x-axis perpendicular to it. The widths of the polar and of the hydrophobic region are denoted by l_{hg} and l_c, respectively. The available cross-sectional area per chain relative to the dividing surface is denoted by a.

We assume the mechanical head group, chain, and electrostatic contributions to the free energy to be independent from each other $F = F_{hg} + F_c + F_{el}$. Here, we are interested in the bending rigidity k and the modulus of mixed deformation τ, both of which are additive: $k = k_{hg} + k_c + k_{el}$ and $\tau = \tau_{hg} + \tau_c + \tau_{el}$.

Chain contribution

A microscopic-level formulation for the chain contribution to the bending elasticity of monolayers in the fluid state has been developed by Szleifer et al. [2, 3]. There, the chain statistics were treated by a mean field (single chain) theory. The sole assumption of this theory is a uniform (liquid-like) density in the hydrophobic region made up of chain segments or solvent molecules in case of a "good" solvent (chain-chain and chain-solvent interactions are the same). The main outcome of the theory is a lateral pressure profile $\pi(x)$ forcing the chains to occupy, on average, that space by which the chains can be packed to an aggregate of a given geometry. After choosing the *rotational isomeric state model* [4] as an appropriate chain approximation, $\pi(x)$ could be evaluated numerically. It depends strongly on the given available cross-sectional area per chain a.

In this work, we use the same formalism with arbitrary position of the dividing surface. Derivatives of $\pi(x)$ with respect to c_+ and a of a flat monolayer allow the determination of k_c and τ_c

$$k_c = - \int_{\delta - l_c}^{\delta} \left(\frac{\partial \pi(x)}{\partial c_+} \right)_{0, 0, a_0} x\, dx, \tag{2}$$

$$\tau_c = - \int_{\delta - l_c}^{\delta} \left(\frac{\partial (a\pi(x))}{\partial a} \right)_{0, 0, a_0} x\, dx. \tag{3}$$

The derivatives $\left(\dfrac{\partial \pi(x)}{\partial c_+} \right)_{0, 0, a_0}$ and $\left(\dfrac{\partial (a\pi(x))}{\partial a} \right)_{0, 0, a_0}$ have to be taken for $c_+ = c_- = 0$ and at the area per molecule a_0. They can be determined from the pressure profile $\pi(x)$ of a flat monolayer using the *pyramid approximation* [2].

Head group contribution

Owing to the lack of a reliable model for the various head group interactions (steric repulsions, attraction due to hydrogen bonding, attraction caused by hydrophobic effects, electric repulsions between charged head groups, and dipolar repulsions) we are dependent on a simple phenomenological approach. Because the width of the head group region is usually small compared with the hydrophobic chain region, we neglect the explicit dependence of the free energy F_{hg} on the curvature and approximate the potential by a $\frac{1}{a(x)}$ dependence at all positions x

$$F_{hg} = \int_{\delta}^{\delta + l_{hg}} \pi(x) \frac{a_0^2}{a(x)} dx. \tag{4}$$

The head group pressure profile, representing the head group structure, can in the simplest case be chosen to be constant $\pi(x) \equiv \pi_{hg} = \dfrac{\Pi_{hg}}{l_{hg}}$ with $\delta \leq x \leq (\delta + l_{hg})$. Then, using the curvature dependence of the area $a(x) = a[c_+ x + \frac{1}{4}c_+^2 - c_-^2 x^2]$, we derive for the mechanical head group contribution of the bending rigidity modulus and the modulus of mixed deformation

$$k_{hg} = 2\Pi_{hg}\left(\delta^2 + \delta l_{hg} + \frac{1}{3} l_{hg}^2 \right), \quad \tau_{hg} = \Pi_{hg}\left(\delta + \frac{1}{2} l_{hg} \right). \tag{5}$$

Electrostatic contribution

A special part of the head group contribution to the elastic moduli is the electric part caused by net surface charges. Its contribution can be evaluated exactly in the framework of the Gouy–Chapman theory of the diffuse double layer using approximate solutions of the nonlinear Poisson–Boltzmann equation. We assume a 1:1 electrolyte and the surface charges directly attached to the head group-water boundary. Then, the distance between the surface charges and the dividing surface is $l_{hg} + \delta$. Generalizing the work of Lekkerkerker [5] to an arbitrary position of the dividing surface, expressions of the electrostatic parts of the bending rigidity k_{el} and the modulus of mixed

Progr Colloid Polym Sci (1994) 97:9–11
© Steinkopff-Verlag 1994

deformation τ_{el} can be derived

$$k_{el} = \frac{1}{2\pi} \frac{k_B T}{Q\kappa} \frac{(q-1)(q+2)}{(q+1)q}$$

$$+ \frac{1}{\pi} \frac{k_B T}{Q\kappa} \left\{ 2(\delta + l_{hg})\kappa \frac{q-1}{q} + ((\delta + l_{hg})\kappa)^2 p \right.$$

$$\left. \times \left[\frac{p}{q} + 2\ln(p+q) \right] \right\}, \tag{6}$$

$$\tau_{el} = \frac{1}{\pi Q} \left\{ \frac{p^2 q}{(1+p^2)(1+q)} - \ln \left\{ \frac{1}{2}(q+1) \right\} \right.$$

$$\left. + (\delta + l_{hg})\kappa \frac{qp^2}{(1+p^2)} \right\}, \tag{7}$$

where $p = \frac{2\pi Q}{\kappa a}$, $q = \sqrt{p^2 + 1}$, $Q = \frac{e^2}{4\pi\varepsilon\varepsilon_0 k_B T}$, and $\kappa^2 = 8\pi n_0 Q$. Here, Q is the Bjerrum length, κ is the inverse Debye length, n_0 is the ionic bulk density, e the elementary charge, ε the dielectric constant, ε_0 the permitivity of the vacuum, T the temperature, and k_B Boltzmann's constant. The electrostatic contribution of the bending rigidity modulus k_{el} consists of two additive parts. The first part $((\delta + l_{hg}) = 0)$ is the pure bending term, the second part takes into account additional area changes during the bending.

Neutral surface

To establish the principal position of the neutral surface and the influence of surface charges on τ and k, we consider a model molecule of the form HG–$(CH_2)_{11}$–CH_3 in a "good" solvent (HG stands for "head group"). The width of the polar region is chosen to be $l_{hg} = 0.5$ nm, which is suited to lipid molecules, the lateral pressure in the polar region is $\Pi_{hg} = 20$ mN/m. The available area per chain relative to the dividing surface is $a = 0.31$ nm^2. Furthermore, the surface charge density is $\sigma = 0.1$ C/m^2 and the Debye length $1/\kappa = 2$ nm. Figure 2 shows the modulus of mixed deformation τ and its chain and head group contribution. The neutral surface ($\tau = 0$) is situated within the hydrophobic region, but less than one C–C bond length (0.153 nm) distant from the polar region. The effect of the rather highly charged head groups is only a small shifting

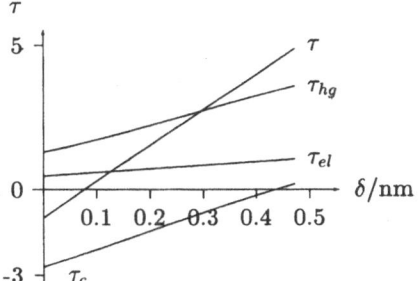

Fig. 2 Modulus of mixed deformation $\tau = \tau_c + \tau_{hg} + \tau_{el}$ and its chain (τ_c), head group (τ_{hg}), and electrostatic (τ_{el}) contribution (in units of $k_B T$/nm) as a function of the position of the dividing surface

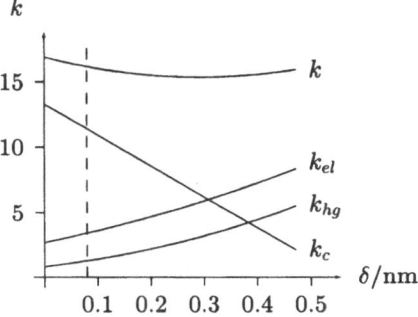

Fig. 3 Dependence of the bending rigidity modulus $k = k_c + k_{hg} + k_{el}$ and its chain (k_c), head group (k_{hg}), and electrostatic (k_{el}) components on the position of the dividing surface (in units of $k_B T$). The position of the neutral surface is marked with a broken line

of the neutral surface towards the polar region. In Fig. 3 the dependencies of the bending rigidity modulus and its head group, chain, and electrostatic components on the position of the dividing surface are displayed. The position of the neutral surface is marked with a broken line. Here, the main contribution of the bending rigidity modulus comes from the chain region. But, the surface charges deliver a considerable contribution to the bending rigidity. This fact comes mainly from the surface area changes during the bending and not from the pure bending part in (6).

Acknowledgement The author gratefully acknowledges the help of Dr. G. Kluge. This work was supported by the Deutsche Forschungsgemeinschaft through Sonderforschungsbereich 197.

References

1. Helfrich W, Kozlov MM (1993) J Phys II France 3:287–292
2. Szleifer I, Ben-Shaul A, Gelbert WM (1990) J Phys Chem 94:5081–5089
3. Szleifer I, Kramer D, Ben-Shaul A, Gelbart WM, Safran SA (1990) J Chem Phy 92:6800–6817
4. Flory PJ (1969) Statistical Mechanics of Chain Molecules, Wiley-Interscience, New York, pp 55–61
5. Lekkerkerker HNW (1989) Physica A 159:319–328

Progr Colloid Polym Sci (1994) 97:12–15
© Steinkopff-Verlag 1994

G. Caminati
E. Margheri
G. Gabrielli

Complexation of metal ions at the monolayer-water interface

Received: 16 September 1993
Accepted: 16 March 1994

Dr. G. Caminati (✉)
E. Margheri; G. Gabrielli
Dipartimento di Chimica
Via G. Capponi, 9
50121 Firenze, Italy

Abstract Spread monolayers of nonadecylpyridine (NDP) were studied using subphases containing two different metal ions, Co^{2+} and Cu^{2+}, at the same concentration, in order to investigate the interactions between the monolayer and the ions in the aqueous subphase and to deduce the behavior of NDP as a ligand.

Surface pressure-area and surface potential-area isotherms were recorded at 20 °C. Langmuir–Blodgett (L-B) films were prepared in different experimental conditions and characterized by spectroscopic techniques: UV-vis spectra were recorded and electron spectroscopy for chemical analysis (ESCA) experiments were performed on the dry films prepared by transferring NDP monolayers from both Co^{2+} and Cu^{2+} ions containing subphases.

The body of the experimental results suggested that NDP forms complexes with different stoichiometries as a function of the nature of the metal ion and as a function of the surface pressure at which the monolayer was transferred onto the solid support, that is to say, as a function of the orientation of the pyridine ring at the interface.

Key words Monolayers – L-B films – metal ion complexes

Introduction

The preparation of Langmuir–Blodgett (L-B) multilayers containing metal ions is particularly important, both for basic scientific research and for applications, especially in the field of advanced materials [1]. The aim of this work was to define the interactions between the monolayer and the transition metal ions in the subphase. The investigation of this process eventually leads to the realization of fiber optic sensors for the detection of transition metals in polluted waters, by transferring monolayers directly on to the fiber [2].

The results obtained in a previous work [3] showed that nonadecylpyridine (NDP) acts as a ligand for Ni^{2+} ions when it is arranged in monolayer at the water-air interface and the metal ions are dissolved in the aqueous subphase. We therefore investigated the behavior of NDP in the presence of different metal ions such as Co^{2+} and Cu^{2+} and compared the results with the previous ones obtained in the presence of Ni^{2+} solutions at the same ionic strength.

The complexation was deduced from changes in the UV-vis spectra and from the analysis of ESCA spectra.

Experimental

Materials

Nonadecylpyridine (NDP), $Ni(ClO_4)_2$, $Co(ClO_4)_2$ and $Cu(ClO_4)_2$ were supplied by Fluka and used without further purification. Water was twice distilled and purified

Progr Colloid Polym Sci (1994) 97:12–15
© Steinkopff-Verlag 1994

with a Milli-Q apparatus (Millipore). Chloroform (Merck) was used as a spreading solvent.

Methods

Surface pressure-area isotherms were recorded using a Lauda filmbalance with a continuous compression at a rate of 7 mm/min. Surface potential-area isotherms were recorded using ^{241}Am electrodes as a function of the monolayer compression [4].

Langmuir–Blodgett films were prepared using a previously described KSV apparatus [5]: all films were transferred at 2 mm/min for upstroke and 10 mm/min for the downstroke, at a compression rate of 7 mm/min. Absorption spectra were recorded using a Perkin-Elmer Lambda5 spectrophotomer. ESCA (electron spectroscopy for chemical analysis) measurements were performed with an ESCA100 instrumentation of V.S.W.

Results and Discussion

Spreading isotherms

In Fig. 1, we report the spreading isotherms of NDP on two different subphases, that is $Co(ClO_4)_2$ and $Cu(ClO_4)_2$ 0.1 mM solutions at 20°C. The spreading isotherm of NDP on $Ni(ClO_4)_2$ 0.1 mM at the same temperature is also shown for comparison.

When Co^{2+} ions are present in the subphase, the NDP monolayer does not show significant changes with respect to the Ni^{2+} case: an inflection point is always present around 13 mN/m [3] and the monolayer is always in a condensed phase, characterized by high values of the

surface compressional modulus C_s^{-1} and collapse pressure π_c (see Table 1). The molecular area value related to the most condensed phase (A_0) is the nearest to the one expected for NDP molecules lying with the pyridine group parallel to the interface and the hydrophobic chains slightly tilted with respect to the normal to the interface. The surface potential-area isotherms show a plateau around 600 mV and reach the maximum value at 800 mV, corresponding to a molecular area of 50 Å^2/molec.

On the contrary, the presence of Cu^{2+} ions in the subphase induces an expansion of the NDP film, as it was deduced from A_0 and C_s^{-1} values (see Table 1); the collapse pressure value is lower than in the Co^{2+} and Ni^{2+} cases, while the inflection point is at the same surface pressure value.

L-B films

Monolayers of NDP were then transferred onto quartz slides from the water-air interface both before ($\pi = 12$ mN/m) and after ($\pi = 30$ mN/m) the inflection point. For L-B films of NDP prepared from subphases containing 0.1 mM $Co(ClO_4)_2$ solutions, the transfer was rather good, even in the low pressure regime, both for the downstroke and the upstroke, as it was deduced from transfer ratio values always close to 1.00.

Table 1 Surface parameters of NDP monolayers on different subphases.

Subphase	A_0 (A^2/molec)	π_c (mN/m)	C_s^{-1} (mN/m)
$Co(ClO_4)_2$	30	48	130
$Cu(ClO_4)_2$	39	40	70

Fig. 1 Surface pressure-area (solid line) and surface potential-area (dashed line) isotherms of nonadecylpyridine on different subphases: (□) $Co(ClO_4)_2$ 0.1 mM solutions: (○) $Cu(ClO_4)_2$ 0.1 mM solutions: (∗) $Ni(ClO_4)_2$ 0.1 mM solutions

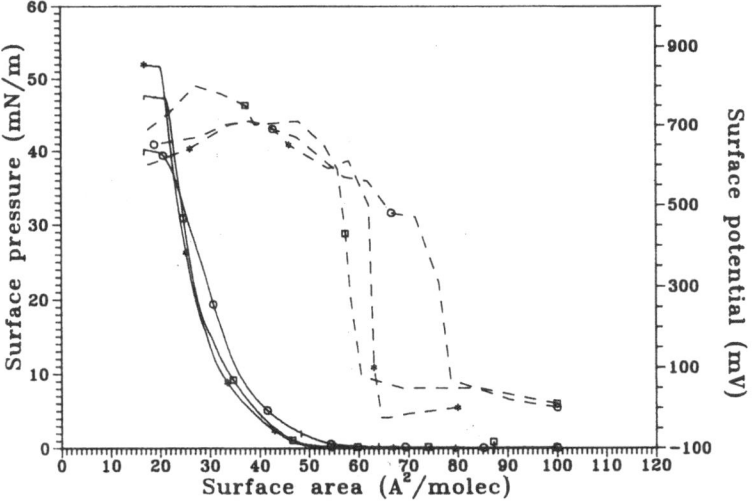

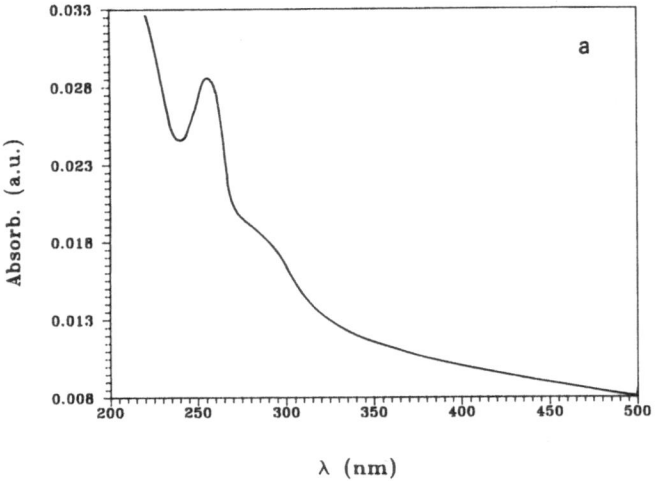

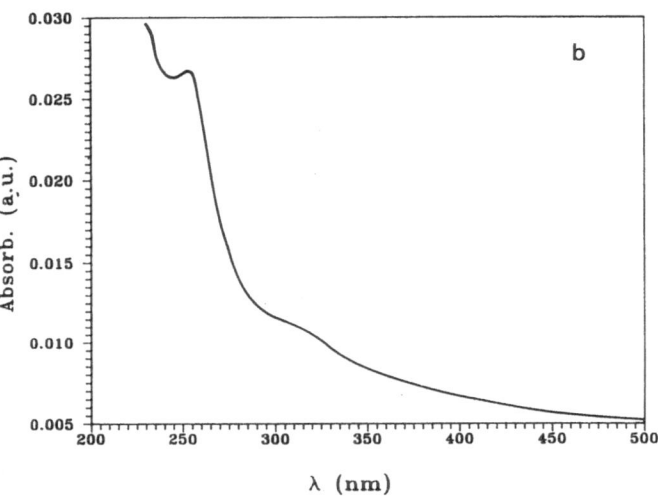

Fig. 2 UV-vis absorption spectra of two-layers L-B films of NDP transferred from Co^{2+} subphase (a) and Cu^{2+} subphase (b)

ESCA experiments were also performed in order to confirm the presence of metal ions in the L-B films and to obtain information about the interactions between NDP molecules and the metal ions.

Figure 3 shows the results obtained in the case of Co^{2+} ions: the presence of the metal ions was confirmed and a nitrogen:cobalt ratio close to 2:1 was found for films transferred at low surface pressure value (15 mN/m) and close to 3:1 for films transferred at high surface pressure value (30 mN/m).

Figure 4 shows the results obtained in the case of Cu^{2+} ions: the presence of the metal ion in the L-B film was confirmed and a nitrogen:cupper ratio close to 2:1 was found both at low and high surface pressure values.

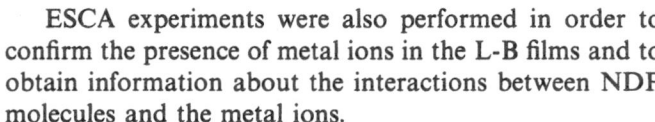

Fig. 3 ESCA spectra for L-B films of NDP transferred from Co^{2+} subphase. The absorption peak of Co^{2+} is shown

Fig. 4 ESCA spectra for L-B films of NDP transferred from Cu^{2+} subphase. The absorption peak of Cu^{2+} is shown

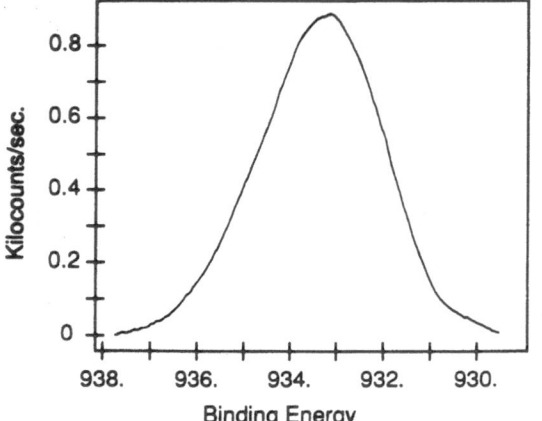

Different results were found for 0.1 mM $Cu(ClO_4)_2$ solutions: the quality of the transfer before the inflection point was poor, whereas at high surface pressure, transfer ratio values close to 1.00 were found both for the upstroke and the downstroke. This behavior was probably due to low interactions between the hydrophobic chains of NDP molecules, thus making the transfer difficult. The L-B films so prepared were then characterized with different spectroscopic techniques; UV-vis spectra are shown in Fig. 2. All the spectra show the band of the pyridine ring at 256 nm. In addition, a new band at 280 and 310 nm was found for the two-layers films prepared by transferring NDP monolayers respectively from cobalt and copper perchlorate 0,1 mM solutions. This result was found for films transferred both at low and high surface pressure.

Progr Colloid Polym Sci (1994) 97:12–15
© Steinkopff-Verlag 1994

Conclusions

In summary the following main conclusions were drawn:

1) NDP forms stable monolayers at the metal ion solution-air interface and their properties depend on the type of metal ion present in the subphase: in particular, Cu^{2+} ions gave evidence of an expanding effect which was not found for Co^{2+} ions.

2) NDP monolayers may be easily transferred onto quartz slides when the subphase contains Co^{2+} ions, while the L-B films may be prepared only from a monolayer at high surface pressure values, when Cu^{2+} ions are present.

3) The body of the results provided evidence that the NDP molecules act as a ligand for Co^{2+} and Cu^{2+} because the metal ions were detected in the two-layer L-B films.

4) It was possible to deduce that the properties of the complex depend on the type of metal ion. In the case of Co^{2+} ions, they depend also on the arrangement of NDP molecules in the monolayer.

5) The obtained results allowed us to employ the NDP system to build sensors for Co^{2+} and Cu^{2+} ions.

Acknowledgements Thanks are due to C.N.R. (Consiglio Nazionale delle Ricerche, Progetto Finalizzato Chimica Fine) for financial support. The authors also acknowledge Dr. M. Galeotti who performed the ESCA measurements and carefully analyzed the data.

References

1. a) Roberts G (ed) (1990) Langmuir–Blodgett films, Plenum Press, New York b) Ulman A (1991) An introduction to Ultrathin Organic Films: from Langmuir–Blodgett to Self Assembly; Academic Press, New York

2. Proceedings of the 1st European Conference on Optical Chemical Sensors and Biosensors, Graz, Austria, 1992

3. Caminati G, Margheri E, Gabrielli G (submitted to Thin Solid Films)

4. Gaines GL Jr., in I. Prigogine Ed., Insoluble Monolayers at Liquid-Gas Interfaces, Interscience, New York, pg. 144, (1966)

5. Gilardoni A, Margheri E, Gabrielli G, (1992) Coll Surf, 68:235–242

Progr Colloid Polym Sci (1994) 97:16–20
© Steinkopff-Verlag 1994

Morphological structures in monolayers of long chain alcohols

S. Siegel
D. Vollhardt

Received: 16 September 1993
Accepted: 25 March 1994

S. Siegel (✉) · D. Vollhardt
Max-Planck-Institut für Kolloid- und
  Grenzflächenforschung
Rudower Chaussee 5
12489 Berlin, FRG

Abstract The morphological
structure of monolayers of straight-
chain alcohols with 14 to 18
C-atoms are studied.
Compression/decompression
isotherms are illustrated by images
taken with a Brewster angle
microscope. Tetradecanol with
a plateau region in the surface
pressure-area isotherm exhibits
circular domains during the phase
transition. Decompression leads to
the formation of a stable two-
dimensional foam in the plateau
region in the isotherm, if the
monolayer was compressed to
a surface pressure of about 10 mN/m
above the kink. The corresponding
point in the isotherm is marked by
a weak kink. If the expansion starts at
lower pressures, the original domain
structure can be attained, and the
compression/decompression cycle is
reversible. Longer alkyl chains or
lower temperatures do not drastically
change this behavior.

Key words Fatty alcohols
– monolayer – Brewster angle
microscopy – isotherm – phase
behavior

Introduction

Monolayers of fatty alcohols are often studied [1, 2] as simple models for various aspects of monolayers and membranes. Recently, the development and the use of new experimental techniques have provided direct information on the morphological structure and phase behavior of monolayers. In this work, the surface pressure-area measurement has been combined with a Brewster angle microscope to study and discuss some features of long chain n-alcohols, as simple amphiphilic molecules, at the water surface.

Experimental

The fatty alcohols with 14 to 18 carbon atoms were purchased from Sigma and used without further purification. They were dissolved in heptane (Merck, for spectroscopy) and spread on double distilled water at an initial area per molecule of about 1.5 nm². A computer-interfaced film balance with a Langmuir float (Lauda) and a Brewster angle microscope (BAM 1, NFT Göttingen) were used for our investigations.

The microscope is sensitive to thickness, density, and molecular orientation of the monolayer. The resolution is about 4 µm. For more details to the BAM see refs. [3–6]. Because of the visual angle, the images appear compressed in one direction. An image-processing system (Data Translation) was used to correct this and to enhance the contrast.

Results and discussion

Myristyl alcohol

In Fig. 1 the surface pressure-area per molecule (π-A) isotherms of myristyl alcohol (tetradecanol) at selected

Progr Colloid Polym Sci (1994) 97: 16–20
© Steinkopff-Verlag 1994

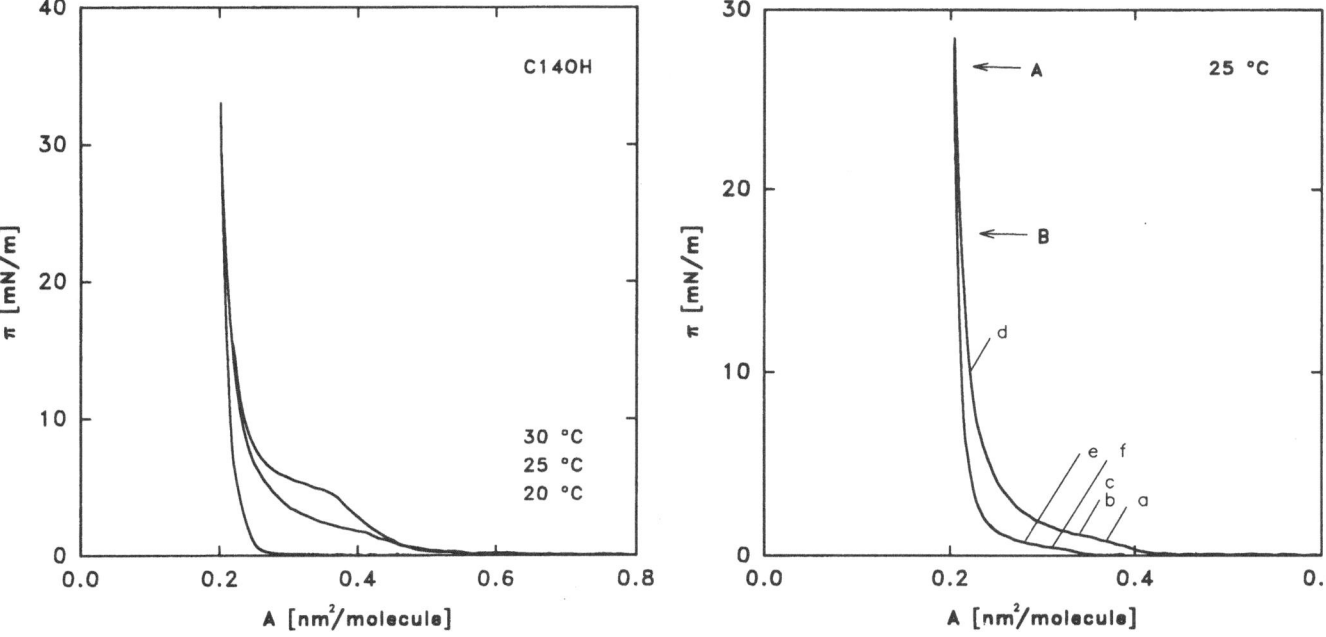

Fig. 1 Pressure-area isotherms of a myristyl alcohol monolayers at different temperatures

Fig. 2 Compression/expansion isotherm of myristyl alcohol at 25 °C. The letters correspond to those in Fig. 3

temperatures are shown. It can clearly be seen that the plateau region is formed with increasing temperature, which is a characteristic for a phase coexistence region of first order between a fluid phase of very low density and a condensed phase forming two-dimensional domains. Therefore, in the literature the characteristic point at which the plateau region begins is denoted as the main phase transition [7]. At 20 °C, however, the beginning of the coexistence region cannot be determined exactly, and it may be at large areas per molecule.

A compression/expansion cycle at 25 °C including the references to the BAM images (Fig. 3) is presented in Fig. 2.

If the myristyl alcohol is spread at temperatures above 24 °C the background of the image is homogeneous and dark. Compressing the monolayer, the formation of circular condensed domains becomes visible at the beginning of the plateau region. These domains grow with further compression (Fig. 3a, b). Using an analyzer, the domains show different contrasts, obviously, because of different orientation of the tilted chains. There also exists a structure within the domains, as reported for other substances [6, 8, 9], but it is hardly to be seen due to the weak contrast (Fig. 3c). The domains have homogeneously reflecting segments with the common point in the middle of the domain or at the edge of it. In some cases, the segments formed are irregular. The segments differ only in the orientation of the tilted molecules [6].

In Fig. 3d, the domains are tightly packed and deformed, but no coalescence can be seen. The domains are

transformed into a nearly hexagonal arrangement because of their same size. An expansion of this monolayer leads to the same structure as shown in Fig. 3b.

The contrast of the image diminishes with further compression and vanishes completely at a film pressure π of about 18 mN/m (point B in Fig. 2). At this pressure, also a kink can be seen in the isotherm (compare Fig. 1). In this monolayer state, the alkyl chains are oriented perpendicular to the water surface, and the compressibility becomes very small. The monolayer seems already to be homogeneous. It is interesting to note that, in this state, decompression leads again to the preceding domain structure (as in Fig. 3b), as, obviously, the hexagonal structure of the monolayer still exists. Here, the compression/decompression cycle is reversible. Only a further compression to $\pi > 25$ mN/m (point A in Fig. 2) leads to a completely homogeneous structure, as can be concluded from a following decompression.

Expanding this highly compressed monolayer, at first irregular structures with uniform molecular orientation are formed. They are similar to that also observed with palmityl alcohol (Fig. 5a), but they are smaller and have a lower contrast. Near the surface pressure plateau small holes arise and grow with further expansion. The size and size distribution of the holes are dependent on the barrier velocity. Figure 3e and f show the expanded monolayer in the plateau region. Note that now not the condensed domains, but the expanded phase has a circular shape and

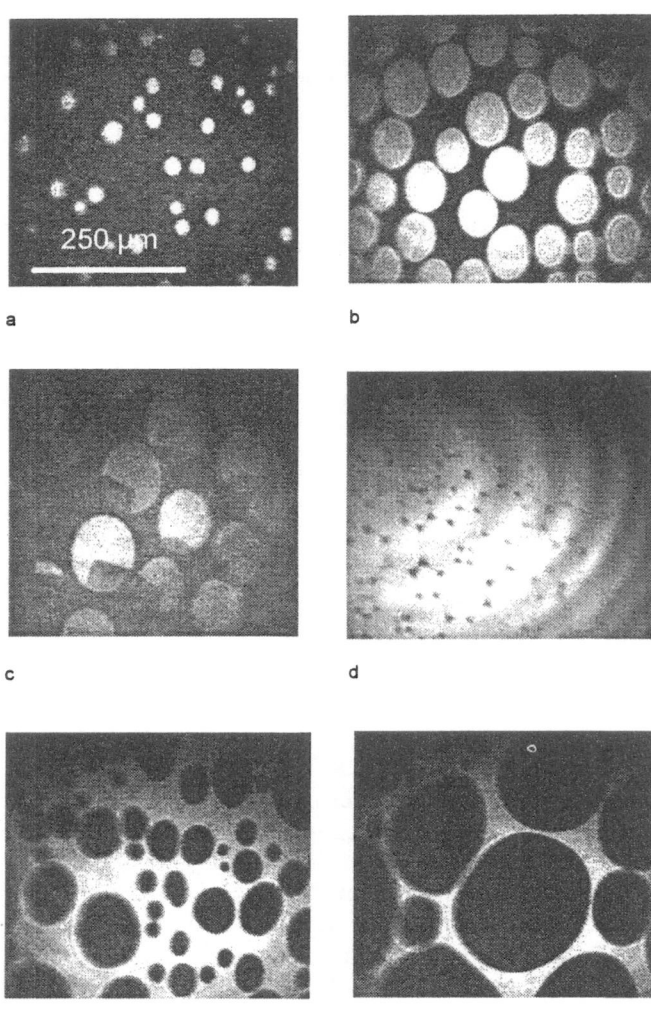

a   b

c   d

e   f

Fig. 3 Brewster angle micrographs of a monolayer of myristyl alcohol at 25 °C. See Fig. 2 and text. All images are on the same scale

Palmityl and stearyl alcohol

Longer molecule chains give a better contrast in the BAM images, however, at room temperature, no plateau region indicating the main phase transition is formed in the isotherm. The plateau pressure is nearly zero and the transition from the fluid phase of low density to the condensed phase is not visible in the isotherm. Even after spreading, the Brewster images show coexisting phases. Therefore, the condensed phase has a complex structure; not only circular domains, but also irregular-shaped domains are formed, such as large areas of condensed material with holes in them or foams. These structures are produced by spreading and evaporating the solvent and they seem to be frozen. Figure 4 illustrates the inhomogeneous distribution of the material and some structures of different shapes and densities. Consequently, in this coexistence state, the size and the structure of the domains of the condensed phase are sensitively affected by temperature changes, spreading solvent, spreading technique, surface instabilities and other factors. Some difficulties in reproducibility [11, 12] can be explained in this simple way.

The steep increase of the surface pressure (see also Fig. 1, 20 °C) indicates the contact and deforming of the condensed structures in addition to a decreasing tilt angle of

Fig. 4 Brewster angle micrographs of alcohol monolayers after spreading. **a, b** palmityl alcohol, 25 °C, 1.2 nm^2/molecule; **c** stearyl alcohol, 20 °C, 1.2 nm^2/molecule; **d** stearyl alcohol, 30 °C, 0.5 nm^2/molecule

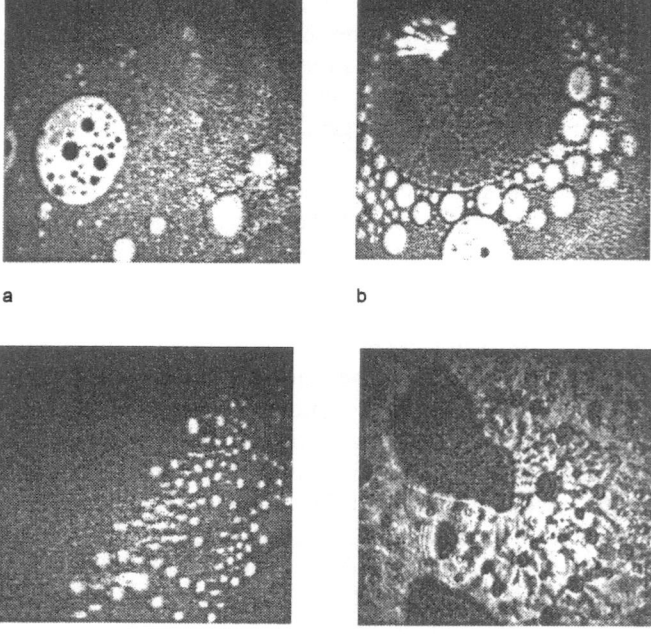

a   b

c   d

thus the continuous phase is the condensed phase. Holes of different sizes indicate a progressive process of hole nucleation. Whereas the compression of the monolayer in the plateau region is characterized by nucleation and growth of domains, the decompression is based on hole nucleation and foam formation.

After stopping the barrier, the two-dimensional foam remains stable. Further expansion leads to a more expanded foam (Fig. 3f). Finally, the thin lamella tear and the condensed phase vanishes at the end of the plateau region. Foam structures known from fatty acid monolayers have been studied mostly under non-equilibrium conditions (10–12, 4).

the molecules. The pressure at which the molecules are oriented perpendicularly, is marked by a kink in the isotherm, typical for fatty alcohols. Additionally, a second weak kink, not reported yet, is visible. At this point the monolayer loses the original structure and becomes homogeneous. The corresponding surface pressure is 6 ... 12 mN/m higher than the surface pressure at the first kink, depending on chain length and temperature.

Decompression of the tight monolayer results at first in irregular formed domains with different chain orientations (Fig. 5a). Such domains are also reported for fatty acids [4, 13] and eicosanol [14]. At $\pi \approx 0$, the condensed monolayer phase tears and forms large floes; additionally, holes and foams can be observed (Fig. 5b). The distribution of the material on the trough area is very inhomogeneous, and the structures are not stable.

General aspects

1) At monolayer compression, the formation of the condensed phase domains begins at the kink point of the plateau region of the π-A isotherm. The corresponding surface pressure depends on the chain length and the temperature and can be about zero.
2) At equilibrium, circular domains are formed. This circular shape due to the effect of the line tension seems to be typical for monolayer substances with one alkyl chain. The kinetics of formation of circular structures depend on the line tension and the shear viscosity of the condensed monolayer phase. Both are dependent on chain length and temperature.
3) The orientation of the tilted molecules can be changed continuously [4] or discontinuously (Fig. 3c and refs. [6, 8]), leading to the inner structure of the domains.

4) At the end of the plateau region, the domains touch one another, and on further compression the domains are deformed increasingly. The phase transition is not complete, but the character of the transition is changed. Now, the condensed phase becomes continuous and determines the properties of the monolayer. A second process is the change of the tilt angle of the molecules, caused by the film pressure.
5) The first kink in the alcohol isotherms indicates a nearly perpendicular orientation of the molecules. There are good reasons to assume that the kinks in the isotherms of fatty acids and the esters of fatty acids represent similar 2D condensed phase transitions. The occurrence of a plateau region instead of such kinks is not probable owing to these morphological images.

A complete homogeneous monolayer exists only at above a second weak kink in the steep part of the alcohol isotherms. The 2D condensed phase transitions end at this point.
6) The morphological structure during decompression depends on the initial state. Expansion starting at lower pressures results in the original domain structure, expansion from higher pressures gives irregular domains with different chain orientations and, at larger areas, formation of holes.
7) Further expansion ends in a two-dimensional foam, i.e., the condensed phase is continuous. This foam is stable and corresponds to the domains in the plateau region during compression.
8) On continuing expansion, the foam rupture results in a homogeneous fluid phase of very low density (gaseous state). Hence, at lower temperatures, the condensed material is frozen in irregular structures and coexists with the gas phase.
9) Foam is formed either after a compression/decompression cycle or as one of the possible shapes of the condensed material in the zero surface pressure region.
10) The study of homologous amphiphiles is necessary to understand the phase behavior of their monolayers. The morphological structure, visible in Brewster angle microscopes, is an important complement to surface pressure measurements.

Fig. 5 a Domains with different chain tilting after compression and expansion. Palmityl alcohol, 30 °C, $\pi \approx 1$ mN/m; b Film after expansion. Palmityl alcohol, 20 °C, 0.4 nm²/molecule

Acknowledgement The authors are indebted to Koordinierungs- und Aufbau-Initiative e.V. Berlin (KAI), the Deutsche Forschungsgemeinschaft (DFG), and the Fonds der Chemischen Industrie for financial support.

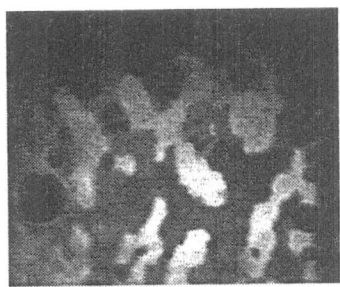
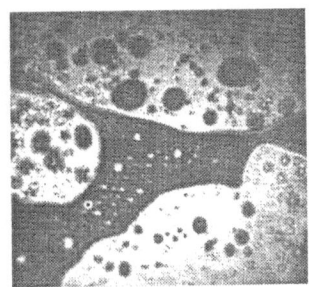

a   b

References

1. Harkins WD (1952) Physical Chemistry of Surface Films, Reinhold, New York
2. Gaines GL (1966) Insoluble Monolayers at Liquid–Gas Interfaces, Interscience, New York
3. Henon S, Meunier J (1991) Rev Sci Instrum 62:936–939
4. Hönig D, Overbeck GA, Möbius D (1992) Adv Mater 4:419–424
5. Hönig D, Möbius D (1991) J Phys Chem 95:4590–4592
6. Vollhardt D, Gehlert U, Siegel S (1993) Colloids Surfaces A 76:187–195
7. Möhwald H (1990) Annu Rev Phys Chem 41:441–476
8. Qui X, Ruiz-Garcia J, Stine KJ, Knobler CM, Selinger JV (1991) Phys Rev Lett:703–706
9. Gehlert U, Siegel S, Vollhardt D (1993) Progr Colloid Polymer Sci 93:247
10. Moore B, Knobler CM, Broseta D, Rondelez F (1986) J Chem Soc, Faraday Trans 2 82:1753–1761
11. Stine KJ, Rauseo SA, Moore BG, Wise JA, Knobler CM (1990) Phys Rev A 41:6884–6892
12. Berge B, Simon AJ, Libchaber A (1990) Phys Rev A 41:6893–6900
13. Hönig D, Möbius D (1992) Thin Solid Films 210/211:64–68
14. Overbeck GA, Hönig D, Möbius D (1993) Langmuir 9:555–560

Progr Colloid Polym Sci (1994) 97:21–26
© Steinkopff-Verlag 1994

É. Kiss
I. Bertóti

Preparation and characterization of PEO grafted surfaces by wettability measurements

Received: 16 September 1993
Accepted: 31 January 1994

É. Kiss (✉) · I. Bertóti
Eötvös University
Department of Colloid Chemistry
P.O. Box 32
Budapest 112, Hungary 1518

Research Laboratory for Inorganic
Chem, Hung. Acad. of Sci.
Budapest, POB 132
Hungary 1518

Abstract Surface modification techniques were developed to graft chemically poly(ethylene oxide) chains on mica and polyethylene substrates. The chemical composition of surface layer was characterized by X-ray photoelectron spectroscopy measurements. Static and dynamic wettability studies were performed to get information on the solid/liquid interaction. The wetting parameters indicate a strong influence of temperature and electrolyte concentration on the hydration of surface PEO layer which correlates to the phase behavior of PEO in solution.

Key words Poly(ethylene oxide) layer – wettability – surface grafting – surface characterization – XPS – theta conditions

Introduction

Poly(ethylene oxide), PEO, has been widely used as an efficient steric stabilizer of colloidal dispersions for several decades [1, 2]. PEO is also a material of growing importance in the biomedical world. It possesses a variety of properties which led to important biochemical and biomedical applications like controlling pharmacodynamics, affecting immunogenicity, as well as developing separation techniques and enzyme or polymer surface modification [3]. The great attention that PEO is receiving is due to its water solubility and nonionic character, the lack of toxicity, and its availability in a wide range of molecular weights [4]. The terminal hydroxyl groups of a PEO molecule provide a site for covalent binding to other molecules or a substrate. Currently, much interest has been shown in PEO containing surfaces prepared either by copolymerization or the grafting of PEO chains to the surface. These surfaces are expected to be biocompatible materials due to their low protein adsorption.

All of the applications mentioned above are related to an aqueous environment. PEO is known to be highly soluble in water and also in many organic solvents including benzene, ethanol, and acetone. Hence, PEO can be described as amphiphilic although it is generally considered to be a hydrophilic polymer. The PEO water interaction has received intensive theoretical and experimental study [5] based on hydrogen bond acceptor character of ether oxygen atoms. The structural similarity of PEO to water and the strong hydrogen bonding explain its unlimited solubility in water at room temperature [6, 7]. The solubility of PEO decreases upon heating, resulting in a lower consolute temperature of approximately 100 °C. (PEO also has an upper consolute temperature but with less practical importance due to its high value.) On raising the temperature above 100 °C a two-phase system is formed except for a narrow low and a narrow high concentration ranges [8]. Salt additives lower the lower consolute temperature to a different extent.

The interaction of surface immobilized PEO molecules with the medium is basically responsible for such important properties as colloidal stability or protein repellency. The solvent-segment and segment-segment interactions have to be taken into account to describe characteristic properties like the thickness and the structure of the layer. The surface grafted PEO layer in an aqueous environment contains hydrated, highly mobile molecules with a large exclusion volume. NMR relaxation time studies show rapid motion of these chains [9]. If it is not fulfilled by any reason (e.g., decreased interaction with water, too high a grafting density) the consequence can be an increased interaction with other molecules dissolved in the medium, for instance, the protein repellent surface can lose its advantageous feature.

The aim of our investigation was to obtain information on the property of surface grafted PEO layers under the different conditions which affect PEO in solution. We wanted to know whether there is an indication of changed water–PEO interaction on approaching the theta condition in the case of surface immobilized PEO molecules. Therefore, the effect of temperature and electrolyte concentration on the wettability of the surface was studied. In general, wettability measurements provide a suitable and highly sensitive method to characterize solid/liquid interaction [10]. The hysteresis which is the difference between advancing and receding contact angles indicates the mechanical and energetic heterogeneity of the surface. Time-dependent processes like swelling of the surface layer or the ability of polymer surfaces to reconstruct themselves in contact with different liquid media can be revealed by dynamic wettability measurements [11].

In this paper, we propose a newly developed method and two others presented earlier [12, 13] for chemical grafting of PEO to polyethylene and mica substrates. The results of both static and dynamic wetting studies are interpreted in connection with the phase behavior of PEO.

Experimental

Materials

Substrates: low density polyethylene (PE, Noax, Sweden) plates and freshly cleaved mica sheets were used as substrate materials for PEO grafting. Smooth and clean PE surfaces were prepared by melting of the polymer sample pressed between glass plates at 120 °C, followed by an ultrasonic rinse in ethanol for 10 min. Mica plates were cut in 20×40 mm pieces and cleaved to a thickness of 0.2 mm immediately before use in order to minimize carbonaceous surface contamination. Non-modified polyethylene and some other polymer surfaces (polystyrene, PVC, PTFE) were used for static wettability measurements for comparison with the PEO grafted surface. The polymer samples were thoroughly rinsed in ethanol in an ultrasonic bath.

Monomethoxy-poly(ethylene oxide) PEO $M_n = 1900$ $M_w/M_n = 1.08$ was obtained from Sigma (USA).

Monomethoxy-poly(ethylene oxide)aldehyde PEO–CHO, was prepared from PEO by partial oxidation according to the procedure of Harris et al. [14] to obtain PEO chains with one terminal aldehyde group which were coupled to the surface amino groups by reductive amination of PEO-CHO.

A branched poly(ethylene imine) PEI, Polymin SN (BASF, FRG) was used for surface amination of the oxidized PE sample. This PEI contains primary, secondary, and tertiary amino groups in the approximate molar ratio $1:2:1$. The PEI was fractionated by ultrafiltration and a fraction with molecular weight between 10^5–10^6 was selected.

All other chemicals used either in surface preparation or wettability measurements were of analytical grade. The purity of wetting liquids was also checked by surface tension measurements.

Methods

Grafting procedures

Figure 1 shows the scheme of chemical modification processes developed in order to obtain PEO grafted surfaces on polyethylene and mica substrates.

I) PEI as an adsorbed cationic polyelectrolyte layer was used to immobilize the PEO chains to the substrate in the method I which can be applied on both substrates. PE substrate had to be chemically oxidized before the adsorption to get a surface with a high density of ionic groups. Freshly cleaved mica surface which provides negative charge in an aqueous environment does not require any pretreatment prior to the PEI adsorption. The pH of the PEI solution was set to obtain a strongly attached polymer layer with a high density of amino groups available for further reaction [15]. The chemical coupling of PEO–CHO molecules to the amino surface was performed by reductive alkylation in the presence of $NaCNBH_3$. Details of our method and the chemical composition of the surface have been given in an earlier paper [12].

II) Grafting of PEO to mica substrate was performed by silylation with 3-isocyanatopropyldimethylchlorosilane, IPS. Silanol groups available for chemical coupling of IPS were produced by plasma activation of mica [16]. Silanization was performed in gas phase at room temperature letting IPS in an evacuated desiccator. Melted PEO was chemically bound to the surface coupled silane molecules by their isocyanate functional groups. The efficiency of PEO grafting was measured by XPS and compared to other modification methods [13]. Because of the highest reaction yield this "melt" method was selected for further sample preparations.

Progr Colloid Polym Sci (1994) 97: 21–26
© Steinkopff-Verlag 1994

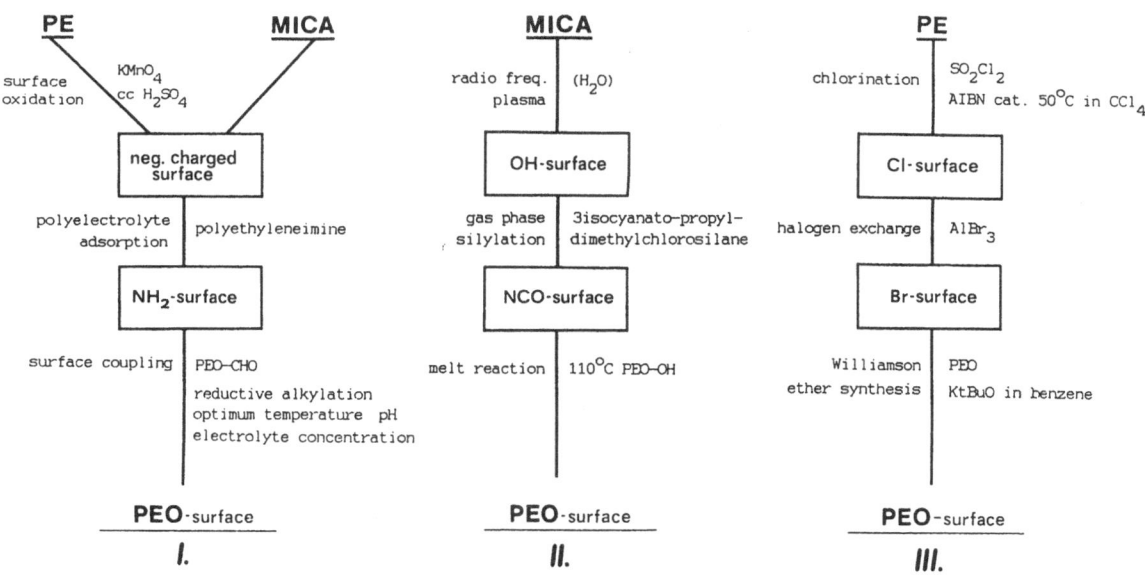

Fig. 1 Scheme of surface modification processes to prepare PEO grafted surfaces

III) In the third reaction route of proposed surface treatment series PE was sulphochlorinated by SO_2Cl_2 in carbontetrachloride in the presence of azobisizobutyronitrile as a catalyst. The surface reaction, like the bulk radical chain chlorination [17], led to a mainly chlorinated rather than a sulphated product. This fact was supported by XPS results of surface composition showing atomic ratios related to carbon 0.053 and 0.007 for chlorine and sulphur, respectively. The chlorine surface was subjected to halogen exchange reaction in order to obtain a more reactive component for PEO coupling reaction. PEO molecules were reacted with the PE-bromine surface according to the Williamson's ether synthesis [14, 18]. This modification step together with the results of surface analysis will be published shortly.

Surface analysis

XPS spectra were recorded on a Kratos XSAM 800 spectrometer operating in the fixed retarding ratio mode using Mg $K_{\alpha1,2}$ X-irradiation (1253,6 eV). Spectra were referenced to the Cls line at a binding energy of 284.6 eV. For quantitative analysis, the relative differential photoionization cross-sections of Evans et al. [19] were used. Peak deconvolution was performed by means of the Kratos DS300 or DS800 software. The surface concentration values of PEO grafted on mica were derived by the method of Herder et al. [20] using the potassium signal originating from mica substrate as an internal standard.

Wettability

Static advancing and receding contact angles of water, diiodomethane and formamide on polymer and PEO grafted surfaces were measured by means of a Reme-Hart Contact Goniometer. Surface free energy data of the solid surfaces were calculated from the cosine average of contact angles. $\cos \Theta = (\cos \Theta_A + \cos \Theta_R)/2$ [21]. Three pairs of liquid contact angles were used for the calculation according to the Wu method [22]. Solid/water interfacial free energies could be obtained from the Young equation. Static contact angles were also measured by a Wilhelmy balance by immersing the vertical suspending solid plate into the liquid phase. The contact angle values were calculated from the capillary forces registered both in advancing and receding positions.

Dynamic wettability behavior of the PEO grafted surface was studied by a dynamic wetting balance equipped with a computer for data analysis. Dynamic advancing and receding contact angles as well as relaxed values were obtained in the velocity range of 0.03–3.5 mm s^{-1}. The conditions of wettability measurements were chosen on the basis of clouding behavior of PEO in aqueous solution as follows: water at 25°, 45°, and 65 °C; 0.63 M K_2SO_4 solution at 45° and 65 °C. The two latter cases represent parameter combinations corresponding to the vicinity of the theta conditions [6, 23]. Using water as a wetting medium the PEO layer is in a good solvent condition at 25 °C, which is a changed towards the theta condition by increasing the temperature up to 65 °C.

Results and discussion

XPS

The chemical composition of the surface layer of different samples was characterized by XPS measurements. A typical example of a carbon 1s spectrum is shown in Fig. 2. For this sample PEO grafting was carried out on PE substrate according to the method III. The C1s envelope decomposed two main components: the higher intensity one corresponds to $-(CH_2)-$type carbon at binding energy (BE) of 284.6 eV originating from the PE substrate, while the other with chemical shift of 1.8 eV is related to the grafted PEO layer. The considerable contribution of the substrate is explained by the fact that only a relatively loosely packed monomolecular layer of PEO can be chemically fixed on PE, which represents a layer thickness under dry condition much less than the usual sampling depth of XPS. Small peaks at 287.6 and 288.6 eV BE could be assigned to some $C=O$ and $C=O-O$ type surface contaminants. Due to the low intensity they are ignored in further analysis of the data.

For mica samples, irrespective of the preparation methods, the PEO grafting densities determined by the internal standard method were in the range of 205 ± 15 Å²/PEO molecule. That surface concentration

Fig. 2 XPS C1s spectrum of PEO grafted surface prepared on PE substrate by method III

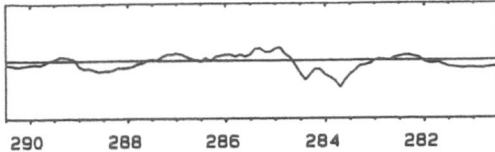

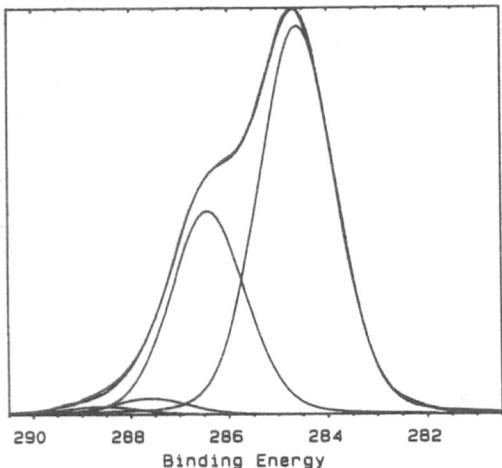

Binding Energy

corresponds to approximately 1.4 mg m^{-2} chemically bound PEO. Taking into account the value of radius of gyration, $R_g = 14$ Å, of a PEO molecule with a relative molecular mass of 1900 [24], the structure of the PEO layer can be described as brush like. Due to the overlapping of neighboring terminally immobilzed PEO molecules the chains are stretched at least to about 50% of their fully extended (meander structure [25]) length. This extended conformation could be observed in force measurements [26], which gave 40 Å as a compressed layer thickness in aqueous solution.

The separation distance of the terminally attached molecules on the surface was found to be 15 Å which is within the range of 13–17 Å given by Jeon et al. [27] as an optimal value for protein repellent surface. Experimental adsorption results are in good agreement with their theoretical prediction [28]. Protein adsorption was highly reduced on PEO grafted surfaces, indicating the practically complete coverage of the substrate material.

Wettability

Surface free energy and solid/water interfacial free energy values calculated from static contact angle data are presented for different polymer surfaces in Fig. 3. Comparing the $\gamma_{s,v}$ values, a moderately hydrophobic character for the PEO grafted surface, similar to PS and PVC, can be deduced. Water does not spread on this surface, rather it forms droplets with contact angles of $50 \pm 2°$ as a characteristic value. The solid/water interfacial free energy data show a different pattern. Those are in the range of 35–50 mN m^{-1} for all the polymers studied except the PEO grafted surface. The highly hydrated surface layer of PEO gives the lowest $\gamma_{s,w}$ value about 10 mN m^{-1}.

Fig. 3 Solid/vapor, γ_{sv}, and solid/water γ_{sw}, interfacial free energies of different polymer and the PEO grafted surfaces calculated from contact angle data by the Wu method

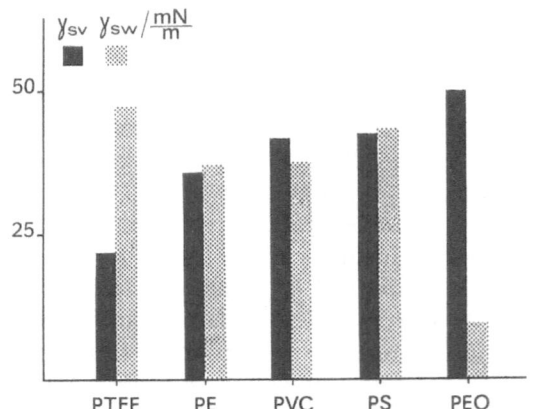

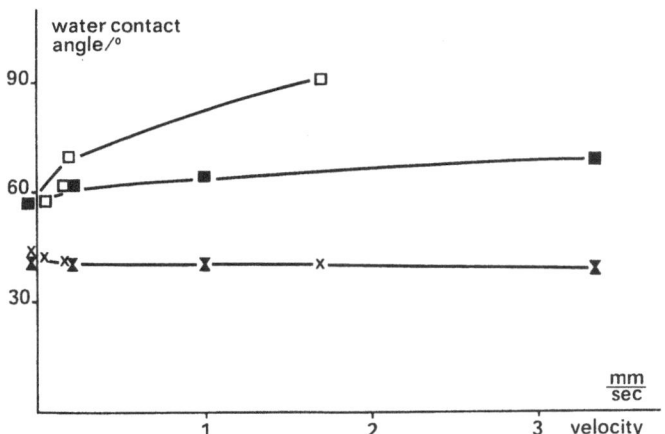

Fig. 4 Dynamic wettability of PEO grafted surfaces prepared on oxidized polyethylene (open s.) and mica (filled s.) substrates. Advancing (□) and receding (×) contact angles of water as a function of contact line velocity

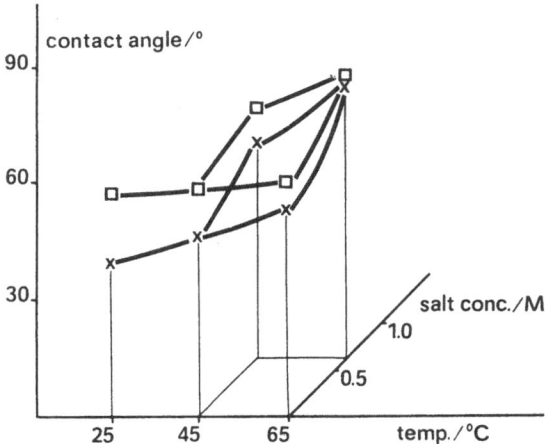

Fig. 5 Static advancing and receding contact angles measured on PEO grafted surface as a function of temperature and K_2SO_4 concentration

Static advancing and receding water contact angles measured on a PEO grafted surface are shown in Fig. 4. Dynamic contact angles are plotted as a function of the velocity of the three-phase contact line. The static contact angles and the static hysteresis are about the same for both PEO surfaces prepared on mica and PE substrates. On the contrary, the dynamic wettability behavior proved to be dependent on the substrate material the PEO was grafted on. The strong increase of advancing contact angle with the increased velocity observed in the case of PE substrate is probably due to the roughness of the oxidized surface. The receding contact angles, which are common, seem to be less sensitive to the mechanical heterogeneity of the surface. The velocity dependence of contact angles was found to be less pronounced on the smooth mica samples. The increase of the advancing and the decrease of receding angles do not exceed 10° over the whole velocity range. From here it can be concluded that hydration must be a very fast process resulting in only a small distortion of moving meniscus even at high velocity of contact line.

The effect of temperature and electrolyte concentration was also studied on the static wettability of the PEO grafted surface (Fig. 5). The water contact angles (especially the receding values) increase with increasing temperature. The presence of K_2SO_4 in the wetting medium further increases the contact angles up to about 75 degrees. A characteristic feature of the wetting behavior is that the hysteresis of 20 degrees observed at 25 °C almost disappears at high temperature and salt concentration. The hydration of the surface layer during contact with the aqueous phase does not have such a major role as at 25 °C. The PEO surface seems to be more hydrophobic and more homogeneous under these conditions which correspond to theta conditions of PEO solution.

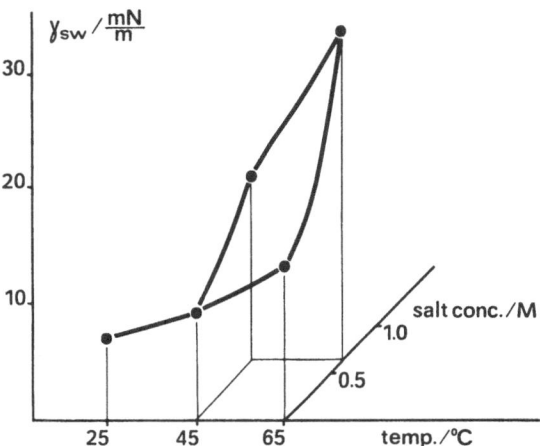

Fig. 6 Solid/water interfacial free energy, γ_{sw}, of the PEO grafted surface as a function of temperature and K_2SO_4 concentration

There is a marked indication of changing character of the PEO surface in the values of solid/water interfacial free energies which were calculated at different temperatures and K_2SO_4 concentrations (Fig. 6). $\gamma_{s,w}$ rose above $30\ mN\ m^{-1}$ at high temperature and electrolyte concentration which is a characteristic property of such hydrophobic polymers as PE. The consequence of the enhanced hydrophobicity of PEO surface appeared in the adsorption properties as well. The amount of protein adsorbed was considerbly higher at elevated temperature than at 25 °C [29].

Details of the structural change within the PEO layer as a result of the various conditions are not fully understood yet. The consideration of a layer as a whole and described by average characteristic height and volume fraction predicts an approximately 20% increase in the volume fraction and corresponding 20% decrease in the height of polymer layer [30] under theta condition compared to the case of a good solvent, like water, for the PEO at 25 °C. These structural quantities cannot be derived experimentally from wettability measurements, but it is clearly shown that the hydration of the PEO layer becomes less favored as the theta condition is approached. The behavior of the surface grafted PEO layer is proved to be similar to the phase behavior of PEO in solution. This similarity may allow an estimate to be made of the character of surface grafted polymer layers and the influence of different conditions on their properties.

Acknowledgement We thank Dr. J. Samu for fruitful discussions. The work was supported by the research projects OTKA 2164/1991, B0036/1992 and a Phare/Accord program 0380/1993.

References

1. Napper DH (1982) In: Goodwin JW (ed) Colloidal Dispersions, R Soc Chem, London
2. Tadros Th F, Vincent B (1979) J Colloid Interface Sci 72:505
3. Harris JM (ed) Poly(ethylene glycol) (1992) Chemistry, Plenum Press, New York
4. Bailey FE Jr, Koleske JV (1976) Poly(ethylene Oxide). Academic Press, New York
5. Lim K, Herron JN (1992) In: Harris JM (ed) Poly(ethylene glycol) Chemistry, Plenum Press, New York, pp 29–56
6. Kjellander R, Florin E (1981) J Chem Soc, Faraday Trans 1 77:2053
7. Karlström G (1985) J Phys Chem 89:4962
8. Saeki S, Kuwahara N, Nakata M, Kaneko M (1976) Polymer 17:685
9. Nagaoka S, Mori Y, Takiuchi H, Yokota K (1983) Polym Prepr, Am Chem Soc Div Polym Chem 24:67
10. Andrade JD (ed) Polymer Surface Dynamics, Plenum Press, New York 1988
11. Andrade JD, Smith LM, Gregonis DE (1985) In: Andrade JD (ed) Surface and Interfacial Aspects of Biomedical Polymers vol 1. Plenum Press, New York, pp 249–291
12. Kiss É, Gölander C-G, Eriksson JC (1987) Progr Colloid Polym Sci 74:113
13. Kiss É, Gölander C-G (1990) Colloids Surfaces 49:335
14. Harris JM, Struck EC, Case MG, Paley MS, Yalpani M, van Alstine JM, Books DE (1984) J Polym Sci Polym Chem Ed 22:341
15. Gölander C-G, Eriksson JC (1987) J. Colloid Interface Sci 119:38
16. Parker JL, Claesson PM, Cho DL, Ahlberg A, Tidblad J, Blomberg E (1990) J Colloid Interface Sci 134:449
17. Ford MC, Waters WA (1951) J Chem Soc 1851
18. Vogel AI (1948) J Chem Soc 616
19. Evans S, Pritchard RG, Thomas JM (1978) J Electron Spectrisc Relat Phenom 14:341
20. Herder PC, Claesson PM, Herder CE (1987) 119:155
21. Wolfram E, Faust R (1978) In: Padday JF (ed) Wetting, Spreading and Adhesion. Academic Press, London, pp 213–222
22. Wu S (1982) Polymer Interface and Adhesion. Dekker, New York
23. Kiss É, Gölander C-G (1991) Colloids Surfaces 58:263
24. Brandrup J, Immergut EH (eds) Polymer Handbook. 3rd ed. Wiley&Sons, New York 1989
25. Schönfeldt N: Oberflachenaktive Anlagerungsprodukte des Athylenoxyds. Wissenschaftliche Verlagsg, Stuttgart 1959
26. Claesson PM, Cho DL, Gölander C-G, Kiss É, Parker JL (1990) Progr Colloid Polym Sci 82:330
27. Jeon SI, Lee JH, Andrade DJ, de Gennes PG (1991) J Colloid Interface Sci 142:149
28. Gölander C-G, Kiss É (1988) J Colloid Interface Sci 121:240
29. Kiss É (1993) Colloids Surfaces A 76:135
30. Birshtein TM, Lyatskaya Yu V (1993) Proc Polymers at Interfaces Conf Univ of Bristol, vol 2 pp 1–20

Progr Colloid Polym Sci (1994) 97:27–30
© Steinkopff-Verlag 1994

Amphiphilic molecules with a structured head on a water surface: a Monte Carlo simulation

H. Stettin
H.-J. Mögel

Received: 16 September 1993
Accepted: 25 March 1994

H. Stettin (✉)
Martin-Luther-University HalleWittenberg
Institute for Physical Chemistry
Mühlpforte 1
06108 Halle/Saale, FRG

H.-J. Mögel
Freiberg University of Mining and
  Technology
Institute for Physical Chemistry
Leipziger Str.29 09596 Freiberg, FRG

Abstract Langmuir films are formed by amphiphilic molecules. These molecules consist of a more or less structured head and one or more tails of different length. The hydrophilic heads are strongly attracted by the water surface, whereas the hydrophobic tails can move in the upper half space. We have carried out athermal MC simulations on a simple cubic lattice using periodic boundary conditions in x- and y-direction.

Key words Monte Carlo simulations
– amphiphilic molecules
– liquids/order behavior

Introduction

In recent years, knowledge about the structure and properties of monomolecular layers of amphiphilic molecules has been developing rapidly. There is a wide range of applications for these quasi two-dimensional systems. We are interested in such liquid-supported monofilm properties which are of great importance for the modeling of biomembranes. From molecular-biological research it has been known for many years that the shape of the hydrophilic part strongly affects the membrane structure. Therefore, we applied Monte Carlo simulations using a simplified head-tail model for the amphiphilics. Additional to our earlier simulations where we only varied the structure of the hydrophobic tails [1–3], in the present paper we address the problem of a structured head. We are interested in the variation of the order behavior due to the changed head geometry. We have carried out calculations with molecules consisting of four head segments and up to 13 tail segments on a simple cubic lattice and obtained the dependence of a number of properties of the system upon the head density.

The Model

We have adopted the cubic lattice model proposed by Harris and Rice [4]. For the simulations, we used an athermal version of their algorithm. A monolayer is built from N single chain amphiphilic molecules consisting of 4 head segments and 3, 8, and 13 hydrophobic chain segments, giving total segment numbers $s = 7, 12$, and 17 (see Fig. 1). The head segments occupy the corners of an elementary square of the lattice. These squares stand perpendicular on the water surface, i.e., segments 1 and 4 belong to the layer $z = 0$ and segments 2 and 3 belong to $z = 1$. The hydrophobic tails join segment 3 in layer $z = 1$. The constraints of chain connectivity require that consecutive elements of a chain lie on adjacent sites. Each site can be occupied by no more than one segment. The hydrophobic tail segments are presumed to be insoluble in the dense liquid substrate. All molecular configurations that would place tail segments below the surface plane ($z < 0$) are forbidden. Every simulation was equilibrated by 10^6 attempted moves. Each of the values presented below is derived from an average of at least $2 \cdot 10^3$ configurations

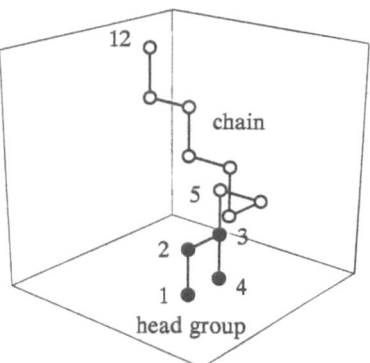

Fig. 1 A typical molecule with 8 tail segments and a total segment number of $s = 12$

with 500 attempted moves between configurations. For each move a chain was selected randomly, the chain was erased, and a trial chain was regrown using a self-avoiding walk in the cage of the surrounding molecules. The trial chain was accepted as the new chain with a probability which equals the ratio of the Rosenbluth weights of the new and old molecules [5]. The surface plane is taken to be a square lattice with n surface sites. The head segment density $\Phi = 2N/n$ (surface coverage) is a measure of the area density of the monolayer and may vary within $0 \leq \Phi \leq 1$. It was varied within $0.005 \leq \Phi \leq 0.9$.

Calculated properties

We were interested in calculating mean values of the molecular geometry and the associated order behavior as well as the lateral pressure. From end-end vectors $r_i = (\Delta x_i, \Delta y_i, \Delta z_i)$ of the molecules ($1 \leq i \leq N$) several mean values have been estimated. The angular brackets indicate the average over at least $2 \cdot 10^3$ system configurations:

The mean end-end distance is a measure of the molecular size which is governed by the equilibrium conformations [1]. The degree of order values the mutual alignment of all end-end vector pairs and detects if a direction is preferred without telling which direction this is [1].

The bond order parameter

$$b(p) = \frac{1}{N} \sum_{i=1}^{N} (3 < (z_{i,p+1} - z_{i,p})^2 > - 1)/2 \tag{1}$$

values the alignment of each bond ($p = 1, \ldots, s - 1$) in reference to the layer normal. It provides additional information to the degree of order. $b(p) = -0.5$ means the p-th bond lies in a plane parallel to the water surface ($z = 0$) and $b(p) = 1$ states that the bond exactly points in z-direction.

Results

The end-end distance was estimated in dependence on the head density for different segment numbers. For fixed number of segments the end-end distance is a growing function of head density. The more segments per molecule, the greater is the end-end distance at constant head density. This qualitative behavior already was found for molecules without structured head [1]. The 7 segment molecules with structured head in their stretched-out configuration reached a height of 4 bond lengths. A comparison with linear 5 segment molecules with a 1 segment head (also 4 bond length in stretched-out configuration) yields a very similar course of $r(\Phi)$. In general, the end-end distance of molecules with a 1 segment head is slightly smaller compared to that of molecules with structured head. The comparison between 12 and 17 segment molecules with structured head and 10 and 15 segment molecules with a simple head, respectively, provides the same results.

The degree of order is strongly influenced by the large head which is fixed in the direction perpendicular to the water surface. For this reason, two different order parameters are required: the usual one for the hydrophobic tails and a second one for the head. The head order parameter must indicate the mutual planar alignment of the heads in the water surface. At high head density the heads are not aligned randomly, but as is seen in Fig. 2, they form domains of mutually aligned heads.

In Fig. 3 the mean bond order parameter $b(p)$ of only the hydrophobic chains for 17 segment molecules is drawn

Fig. 2 Arrangement of head bonds and tail segments in layer $z = 1$ for molecules with $s = 17$ at a head density $\Phi = 0.90$; bond lines connect the second and third segment of the heads, the circles are tail segments, vertically aligned heads are gray, whereas horizontally ones are black

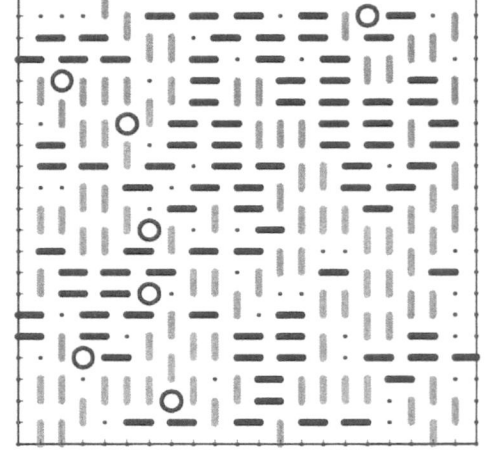

versus the bond number p for different head densities Φ. All curves start off at $b(3)$ because the first three bonds have fixed bond order parameters. All curves approach $b \approx 0$ with growing bond number. The higher the head density the greater is the bond order parameter for a determined bond. However, bond number 5 disturbs the monotonous course. At low head densities it is more aligned in the z-direction than the adjoining bonds 4 and 6, whereas at high head densities it is less aligned. That means the bond order parameter of the fifth bond increases less than for the neighbor bonds. The same holds for the 5th bond of the 12 segment molecules whereas the bond order parameter for the chain bonds of the 7 segment molecules is a monotonous decreasing function of bond number. Molecules with 4, 9, and 14 tail segments and 1 head segments show very similar bond order parameter behavior: at fixed head density the first bond has a certain order parameter. The second bond in general is most aligned in z-direction. With further growing bond number the order parameter

decreases, approaching zero. The higher the head density the greater is the bond order parameter.

The density profile of 17 segment molecules including all head segments is shown in Fig. 4. Φ is the head segment fraction in layers $z = 0$ and $z = 1$ and $\Phi(z)$ is the tail segment fraction of occupied sites in layer z. Each curve has a maximum in the first layer. With increasing height all curves approach zero density. However, with increasing head density the monotonous behavior diminishes, changing into a slight minimum at $z = 2$. The density profiles above $z = 1$ qualitatively agree with those of molecules with unstructured 1 segment heads [1].

Discussion

Lattice MC simulations have been carried out for molecules with a structured head. The hydrophobic chains were connected at the head and varied in length from 3 to 13 tail

Fig. 3 Bond order parameter $b(p)$ for the hydrophobic tails of 17 segment molecules in dependence on the bond number p for different head densities Φ

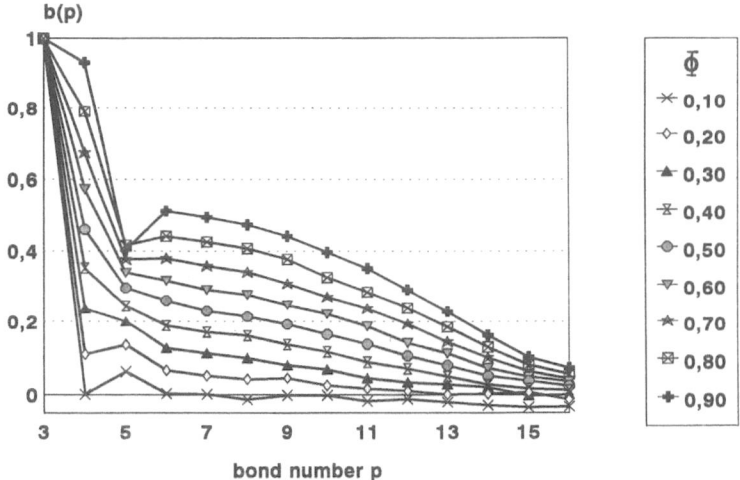

Fig. 4 Density profiles for 17 segment molecules at different head densities, Φ is the head density, and $\Phi(z)$ is the tail segment layer fraction in layer z

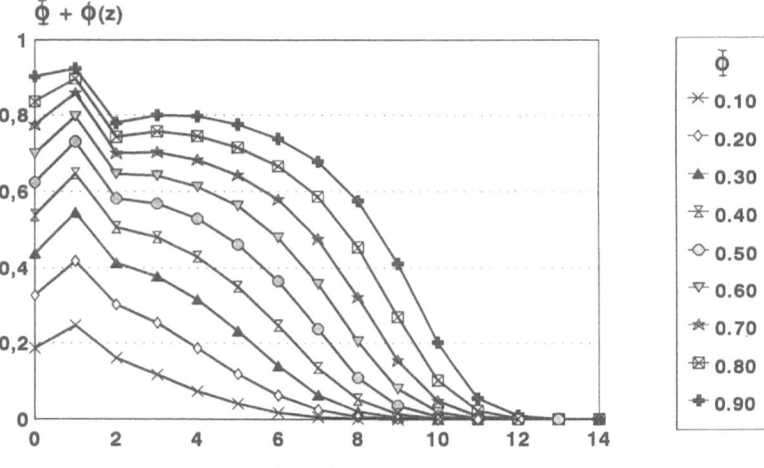

segments. The mean end-end distance of the molecules increased with growing head density in a monotonous manner for all chain lengths. For the mean radius of gyration qualitatively the same course was estimated.

The domains in Fig 2 are due to the repulsive interaction of the 2 segment head bonds at high head densities. With decreasing head density the correlation decreases and vanishes completely only at very low head densities. The longer the hydrophobic tails the stronger is the alignment correlation at fixed head density. At low head densities lattice sites which are occupied by tail segments additionally support the head alignment.

The bond order parameter at moderate and high head densities behaves differently in comparison to molecules with 1 head segment. The fifth segment at high head densities shows a greater tendency to align planar in comparison to the neighboring bonds (see Fig. 3). This can be explained by the different cross-sections: the cross-sectional area of a head is two lattice sites whereas a tail needs only one. At high head densities the first tail segment with high probability occupies a lattice site above the head in layer $z = 2$. The fifth segment can occupy a site in layer $z = 2$ with high probabilty. Obviously, this is done and gives rise to a preferred planar alignment.

The first maximum of the density profiles of Fig. 4 arises from the high number of tail segments in this layer. The tail segments are placed in this layer by lateral alignment and downwards and upwards backfolding. The effect of backfolding is more pronounced in the low density region.

Acknowledgements I wish to thank the Deutscher Akademischer Austauschdienst (DAAD, FRG) for giving the grant no. 517 009 505 2. The investigations also were supported by the Deutsche Forschungsgemeinschaft (DFG) within the Sonderforschungsbereich 197: Lipidorganisation und Lipid-Protein-Wechselwirkung in Bio- und Modellmembranen.

References

1. Stettin H, Mögel H-J, Friedemann R (1993) Ber Bunsenges Phys Chem 97(1): 44–48
2. Stettin H, Care CM (submitted) J Chem Soc Far Trans
3. Stettiln H, Mögel H-J (1994) Prog Colloid Polym Sci 97:31–34
4. Harris J, Rice SA (1988) J Chem Phys 88(2): 1298–1306
5. Rosenbluth MN, Rosenbluth AW (1955) J Chem Phys 23(2): 356–59

Progr Colloid Polym Sci (1994) 97:31–34
© Steinkopff-Verlag 1994

H. Stettin
H.-J. Mögel

Branched amphiphilic molecules on a water surface: a Monte Carlo simulation

Received: 16 September 1993
Accepted: 25 March 1994

H. Stettin (✉)
Martin-Luther-University Halle-Wittenberg
Institute for Physical Chemistry
Mühlpforte 1
06108 Halle/Saale, FRG

H.-J. Mögel
Freiberg University of Mining and
   Technology
Institute for Physical Chemistry
Leipziger Str. 29 09596 Freiberg, FRG

Abstract Amphiphilic molecules consist of a hydrophilic head and one or more hydrophobic tails. The heads are strongly attracted by the water surface, whereas the tails can move in the upper half space. We investigated the behavior of a monomolecular film consisting of branched amphiphilic molecules in dependence on the head density and the side chain length. The Monte Carlo simulations were carried out on a simple cubic lattice using periodic boundary conditions in x- and y-direction.

Key words Lattice Monte Carlo simulations – amphiphilic molecules – surfaces/liquids – order behavior

Introduction

Assemblies of amphiphilic molecules are of fundamental interest. The possible application range extends from electronic and electro-optical devices to drug transport within the human body, and lipid mono- and bilayers serve as model membranes in biology. The molecules of these layers mainly consist of a hydrophilic head and two or more hydrophobic chains [1]. Extensive theoretical and experimental work has been carried out on liquid supported Langmuir films and on Langmuir–Blodgett films transferred onto solid surfaces. The standard method for investigations of amphiphilic molecules on liquid supported surfaces (usually water) is the estimation of π-A isothermes with a Langmuir trough [1,2].

Monomolecular films show a rich polymorphism with phase transitions in dependence on the temperature, lateral pressure, and the molecular species [2]. These effects require a statistical interpretation on a molecular level. Monomolecular layers have been studied by Monte Carlo

simulations within several model systems [3–6]. Our athermal Monte Carlo simulations of branched chain molecules were carried out on a simple cubic lattice. The results are compared with simulations of linear (nonbranched) molecules on the same lattice [7]. Our goal is to study how a side chain connected near the head segment influences the behavior of linear molecules.

Model

We have used the cubic lattice model proposed by Harris and Rice [3]. A monolayer is built up from N amphiphilic chain molecules consisting of $s = s_1 + s_2$ segments. s_1 segments form the main chain. The first segment is the hydrophilic head which is strongly attracted to the liquid surface. All other units are hydrophobic. At the second unit a further hydrophobic chain is connected with s_2 segments ($s_1 \geq s_2$). The number of segments in the main chain was maintained at $s_1 = 10$. The number of side chain segments was varied in $0 \leq s_2 \leq 8$. For the simulations we use an athermal version of the algorithm proposed by Harris and Rice. The constraint of chain connectivity requires that consecutive segments of a chain lie on adjacent sites and that no more than one segment can occupy a site. Within this model we already investigated linear molecules with varying chain length and molecules with a structured head. The surface is taken to be a square lattice with n surface sites. The head or molecular fraction $\Phi = N/n$ (surface coverage) is a measure of the density of the system. It may vary in $0 < \Phi < 1$.

Calculated properties

We are interested in calculating average values of the molecular geometrical properties, the associated order parameter, and the π-A isotherms. From end-to-end vectors $r_{i,k} = (\Delta x_{i,k}, \Delta y_{i,k}, \Delta z_{i,k})$ of the molecules several mean values have been estimated as the end-end distance of the main chains and of the side chains, as well as the total degree of order by taking into consideration all end-to-end vectors [7]. The index i labels the number of the molecules ($1 \leq i \leq N$), whereas k labels the main and the side chains ($k = 1$ for main chain, $k = 2$ for side chain). Furthermore, the radius of gyration was estimated which is a measure of the total mean molecular size:

$$r_{\mathrm{gyr}} = \left\langle \frac{1}{Ns} \sum_{i=1}^{N} \sum_{p=1}^{s} (r_{i,p} - r_{i,\mathrm{com}}) \right\rangle^{1/2}, \tag{1}$$

with $r_{i,p}$ the chain vector (actually p vectors) containing all segment coordinates and $r_{i,\mathrm{com}}$ the center of mass of the i-th chain molecule. The angular brackets indicate the average of the property over at least $2 \cdot 10^3$ system configurations.

The lateral pressure π can be estimated using the virial theorem of Clausius. Within an athermal lattice system there is only a contribution to the virial if two segments lie on adjacent lattice sites. For the *lateral* pressure, due to pure repulsive interactions, we have only to count the number of nearest neighbor segment pairs in x and y direction. Thus, we get Eq. (2) with K being the force between two neighbor segments (typically 10^{-10} N) and L being the lattice constant (typically 5 Å). The upper term counts the intermolecular nearest-neighbour segment pairs which lie in the same z plane, while the lower term counts the intramolecular nearest neighbor segment pairs.

$$\pi A = K \cdot L \left\langle \sum_{i=1}^{N-1} \sum_{j>i}^{N} \sum_{p=1}^{s} \sum_{q=1}^{s} \delta(|r_{i,p} - r_{j,q}| - 1) \right.$$
$$\times \delta(z_{i,p} - z_{j,q}) \Big\rangle$$
$$+ K \cdot L \left\langle \sum_{i=1}^{N} \sum_{p=1}^{s} \sum_{q>p}^{s} \delta(|r_{i,p} - r_{i,q}| - 1) \right.$$
$$\times \delta(z_{i,p} - z_{i,q}) \Big\rangle \tag{2}$$

Results

Every simulation was equilibrated by 10^6 attempted moves starting with uniformly stretched-out chains. Each of the values presented below is derived from an average of at least $2 \cdot 10^3$ configurations with 500 attempted moves between configurations. For each move a molecule was selected randomly, both chains were erased from the lattice and a trial molecule was regrown using a self-avoiding walk. The trial molecule was accepted as the new one with a probability which equals the ratio of the Rosenbluth weights of the new and old molecules [3, 4, 9]. The head density was varied within $0.0025 \leq \Phi \leq 0.45$.

Figure 1 shows a molecule with $s_2 = 4$. In the case of $s_2 = 8$ both chains would have the same length.

The mean end-to-end distance $r_1(\Phi)$ of the main chains increases with growing head density Φ. The onset of $r_1(\Phi)$ at small head density (Φ nearly 0) hardly depends on the number of side-chain segments. For $0 \leq s_2 \leq 6$ holds: the more side-chain segments the greater is the end-to-end distance of the main chains at fixed head density. The end-to-end distance r_1 for $s_2 = 8$ has slightly diminished in comparison to molecules with $s_2 = 6$.

The mean end-to-end distance $r_2(\Phi)$ of the side chains in dependence on the head density Φ exhibits a different behavior. In general, the end-to-end distance r_2 is smaller than r_1. For $s_2 = 1$ the end-to-end distance weakly decreases with increasing Φ ($r_2(\Phi = 0.005) = 1.57$, $r_2(\Phi = 0.45) = 1.54$). For $s_2 = 4$ $r_2(\Phi)$ decreases till $\Phi = 0.1$ and

Fig. 1 A typical molecule with 4 side-chain segments

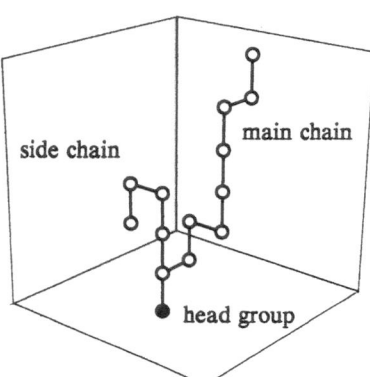

side chain

main chain

head group

Progr Colloid Polym Sci (1994) 97:31–34
© Steinkopff-Verlag 1994

then monotonously increases with Φ. If $s_2 = 8$ then the end-to-end distance r_2 is a monotonous growing function of Φ. The more side chain segments the higher is the onset of $r_2(\Phi)$ at small head densities. For $s_2 = 8$ the curves $r_1(\Phi)$ and $r_2(\Phi)$ agree within the error bars as expected. The radius of gyration r_{gyr} of the entire chain molecules in dependence on the head density for different numbers of side chain segments is shown in Fig. 2. For fixed number of side chain segments r_{gyr} first decreases with increasing head density, reaching a minimum at about $\Phi = 0.10$ and finally increases. The more side chain segments in the molecule the greater is the radius of gyration at constant head density. With increasing number of side chain segments the minimum of r_{gyr} is shifted to higher head densities.

The total degree of order of all end-end vectors is a growing function of the head density Φ. However, for small head densities ($\Phi < 0.05$), we could not detect an alignment, neither for the main chains, nor for the side chains, no matter if the side chains were short or long. With increasing head density the unbranched 10-segment molecules start to align in z direction. Adding 1 and 2 side-chain segments strongly decreases the degree of order in comparison to molecules without side-chain segments. Molecules with 4 side chain segments behave similar to unbranched ones. Molecules with 6 and 8 side-chain segments yield a degree of order which more strongly increases in dependence on head density in comparison to unbranched molecules.

In Fig. 3 the lateral pressure π is shown in dependence on the lattice sites per molecule ($= 1/\Phi$). For fixed number of side-chain segments the lateral pressure strongly in-

Fig. 2 Radius of gryration r_{gyr} of the entire molecules in dependence on the head density Φ for different numbers of side-chain segments s_2

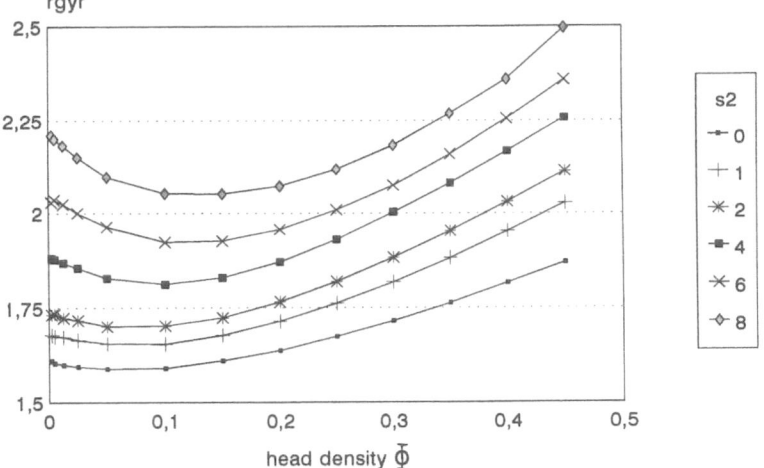

Fig. 3 Lateral pressure π in arbitrary units in dependence on the number of lattice sites per molecule ($1/\Phi$) for different numbers of side-chain segments s_2

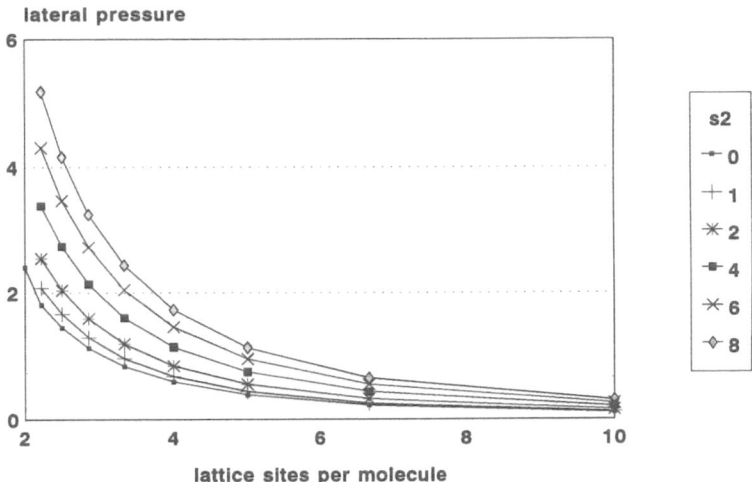

Fig. 4 Pressure profile for molecules with 4 side-chain segments: contribution of each z-layer to the lateral pressure π for different head densities Φ

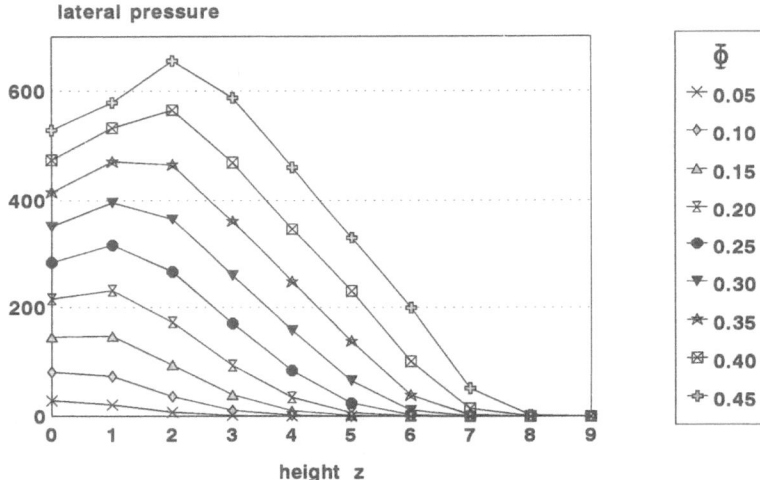

creases with decreasing area per molecule. The more side-chain segments, the higher is the lateral pressure at constant area per molecule.

Discussion

The main chains show a different behavior in comparison to the side chains. In the case of short side chains ($s_2 < 4$) the end-to-end distance of the side chains weakly decreases with increasing head density. The side chains are pushed down, giving rise to smaller end-to-end distances because the occupancy of lattice sites by main chain segments in upper z-layers ($z = 3$ and 4) increases with growing head density.

The mean size of the entire molecules is described by the radius of gyration. Figure 2 clearly shows a minimum behavior in dependence on the head density. This can be explained by a competition between *lateral compression* and *normal extension*. At low head densities the lateral compression predominates until the molecular excluded volume is minimized, whereas at high head densities the extension prevails.

The total order behavior clearly shows that short side

chains strongly destroy the order tendency, but long side chains again support the alignment. The lateral pressure integrates over the layer normal. In Fig. 4, we present the contribution of each layer to the lateral pressure for molecules with 4 side-chain segments. It is clearly seen that the pressure changes with height. For all head densities the pressure approaches zero for $z = 9$. The largest contribution at highest head density arises from layer $z = 2$. However, the maximum of segments is located in layer $z = 1$, whereas the maximum pressure (and together with it, the maximum number of non-bonded nearest-neighbor segment pairs) lies in layer $z = 2$. This is due to the connectivity of the chain molecules. Weitzel and coworkers estimated π-A isotherms of branched carbonic acids and the dependence on the side-chain length [8]. (We found the same sequence of isotherms, as can be seen in Fig. 3.) They estimated further effects connected with the π-A isotherms which cannot be investigated with our coarse-grained model.

Acknowledgements The investigations were supported by the Deutsche Forschungsgemeinschaft (DFG) within the Sonderforschungsbereich 197: Lipidorganisation und Lipid-Protein-Wechselwirkung in Bio- und Modellmembranen.

References

1. Gaines Jr, GL (1965) Insoluble Monolayers at Gas–Liquid Interfaces. Intersciences, New York
2. Dietrich A, Möhwald H, Rettig W, Brezesinski G (1991) Langmuir 7(3):539–46
3. Harris J, Rice SA (1988) J Chem Phys 88(2):1298–1306
4. Siepmann JI, Frenkel D (1992) Mol Phys 75(1): 59–70
5. Scheringer M, Hilfer R, Binder K (1992) J Chem Phys 96(3):2269–77
6. Haas F, Lai P-Y, Binder K (in press) Makromol Chem
7. Stettin H, Mögel H-J, Friedemann R (1993) Ber Bunsenges Phys Chem 97(1):44–48
8. Weitzel G, Fretzdorff A, Savelsberg W (1950) Hoppe-Seylers Z Physiol Chem 285:230–37
9. Rosenbluth MN, Rosenbluth AW (1955) J Chem Phys 23(2): 356–59

Progr Colloid Polym Sci (1994) 97:35–39
© Steinkopff-Verlag 1994

C. Johner
C. Graf
U. Hoß
H. Kramer
C. Martin
E. Overbeck
R. Weber

Static light scattering by aqueous, salt-free solutions of charged polystyrenesulfonate at different molecular weights

Received: 16 September 1993
Accepted: 25 March 1994

C. Johner (✉) · C. Graf · U. Hoß ·
H. Kramer · C. Martin · E. Overbeck ·
R. Weber
Fakultät für Physik
Universität Konstanz
78464 Konstanz, FRG

Abstract Static light-scattering measurements on aqueous solutions of charged polystyrenesulfonate (PSS) with five molecular weights between 1 100 000 g/mol and 1 132 000 g/mol are presented. All experiments were performed in the dilute/semidilute regime at minimum ionic strengths ($\approx 10^{-6}$ M) in order to maximize the electrostatic interaction between monomers of different polymers or monomers belonging to the same polymer. A single broad peak in the scattered intensity was always found. The scattering vectors of these peaks increase with increasing concentrations c and scale either with $c^{1/3}$ or with $c^{1/2}$, depending on concentration: Below about $20 c^* \{1 c^* := 1$ particle/(contour lengths $l_c)^3\}$, we found the $c^{1/3}$ dependence of the scattering vector; above $20 c^*$, the $c^{1/2}$ law is valid. A very similar behavior has been observed for rigid rods [1]. Furthermore, our results are compared with previous light-[2, 3], small-angle neutron-[4] and small-angle x-ray [5]-scattering investigations. Nearly all of these studies confirm the validity of the $c^{1/3}$- and $c^{1/2}$-law, respectively. It is claimed that these laws can be explained by a flexibility of more or less strongly elongated PSS-rods which decreases with increasing molecular weight. Theories calculating the persistence lengths of charged linear polymers [6], and measurements of the electric birefringence being performed at present confirm our findings.

Key words Static light scattering
– flexible polyelectrolyte
– polystyrenesulfonate – persistence length

Introduction

Polyelectrolytes have been developed into one of the most examined objects in colloidal physics. While both solutions consisting of charged spheres [7] and solutions of rigid rods [1] are quite well investigated and theoretically understood, the structural properties of solutions of polyelectrolytes like polystyrenesulfonate are studied under various experimental conditions. Previous SANS [4] and SAXS [5] investigations dealt with concentrations between 5 g/l and 300 g/l. These regimes are partially covered theoretically by Odjik [6], de Gennes [8], and Koyama [9], who calculated the persistence length and transitions concentration between different conformations of the polymers.

This work is concerned with solutions of polystyrenesulfonate in order to get more information about the conformation and shape of the particles in the dilute/semidilute regime (0.005 g/l < c < 0.05 g/l). We are especially interested in the changes due to the flexibility of

36

C. Johner et al.
Static light scattering of salt-free solutions of polystyrene sulfonate

these chains which is determined by the contour length l_c ($= N \cdot$ monomer length), the particle concentration, and the ionic strength. According to Odijk [6], this may be expressed by the persistence length

$$l_p = l_i + l_e , \qquad (1)$$

where l_i is the intrinsic part of the persistence length (1.2 mm for PSS) and l_e the electrostatic part. In order to maximize the Coulomb interaction of the charged monomers, deionized ("salt-free") solutions were chosen. With a special experimental setup, we reached minimum ionic strengths of about 10^{-6} M. Until now, only the data of Drifford et al. [3] being measured on samples of MW = 780 000 g/mol (l_c = 950 nm) and Krause et al. [2] on samples of MW = 354 000 g/mol (l_c = 430 nm) and MW = 1 060 000 g/mol (l_c = 1286 nm), respectively, were available in the light-scattering regime, both in relative contradiction to each other. The aim of our work was to extend the range of molecular weights from MW = 1 100 000 g/mol (l_c = 121 nm) to MW = 1 132 000 g/mol (l_c = 1374 nm) and to reconcile all scattering experiments mentioned above. All these results agree in that a peak in the scattered intensity is observed which is caused by intermolecular electrostatic correlations. It is observed above and below the overlap concentration. Kaji et al. [5] suggest a critical concentration c_R^* which separates the dilute and the semidilute regime:

$$c_R^* = \frac{N}{\frac{4}{3} \cdot \pi \cdot \left(\frac{\langle R^2 \rangle^{1/2}}{2} \right)^3} , \qquad (2)$$

where $\langle R^2 \rangle$ means the mean-square end-to-end distance:

$$\langle R^2 \rangle = 2 \cdot l_p^2 \cdot (\alpha - 1 + e^{-\alpha}) , \qquad (3)$$

where $\alpha = l_c / l_p$. With decreasing concentration the peak indicating a liquid-like phase disappears, since the intermolecular distances are getting larger than the electrostatic screening length κ^{-1}, the intermolecular correlations are lost, and a gas-like phase is obtained. This crossover concentration c_κ can be estimated [5] as

$$c_\kappa^* = \frac{1}{4 \cdot \pi^3 \cdot a^3 \cdot N^2} , \qquad (4)$$

where a is the length of a monomeric unit being equal to 0.25 nm.

The measurements using the transient electric technique apparatus have been started in order to determine the rotational diffusion constant D_R. There are some theories both for rigid rods and for flexible chains calculating D_R and the time constant τ of the decay of the birefringence signal. The relation for the former quantity is given by Newman et al. [10], and, quite recently, the validity was

again confirmed by Kramer et al. [11] for the rigid tobacco mosaic virus (TMV) and the only slightly flexible fd-virus:

$$D_{R,\text{rod}} = \frac{3 \cdot k_B \cdot T}{\pi \cdot \eta \cdot L^3} \cdot$$

$$\left[\ln \left(\frac{L}{d} \right) - 0.76 + 7.5 \cdot \left\{ \frac{1}{\ln \left(2 \cdot \frac{1}{d} \right)} - 0.27 \right\}^2 \right] , \qquad (5)$$

where kT is the thermal energy, η is the solvent viscosity, L and d the length and the diameter of the rod. The relation for τ is calculated by Yoshizaki et al. [12]:

$$\tau(x) = \frac{1}{6 \cdot D_{R,\text{rod}}} \cdot \left[x + \frac{e^{-2 \cdot x} - 1}{2} \right]^{1.5}$$

$$\cdot \frac{[1 + 0.539\,526 \cdot \ln(1 + x)]}{x^3} , \qquad (6)$$

where $x = l_c / l_p$. This equation enables the estimation of the persistence length of our polymers: In the case of $x \ll 1$, the rigid rod limit is reached; in the case of $x \approx 1$, we have distinct flexible polyelectrolytes, and in the case of $x \gg 1$ the polymers can be considered as Gaussian coils.

Experimental

The sodium salt of polystyrenesulfonate (NaPSS) (supplied by Polyscience) is, according to the manufacturer, characterized by $M_w / M_n \le 1.1$, and it was used without further purification. In order to obtain polyelectrolytes with various contour lengths l_c, we used five different molecular weights: MW1 = 100 000 g/mol (l_c = 121 nm), MW2 = 200 000 g/mol (l_c = 243 nm), MW3 = 400 000 g/mol (l_c = 485 nm), MW4 = 780 000 g/mol (l_c = 946 nm) and MW5 = 1 132 000 g/mol (l_c = 1373 nm). First, for each molecular weight a stock solution of 1 g/l was prepared by dissolving the carefully weighted salt in deionized water ($R > 18$ MΩ), then the solution was checked by an absorption measurement. The absorption coefficient (measured with a Beckmann spectrometer DU 64, Darmstadt, FRG) at the absorption maximum ($\lambda = 224$ nm) was in a good agreement with that of other authors [13, 14]. The desired concentrations were obtained by dilution. In order to avoid any dirt or dust all vessels (for example the scattering cells) were cleaned with acetone following ethanol p.a. and finally excessively with highly purified water. Every step of preparation occurred in a dust-free flow box. In order to reach the minimum ionic strength, a cleaned mixed-bed ion exchange resin (Serva, lot No. 45500) (but no neutralizing agent such as NaOH) was added to each sample. As described in [2] the dissolved NaPSS is nearly completely (80%) converted to HPSS. To be sure that there were not any dust particles left in the scattering volume, all samples were centrifuged at 6000 rpm for at least 3 h before each measurement.

The light-scattering apparatus is a commercial instrument (ALV, Langen, FRG) consisting of a computer-controlled goniometer table with focusing and detector optics, a power stabilized 3 W argon laser (Spectra Physics), a digital rate meter, and a temperature control which stabilizes the temperature of the sample cell at $21 \pm 0.1°$ C. The available scattering vector

$$q = \frac{4 \cdot \pi \cdot n}{\lambda_0} \cdot \sin\left(\frac{\vartheta}{2}\right) \qquad (7)$$

ranges from $0.01 \text{ nm}^{-1} \leq q \leq 0.033 \text{ nm}^{-1}$, where $\lambda_0 = 488$ nm is the vacuum wavelengths of the incident beam, $n = 1.33$ the refractive index of the solution, and ϑ the scattering angle. The intensities were measured in steps of 5° and normalized to a reference sample (toluene) to correct power fluctuations and to get a standard of the incident laser intensity. Furthermore, the background scattering due to water and the dark rate of the photo multiplier were subtracted from the measured count rate. Also, the usual correction due to the geometrical scattering volume ($\sin \vartheta$) was made.

The used birefringence apparatus, described in [11], partially orientates the PSS chains in an electric field, which leads to a birefringence signal. After the applied electric field has been switched off the birefringence signal decays. In the case of a monoexponential decay the signal can be fitted by

$$\Delta n(t) = \Delta n_0 \cdot e^{-\frac{t}{\tau_1}}, \qquad (8)$$

where $\Delta n(t)$ is the decay of the electric birefringence and Δn_0 is the steady-state value. The time constant τ_1 is related to D_R via

$$\tau_1 = \frac{1}{6 \cdot D_r}. \qquad (9)$$

Results and discussion

The plots of the normalized intensities $I(q)$ of all molecular weights and measured concentrations c exhibit a single broad but well defined peak at certain q_m. In addition, $I(q)$ increases with decreasing q as observed previously [2, 3]. The peaks for polystyrenesulfonate with

MW1 = 100 000 g/mol were hardly detectable because of a very low scattering intensity. Figure 1 shows $I(q)$ for four different concentrations of MW5.

In Fig. 2 the position of the peak maxima for all molecular weights are plotted versus concentration.

Very similar to the results found by Krause et al. [2], a gap between the peak positions of the higher (MW $\geq$ 780 000 g/mol) and the lower (MW $\leq$ 400 000 g/mol) molecular weights appears. The data obtained by Drifford et al. [3] (MW = 780 000 g/mol) lie somewhat below ours.

Always the position of the maxima scale with a certain power law: For the lower molecular weights, we found $q_m \propto c^{1/3}$, for the higher ones the exponents increase, pointing to a "non-Gaussian" slope of our polyelectrolytes: It would be expected that in the dilute/semidilute regime Gaussian coils behave very similar to spheres with a well known $c^{1/3}$-dependence of the scattering vector of

Fig. 1 Normalized scattering intensities $I(q)$ of PSS (molecular weight = 1 132 000 g/mol) with following concentrations $c = 0.015$ g/l, $c = 0.031$ g/l, $c = 0.040$, $c = 0.050$ g/l (from top to bottom)

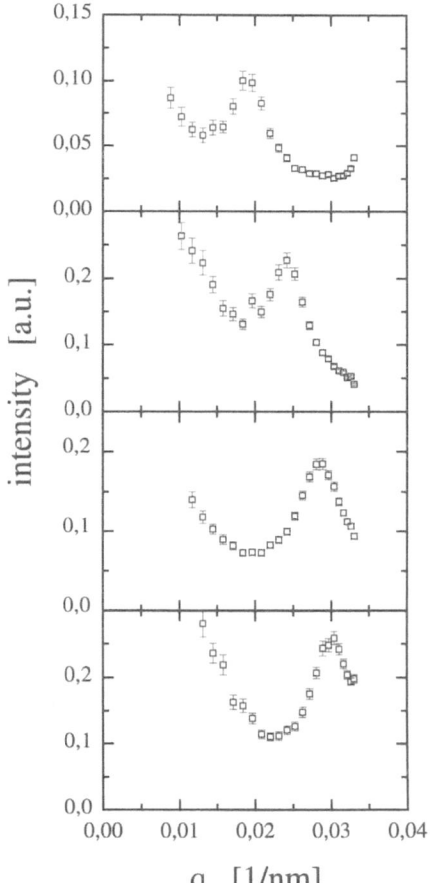

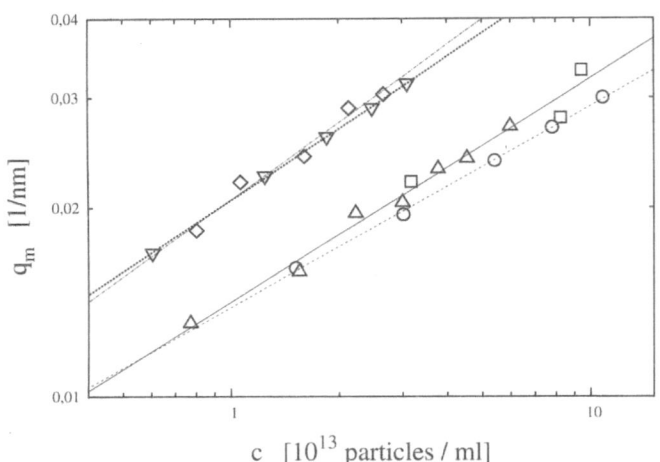

Fig. 2 Position of q_m of the samples of MW1 = 100 000 g/mol ($\square$), MW2 = 200 000 g/mol ($\bigcirc$), MW3 = 400 000 g/mol ($\triangle$), MW4 = 780 000 g/mol ($\triangledown$) and MW5 = 1 132 000 g/mol ($\diamondsuit$) in dependence of concentration is shown in a double logarithmic scale. Also, the accompanying fits are plotted for MW2 (– – – –), MW3 (———), MW4 (· · · ·), and MW5 (– · – · –). The results are listed in Table 1. The lines fits the $q_m(c)$-values

the peak maximum. The exact values of our exponents are listed in Table 1.

Koyama et al. [9] predict for very high $c \cdot N$ values

$$q_m \propto \left(\frac{c}{N}\right)^{\frac{1}{2}}, \tag{10a}$$

and

$$q_m \propto (c)^{\frac{1}{3}}, \tag{10b}$$

for very small $c \cdot N$ values (dilute solutions) which are obtained for particles reduced to their centre of mass. The molecular weights MW1, MW2, and MW3 fulfill the condition of the latter case and are in agreement with Koyamas' prediction. The higher ones (MW4 and MW5) seem to be in the transition regime between these two extrema. In Fig. 3 results from light-scattering, SANS-, SAXS experiments performed at PSS are "rescaled" by plotting $q_m \cdot l_c$ versus c^* ($= 1/l_c^3$).

Table 1. The results from a least square fit of $q_m(c)$ are listed for all molecular used weights. (See Fig. 2).

Molecular weight MW [g/mol]	Linear fit for $q_m(c)$: $A + B \cdot x$	
	A	B
100 000	−1.80	0.29 ± 0.11
200 000	−1.89	0.32 ± 0.02
400 000	−1.85	0.35 ± 0.02
780 000	−1.69	0.38 ± 0.04
1 1132 000	−1.69	0.41 ± 0.04

Obviously, Eqs. (10a) and (10b) are generally valid over seven orders of magnitude, and there is no dependence recognizable on the contour length l_c or the absolute concentration (for example in mg/ml) and all samples have a critical concentration-expressed in units of c^*-in common; below this the 1/2-exponent is found, above the 1/2-exponent is valid. These facts are comparable to the behavior of rigid rods like TMV or fd [1]; the power laws obtained by scattering experiments for these rods are shown in Fig. 3 as dotted lines. Assuming an effective length l_{eff} allows us to attribute our $q_m \cdot l_c$ values to the already known "rigid rod values", that means in static scattering experiments the polyelectrolytes are behaving like rigid rods with new lengths l_{eff} which are shown in Fig. 4.

With increasing concentrations (units in c^*!) l_{eff} is getting smaller until a steady state value of about 0.35 times l_c is reached. Thus, we have a first hint that the short PSS chains (MW1 and MW2) are more or less stretched while the longer chains are becoming distinctly bent.

Odijk et al. [6] calculated the persistence length as follows

$$l_p = \frac{1}{4 \cdot Q \cdot \kappa^2} + l_{i;} \quad \sigma \leq Q \tag{11a}$$

$$l_p = \frac{Q}{4 \cdot \kappa^2 \cdot \sigma^2} + l_i \quad \sigma > Q, \tag{11b}$$

where $Q = \dfrac{e^2}{4 \cdot \pi \cdot \varepsilon_0 \cdot \varepsilon \cdot k \cdot T} \approx 0.7$ nm is the Bjerrum length in water ($T = 21\,°C$) and σ is the mean distance of two neighboring charges, and it is assumed to be about the

Fig. 3 Wave vector of the maximum of intensity times the contour length l_c of the polyelectrolytes versus relative concentration c in units of c^* ($1c^* = 1$ particle/l_c^3). The symbols are defined in Fig. 2. Also, the results from other authors are shown: ($\times$) from [5], ($*$) from [2], and ($|$) from [4]. The dotted lines represent the $c^{1/3}$- and $c^{1/2}$-law, obtained by scattering experiments on rigid rods [1]

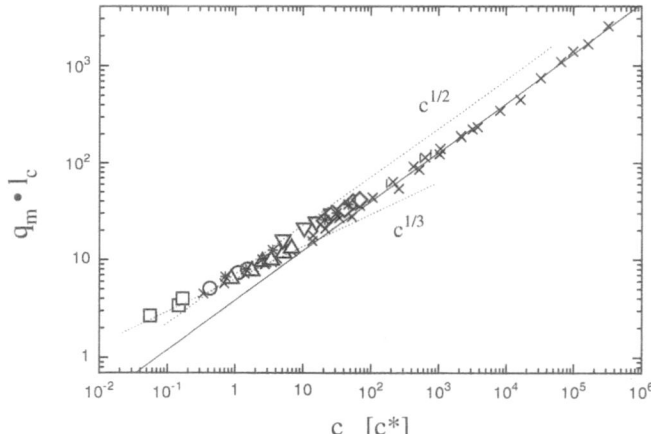

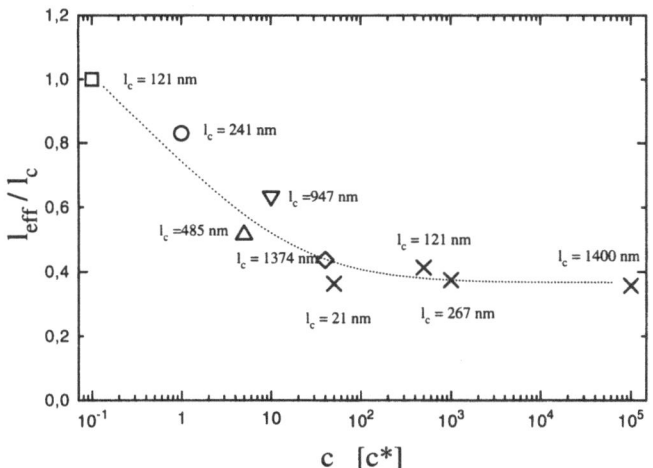

Fig. 4 Effective length of the particles divided by their contour length l_c versus relative concentration c in units of c^*. The symbols are the same as used in Fig. 3. The dotted line is only a guide to the eye

length of a monomeric unit $a = 0.25$ nm. Therefore the formula (11a) has to be taken in order to calculate the persistence lengths of PSS. In the case of no salt (minimum ionic strength), we obtain for our concentration regime for MW2 (as an example) a persistence length which is 10 times larger than the contour length; that means these rods are rigid. On the other hand, for MW5 a persistence length which is about the same as the contour length is found, in this case the PSS molecular is clearly bent. This result is in a very good agreement with our first hint.

Furthermore, the decay of the birefringence signal after a rectangular electric pulse was measured for MW2, MW3,

and MW5. In the case of MW2 and MW3, this decay was exactly monoexponential, signifying that only one relaxation process – the rotational diffusion-occurs. In the case of MW5, however, the decay seems to be composed of two exponentials. There is a second relaxation process, indicating internal modes.

Conclusions

Static light-scattering experiments on aqueous solutions of polystyrenesulfonate show a single broad but well defined peak. The position q_m of this peak scales with the concentration c as $c^{1/3}$ for diluted solutions and like $c^{1/2}$ for semidiluted ones. The position where the transition occurs depends only on the concentration measured in units of c^*. This law is also valid for the results of SAXS- and SANS- measurements. It was demonstrated that the flexibility (the persistence length) of the polymers in the dilute/semidilute regime increases with increasing molecular weights and contour length, respectively.

Now, we are interested in more structural details, for example in the optical and electrical polarizibility, in the ion cloud surrounding the particles, and in the flexibility of chains of other molecular weights. These investigations have been started.

Acknowledgment This work was financial supported by the Deutsche Forschungsgemeinschaft (SFB 306).

References

1. Hagenbüchle M, Weyrich B, Deggelmann M, Graf C, Krause R, Maier EE, Schulz SF, Klein R, Weber R (1990) Physica A 169:29–41
2. Krause R, Maier EE, Deggelmann M, Hagenbüchle M, Schulz SF, Weber R (1989) Physica A 160:135–147
3. Drifford M, Dalbiez J-P (1984) J Phys Chem 88:5375
4. Nierlich M, Wiliams CE, Boué F, Cotton JP, Daoud M, Farnoux B, Jannink G, Picot C, Moan M, Wolff C, Rinaudo M, de Gennes PG (1979) J Phys France 40:701–704
5. Kaji K, Urakawa H, Kanayaa T and Kitamuru R (1988) J Phys Frace 49:993-1000
6. Odijk T (1979) Macromolecules 12:688–693
7. Krause R (1991) Dissertation, Universität Konstanz, pp 67
8. de Gennes PG, Pincus P, Velasca RM, Brochard FJ (1976) J Phys France 37:1461
9. Koyama R (1985) Macromolecules 19:178–182
10. Newman H, Swinney H, Day LA (1977) J Mol Biol 116:593
11. Kramer H, Deggelmann M, Graff C, Hagenbüchele M, Johner C, Weber R (1992) Macromolecules 25:4325–4328
12. Yoshizaki T, Yamakawa H (1984) J Chem Phys 81:982
13. Vink H (1981) Makromol Chemie 182:279
14. Reddy M, Marinsky YA (1970) J Phys Chem 74:3884

Progr Colloid Polym Sci (1994) 97:40–45
© Steinkopff-Verlag 1994

H. Kramer
C. Martin
C. Graf
M. Hagenbüchle
C. Johner
R. Weber

Electro-optic effects of electrostatically interacting rod-like polyelectrolytes

Received: 16 September 1993
Accepted: 16 March 1994

H. Kramer (✉) C. Martin
C. Graf · M. Hagenbüchle · C. Johner
R. Weber
Fakultät für Physik
Universität Konstanz
78464 Konstanz, FRG

Abstract Orientation behavior in external electric fields of rod-like fd-virus particles (length $l = 895$ nm, diameter $d = 9$ nm) in aqueous suspensions is examined by the electric birefringence method and static light-scattering measurements. In aqueous suspensions the negatively charged fd-particles are surrounded by a diffuse Debye cloud of counterions, which is characterized by the Debye–Hückel parameter κ. A special experimental setup is used to vary the ionic strength of the suspension, i.e., the Debye–Hückel parameter κ and, therefore, the electrostatic interparticle interaction. The birefringence signal Δn and the relative change of the scattered light intensity $\Delta I/I_0$ is measured as a function of the strength and frequency of the applied electric field in suspensions of very low ionic strength (10^{-6}–10^{-4} M). At low field strengths Kerr-behavior is found. From the dependence of the electric anisotropy $\Delta \alpha_{el}$ on the Debye–Hückel parameter κ it is concluded that the orientation of the fd-particles is correlated to an induced dipole due to a deformation of the diffuse Debye cloud. Saturation electric birefringence and electric light-scattering data are far from that theoretically expected. This can be interpreted as a destruction of the diffuse Debye cloud at high electric fields. At low field strengths the frequency dispersions below 1 kHz of Δn and $\Delta I/I_0$ of the electrostatically interacting fd-virus suspensions show anomalous behavior. This negative electro-optic effect is an evidence for the orientation of the particle long symmetry axis perpendicular to the applied electric field. The dispersion has a positive maximum at about 3 kHz. This maximum could be explained by different frequency dependencies of the electric polarizabilities parallel and perpendicular to the long symmetry axis of the fd-rods.

Key words Electric birefringence – electric field light scattering – fd-virus – electric polarizability

Introduction

Aqueous suspensions of charged macromolecules are very interesting two-component systems, built up by the charged macromolecular particles and the surrounding aqueous media. The particle charge can lead to electrostatic interparticle interaction at relatively small particle concentrations where short range interparticle interactions are still negligible. This allows the formation of ordered states of the macromolecules. These ordered states

Progr Colloid Polym Sci (1994) 97:40–45
© Steinkopff-Verlag 1994

are dependent on the electrostatic interaction a short-range order of the particles' centers of mass, called liquid-like phase, and at stronger interaction a long-range order of the centers of mass called liquid-crystalline phase, and for anisometric particles a long-range order of the particle orientations.

In this paper, the changing optical properties of aqueous fd-virus-suspensions under the influence of an external electric field are examined by the electric birefringence method and static light scattering. The fd-virus is a filamentous bacteriophage of length $l = 895$ nm, diameter $d = 9$ nm, and very low polydispersity [1]. The virus has a rod-like shape and is due to its molecular structure optically anisotropic. On its surface approximately 10^4 ionizible acid groups are located. In aqueous solutions these give rise to a negative charge on each virus which, in highly deionized water, leads to long-range electrostatic interaction. It can be screened by adding salt to the solution. The particle concentrations of the fd-suspensions were fixed such that no or liquid-like order occurred without external electric field; this means that the particles are isotropically oriented and the suspension is optically isotropic. Under the influence of an external field the fd-particles become oriented, the suspension is optically anisotropic and, therefore, birefringent. From the magnitude of the birefringence conclusions about the orientation of the fd-particles, the properties and the relevant processes which lead to this orientation behavior can be concluded. The fieldfree relaxation after switching off the external electric field contains information about the rotational diffusion of the particles [2]. Particularly the intensity change of the scattered light in an external electric field contains information about the orientation of the particles with respect to the electric field direction [3]. Here, we are interested in fd-virus-suspensions at very low ionic strength. Under this condition the charged particles are surrounded by diffuse counterion clouds with extensions of several hundred nanometers. To our knowledge, electro-optic measurements on charged particles with very extended counterion clouds have not yet been made. We can show that the extended counterion clouds are dominating the electro-optic properties of the charged particles. The extension of the counterion clouds determines the interparticle interaction and the interaction of the single particle with the external electric field and, therefore, the orientation behavior of the particle.

There are a few theoretical publications which try to describe the influence of electric fields on diffuse counterion clouds, which could not be tested yet because of the lack of experimental data. This theoretical work quantitatively predicts the polarizability of the counterion clouds [4, 5] and the destruction of the counterion clouds in high electric fields [6].

In the literature anomalous birefringence effects are described [7–10]. These phenomena have been, for over 50 years, an object of controversial speculation which is mainly caused by insufficient characterization of the studied systems. The well characterized fd-virus in combination with our electro-optic methods allows measurements at well defined very low ionic strengths and is therefore very appropriate to study these effects.

Experimental section

Materials

A stock solution (9 mg/ml) of fd-virus particles (length $l = 895$ nm, diameter $d = 9$ nm, MW = $1.64 \cdot 10^7$, about 10^4 ionizable groups on the surface) was prepared with the help of Professor I. Rasched (University of Konstanz) following a method of Marvin et al. [11]. *Escherichia coli* bacteria were infected with the fd virus. After 8 h at 37 °C the fd virus bred rapidly by a factor of 1000. The reproduction was stopped, and the viruses were separated from the bacteria by several steps of precipitation and centrifugation. Finally, the fd-virus solution was ultracentrifuged in a CsCl gradient and dialyzed against a solution of 0.01 M Tris/HCl to obtain a pure stock solution. The samples were prepared by diluting the stock solution with highly purified water ($R = 18$ MΩ). The actual concentrations of the samples were determined by their UV absorption at $\lambda = 269$ nm (extinction coefficient $\varepsilon = 3.84$ cm^2/mg), using a Beckmann spectrometer (DU-64, Darmstadt, FRG).

Tube pump system

In a closed circuit, including the birefringence resp. the light-scattering cell containing the electrodes, the samples were deionized by pumping them with a tube pump through mixed-bed-ion-exchange resin (MB3, Serva Diagnostics, Heidelberg, FRG) until the desired conductivity was achieved [12]. The minimum ionic strength depends on the particle concentration. At this ionic strength the suspension could be considered free of small ions other than H$^+$ or OH$^-$. NaCl was added to the suspensions to obtain higher conductivities, i.e., smaller Debye clouds. All measurements were carried out at 20.0 $\pm$ 0.2 °C. Conductivity measurements were performed using a Knick conductometer (Knick, Berlin, FRG).

42   H. Kramer et al.

Electro-optic effects of rod-like polyelectrolytes

Electric birefringence apparatus

The electric birefringence apparatus is a commercial instrument (spectrometer DB 10, Suck, Siegen, FRG) and similar to that described elsewhere [2,13,14]. The birefringence was measured at the wavelength $\lambda = 633$ nm.

Electric field light scattering apparatus

The intensity change of the scattered light in an external electric field ΔI was measured with a commercial light-scattering apparatus (ALV, Langen, FRG) and a self-constructed light-scattering cell containing the electrodes. The electric field direction is perpendicular to the observation plane. The observation angle was fixed at $\theta = 140°$, i.e., the magnitude of the scattering vector is

$$q = \frac{4\pi n_s}{\lambda} \sin\left(\frac{\theta}{2}\right) = 0.0323 \text{ nm}^{-1} , \tag{1}$$

with a wavelength $\lambda = 488$ nm and a refractive index of the aqueous suspension $n_s = 1.33$.

Theory

The parameter describing the extension of the diffuse counterion cloud is the inverse Debye–Hückel parameter, the Debye–Hückel length κ^{-1}, which is given by [15]

$$\kappa^{-1} = \left(\frac{e_0^2}{\varepsilon_r \varepsilon_0 k_B T} \sum_i c_i z_i^2\right)^{-1/2} , \tag{2}$$

where z_i and c_i are the charge number and the concentrations of the charged particle class i. ε_0 is the permittivity of vacuum, ε_r the relative permittivity, k_B the Boltzmann constant, e_0 the elementary charge, and T the temperature. The ionic strength IS is defined by

$$IS = \frac{1}{2} \sum_i c_i z_i^2 . \tag{3}$$

For small NaCl concentrations the ionic strength is proportional to the measured conductivity and the Debye–Hückel length κ^{-1} can be calculated in good approximation by

$$\kappa^{-1} = \frac{9.6 \text{ nm}}{\sqrt{IS[\text{mM}]}} . \tag{4}$$

The electric polarizability $\Delta\alpha_{el}$ of the fd-virus particles can be calculated from the Kerr behavior at small field

strength by the Kerr law

$$\Delta n_0 = \frac{\Delta n_{sat} \Delta\alpha_{el}}{15 k_B T} E^2 , \tag{5}$$

where Δn_0 is the measured birefringence, Δn_{sat} is the birefringence at infinite high field strength, and E is the electric field strength. Normally, when all particles are totally oriented in the field direction, from the saturation value Δn_{sat} the optical anisotropy of the particles can be calculated, which is dominated by the molecular structure of the virus.

The relative intensity change of light scattered by anisometric particles in an external electric field is given by:

$$\frac{\Delta I}{I_0} = \frac{I_E - I_0}{I_0} , \tag{6}$$

where I_E is the steady-state intensity of the scattered light when the system is subjected to an electric field of strength E, and I_0 is the intensity of scattered light when there is no electric field applied to the system. For rod-like particles (in the RDG approximation) oriented parallel to the external field the saturation electric field light-scattering effect is [16]:

$$\left(\frac{\Delta I}{I_0}\right)_\infty = \frac{1}{P_0(q, l)} - 1 , \tag{7}$$

with the formfactor P_0 [17] which is proportional to the scattered light intensity without interparticle interaction of isotropically oriented rods. For suspensions of interacting particles formula (7) is valid for a static structure factor $S(q, l) \equiv 1$.

Results and discussion

One aim of this paper is to measure the dependence of the electric polarizability of the fd-virus particle dependent on the Debye–Hückel length κ^{-1}. Therefore, the field strength dependence of the transient electric birefringence signal Δn_0 was analyzed at a concentration well below c^* (c^* is the overlap concentration 1 particle/(length 1)^3 which corresponds to 0.04 mg/ml) where anomalous birefringence effects do not occur [2]. Figure 1 shows Δn_0 under these conditions. A significant result of Fig. 1 is the strong dependence of the high field saturation value of Δn_0 for the different conductivities. For fd-virus suspensions Torbet and Maret [18] measured a saturation value of $\Delta n_{sat}(1c^*) = 2.4 \cdot 10^{-6}$ at complete orientation of the particles in a magnetic birefringence experiment. This value is much larger than the values measured in Fig. 1. The saturation value is given by the optical anisotropy of the virus which is not influenced by the conductivity of the suspension at these low ionic strengths. The theoretically predicted value for the saturation electric field light scattering (7) is about four times larger than the measured one. Therefore, we must conclude that the measured saturation values in Fig. 1 do not correspond with totally oriented particles. Kerr-behavior is found at low electric field strengths, and with the help of the given saturation value of Torbet and Maret the electric anisotropy $\Delta\alpha_{el}$ is calculated as a function of the Debye–Hückel length κ^{-1}. The result is shown in Fig. 2. A strong increase of $\Delta\alpha_{el}$ with κ^{-1} of over one order of magnitude in the κ^{-1} range from 40 to

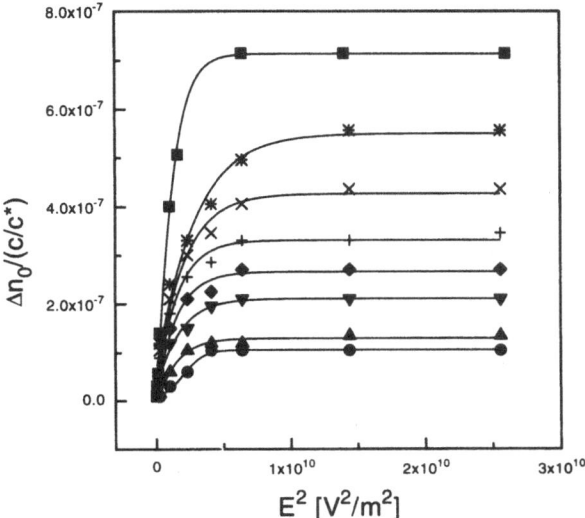

Fig. 1 The birefringence of a fd-virus suspension with a concentration of 0.4 c^* for different electric conductivities. The various conductivities are (starting from the top): 0.22 μS/cm; 1.3 μS/cm; 1.61 μS/cm; 1.97 μS/cm; 2.4 μS/cm; 3.15 μS/cm; 5.2 μS/cm; 6.7 μS/cm

Fig. 2 The electric anisotropy $\Delta\alpha_{el}$ of the fd-virus in aqueous suspension

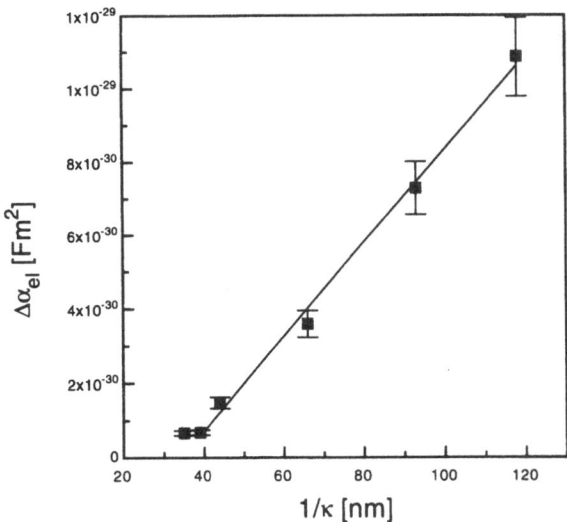

120 nm is found. A similar linear behavior was measured by Hogan et al. [19] and Rau and Bloomfield [20] on charged DNA rods at higher ionic strengths, and interpreted in the way that the electric anisotropy is dominated by the diffuse Debye cloud and not by counterions bound to or adsorbed on the macromolecular surface. Theoretical considerations of Rau and Charney [21] concerning DNA particles led them to the result that in the κ^{-1} range below 10 nm adsorbed counterions are dominating the electric polarizability, at κ^{-1}-values above 20 nm the diffuse counterion cloud begins to dominate the electric polarizability. Rau and Charney could compare their results with experimental values up to $\kappa^{-1} = 27$ nm. The results presented in this paper show the strong influence and domination of the diffuse counterion cloud on the electric polarization at higher Debye–Hückel lengths.

Our results yield that the fd-virus particles at low ionic strength are oriented in electric fields by an induced dipole mechanism caused by the polarization of their surrounding diffuse counterion clouds. This orientation mechanism explains why the particles cannot be fully oriented at higher electric field strengths. In the literature, several authors theoretically discuss the influence of high electric fields on the diffuse ion cloud [22–24, 6]. Considerations of Fixman and Jagannathan on spherical polyelectrolytes show that at moderate electric field strengths of several hundred V/cm the induced dipole shows no linear behavior, and the ion cloud starts to be stripped away. Yoshida et al. showed by Monte Carlo simulations similar behavior for rod-like particles. A destruction of the diffuse ion cloud leads to a destruction of the orientation mechanism and therefore to an incomplete orientation for rod-like particles at high electric fields.

In the following measurements of fd-virus-suspensions in alternating electric fields are discussed. In the case of a sinusoidal field the frequency dependent electric birefringence signal $\Delta n(\omega, t)$ is given by

$$\Delta n(\omega, t) = \Delta n_{sta}(\omega) + \Delta n_{alt}(\omega) \cos(2\omega t - \varphi) , \quad (8)$$

where $\Delta n_{sta}(\omega)$ is a stationary, time-independent component, $\Delta n_{alt}(\omega)$ is the amplitude of the alternating component, and φ is the phase shift between the alternating component and the sinusoidal electric field. In the birefringence experiment the property $\Delta n_{max}(\omega) := \Delta n_{sta}(\omega) + \Delta n_{alt}(\omega)$ is measured. In the static light-scattering experiment $\Delta I/I_0$ is measured in a square wave field. Figure 3a resp. 3b show Δn_{max} and $\Delta I/I_0$ as a function of the frequency $f = \omega/2\pi$ of the electric field for various fd-virus concentrations at minimum ionic strength. At low frequencies of about 10 Hz the behavior known from the literature is found [2]. In both experiments suspensions of smaller fd-virus concentration show normal behavior which means a positive signal, suspensions of concentration above the overlap

44

H. Kramer et al.
Electro-optic effects of rod-like polyelectrolytes

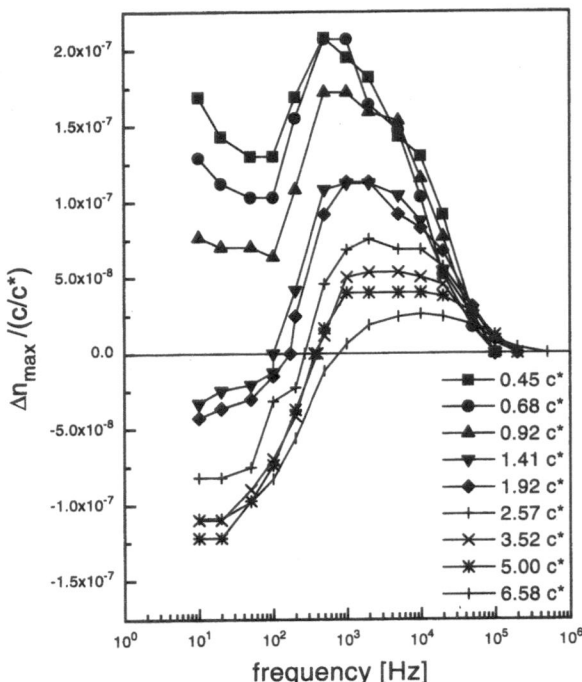

Fig. 3A The birefringence of fd-virus suspensions at minimum conductivities and an amplitude of the sinusoidal electric field of $2.4 \cdot 10^4$ V/m

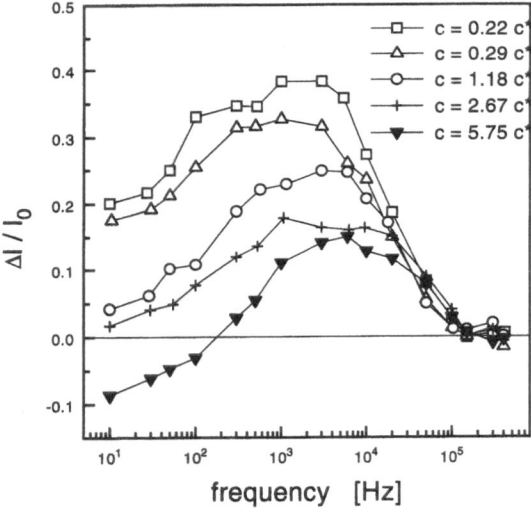

Fig. 3B The intensity change of the scattered light $\Delta I/I_0$ at minimum conductivities and an amplitude of the square wave electric field of $2.0 \cdot 10^4$ V/m

concentration show anomalous, negative behavior. Suspensions showing anomalous behavior at low frequencies lose this with increasing frequency. The negative signal measured in the electric field light-scattering experiment for the sample with a particle concentration of 5.75 c^*

states that the particles tend to orient perpendicular to the electric field. The decreased intensity I_E can be led back to a decreased formfactor P_E in the external field compared to the isotropic formfactor P_0. A decrease of the formfactor can only be achieved if the particles are oriented parallel to the observation plane, i.e., in our case, perpendicular to the electric field. This consideration corresponds to the derived formula for complete particle orientation perpendicular to an external field by Stoimenova [25].

For all suspensions a positive maximum of the signal is found at frequencies of about 10^3 Hz. The disappearance of negative signals and the positive maximum of the signal at about 10^3 Hz can be explained by the frequency dependencies of the electric polarizabilities parallel and perpendicular to the rod length axis. The measured signal at small electric field strength is proportional to the electric anisotropy $\Delta\alpha_{el}$ (see Eq. (5)), which is the difference of the polarizabilities parallel and perpendicular to the rod length axis:

$$\Delta\alpha_{el} = \alpha_{el}^{pa} - \alpha_{el}^{pe} . \tag{9}$$

A maximum of the signal with frequency at constant anisotropy and at constant electric field amplitude can only be explained with a maximum in the dispersion characteristic of the electric anisotropy $\Delta\alpha_{el}$. This can be caused by different dispersion behaviors of the polarizabilities parallel and perpendicular to the rod, in the case of a significant perpendicular polarizability. Fixman and Jagannathan showed that the perpendicular polarizability gets a considerable magnitude in the case of extended diffuse counterion clouds [5]. Secondly, the existence of anomalous birefringence is not explainable without such a significant polarizability [9, 14]. A maximum in the dispersion of the electric anisotropy $\Delta\alpha_{el}$ occurs when the perpendicular polarizability decreases earlier, and in a certain frequency range faster with the frequency than the parallel polarizability. The strong decrease of the perpendicular polarizability leads also to the disappearance of the anomalous birefringence.

One idea to understand the frequency dependent behavior of the parallel and perpendicular polarizabilities is the following. A shift of the counterion cloud perpendicular to the rod length axis causes that large parts of the cloud come closer to the rod in regions of a stronger rod potential. Shifts of the counterions closer to the rod cause that the ions are stronger bound and that their translation diffusion constant is lower. Shifts of the counterion cloud parallel to the rod length axis cause that large parts of the cloud move at constant distance of the rod on equipotential surfaces. This fact can cause that with increasing frequency the perpendicular polarization can only with more difficulty follow the stimulating electric field than can the parallel polarization. Therefore, the two components of

Progr Colloid Polym Sci (1994) 97:40–45
© Steinkopff-Verlag 1994

the polarizability show different behavior with increasing frequency.

At frequencies in the range between 10^5 and $3 \cdot 10^5$ Hz the birefringence signal vanishes. The vanishing of the signal and therefore the impossibility to orient the fd-particles at frequencies above $3 \cdot 10^5$ Hz can be easily understood. At minimum ionic strength the diffuse counterion cloud surrounding the particle is mainly built up by H^+ counterions originating from the acid groups located on the virus surface. The orientation of the particle is therefore evoked by the polarization of the H^+ counterion cloud, which means a shifting of the cloud to the negatively charged rod. The relaxation time and the reorientation time of such an induced dipole can be approximately calculated by a formula given by Oosawa [3, 26]:

$$\tau = \frac{l^2}{(2\pi)^2 D},$$ (10)

where l is the rod length and D the translation diffusion constant of the counterions. Using for D the value for free H^+ ions at 20 °C of about $8.5 \cdot 10^{-9}$ m^2/s [27], which is possible because the ions are very weakly bound, leads to a time τ of about $2.4 \cdot 10^{-6}$ s. This corresponds to a frequency of about 400 kHz, which is very close to the frequency at which the signals vanish, i.e., the induced dipole can no longer follow the stimulating electric field and the rod is no longer oriented.

Conclusions

The orientation behavior of rod-like fd-virus particles in aqueous suspensions of very low ionic strength has been examined. It was demonstrated that at very low ionic strength (10^{-5} M–10^{-6} M) the particles are oriented by an induced dipole mechanism due to the anisotropic polarization of their diffuse Debye cloud. This is in good agreement with predictions by Rau et al. [4] and Fixman et al. [5]. It is shown that the counterion cloud is stripped away from the negatively charged rod at electric field strengths of about 10^5 V/m. From our measurements we conclude that the rods possess a significant perpendicular polarizability. The different dispersion behavior of the polarizabilities parallel and perpendicular to the rod length axis leads to a maximum of the measured signals at about 10^3 Hz and to the disappearance of anomalous orientation at frequencies above 10^2 Hz. Electric field light-scattering measurements prove the perpendicular orientation of the fd-particles with respect to the electric field direction. At alternating electric fields above $3 \cdot 10^5$ Hz no orientation of the particles is possible, which can be explained by the polarization of the diffuse counterion cloud which is limited by the diffusion of the H^+ counterions.

Acknowledgment This work was supported by the Deutsche Forschungsgemeinschaft (SFB 306)

References

1. Newman J, Swinney H, Day LA (1977) J Mol Biol 116:593
2. Kramer H, Deggelmann M, Graf C, Hagenbüchle M, Johner C, Weber R (1992) Macromolecules 25:4325
3. Stoylov SP (1991) Electro-Optics Theory, Techniques, Applications, Academic Press, London
4. Rau DC, Charney E (1981) Biophys Chem 14:1
5. Fixman M, Jagannathan S (1981) J Chem Phys 75:4048
6. Yoshida M, Kikuchi K, Maekawa T, Watanabe H (1992) J Phys Chem 96:2365
7. Lauffer MA (1993) J Am Chem Soc 61:2412
8. Asai H, Watanabe N (1976) Biopolymers 15:283
9. Hoffmann H, Krämer U, Thurn H (1989) J Phys Chem 94:2027
10. Krämer U, Hoffmann H (1991) Macromolecules 24:256
11. Marvin DA, Wachtel EJ (1975) Nature 253:19
12. Deggelmann M, Palberg T, Hagenbüchle M, Maier EE, Krause R, Graf C, Weber R (1991) J Col Int Sci 143:318
13. Fredericq E, Houssier C (1973) Electric Dichroism and Electric Birefringence, Clarendon Press, Oxford
14. Kramer H (1993) Elektrooptische Untersuchungen an wässerigen Suspensionen geladener, linearer Makromoleküle, Hartung-Gorre Verlag, Konstanz
15. Hunter RJ (1981) Zeta Potential in Colloid Science Principles and Applications, Academic Press, London
16. Stoylov SP (1971) Advan Col Int Sci 3:45
17. Hagenbüchle M, Weyerich B, Deggelmann M, Graf C, Krause R, Maier EE, Schulz SF, Klein R, Weber R (1990) Physica A 169:29
18. Torbet J, Maret G (1981) Biopolymers 20:2657
19. Hogan M, Dattagupta N, Crothers DM (1978) Biochemistry 75:195
20. Rau DC, Bloomfield VA (1979) Biopolymers 18:2783
21. Rau DC, Charney E (1983) Biophys Chem 17:35
22. Fixman M, Jagannathan S (1983) Macromolecules, 16:685
23. Rau DC, Charney E (1983) Macromolecules 16:1653
24. Altig JA, Wesenberg GE, Vaughan WE (1986) Biophys Chem 24:221
25. Stoimenova MV (1975) J Col Int Sci 53:42
26. Oosawa F (1970) Biopolymers 9:677
27. American Institute of Physics Handbook (1972) Mc-Graw-Hill Book Comp., New York

Progr Colloid Polym Sci (1994) 97:46–50
© Steinkopff-Verlag 1994

APPLICATION OF THE PRINCIPLES OF COLLOID SCIENCE

B. Biliński
A. L. Dawidowicz
W. Wójcik

The surface properties of controlled porosity glasses of various porosity

Received: 16 September 1993
Accepted: 1 March 1994

B. Biliński (✉) · W. Wójcik
Department of Physical Chemistry
Faculty of Chemistry
M. C. Skłodowska University
20–031 Lublin, Poland

A. Dawidowicz
Department of Chemical Physics
Faculty of Chemistry
M. C. Skłodowska University
20–031 Lublin, Poland

Abstract The surface free energy of controlled porosity glasses was correlated with "geometrical" factors, i.e., specific surface area, mean pore diameter, and total pore volume. A new parameter related to the specific surface area was evaluated based on mean pore diameter and total pore volume and assuming various capillary models. Two CPG's differing in their total pore volume were selected for further experiments

Key words Controlled porosity glasses – surface free energy – porosity – surface rehydroxylation

Introduction

The surface free energy may be considered as a very important parameter determining the course of many interfacial processes. It used to be expressed as a sum of two components [1]:

$$\gamma_S = \gamma_S^d + \gamma_S^p, \tag{1}$$

where γ_S^d is the dispersion component and γ_S^p is the polar component of surface free energy.

The dispersion component may be calculated based on the adsorption data of a substance interacting by dispersive (non-specific) forces, e.g., n-octane [2, 3]. The polar component may be calculated from the adsorption data of a substance which interacts also by polar (specific) forces (e.g., toluene), however, the value of γ_S^d must be known [2, 3].

The individual surface properties of porous materials used to be attributed to the geometrical factors as specific surface area, total pore volume or pore diameter. This is not necessarily true, because surface chemistry appears

also very important [4]. The preparation of controlled porosity glasses (CPGs) differing in their geometrical factors may imply various surface composition [3, 5]. As the specific surface area, the total pore volume, and the pore diameter are strongly dependent on each other, it is practically impossible to investigate the influence of only one factor on surface properties [4]. On the other hand, the final geometry of porous material seems correlated with surface composition, especially in the case of porous glasses.

The aim of this paper was to seek any correlations between the geometry and the surface free energy of controlled porosity glasses. However, not only the geometrical factors themselves should be considered as fundamental, but also the differences in surface chemistry (e.g., the distribution of surface active sites). Two porous glasses were selected for further experiments, which consisted of the determination of the components of surface free energy for glasses thermally treated (and dehydroxylated) and subsequently rehydroxylated in different ways.

Progr Colloid Polym Sci (1994) 97: 46–50
© Steinkopff-Verlag 1994

Experimental

Seven CPGs of different specific surface area (A_{BET}), mean pore diameter (d_p), and total pore volume (V_p) were taken for the experiments. They were prepared from Vycor glass composed of 55% SiO_2 35% B_2O_3 and 10% Na_2O, by thermal treatment and a leaching process described in detail elsewhere [5, 6].

The specific surface area of these materials was determined by thermal desorption of nitrogen [7].

The total pore volume and the mean pore diameter were determined by mercury intrusion emloying the mercury porosimeter Type 4000 (Carlo Erba, Milan, Italy).

Two porous glasses were subjected to thermal treatment, and then to rehydroxylation in different ways [2, 3].

The components of surface free energy of all materials was determined from the adsorption data of n-octane and toluene. The procedure for the surface free energy calculation was presented previously [2, 3].

Results and discussion

Table 1 contains the values of the specific surface area (A_{BET}), the mean pore diameter (d_p), and the total pore volume (V_p) of the investigated CPGs. As can be seen from this table both A_{BET} and d_p vary over a relatively wide range (from 28.5 to 360 m^2/g and from 99.8 to 6.2 nm, respectively). Therefore, this series of CPGs may be considered as representative for such kind of materials.

The first attempt at searching for a relationship was to plot the values of γ_S as a function of A_{BET}. It can be seen from Fig. 1 that a regular curve may be drawn for all materials except CPG4 and CPG5. Those two seem to possess either too high a surface free energy or too high a value of A_{BET}. On the other hand, as can be seen from Table 1, both materials are characterized by the highest values of V_p, significantly exceeding 1 cm^3/g.

The second attempt was to correlate the values of γ_S with pore diameter (Fig. 2). As can be seen from this figure,

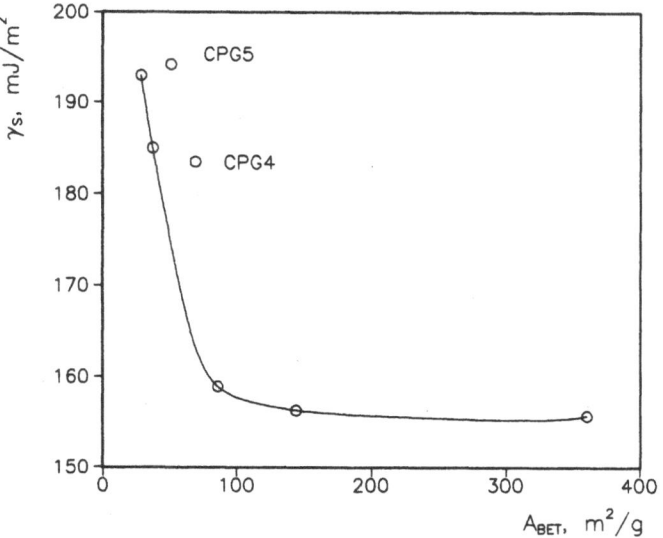

Fig. 1 The relationship between the value of γ_S and the specific surface area A_{BET}

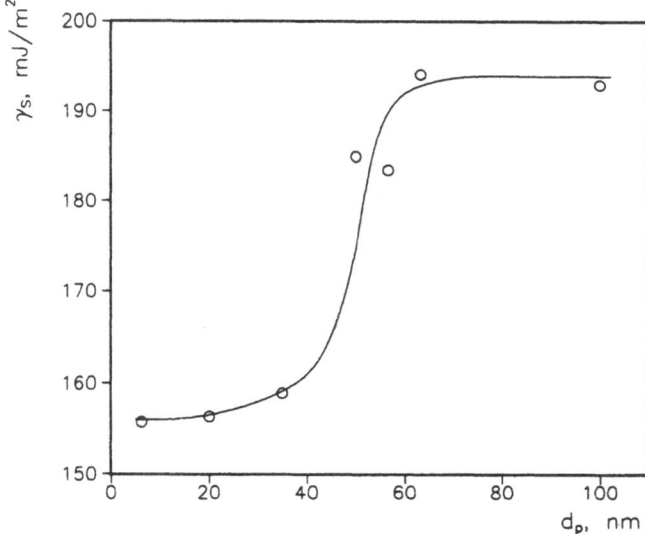

Fig. 2 The relationship between the value of γ_S and the mean pore diameter d_p

an important difference in surface properties takes place for the investigated materials between 35 and 63 nm.

It is obvious that the geometric factors determined experimentally (i.e., A_{BET}, d_p and V_p) cannot strictly describe the real geometry of the material. Of course, the surface properties (as energy of interactions) must be correlated with these parameters, and they should not be considered separately from each other. Therefore, our next attempt was to find such a parameter which could generalize A_{BET}, d_p, and V_p, and might be used for correlation with

Table. 1. The specific surface area (A_{BET}), the pore diameter (d_p) and the total pore volume (V_p) for all investigated CPGs.

CPG	$A_{BET} [m^2/g]$	$d_p [nm]$	$V_p [cm^3/g]$
CPG1	360	6.2	0.73
CPG2	144	20.0	0.70
CPG3	85.7	35.0	0.94
CPG4	68.5	56.6	1.65
CPG5	50.6	63.2	1.34
CPG6	37.0	50.0	0.79
CPG7	28.5	99.8	0.99

surface energy. The simplest model is a file of cylindrical capillaries possessing certain values of A (geometric area), r (pore radius) and V (pore volume). According to this model;

$$A = n \cdot 2 \cdot \pi \cdot r \cdot h \qquad (2)$$

$$V = n \cdot \pi \cdot r^2 \cdot h , \qquad (3)$$

where n is the number of capillaries and h is the length of a capillary.

Combining these equations, one can obtain:

$$A = \frac{2 \cdot V}{r} , \qquad (4)$$

This is a strictly geometric dependence between surface area, pore radius, and pore volume. Dividing A by the experimentally determined A_{BET}, replacing r with diameter d, and introducing the experimental values of V_p and d_p a new parameter may be evaluated [8]:

$$A^0 = \frac{4 \cdot V_p}{A_{BET} \cdot d_p} . \qquad (5)$$

This parameter represents a relative surface area determined by experimental data.

The relationship between γ_S and A^0 is plotted in Fig. 3. As can be seen from this figure the regular curve may be drawn through the points, except that one for CPG7. The glass CPG7 possesses the lowest value of A_{BET} and the highest value of d_p. This suggests that the model of open capillaries is not fully adequate for the investigated materials [8]. Especially for a CPG of low specific surface area (corresponding to large pore diameter) the contribution of

the area of "bottom" closing capillaries may be significant. Therefore, the model of closed capillaries was taken into account. According to this model [8]:

$$A = n \cdot 2 \cdot \pi \cdot r \cdot h + n \cdot \pi \cdot r^2 . \qquad (6)$$

Combining this equation with Eq. (3), one can obtain:

$$A = \frac{V}{d}(2 + r/h), \qquad (7)$$

where the factor r/h results from the surface area of the "bottom" closing of the pores. Hence by analogy to Eq. (5):

$$A^0 = \frac{2 \cdot V_p}{A_{BET} \cdot d_p}(2 + r/h) . \qquad (8)$$

As can be seen, the ratio r/h has to be estimated for each glass. This may be done from the dependence between V_p and d_p, by transformation of Eq. (3);

$$r/h = \frac{\pi \cdot d_p^3}{8 \cdot V_p} . \qquad (9)$$

These values correspond to the ratio between r and h for single long capillary and they are completely unrealistic (order of magnitude 10^{-16}–10^{-19}). However, the proportions between them should estimate the ratio r/h for a series of CPGs. In order to reduce the ratio r/h to more reasonable physicochemical values, they were divided by the value of r/h obtained for CPG4. This was chosen arbitrarily and used for calculations of the parameter A^0 according to Eq. (8). The values of γ_S were plotted against $lg A^0$ in Fig. 4 (curve 1). As can be seen from this figure, the

Fig. 3 The relationship between the value of γ_S and the value of parameter A^0 calculated from Eq. (5) (the open pores model)

Fig. 4 The relationship between the value of γ_S and the logarithm of parameter A^0 calculated from Eq. (8) (the closed pores model); curve 1-CPG4, curve-CPG5 and curve 3-CPG3 as reference glasses

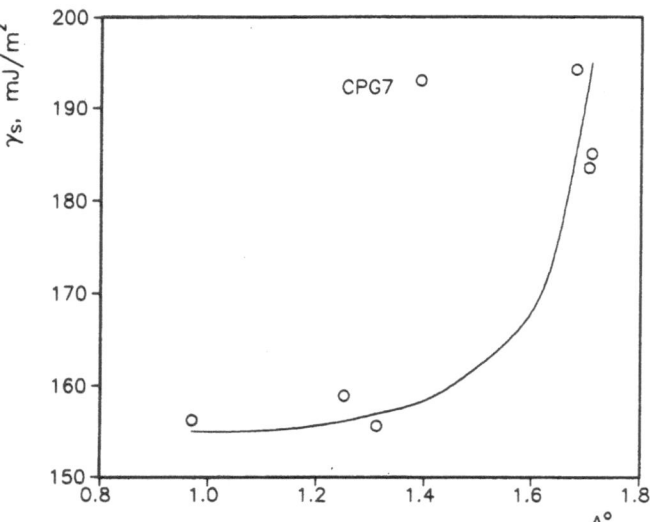

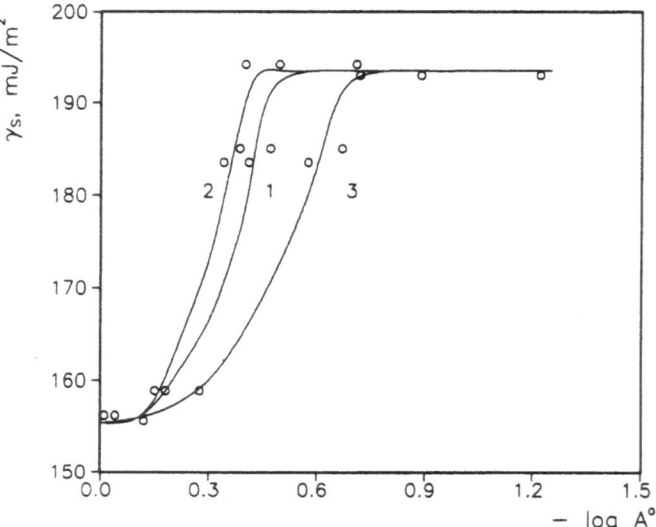

Progr Colloid Polym Sci (1994) 97:46–50
© Steinkopff-Verlag 1994

Table 2.

Glass	Manner of preparation	Specific surf. area [m²/g]
CPG5	bare glass	50.6
CPG5.2	CPG5 heated for 2 h at 600°C	53.9
CPG5.20	CPG5 heated for 20 h at 600°C	52.5
CPG5.100	CPG5 heated for 100 h at 600°C	46.7
CPG5.20–5	CPG5.20 rehydroxylated with water vapor for 5 h	50.2
CPG5.20–10	CPG5.20 rehydroxylated with water vapour for 10 h	50.6
CPG5.20–20	CPG5.20 rehydroxylated with water vapor for 20 h	47.2
CPG5.100–5	CP5.100 rehydroxylated with water vapor for 5 h	46.4
CPG5.100–10	CPG5.100 rehydroxylated with water vapor for 10 h	44.9
CPG5.100–20	CPG5.100 rehydroxylated with water vapor for 20 h	44.1
CPG5.100–100	CPG5.100 rehydroxylated with water vapor for 100 h	42.9
CPG5.100–W	CPG5.100 rehydroxylated with liquid water	44.0
CPG5.100–S	CPG5.100 rehydroxylated with NaOH solution	49.1

Table 3.

Glass	Manner of preparation	Specific surf. area [m²/g]
CPG6	bare glass	37.0
CPG6.2	CPG6 heated for 2 h at 600°C	40.3
CPG6.2–20	CPG6.2 rehydroxylated with water vapor for 20 h	42.4
CPG6.20	CPG6 heated for 20 h at 600°C	39.0
CPG6.20–20	CPG6.20 rehydroxylated with water vapor for 20 h	38.9
CPG6.20–S	CPG6.20 rehydroxylated with NaOH solution	37.3
CPG6.120	CPG6 heated for 120 h at 600°C	36.4
CPG6.120–20	CPG.120 rehydroxylated with water vapor for 20 h	36.7
CPG6.120–S	CPG6.20 rehydroxylated with NaOH solution	39.4

shape of the curve is very similar to that in Fig. 2. For comparative purposes, glasses CPG5 and CPG3 were also taken as a reference in the calculation of A^0; the proper curves are plotted in Fig. 4 (curves 2 and 3, respectively). It should be emphasized that the assumption of another reference glass does not change the shape of the general relationship.

Table 2 contains the identification, the manner of preparation, and the specific surface area of CPG5 while Table 3 contains the same data for CPG6. The behavior of both controlled porosity glasses with thermal treatment is very similar [3]. The dispersion component of the surface free energy remains almost constant. However, the highest value of γ_S^d occurs after 2 h of heating. The most important changes in γ_S^p take place after 20 h of heating (Table 4).

A certain difference appears in the possibility of rehydroxylation with water vapor of both kinds of glasses, after the thermal treatment. The general conclusion may be drawn that CPG5 requires longer time of preliminary thermal treatment for effective rehydroxylation with water vapor, when compared to CPG6 (Table 4) [3].

The previous investigation [2] demonstrated that the rehydroxylation with water vapor was strongly dependent on the content of alkali-borate constituents on the surface.

Table 4 The values of the maximal extrapolated film pressure π_{max} for n-octane and toluene as well as the calculated values of γ_S^d, γ_S^p and γ_S for controlled porosity glasses [mJ/m²].

CPG	Octane		Toluene		γ_S
	π_{max}	γ_S^d	π_{max}	γ_S^p	
CPG5	11.7	35.0	33.7	159.2	194.2
CPG5.20	14.8	39.1	23.6	44.8	83.9
CPG5.20–5	14.9	39.3	25.6	56.4	95.7
CPG5.20–10	14.6	38.8	25.3	56.7	95.5
CPG5.20–20	15.5	40.0	27.0	61.7	101.7
CPG5.100	14.0	38.1	22.0	40.8	78.9
CPG5.100–5	13.4	37.3	22.8	49.4	86.7
CPG5.100–10	13.5	37.4	25.1	63.9	101.3
CPG5.100–20	13.6	37.5	28.9	86.8	124.3
CPG5.100–100	13.6	37.5	29.8	99.3	136.8
CPG5.100–W	11.8	35.3	33.3	153.1	188.4
CPG5.100–S	10.5	33.6	31.3	146.5	180.1
CPG6	13.5	37.3	34.6	147.7	185.0
CPG6.2	15.6	40.2	26.2	55.8	96.0
CPG6.2-20	14.2	38.3	28.0	79.8	118.1
CPG6.20	15.1	39.5	22.6	37.5	77.0
CPG6.20–20	14.2	38.4	27.3	73.5	111.9
CPG6.20–S	11.6	34.9	31.7	138.2	173.1
CPG6.120	13.0	36.8	20.3	37.4	74.2
CPG6.120–20	11.2	34.4	27.3	100.8	135.2
CPG6.120–S	10.1	33.0	29.2	130.4	163.4

The CPG preparation procedure involves thermal treatment of Vycor glass. During the heating of this material, the demixion of glass components takes place. This results from the diffusion of alkali and boron oxides from silica phase to the nucleus of heterogeneity [5]. Thus, the thermal treatment provides a significant decrease of the alkali-borate content in the silica phase.

The leaching in alkaline solution, which increases the pore volume of CPG, provides the dissolution of the surface layer of silica skeleton (mostly enriched in B and Na) [5, 6]. Thus, the CPG possessing the high pore volume is characterized by lower residue of alkali-borate, and the heating of such material results in a smaller surface enrichment in boron and sodium compounds when compared to that for glasses of smaller porosity [3, 5].

References

1. Wu S (1978) In: Paul DR, Newman S (eds) Polymer Blends, Vol. 1, Academic Press, New York, p 243
2. Biliński B, Wójcik W, Dawidowicz AL (1991) Appl Surf Sci 54:125–131
3. Biliñski B, Dawidowicz AL (1994) Appl Surf Sci 74:277–285
4. Nawrocki J (1991) Chromatographia 31:177–192
5. Haller WJ (1965) J Chem Phys 42:686–697
6. Dawidowicz AL, Waksmundzki A, Deryło A (1979) Chem Anal 24:811–820
7. Nelsen FM, Eggersten FT (1958) Anal Chem 3:1387–1390
8. Biliñski B, Dawidowicz AL (1993) Colloids Surf 70:61–67

Progr Colloid Polym Sci (1994) 97:51–58
© Steinkopff-Verlag 1994

SUSPENSION

J. B. Rosenholm
F. Manelius
H. Fagerholm
L. Grönroos
H. Byman-Fagerholm

Surface and bulk properties of yttrium stabilized ZrO$_2$ powders in dispersions

Received: 20 September 1993
Accepted: 13 June 1994

J. B. Rosenholm (✉)
F. Manelius · H. Fagerholm
L. Grönroos · Byman-Fagerholm
Department of Physical Chemistry
Abo Akademi University
20500 Abo, Finland

Abstract The electrophoretic mobility of ZrO$_2$ particles stabilized with 3 mol % of yttria recorded at different pH is known to vary as a function of time. The cause of the shift in the surface charge was investigated since it is known to influence a range of important properties controlling the slip casting of ceramic green bodies. It was found that yttrium and some zirconium ions are leached out from the powder surface both in the acid and the basic range, but that the dissolution of yttrium is particularly pronounced in acidic dispersions. As a result, the pH of the dispersions are buffered to values around 5.5. The buffering effect is found to be effective over a period of a few days. The leached yttrium ions influences the electrophoretic mobility, the particle size, the stability of the dispersions against coagulation and the viscous flow. The results are compared with pH and EM values measured on some other commercial ZrO$_2$ powders.

Key words Zirconia (ZrO$_2$), Yttria (Y$_2$O$_3$) – Electrophoretic mobility – Particle size, ESCA (XPS) Solubility – Viscosity, Dispersion stability – pH-dependency

Introduction

Dispersing commercial powders "as received" into water results in a characteristic shift of the pH of the slurry. This reaction is not surprising since it is well known that most metal oxides [1] and the other powders of interest in manufacturing of ceramic materials (e.g. SiO$_2$, Si$_3$N$_4$, and SiC) are quite reactive in aqueous dispersions due to leaching of the constituent ions from the solid surface [2–4]. It is, however, surprising to find that large variations of surface properties are also found for ceramic powders of formally the same composition. Most of the differences are, of course, explained by the different manufacturing processes used or organic processing aids added, resulting in variable amounts of trace impurities released from the powders.

An additive-free stabilization of slips used to cast green ceramic bodies is an attractive alternative since it would remove the burning step of the organics prior to sintering. This goal may be achieved by a careful control of the pH giving rise to a charge stabilized slip. Using this option, however, any shift of the pH may have a very detrimental influence on the dispersion behaviour. Since the buffering capacity due to leaching may be several pH-units and may persist over a period of several days, it is of importance to find ways to minimize or remove this effect. The pH shift may also seriously influence any measurement made within the buffering period. The error may be introduced both as the shift in pH and as an error in the ionic strength due to the leaching of the ionic species In the latter case salting-in and salting-our effects of both the powder and of, for example, surfactants, polymers and/or sintering aids added to the dispersion, may seriously distort the results. It was thus considered of importance to investigate if the reactivity is simply due to the dissolution of the main components or whether the additives introduced (as non-reactive components) had to be considered as well.

In the present context, we report on the surface properties of one commercial model powder, ZrO_2 stabilized with 3 mol % Y_2O_3. This powder is compared with the natural pH and electrophoretic mobility of some other typical commercial ZrO_2 powders. The parameters chosen to record the dispersion behaviour are the electrophoretic mobility (zeta potential), the particle size, the dispersion stability (sedimentation) of dilute dispersions and the viscosity of concentrated slips. In this way the properties of the dilute dispersion can be correlated with true ceramic slips.

Experimental

Materials

The powder used was of the type TZ-3YS and supplied by Toyo Soda (now Tosoh Co.), Japan. According to the manufacturer the powder is especially suitable for slip casting and consists of spherical particles of 0.5 μm size with the composition of 94.7 wt% ZrO_2 and 5.2 wt% Y_2O_3. This corresponds to 32.4 mol % Zr, 66.5 mol % O and 1.2 mol % Y. The raw powder was analyzed by ESCA (Perkin-Elmer PHI 5400) which gave the following composition (neglecting the carbon fraction): 26.4 at% Zr, 70.9 at% O and 2.6 at % Y. The average particle size was determined by x-ray sedimentator (Sedigraph 5000 ET, Micrometrics) to be $d_{50} = 0.38$ μm and with Light Scattering (Malvern 4700 c) to be $d_{50} = 0.30$ μm. The surface area was reported by the manufacturer to be ca. 8 m^2/g. The surface area was determined by a single point BET (Flowsorb II 2300, Micromeritics) to be 6.1–6.6 m^2/g and by sorptometry (Carlo Erba 200) to be 6.2 m^2/g.

The KCl (pure), HCl (99%) and NaOH (99%) were all from Merck AG and used without further purification. The water was distilled and purified through a Milli-Q system. The conductivity was then less than 5.56 10^{-8} S/cm at 25 °C.

Instrumentation and experimental methods

The solubility of the ions in the supernatant was investigated with a Spectra Span IIIB Plasma Emission Spectrometer. The leaching of the ions from the powders was investigated with a Perkin-Elmer PHI 5400 ESCA (XPS) spectrometer [5]. The powders were first dispersed in water and dried on molybdenum plates before analysis. All the values are given in corrected atomic percent values.

The electrophoretic mobilities (EM) were measured with a Zetasizer IIc instrument from Malvern Instruments. Prior to the measurements the samples were conditioned for 5 min. in an ultrasonic bath and stabilized for 2h to 2d.

The pH given is the final pH_{fin}. The time dependency of the EM was measured from 20 wt% dispersions conditioned in pH-adjusted water for a variable time. After recording the pH_{fin} the dispersions were diluted according to the requirements of the instrument. Since the leached ions may change the ionic strength and since the size/diffuse layer thickness ratio is outside both the Huckel and the Smoluchowski limits [6], only the EM is reported.

The size of the particles have been measured with a high resolution light scattering instrument (4700 c) from Malvern Instruments at a 90° scattering angle. The dilution was optimized for an optimal response from the instrument.

The stability was observed through visual inspection after mixing 1 g of powder in a graded measuring glass containing 25 ml of water which was pH adjusted with HCl and NaOH accordingly. Before the initial pH_{ini}-readings were taken the samples were placed in a ultrasound bath for 5 min to ensure maximum dispersibility.

The viscosity was measured on dispersions containing 80 wt % or 40 vol % of powder. The measurements were made on slips freshly made, after 15 min treatment with an ultrasonic rod (Branson Sonifer 450) and after further mixing the dispersion with a magnetic stirred for 24 h. The shear stress and viscosity measurements were made using a Bohlin VOR Rheometer System.

All the measurements were carried out at 25 °C.

Results and discussion

The native pH-values found for a few commercial ZrO_2 powders (Table 1) varied surprisingly little around pH = 5.5 $\pm$ 1 although a range of sintering and processing aids had been mixed into the matrix. However, the variable composition was clearly evident from the alternating electrophoretic mobilities found. The EM-values were dependent both on the composition and the conditioning of the sample. For the model powder TZ-3YS a shift of roughly one pH-unit was recorded in the acidic range (pH = 2–5.8) while a pH-shift of up to four units was observed in the alkaline range (pH = 5.8–12). As shown in Fig. 1, the strongest effect was observed at pH = 9. In general, the pH did stabilize within 2 days, but close to the native pH (pH = 5.5–7) the stabilization requires an even longer time.

When considering the cause of this pH-effect it seems natural to first investigate the dissolution of the constituent metal ions (Zr and Y) from the powder. It has been suggested that the surprisingly high surface charge density obtained for typical ceramic powders as compared to their electrophoretic mobility is due to a high surface porosity

Progr Colloid Polym Sci (1994) 97 : 51–58
© Steinkopff-Verlag 1994

Fig. 1 The pH change found after the initial pH adjustment during a 2-day period plotted against the (initial) pH recorded after the conditioning time

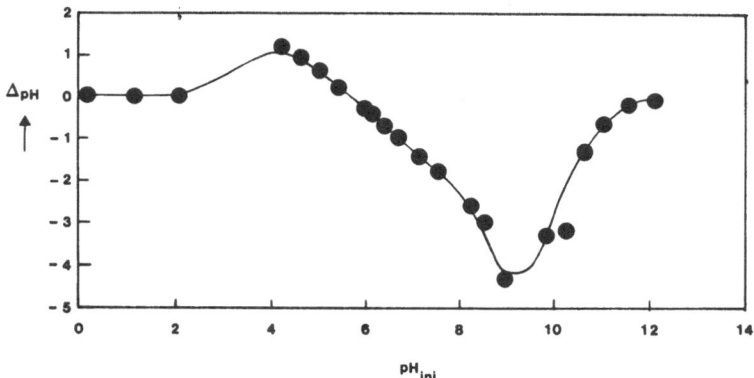

enabling ions to diffuse relatively deep from the particle interior [2]. However, the diameters calculated from the surface area values, using the bulk density value of 6.03 g cm³ provided by the manufacturer, are very close to the measured ones; d(BET) = 0.15–0.16 μm, d(sorpt.) = 0.16 μm and d(manuf.) = 0.12 μm. This indicates that the particles are uniform in size and dense. The small difference found with the two particle sizing instruments used may, at least in part, be due to the inherent properties of the measuring techniques employed [7].

Figure 2 presents the change in surface composition when the powder has been conditioned at a constant pH over a period from 1 day to 1 week. Only a small amount of Zr ions seem to be leached out from the powder surface over a 1 week period in the pH range of 2 to 11 (Fig. 2a). As expected from the higher solubility of its metal oxide [1] a considerable amount of yttrium is dissolved (probably as hydrated ion species) in both acid and alkaline solutions (Fig. 2b). The effect is, however, particularly pronounced in the acidic range.

Table 1. The native pH and the electrophoretic mobility (EM) in (μm/s)/(V/cm) for a range of commerical powders in 1 mmol/dm³ KCl solutions at 25 °C.

Powder	Type	Native pH	EM
ZrO$_2$	TZ-O	5.37	2.6
− " − + Y$_2$O$_3$	TZ-3YA	4.91	0.66
− " − + − " −	TZ-3YS	5.51	1.63
− " − + − " −	TZ-3Y	5.48	2.67
− " − + − " −	TZ-3Y20A	5.03	− 0.34
− " − + − " − + Al$_2$O$_3$	F-5Y	4.58	1.67
− " − + U$_x$O$_y$	HSY-3u	5.5	− 0.95
− " − + MgO	MSZ-8	6.7	− 1.65

The F-5Y and the MSZ-8 powders were from Dynamite Nobel (Sweden), the other samples were supplied by Toyo Soda/Tosoh Co. (Japan).
The dispersions were conditioned in a ultrasonic bath for 5 min. and allowed to equilibrate for 2 h before the EM was recorded.

Fig. 2 The amount of constituent ions extracted from the particle surface expressed as the relative atomic ratio of Zr/O (circle, a) and Zr/Y (triangle, b), respectively, measured with ESCA spectrometry after a 1-day (filled symbols) and 1-week (open symbols) conditioning time at different pH. The theoretical Zr/O-value (squares) is given as a reference (a)

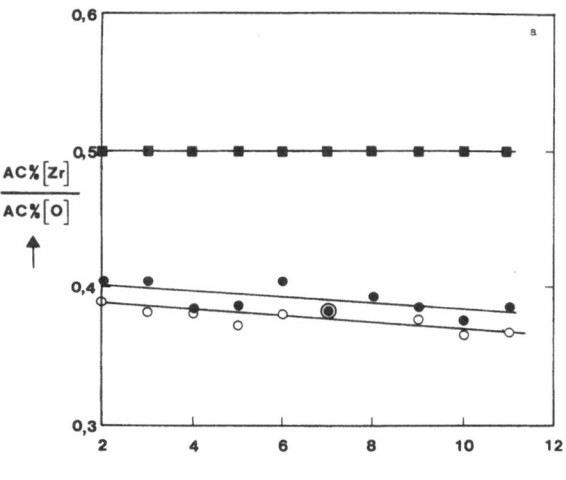

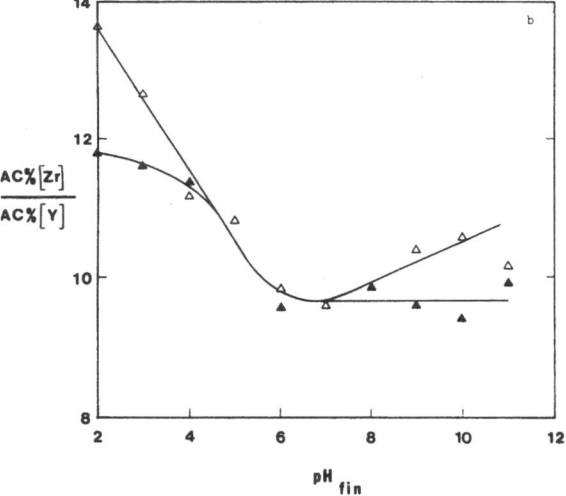

In order to quantify the leaching the amount of Zr and Y was also measured on the supernatant. As presented in Fig. 3, the general trend found with ESCA is confirmed with 100–150 ppm yttrium dissolving from 1 day to 1 week in the acidic range. However, as judged from the plasma analysis, some 25–30 ppm Zr is dissolved in the alkaline range while the amount of yttria in solution remains small. The release of some Zr ions is thus supported by both methods. The rise of the Zr/Y-ratio in Fig. 2 is then suggested to be due to a redeposition of solvated Zr-gel known to precipitate from alkaline solutions [8].

It is well known that neutral and charged molecules may adsorb onto powders, especially when the surface charge is low [9]. This indicates that particle surfaces also contain sites which do not necessarily ionize in aqueous solutions, but which are able both in aqueous and organic solvents to act as binding sites for surface active molecules. In order to fully characterize the solid surfaces it is therefore necessary to detect both the number and the strength of all the potential adsorption sites. In a recent review Jensen has discussed the development of the Lewis acid-base concept as a general scale for non-hydrophobic and non-charged interactions [10]. Fowkes, on the other hand, has introduced this scale as a substitute for the obscure concept of polarity in surface and colloid science [11]. Although a range of experimental methods have been offered to determine the Lewis acid-base forces van Oss et. al. have pioneered the development of a formal framework to account for multi-site interactions [12] and introduced a decay length to be combined with the standard (hydrophobic-electrostatic) DLVO-theory [13].

We have recently determined the strength and the number of Lewis acid-base sites of some metal oxides by a titration method involving so-called Hammet indicators in cyclohexane (and in benzene) [14]. It was found that the number of basic sites increases due to the presence of yttria shifting the equilibrium point, $H_{0,max}$, towards the basic range. This value gives the acid-base equilibrium in terms of the pK_a of the indicators (determined in aqueous solutions). The $H_{0,max}$-values found were 4.5–5.0 for yttria-free ZrO_2 (TZ-O), 5.0–5.5 for yttria stabilized ZrO_2 (TZ3YS) and around 8 for Y_2O_3. Consequently, the small amount of yttria is shown to introduce some basicity to the zirconia particles. However, washing (multiple conditioning in water at the native pH) may change the surface acid-base equilibrium on the particle surface [14].

The zeta potential has been claimed to represent the practically important surface charge of the particles [2]. The charging of the zirconia powder is indicated by the electrophoretic mobility measured at three ionic strengths (Fig. 4) as a function of the final (conditioned) pH_{fin} of the solution. In the pH range of 5.5–7.5 very high EM-values of approximately 7 $(\mu/s)(cm/V)$ was obtained. The increased (indifferent) electrolyte concentration was found to reduce the buffering capacity of the powder. The dispersions with a high content of added salt have thus a stable pH and a less distorted particle surface composition, but only the salt free dispersions mimic the true (dilute) slip conditions. As shown in Fig. 4 the isoelectric point of the zirconia powder appears at roughly pH = 4–5. This value is only slightly lower than the $H_{0,max}$-value referred to above. However, as shown in Fig. 4b, when the zirconia

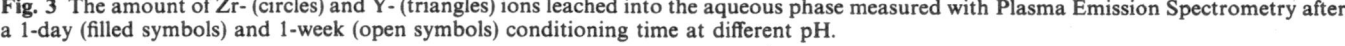

Fig. 3 The amount of Zr- (circles) and Y- (triangles) ions leached into the aqueous phase measured with Plasma Emission Spectrometry after a 1-day (filled symbols) and 1-week (open symbols) conditioning time at different pH.

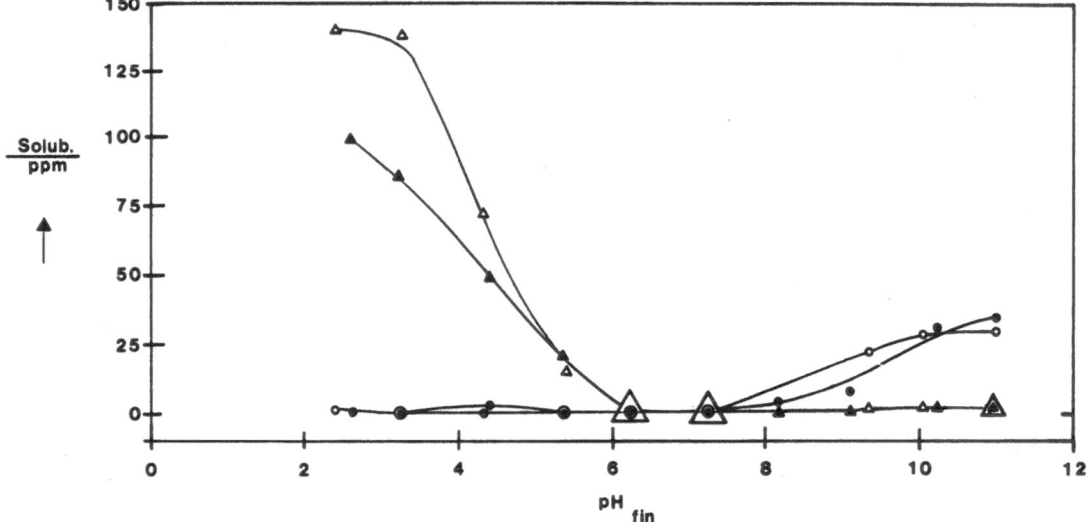

Progr Colloid Polym Sci (1994) 97: 51–58
© Steinkopff-Verlag 1994

powder was conditioned for 1 week in acidic and alkaline solutions the iep shifted to around 7 due to the leaching of yttria. Y_2O_3 was found to have an iep around 8 which compares favorably with the $H_{0,max}$-values found, as well as with previously reported values [14, 15]. It thus seems that the sites on the ZrO_2 particles do charge up in aqueous solutions and may therefore all be ascribed to normal hydroxy groups. However, since some of the powders compared in Table 1 showed considerably different particle mobilities, and since the EM was found to strongly depend on the manufacturing and storage conditions, it remains somewhat unclear whether the iep coincides with the $H_{0,max}$-values given above. Of course, the surface charge density and the point of zero charge (pzc) values should represent the true charged surface sites but the values may be considerably exaggerated due to the substantial variation in the absolute values found for the electrophoretic mobility. If only indifferent ions are present the iep should in general equal the pzc [2, 6].

The influence of the particle size on the pH_{fin} is reported in Fig. 5. As shown, the particle size remains quite unstable in the acidic range at low ionic strengths, especially in the range of acid buffering or enhanced yttrium leaching. The dissolved ions should not be responsible for the particle growth since > 0.04 mol/dm³ is needed to induce a substantial aggregation (Fig. 5b). If the yttrium ions would remain as free Y^{3+} ions in the acidic solutions the aggregation should be enhanced according to the Schulze–Hardy principle [6]. This does not seem likely since the particles are positively charged in this range (Fig. 4) and should repel the ions. If the cause of the instability is to be sought in the dissolved ions their oxo- or hydroxy- species [8] must adsorb specifically onto the particle surface (cf. Hofmeister series) and modify the surface charge accordingly [9]. The ability of yttria sol to neutralize and recharge Si_3N_4 powder has been demonstrated recently [15] and this ability may have a dramatic influence on the particle growth [5]. In the native and slightly alkaline range the zirconia powder remains as single particles, but at both extremes of the pH-scale the instability seem to be reintroduced. The addition of indifferent electrolyte seem to stabilize the dispersion both in the weakly acidic and strongly alkaline ranges. In both cases it is suggested that the electrolyte has a salting-out influence on the dissolved yttrium and possible Zr-gel, respectively. The influence of dissolved yttrium on the particle size, through an increased ionic strength, readily explains the problems encountered in obtaining truly reproducible absolute EM and iep (as well as charge density and pzc) values at low ionic strengths. These effects must be considered when evaluating the results.

The stability of the dilute dispersions was measured visually at two time intervals, 48 h and 72 h (Fig. 6). The pH_{ini} was, however, read shortly after the mixing procedure. The coagulation is almost complete for dispersions with a pH < 2 as well as in the range pH $= 6–9$. In the

Fig. 4 The electrophoretic mobility measured at three different ionic strengths of no KCl added (circles), 0.001 mol/dm³ KCl (triangles) and 0.01 mol/dm³ KCl (squares) as a function of the final pH (a) and in a 0.001 mol/dm³ KCl medium with the time as a parameter; raw powder (circles), 1 day (triangles) and 1 week (squares)-(b). The EM-values found for KCl-free dispersions at pH = 5.5–7.5 (7 (μm/s) (cm/V)) and those outside the pH-range are omitted from the figure

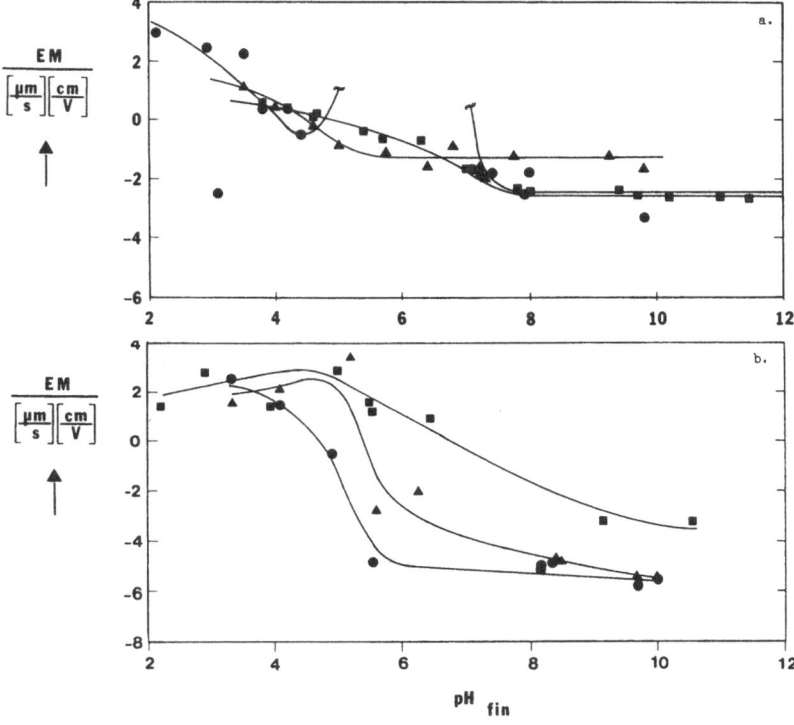

Fig. 5 The mean particle size (d$_{50}$) measured at two ionic strengths of no KCl added (circles) and 0.001 mol/dm^3 KCl as a function of the final pH (a) and as a function of the KCl concentration (diamonds)–(b)

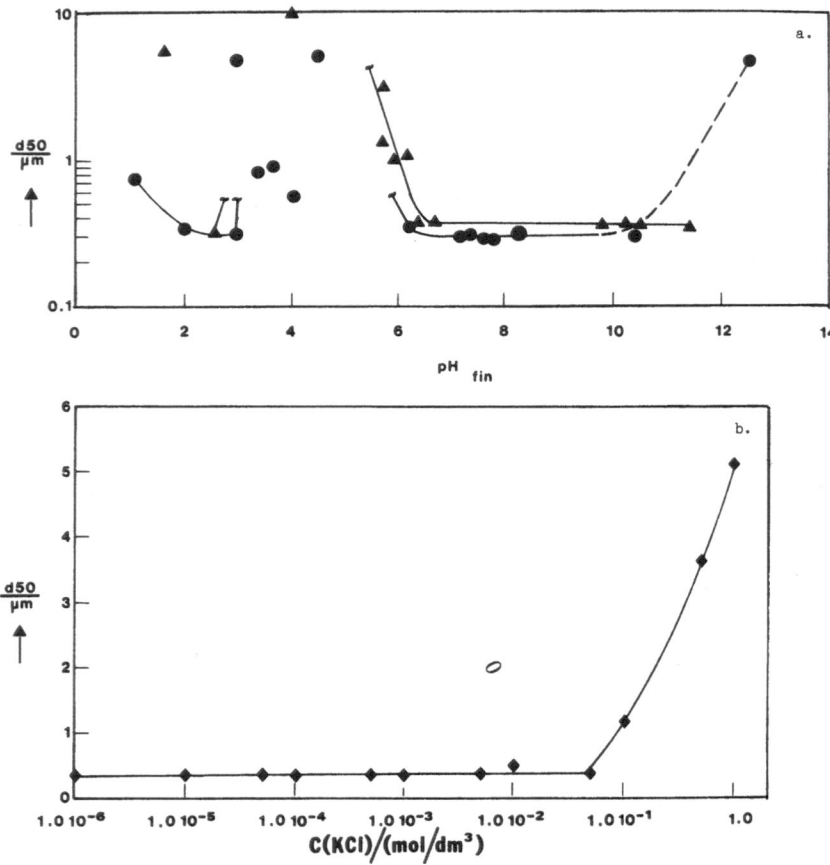

latter case the sedimentation is instant, producing an exceedingly compact coagulate. As shown in Fig. 6 the pH of the dispersion adjusts itself (buffers) to a pH = 5.8 producing a very stable dispersion. In the pH-range of 7.0–9.3 the instability of the pH was too large to remain constant. At pH > 12 the dispersions are again destabilized. Stable dispersions are thus obtained in the pH-range of 2–6 and 9.4–11.3. The range of stability/instability does not coincide with the range of minimum/maximum particle growth (Fig.5) due to the incomplete pH-stabilization in the sedimentations studies. The strongest coagulation after long-term stabilization is, however, found at the iep for the thoroughly conditioned zirconia powder (Fig. 4b). The instability at the extremes of pH seems also to be due to the long-term leaching (> 1 week) indicated in the leaching (Fig. 2) and dissolution (Fig. 3) experiments.

On the basis of preliminary slip casting experiments the concentration of the dispersions for the viscosity measurements were chosen to be 80 wt% or 40 vol%. This concentration may be considered as quite high and it requires careful preparation procedures, but it secured a sufficiently high density of the cast green ceramic body.

The casting procedures and the sintering will be reported together with the final strength features in a subsequent paper [16]. The viscosity of the slips was measured both in the acidic and the alkaline range of maximum stability immediately after the preparation of the slip, after mixing the slip with a ultrasonic bar for 15 min and after further mixing of the slip for 24 h. The measurements of the shear stress and viscosity was measured over shear rates of 0–1470 1/s. Figure 7a presents the dependency of the shear stress on the shear rate for the slips prepared at pH = 2.2 and at the native of pH = 5.4. All the slips prepared in the alkaline range resulted in viscosities being too high for the slip casting purpose and were thus rejected in this context. This feature also explains the use of dispersing aids when alkaline slips are used in colloidal processing [17]. As indicated in Fig. 7 the slips maintained at the native pH produced slips with no observable yield stress, while the acidic slips showed an initial flow resistance. Both freshly prepared slips had a considerable thixotropy, which however disappeared upon conditioning with only a small concomitant reduction in the yield stress. Some of the effects recorded at pH = 2.2 are clearly

Progr Colloid Polym Sci (1994) 97:51–58
© Steinkopff-Verlag 1994

Fig. 6 The stability of powder
dispersions in per cent of
original volume read after 48 h
(dashed line) and 72 h (full
drawn line), respectively,
plotted against the initial pH

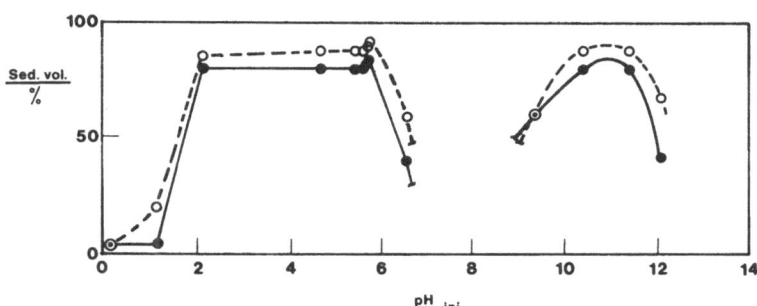

Fig. 7 The shear stress of 80 wt% or 40 vol% powder dispersion at
its native pH (A1 – A3) and adjusted to ph = 2.2 immediately after
mixing (B1), after ultrasonic rod treatment for 15 min (B2, pH = 2.6),
and after further mixing for 2 h (B3, pH = 3.4) plotted against the
shear rate (a). The viscosity and the shear stress of (2h) conditioned
powders at the native pH = 5.4 (A3) and at pH = 3.4 (B3) plotted
against the shear rate (b). The lines from the identification lettering
point to the scale to be read.

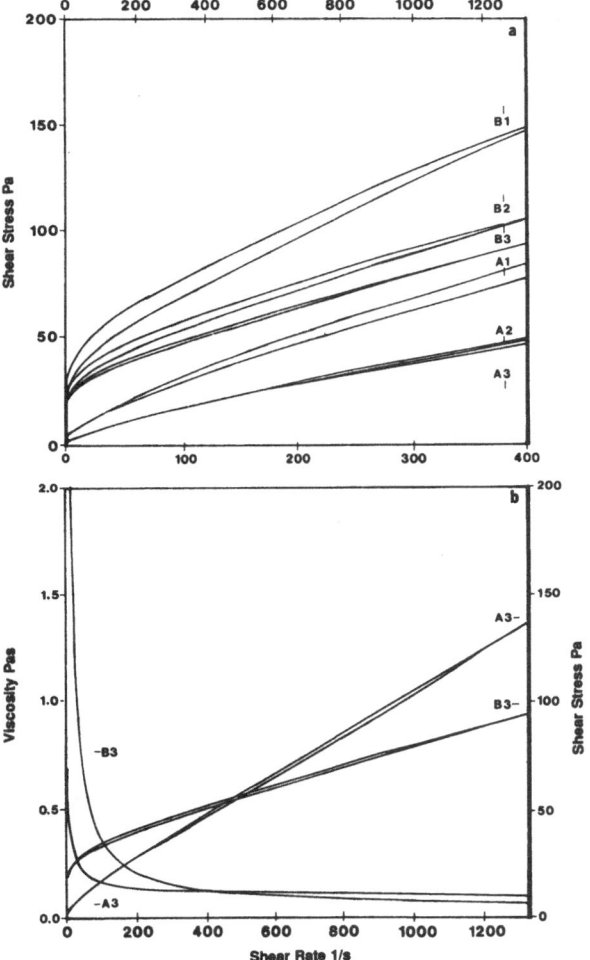

due to the release of yttrium which is shown as a shift of
the pH close to the range of high instability during the time
of the conditioning. In both cases the mixing reduces the
viscosity. No aggregation of the particles was observed
during the conditioning. The particle sizes were measured
for a fraction of the slips and were found to be 0.34 μm
(pH = 2.2), 0.34 μm (pH = 2.6), 0.32 μm (pH = 3.4) and
0.42 μm, 0.40 μm 0.34 μm for the successive samples at the
native pH = 5.4, respectively. Rather a slight reduction of
the average particle size may be observed as a function of
the mixing time. The viscosity and the shear stress of both
systems are plotted on the same shear rate scale in Figure 7b. As shown, the native pH offers slips with superior
gravitational flow properties (used in slip casting) while
a lower pH should be considered if higher shear rates are
used. It seems probable that the higher charge of the
particles (Fig. 4) and the released ions give rise to a sufficiently soft but elastic repulsion in the acidic range resisting the increased shear.

Conclusions

It has been clearly demonstrated that the yttria stabilized
ZrO_2 powder investigated contained matrix components
(Y and Zr) which dissolve into the aqueous solution and
influence the surface properties of the particles as well as
the bulk properties of the dispersion. For the powder
investigated no other compound seem to significantly contribute to the properties of the system.

The dissolution of the matrix components continues
over a period of days and may seriously distort any
measurement performed during this period of time.

The Lewis acid-base characterization of the particle
surface in organic liquids may produce new information of
the active sites which are not ionized in aqueous media but
are able to bind surface active substances.

The optimum alkaline conditions found for the dilute dispersions (EM, particle size, stability) is not suitable for slips of high dry content since the flow resistance at low shearing rates was too high. Dispersing acids have then to be added to ensue a low viscosity. An additive free slip casting is possible at the native pH while acidic conditions provides more favorable slip properties at high shear rates.

References

1. Rich RL (1985) J Chem Educ 62:44
2. Bergström L, Pugh RJ (1989) J Am Ceram Soc 72:103
3. Bergström L, Bostedt E (1990) Colloids and Surfaces 49:183
4. Persson M (1989) Thesis Chalmers University of Technology, Göteborg, Sweden
5. Fagerholm H, Johansson LS, Rosenholm JB, J European Ceramic Soc, in press
6. Hiemenz PC (1986) "Principles of Colloid and Surface Chemistry", 2nd Ed., Ch. 13 p. 737, Marcel Dekker Inc, New York
7. Allen T (1975) "Particle Size Measurements", 3rd Ed. Chapman and Hall, London
8. Merck Index, 10th Ed., p. 1460 Merck Co Inc Rachway, New Jersey
9. Collins KD, Washabaugh MW (1985) Quart Rev Biophysics 18:323
10. Jensen WB (1991) in "Acid-Base Interactions: Relevance to Adhesion Science and Technology", (KL Mittal and HR Anderson, Jr, Eds), p. 3, VSP BV Utrecht, The Netherlands
11. Fowkes FM (1987) J Adhesion Sci & Tech 1:7
12. van Oss, CJ, Chaudhury MK, Good RJ, (1988) Chem Rev 88:927
13. van Oss Giese RF, Costanzo PM (1990) Clays & Clay Min 38:151
14. Pettersson ABA, Byman-Fagerholm H Rosenholm JB (1992) In: "Ceramic Materials and Components for Engines", (Carlsson R, Johansson T, Kahlman L, (Eds), Elsevier Appl Sci Publ Essex, England
15. Lidén E, Persson M, Carlström E, Carlsson R (1991) J Am Ceram Soc 74:1335
16. Grönroos L, unpublished results
17. Byman-Fagerholm, Rosenholm JB, Lidén E, Carlsson R (to be published)

Progr Colloid Polym Sci (1994) 97:59–64
© Steinkopff-Verlag 1994

SUSPENSION

One-phase and two-phase regions of colloid stability

M.V. Smalley

Received: 15 September 1993
Accepted: 20 December 1993

Abstract The Coulombic attraction theory of colloid stability, first proposed by Sogami in 1983, is shown to be well adapted to explain the existence and extent of both the one-phase (suspension) and two-phase (gel) regions of colloid stability which are observed in clay swelling. The central prediction of the Sogami theory is that there is a weak attractive tail in the thermodynamic electrostatic interaction potential between colloidal particles in electrolyte solutions. The position of the minimum in the pair potential between parallel clay plates is given as a function of the plate thickness and the electrolyte concentration (c) and is used to estimate how much solvent the clay will absorb as a function of its initial volume fraction (r). This yields a prediction for the position of the r-c phase boundary, whose curvature is in excellent agreement with recent studies of vermiculite swelling. The standard theory of colloid stability, the DLVO theory, is a limiting case of the Coulombic attraction theory, in the one-phase region. In the two-phase region, Sogami theory combined with the Dirichlet boundary condition (constant surface potential) predicts that the ratio (s) of the salt concentration in the supernatant fluid to the average salt concentration in the gel phase will be constant. For a surface potential of 70 mV, s is equal to 2.8, in excellent agreement with the experimental results on n-butylammonium vermiculite gels.

Key words Clay swelling – colloid stability – Sogami potential – DLVO theory – salt fractionation

M.V. Smalley
Polymer Phasing Project
ERATO, JRDC
Keihanna Plaza
1-7 Hikari-dai
Seika, Kyoto 619-02, Japan

Introduction

The interaction between the charged particles in ionic colloidal solutions is usually treated in terms of the DLVO theory, which was developed independently in the 1940s by Derjaguin and Landau [1] and Verwey and Overbeek [2]. According to this theory, the thermodynamic pair potential that describes the Coulombic interaction between the charged particles is a pure repulsion. The stability of lyophobic colloids is then attributed to the van der Waals force.

The DLVO theory has long been one of the foundations of colloid chemistry, but the basic intuitive concept on which it is based, namely the repulsive nature of the Coulombic interaction between the like charged macroions, has been questioned by Ise and co-workers [3, for a recent review]. They have argued that the ordering observed in highly charged macroionic solutions is due to Coulombic attraction between the like charged particles through the intermediate counterions. This new concept was first given theoretical expression by Sogami in 1983 [4] and generalized by Sogami and Ise in 1984 [5]. The Coulombic attraction theory first received attention in the

West when Overbeek disputed it in 1987 [6]. It has already been shown that Overbeek's criticism of the new theory is fundamentally flawed and that the Coulombic attraction theory provides a logical and self-consistent basis for describing the interparticle interactions in colloidal systems [7]. The Coulombic attraction theory has since been placed on a rigorous basis by calculations of the Helmholtz free energy for the case of the interaction of two flat double layers [8, 9]. In ref. [8] the one-dimensional colloid problem was solved at the mean field theory level subject to the hypothesis of "counterion dominance" and in ref. [9] the exact mean field theory solution to the problem was given. The latter paper has proved rigorously the existence of effective long-range attraction between highly charged plates in an electrolyte solution and provides a new basis for the analysis of a variety of phenomena in macroionic solutions.

The work in refs. [7–9] on the interaction of highly charged plates in an electrolyte has been criticized by Levine and Hall [10] and by Overbeek [11]. Which of us turns out to be right will, of course, be decided by experiment.

The three component system

We shall study a three-component system of a monodisperse colloid, electrolyte and solvent. The best characterized of the one-dimensional colloids, the n-butylammonium vermiculite system [12–16], will be used as an example. There are four constituents in the macroionic solution, the negatively charged clay plates, n-butylammonium ions (counterions), chloride ions (co-ions) and water, but these may not vary independently because they are subject to the restriction that

$$[n\text{-}Bu^+] = [\text{Plate}^-] + [Cl^-]$$

in an obvious notation. Hence, the number of components is $4 - 1 = 3$. In the following we shall refer to the solvent as water and to the electrolyte as salt.

The raw phenomenon of the clay swelling in the n-butylammonium vermiculite system is represented schematically in Fig. 1.

In the cases studied in refs. [12–16] V^* was always much greater than V, the volume occupied by the macroions. We now define V_m to be the volume occupied by the macroions in the coagulated state, as in Fig. 1(a) in the vermiculite system. This is an experimentally controlled variable. We define the sol concentration r by

$$r = V_m/V^* \,,$$

where V^* is the total volume of the condensed matter system. In the case of swelling illustrated by Fig. 1(a) and

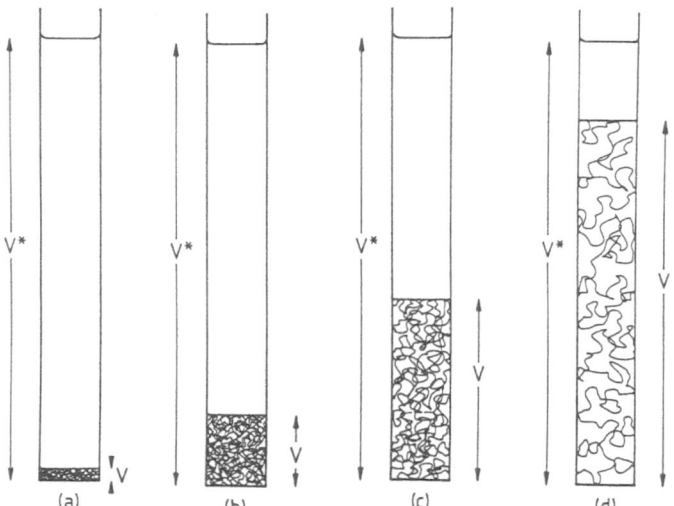

Fig. 1 Schematic illustration of the swelling of n-butylammonium vermiculite. a) shows the unexpanded crystal in a 1.0 molar n-butylammonium chloride solution. b), c) and d) show the gels formed in 0.1, 0.01, and 0.001 M solutions, respectively

(b) V^* decreases by approximately 0.1% [16]. This is a very small fractional volume change compared to that observed in V, so in the following we ignore the electrostriction of the solvent which accompanies swelling, that is, we take $V^* =$ constant. Although the phase boundary has been investigated with respect to temperature and hydrostatic pressure [16], we now restrict attention to P and T constant, so that we can represent the phase behavior of the system on triangular graph paper.

N-butylammonium chloride salts out from simple electrolyte (macroion free) solutions at about 4.5 molar [17]. This value does not depend significantly on the presence of macroions in the beaker and so is independent of r. This enables us to draw in the left-hand wedge in Fig. 2(a), which shows the phase diagram in mass fractions.

The left-hand wedge of Fig. 2(a) represents a three-phase region of crystalline clay, solid salt and saturated salt solution and is labelled region IV in the schematic Fig. 2(c). In Fig. 2(c), the $r = 0.1$, $c = 0.01$ M point has been placed at the center of the triangle and the scale has been distorted to show all four regions clearly on the same plot.

As the salt concentration is decreased below 4.5 M the solid salt phase disappears, but the clay crystals do not swell until c is decreased below 0.2 M (at $T = 4$ °C, $P = 1$ atm.) [15]. To a first approximation, this value is also independent of r, which enables us to draw in the central wedge (region III) in Fig. 2. This represents a two-phase region of salt solution and crystalline clay. In both of regions III and IV, the macroions are in their primary minimum (crystalline, coagulated, flocculated) state. To

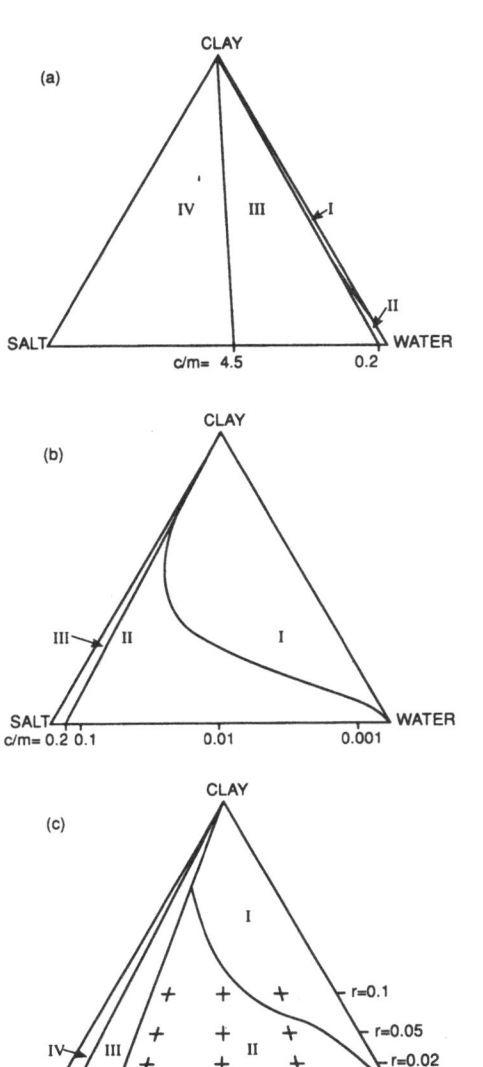

Fig. 2 Phase diagram of the three-component system of clay (*n*-butylammonium vermiculite), salt (*n*-butylammonium chloride) and water at $T = 4\,°C$, $P = 1$ atm. a) shows a straight mass fraction plot. b) shows the mass fraction plot when the molecular weight of the salt is rescaled by 1000. The curved phase boundary is calculated in section 3. c) is a schematic plot. The $r = 0.1$, $c = 0.01$ M point has been placed at the center of the triangle and the scale has been distorted to show all four regions clearly. The labeling of the regions is explained in the text. The crosses indicate the points studied by Williams et al [17]

the right of region III, in electrolyte concentrations $c < 0.2$ molar, the clay absorbs water macroscopically and swells "osmotically" into the secondary minimum (gel, sol) state, giving us the one-phase (I) and two-phase (II) regions of colloid stability.

If we plot the phase diagram in ordinary weight percentages, as shown in Fig. 2(a), the most interesting chem-

istry, that of gel formation, is confined to too small a region on the right-hand edge. We gain more insight into the phase behavior of the system if we re-scale the molecular weight of the salt by a factor of 1000, as shown in Fig. 2(b). This has the effect of fanning out the plot around the $c = 0.01$ M line and clearly shows the curved phase boundary between the one-phase and two-phase regions of colloid stability. It has been calculated using the method described in Section 3.

Coulombic attraction theory

The central prediction of the Sogami theory is that there is a weak attractive tail in the thermodynamic electrostatic interaction potential between colloidal particles in electrolyte solutions [3–9]. In the linearized theory, which we now pursue, the position of the minimum in the thermoelectric potential between macroionic plates is given approximately by the equation

$$\kappa X_{\min} = 4\,, \tag{1}$$

where $X_{\min}$ is the equilibrium separation of the particles and κ is the inverse Debye screening length. This localizes the plates at a distance of four Debye sceening lengths, where, for monovalent ions in water at 25°C the inverse Debye sceening length is defined by

$$\kappa^2 = 0.107\,c\,, \tag{2}$$

where κ is expressed in units of $Å^{-1}$ and c is the concentration of the electrolyte solution in moles/liter. The equilibrium separation of the particles is therefore roughly inversely proportional to the square root of the concentration of the salt solution. Equation (2) suffices to define kappa in a simple ionic solution, but there is no a priori reason why the average salt concentration in the gel phase should be equal to that in the supernatant fluid. It has recently been proved that if we take Eq. (1) in conjunction with the Dirichlet boundary condition (constant surface potential, ψ_0), then the supernatant fluid always contains a constant multiple of the average electrolyte concentration in the gel phase, irrespective of the absolute value of the salt concentration [18]. For the *n*-butylammonium vermiculites, the effect of uniaxial stress on the gels has shown that ψ_0 is constant in the range of electrolyte concentrations between 0.001 M and 0.1 M, its average value being equal to 70 mV [19]. For $\psi_0 = 70$ mV, the salt ratio has been calculated to be 2.8 [18]. Recent studies of *n*-butylammonium vermiculite swelling have shown that the ratio is constant in the range of electrolyte solutions between 0.001 M and 0.1 M, its average value being equal to 2.6 [17]. This confirms that clay swelling is governed by the Coulombic attraction theory with the Dirichlet boundary condition.

Kappa defines one of two independent length scales in the colloid problem. The other independent length variable is the thickness, 2a, of the (supposed monodisperse) colloidal particles. For the n-butylammonium vermiculite plates we take the observed c-axis spacing of 19.4 Å for 2a [15].

Equation (1) is only the small kappa approximation to the equilibrium separation of plate macroions. For plate macroions, the position of the secondary minimum X_{min} is given in full by [7]

$$X_{min} = \frac{1}{\kappa}\left[4 + 2a\kappa \frac{\sinh(2a\kappa)}{1 + \cosh(2a\kappa)} \right]. \tag{3}$$

The two terms in Eq. (3) are given separately in the second and third columns of Table 1, where the second term has been labeled $f(a, \kappa)$.

Table 1 covers the experimentally best characterized salt range, over which this function for X_{min} works qualitatively well to explain the behavior of the vermiculite gels in the two-phase region [7, 18]. Its essential feature is that for separations greater than about $4/\kappa$ (first length scale in the problem) the plates attract each other so that the gel does not soak up any more solvent beyond this point: osmotic swelling stops and the two-phase region begins. X_{min} should therefore give a reasonable approximation to the position of the phase boundary. In order to express this in terms of the sol concentration, r, we have to introduce the thickness of the plates 2a (second length scale in the problem). We define

$$r^* = 2a/X_{min},$$

where r^* is the theoretical prediction for the sol concentration at the phase boundary, given in the fifth column of Table 1. The experimental approach to the phase boundary from the two-phase region in the n-butylammonium vermiculite system [17] shows that the way is cuts across the r-c space is qualitatively well predicted by the theory in the experimentally accessible range between 0.001 M and 0.1 M.

DLVO theory

In DLVO theory the two-phase region of colloid stability can only be created by the van der Waals force, which is independent of the salt concentration across the concentration range 0.001 M < c < 0.1 M [2]. This force has to be balanced with a force which decays exponentially as a function of kappa, which means that it decays by a factor $\exp(-10)$ across this range. The unhappy consequence of this prediction is that the position of the secondary minimum, and therefore the position of the phase boundary, varies very rapidly as a function of c, in contradiction to the experimental results. A further unhappy consequence of this balance is that it always produces a primary minimum much deeper than the secondary minimum. This renders it unable to explain the raw phenomenon of osmotic swelling, in which a primary minimum material develops spontaneously into the secondary minimum, and unable to explain the thermodynamic character of this transition [15, 16]. Such subtle effects as the salt fractionation effect are way beyond its scope.

It is noteworthy that in the one-phase region of colloid stability the net interaction between the plates is a repulsive function which decays approximately exponentially with the separation between the plates (see Fig. 3). This is the prediction of DLVO theory [1, 2]. In this region, therefore, the electrostatic part of DLVO theory still applies. This is the potential which governs the one-phase region of colloid stability, because at high sol concentrations the plates do not have sufficient solvent available to reach their equilibrium separation of about four Debye screening lengths. In this sense, the DLVO theory can be seen to be a limiting case of the Coulombic attraction theory, in the one-phase region.

Temperature dependence of the phase boundary

It is instructive to study the temperature dependence of the phase boundary between the two-phase region of colloid stability (II) and the primary minimum state (III) in the n-butylammonium vermiculite system [15, 17]. The phase transition is thermodynamic, so it will occur when the primary and secondary minima are equal in depth. We might expect the free energy of the crystalline state to be relatively insensitive to temperature. In this case, variations of the phase transition temperature T_c correspond to variations in the depth of the secondary minimum, which in turn is sensitive to the surface potential, ψ_0. The effect therefore gives us a measure of the way in which ψ_0 varies with c. The exact functional dependence of ψ_0 on T_c is unknown, but it seems reasonable to assume that the

Table 1 The r-c phase boundary in n-butylammonium vermiculite swelling

c(M)	$4/\kappa$(Å)	$f(a, \kappa)$(Å)	X_{min}(Å)	r^*
0.001	400	2	402	0.05
0.003	231	3	234	0.08
0.01	126	6	132	0.15
0.03	73	9	82	0.24
0.1	40	15	55	0.36

Progr Colloid Polym Sci (1994) 97:59–64
© Steinkopff-Verlag 1994   63

ratio $e\psi_0/kT_c$ will remain constant along the c, T phase boundary if the behavior of the system is dominated by electrostatic forces. As the temperature of the transition decreases with increasing salt concentration, this suggests that the magnitude of the surface potential is also decreasing with c.

If we now look for Nernstian behavior in the system, then

$$\psi = \psi_0 + (RT/zF)\ln c \qquad (4)$$

suggests that a plot of the surface potential against log c should be linear. This further suggests that a plot of T_c against $\log c$ might also be linear, and experimentally this is indeed the case [17]. The gradient of the experimental plot of T_c against $\log c$ is $0.077\ \mathrm{K}^{-1}$, which corresponds to 13 K per log unit. In electrical terms this corresponds to a decrease of only 1 mV per decade of salt concentration compared to 58 mV as predicted by the Nernst equation. This again shows that the n-butylammonium vermiculite gels form a more or less constant surface potential system and are therefore governed by the Dirichlet boundary condition rather than the Nernst equation.

The weak temperature dependence of the phase boundary also shows that the primary and secondary minima are very delicately balanced in this system, as sketched in Fig. 3(a), which is to be compared with the DLVO potential sketched in Fig. 3(b).

How does the experimentally observed situation come about in the Coulombic attraction theory? The clue to the answer is sketched in Fig. 3(c), in which we have assigned the van der Waals potential its proper role in colloid science, namely as a negligible contribution at the secondary minimum separation, but as a constituent in understanding flocculation into the primary minimum.

The Coulombic attraction theory curve to the right of the point marked by the arrow on the solid (electrostatic) curve in Fig. 3(c) has been taken from ref.[7], its functional form being given by

$$U^G = \frac{2\pi e^2}{\varepsilon} Z^2 \exp(-\kappa X)\left\{[1 + \cosh(2a\kappa)]\left(\frac{3}{\kappa} - X\right) + 2a \sinh(2a\kappa)\right\}, \qquad (5)$$

where Z is the number of charges per unit area on the macroionic plates, X is the interplate separation, e the electronic charge, and ε the permittivity. U^G is the electrostatic contribution to the total pair potential V_T. For plate separations greater than about 100 Å, the exact mean field theory solution to the one-dimensional colloid problem [9] has shown that Z is constant in this region, once the salt concentration has been fixed. An example is shown in Fig. 4.

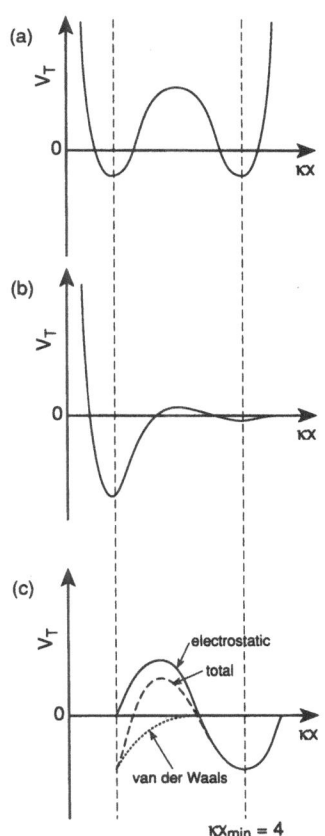

Fig. 3 The phase transition in n-butylammonium vermiculite swelling is thermodynamic so a) the primary and secondary minima must be of equal depth, irrespective of the shape of the total interaction potential, V_T. This cannot be accounted for by the standard DLVO potential, shown in b). c) shows V_T in the Coulombic attraction theory (dashed curve), composed of an electrostatic part (solid curve) and a short-range potential (dotted curve), which includes a weak contribution from the van der Waals potential. The vertical dashed lines indicate the positions of the two minima

In Fig. 4, Z_0 is the charge density on an isolated plate and the ordinate $|Z_i|$ is the magnitude of the charge density on the inner surface of a charged plate as two isolated plates are brought together from infinity. Of course, $Z_i \rightarrow 0$ as $X \rightarrow 2a$ ($d \rightarrow 0$) because in this limit the plates have coalesced and there is no charge separation in the system. This means that $U^G \rightarrow 0$ as $X \rightarrow 2a$ and the negative values of U^G at large particle separations therefore represent a state of lower free energy than the primary minimum. The inner part of the electrostatic curve in Fig. 3(c), in which the decay of the surface charge overwhelms the exponential screened Yukawa potential, has been taken from Fig. 4, but its exact form should not be taken too seriously because in this regime of spacings mean field theory breaks down due to many effects such as dispersion forces, the size of the small ions and the molecular degrees of freedom of the solvent.

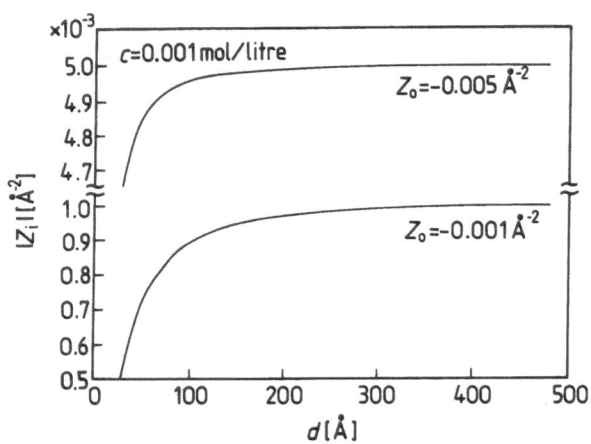

Fig. 4 Dependence of the magnitude of the charge density on the inner plate surface $|Z_i|$ on the interplate distance d for $\Psi_0 = -110$ mV and -190 mV in the Dirichlet model ($Z_0 = -0.001$ and -0.005 Å^{-2}) when $c = 0.001$ M. The density $|Z_i|$ vanishes rapidly to zero as $d \to 0$ ($X \to 2a$), and it saturates quickly to the density of an isolated plate $|Z_0|$

If the short range forces were purely electrostatic then the Coulombic attraction theory would predict that the secondary minimum was always lower in energy than the primary minimum. However, other short range forces,

including the van der Waals potential, do lower the depth of the primary minimum and so bring it into balance with the secondary minimum, as observed experimentally.

Conclusion

These considerations show that the Coulombic attraction theory is well adapted to explain the existence and extent of both the one-phase and two-phase regions of colloid stability which are observed in clay swelling. At low sol concentrations, many important practical problems, such as sedimentation problems in lakes and the rheology of drilling muds, will have to be re-appraised because the interaction between the charged particles is not that envisaged by DLVO theory, which is unfortunately still common currency among many experimental workers in the field.

Acknowledgments I wish to thank the SERC for provision of an Advanced Fellowship to support this work. I also wish to thank Dr. R. K. Thomas and Professor S. Levine for helpful suggestions and Professors N. Ise and I. Sogami for their encouragement.

References

1. Derjaguin BV, Landau L (1941) Acta Physicochimica 14:633–662
2. Verwey EJW, Overbeek JThG (1948) Theory of the Stability of Lyophobic Colloids, Elsevier, Amsterdam
3. Ise N, Matsuoka H, Ito K, Yoshida H, Yamanaka J (1990) Langmuir 6:296–302
4. Sogami I (1983) Phys Lett A 96:199–203
5. Sogami I, Ise N (1984) J Chem Phys 81:6320–6332
6. Overbeek JThG (1987) J Chem Phys 87:4406–4408
7. Smalley MV (1990) Molec Phys 71:1251–1267
8. Sogami IS, Shinohara T, Smalley MV (1991) Molec Phys 74:599–612
9. Sogami IS, Shinohara T, Smalley MV (1992) Molec Phys 76:1–19
10. Levine S, Hall DG (1992) Langmuir 8:1090–1095
11. Overbeek JThG (1993) Molec Phys 80:685–694
12. Garrett WG, Walker GF (1962) Clays Clay Min 9:557–567
13. Walker GF (1960) Nature 187:312–313
14. Norrish K, Rausel-Colom JA (1963) Clays Clay Min 10:123–149
15. Braganza LF, Crawford RJ, Smalley MV, Thomas RK (1990) Clays Clay Min 38:90–96
16. Smalley MV, Thomas RK, Braganza LF, Matsuo T (1989) Clays Clay Min 37:474–478
17. Williams GD, Moody KR, Smalley MV, King SM, Clays Clay Min (in press)
18. Smalley MV, Langmuir (in press)
19. Crawford RJ, Smalley MV, Thomas RK (1991) Adv Colloid Interface Sci 34:537–560

Progr Colloid Polym Sci (1994) 97:65–70
© Steinkopff-Verlag 1994

SUSPENSION

E. Pefferkorn
L. Ouali

Polymer induced fragmentation of colloids: Mechanism and kinetics

Received: 28 September 1993
Accepted: 30 October 1993

Dr. E. Pefferkorn (⊠)
Lahoussine Ouali
Institut Charles Sadron
6, rue Boussingault
67083 Strasbourg Cedex, France

Abstract The particle counter technique is used for studying the rate of aggregate fragmentation induced by polymer adsorption. The influences of the hydrodynamic forces during aggregation and the degree of surface coverage before fragmentation were investigated. We determined that a scaling law $a(x) \cong x^\lambda$ described the rate of break-up $a(x)$ of aggregates of size x. Values of λ equal to 0.65 and 0.90 characterized the process for aggregates formed with and without stirring respectively. The establishment of an adsorption equilibrium for facing surfaces of adhering colloids was achieved by a reptation process in the restricted interfacial zones. The time prior to the beginning of the fragmentation decreased with the degree of coverage of the starved surface at the time of aggregation. The polymer concentration of the fragmentation medium influenced the fragment size distribution, which was described by the scaling law $\phi(x) = x^v$. The exponent v was found to be equal to -1.2 in concentrated polymer solutions and equal to -1.75 in the more diluted ones. The rate of fragmentation determined from the variations of the weight $S(t)$ and number $N(t)$ average sizes, was much slower than the rate of break-up calculated from the theory, indicating that fragmentation is retarded by a concomitant aggregation.

Key words Aggregate fragmentation – aggregate break-up – polymer adsorption – colloid size distribution – fragmentation rate

Introduction

The stability of polymer-coated colloids is related to the polymer solubility parameters and relations between the critical flocculation temperature and the theta temperature of solubilized macromolecules have been established. This shows that the stability of polymer-colloid complexes strongly depends on the macromolecular characteristics [1]. Our investigation was directed to the conditions of achievement of such stabilized colloids. This problem appeared to be non trivial In fact, polymers predominantly behave like destabilizing agents and polymer colloid mixing generally leads to the formation of large aggregates [2]. Colloid stabilization was expected to be enhanced when the colloidal suspension is mixed with a polymer solution of high concentration favoring the polymer adsorption rate. It is, however, well known that concentrated solutions also induce colloid instability by a depletion phenomenon [3, 1]. All of these problems arise because instantaneous mixing produces places where polymer is i) present at a sufficient dosage and induces particle stabilization, ii) present in great excess and induces colloid instability, and iii) in default and induces aggregation by a bridging mechanism.

The objective of this work was to determine the kinetics and mechanism of the aggregate fragmentation in the presence of polymer. Insofar as limited particle aggregation cannot be avoided in concentrated suspensions during mixing of colloids and polymers, the resulting aggregates are expected to fragment with the completion of polymer adsorption. Our experimental investigation consisted in preparing different types of aggregates and in studying the corresponding rate and mechanism of fragmentation. The influence of slow stirring of the suspensions during aggregation was also investigated to determine the impact of hydrodynamic forces on the strength of the interparticle bonds. Our interpretation of the fragmentation processes is based on scaling laws. In the following theoretical part, we briefly present elements of the theory to which we refer.

Materials and methods

Colloids

Monosized polystyrene latices of diameter equal to 1730 nm, kindly supplied by the Laboratoire des Matériaux Organiques (LMO, Lyon, France) were used as model colloids. Sulfonate surface groups at a density of 1.45 $\mu C/cm^2$ constitute the active sites. The density of the latex is equal to 1.045 g/ml.

Polymer

Polyvinylpyridine of molecular weight equal to 580 000 was used as model for a polyelectrolyte. All experiments were performed at 25 °C and at pH 3.0, where the degree of protonation of the pyridine group is equal to 0.5.

Determination of the colloid size distribution

The cluster size distribution $c(x, t)$ is deduced from the histogram given by the Coulter Counter TAII and, the number $N(t)$ and weight $S(t)$ average sizes of the aggregates are calculated from the distribution. The device and analysis technique are described elsewhere [4].

Methods of aggregate formation

Two typical procedures were implemented to obtain aggregates of characteristic size and morphology. Firstly, in the case of instantaneous mixing of polymer and colloids, the aggregation slowly progresses by the sticking together of particles according to the scaling law $S(t) \cong N(t) \cong t^{0.35}$ and the particles forming the aggregates are characterized by a constant degree of coverage. Secondly, in the case of progressive polymer addition, the surface coverage increases with time, so that the degree of coverage of aggregates of large size which are sampled after a long aggregation period is larger than that of smaller ones which are taken from the suspension earlier. To quantitatively correlate the surface coverage with the aggregation period, the following method was implemented. The colloid suspension was introduced into a closed reactor of volume equal to 50 ml and stirred to ensure homogeneity. The polymer solution was injected at a constant rate at the reactor input and the corresponding volume of the suspension (containing the aggregates) was collected at the reactor output, at given periods, to start the fragmentation experiments. Insofar as the adsorption rate on the suspended colloids is instantaneous at low surface coverage [5], this procedure allows the surface coverage A_s to be a linear function of the injection time t.

Method of fragmentation

To induce the fragmentation of these two classes of aggregates, portions of the aggregate suspensions were withdrawn and transferred to polymer solutions of different concentrations. The reference concentration corresponding to the beginning of the coil overlap $C*$ was equal to 3.06 g/l, (the addition of a colloid to this solution provoked phase separation and the corresponding mechanism may be that described by Vincent, Luckham and Waite [3]). The aggregate fragmentation was investigated in the presence of the following concentrations: $C*/5$, $C*/10$, $C*/30$, $C*/50$ and $C*/70$. During fragmentation, the suspension was slowly rotated (one revolution per 2 min) in order to keep the suspension homogeneous.

Theoretical part

Equation (1) may be used to describe the variation of the size distribution function $c(x, t)$, the concentration of aggregates composed of x elementary particles as a function of time [6]:

$$\frac{\partial}{\partial t} c(x, t) = -a(x) c(x, t) + \sum_{y=x}^{\infty} c(y, t) a(y) f(x/y), \qquad (1)$$

where $c(x, t)$ is the concentration of the cluster of size x at time t, $a(x)$ is the rate of break-up, and $f(x/y)$ is the rate at which clusters of size x are produced from the break-up of aggregates of size y

Homogeneous kernels of the rate of aggregate break-up of the form

$$a(x) = x^\lambda \qquad (2)$$

were considered. The fragmentation rate is correlated to the variation of the mean cluster size $S(t)$ and $N(t)$ which decrease as indicated in the theory by:

$$S(t) \approx N(t) \approx t^{-1/\lambda} \qquad (3)$$

Treatment of Eq. (1) is carried out by assuming the cluster size distribution to be self-similar [7]:

$$c(x, t)S^2(t) \approx \varphi[x/S(t)] , \qquad (4)$$

where $S(t)$ is the weight average size of the colloid. λ is related to the slope of $\varphi(x/S(t))$ for large values of the variable in the case of linear fragmentation, as follows

$$\varphi(x) \approx x^{-2} \exp(-ax^\lambda) \qquad (5)$$

Homogeneity implies that the rate of the fragmentation of y to x is described by:

$$f\left(\frac{x}{y}\right) = y^{-1} b\left(\frac{x}{y}\right) . \qquad (6)$$

The rate $b(x)$ of formation of cluster of size x is given by

$$b(x) \approx x^v . \qquad (7)$$

The exponent v is related to the slope of the fragment size distribution $\varphi(x)$ at small values of the size x as indicated by:

$$\varphi(x) \approx x^v \qquad (8)$$

We attribute the deviation from the similar behavior of $S(t)$ and $N(t)$ in the fragmentation induced by the polymer to supplementary and concomitant aggregation processes. To this aim, we introduce the following Eq. (9) to characterize the variation of $N(t)$:

$$N(t) \cong t^{-1/\mu} \qquad (9)$$

Results and discussion

I) The fragmentation of aggregates formed in the presence of polymer without stirring

Figure 1 represents the colloid number $N(t)$ and weight average size $S(t)$ as a function of the fragmentation period for three typical polymer concentrations. The log-log representation is clearly evidence of the validity of the scaling

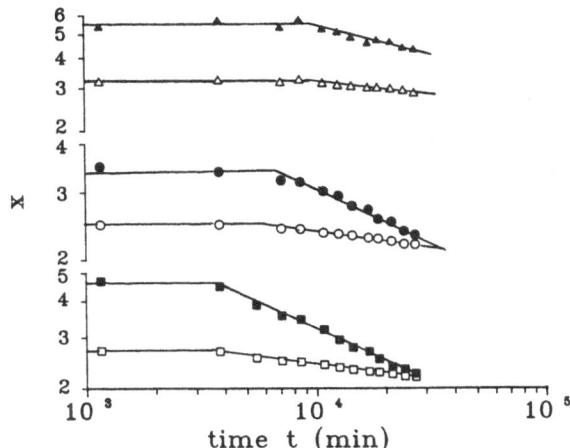

Fig. 1 Fragmentation rate of aggregates formed without stirring. Representation of the fragment weight $S(t)$ (solid symbols) and number $N(t)$ (open symbols) average sizes (measured by x, the number of elementary particles) as a function of the fragmentation time. Time $t = 0$ corresponded to the moment of immersion of the aggregates in polymer solutions of concentration $C^*/5$, (triangle), $C^*/30$, (circle) and $C^*/70$, (square)

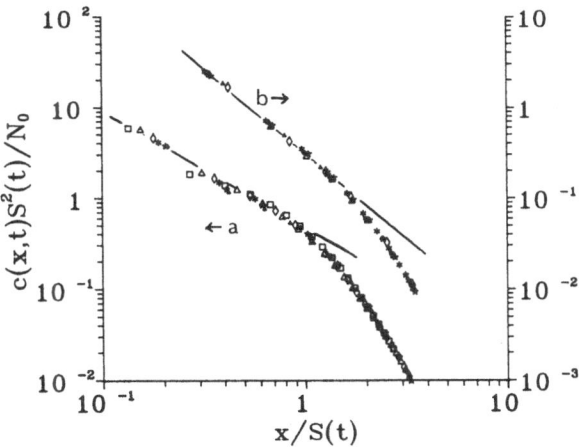

Fig. 2 Fragmentation of aggregates formed without stirring. Reduced size distribution curves of the fragments according to Eq. (4), determined in the concentrated polymer solutions $C^*/5$ and $C^*/10$, (curve a) and in the dilute polymer solutions $C^*/30$, $C^*/50$ and $C^*/70$, (curve b)

laws (3) and (9). Figure 1 shows that both $S(t)$ and $N(t)$ tend towards a unique value which is more rapidly reached in dilute polymer solutions.

Figure 2 represents the cluster size distribution function (4), on a log-log scale, for fragmentation in different polymer solutions. In the concentrated solutions $C^*/5$ and $C^*/10$, a unique curve is obtained characterized by a linear variation with a slope v equal to -1.20 for values of $x/S(t)$ smaller than 1 (Fig. 2, curve a). In the more diluted solution $C^*/30$, $C^*/50$ and $C^*/70$, a unique master curve

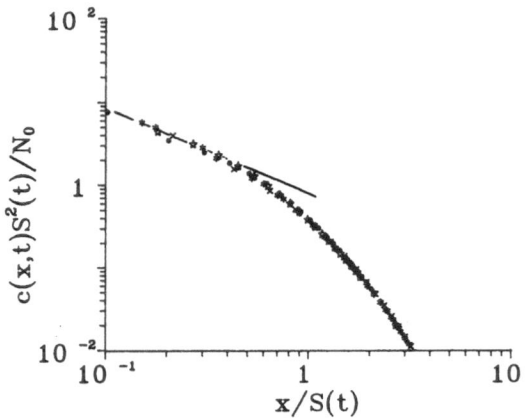

Fig. 3 Reduced size distribution according to Eq. (4) for the aggregates obtained without stirring (prior to the immersion in the polymer solutions)

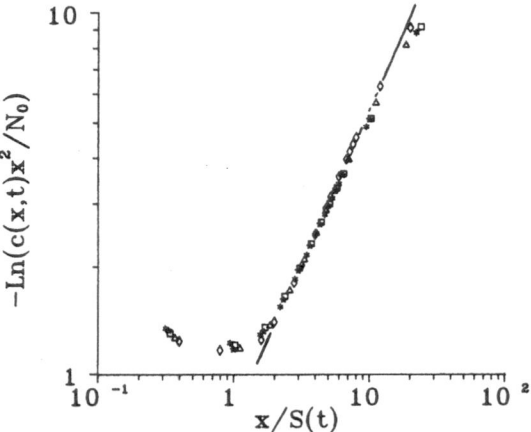

Fig. 4 Fragmentation of aggregates formed without stirring. Reduced size distribution curves of fragments according to Eq. (5) determined at different moments of the fragmentation and valid for the different media from $C^*/5$ to $C^*/70$

with an initial slope equal to -1.75 is determined (Fig. 2, curve b). These colloid size distributions during fragmentation should be compared to that of the initial aggregates represents in Fig. 3, where the slope at small values of x, equal to -1, smaller than the value of -1.5 which characterizes the irreversible aggregation induced by interparticle polymer bridging, may be characteristic of a reversible perikinetic aggregation process resulting from a very small polymer dosage. The finding that $S(t) \cong N(t)$ agrees with this interpretation [8].

In the range of large values of the variable, the exponent λ of Eq. [5] is calculated on the basis of results of Fig. 4 and the following Eq. (10) is valid:

$$a(x) \cong [x/S(t)]^{0.9} \qquad (10)$$

for fragmentation in dilute and concentrated polymer solutions.

Scaling of the colloid size distribution at small and large values of the reduced variable $x/S(t)$ shows that the theoretical assumption of the self-similarity of the aggregate size distribution during fragmentation is valid.

II) Fragmentation of aggregates formed in the presence of polymer under stirring.

1) *Influence of the colloid surface coverage established during aggregation*: By performing this second set of experiments, we tried to investigate the influence of the initial colloid surface coverage established during the aggregate formation in the reactor. We determined that the orthokinetic aggregate formation is a very fast process and the aggregate were sampled in the period where the scaling law $S(t) \cong t^{3.5}$ holds. The following aggregates having initially different sizes $S(t = 0)$ close to 6, 9, 13, 25 and 43 were fragmented in polymer solutions of concentration $C^*/30$. The variations of $S(t)$ with t is given in Fig. 5. Firstly, the period characterized by a constant value of the size $S(t)$ decreases when the initial aggregate size increases. The fragmentation rate in the first domain is described by Eq. (3), where λ is relatively constant and close to 4 for all sizes. In the second domain, the rate increases with the initial aggregate size $S(t = 0)$. The following values of λ, 1.8, 1.6, 1.5, 1.2 and 1.0 characterize the fragmentation rate of colloids of sizes 6, 9, 13, 25 and 43. Equation (11) describes the fragmentation acceleration as a function of

Fig. 5 Fragmentation rate of aggregates formed under stirring. Representation of the weight average size $S(t)$ of the fragment as a function of time (min), (log-log scale). Time $t = 0$ corresponded to the moment of immersion of the aggregates in the polymer solution of concentration $C^*/30$. Initial size $S(t = 0)$ of the aggregates before immersion: 6, ($\diamond$); 9, ($\star$); 13, ($\triangle$); 25, ($\square$) and 43, ($\bigcirc$)

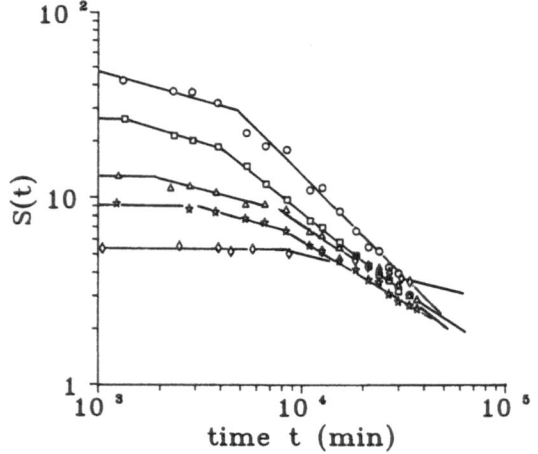

Progr Colloid Polym Sci (1994) 97:65–70
© Steinkopff-Verlag 1994

the initial size of the colloid:

$$\lambda \cong S(t = 0)^{-0.3} . \qquad (11)$$

Taking into account that the aggregation procedure implies that $S(t) = S(t = 0) \cong t^{3.5}$, Eq. (12) demonstrates that the exponent of the fragmentation rate in this second domain is a function of the initial degree of surface coverage A_S of the colloid:

$$\lambda^{-1} \cong A_S . \qquad (12)$$

Using Eq. [5] to calculate the exponent λ characterizing the rate of the aggregate break-up, we found a unique value equal to 0.65 ± 0.03.

2) *Influence of the polymer concentration in the fragmentation medium*: The temporal variation of the average sizes displays three domains. As previously, there was an initial period where the mean sizes remained constant: the time lag prior to fragmentation decreased with the dilution of the polymer solution. Then, a domain of slow fragmentation rate, for which the slope of the temporal variation of the average sizes does not depend on the characteristics of the medium. This domain is followed by a more rapid process, where $S(t)$ and $N(t)$, decreasing at different rates, finally attain the value 2. However, the polymer concentration influenced the fragmentation rate in a rather complex manner. To get over this problem, we used the theory of Cheng and Redner to calculate the rate of aggregate break-up $a(x)$ and the rate of fragmentation $b(x)$ [6]. From the slope of the function at large values of the variable x, one obtains the rate of break-up, which is independent of the polymer concentration in the fragmenting medium as well as of the initial size (or coverage) of the aggregates. The scaling law (13) emerges from these experiments:

$$a(x) \cong [x/S(t)]^{0.65} \qquad (13)$$

From the slope of the variation of the reduced size distribution at small values of $x/S(t)$ represented in Fig. 6, we determined that the exponent v of the scaling law of production of fragments of size x is equal to -1.65. This continuously decreasing function also should be compared to the bell-shaped initial size distributions of the aggregate before fragmentation which is also represented in Fig. 6.

The main results of the study are relative to:

1) *The influence of hydrodynamic forces in aggregation and their influence on the aggregate fragmentation rate.*

The reduced size distribution of the fragmented colloids indicates that fragmentation does not produce fragments of a typical size: no size emerges from the distribution so

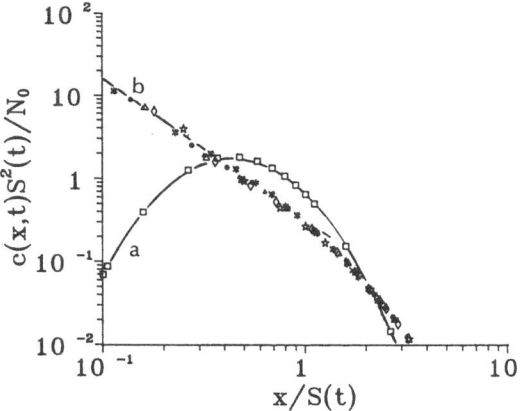

Fig. 6 Reduced size distribution according to Eq. (4) of aggregates formed under stirring (recorded during the aggregation phase prior to immersion i the polymer solution): curve a (bell-shaped curve); reduced size distribution recorded during the fragmentation in the polymer solution (curve b)

a bell-shaped distribution is never formed. This implies that attrition, which releases isolated particles and doublets and triplets in the suspension, is the main fragmentation process. Internal rupture of colloids may be a much slower process. Thus, we assume that the unique exponent λ in Eq. (5) characterizes the main process of release of small fragments. Obviously, the break-up of colloids formed without stirring is faster than that of colloids formed during stirring. This implies that the stirring of the suspension and the resulting multiple interparticle collisions, not only strongly accelerates the aggregation process as revealed by the equation $S(t) \cong t^{3.5}$ in comparison to the temporal variation $S(t) \cong t^{0.35}$ in our perikinetic aggregation process, but essentially consolidates the aggregate structure. We previously determined that the fragmentation mode strongly depends on the aggregation mode [9]. Our interpretation of the different fragmentation rates is also based on the mechanism of the corresponding aggregation modes. It the orthokinetic aggregation which develops in the scaling domain of $S(t)$, the aggregate size distribution is represented by a bell-shaped curve. This implies that small colloids have an unusual reactivity in the orthokinetic process when compared to the reversible perikinetic aggregation induced by the interparticle polymer bridging [10]. These considerations lead us to conclude that intense stirring during the colloid and polymer mixing step leads to a non desirable aggregate stability against fragmentation. On the contrary, gentle mixing prevents this phenomenon and the resulting aggregates fragment more easily,

2) *The interpretation of the lag time before fragmentation*

The time-lag preceding the onset of fragmentation was determined to decrease with the initial surface coverage of the colloid. The fragmentation starts when the polymer layer is attaining its equilibrium value of full coverage at the level of the stuck particles. This suggests that the rate limiting process is the chain reptation in the starved diffuse layer. Thus, the time lag can be shortened by allowing each elementary particle to be covered nearly to its equilibrium value before it collides with other particles. Therefore colloids may be stabilized more easily in dilute suspension.

3) *The aggregate size distribution at low values of the reduced size $x/S(t)$*

The fragmentation of aggregates formed without stirring leads to well characterised size distributions at low and high polymer concentrations. At high concentration, the lower slope $v = -1.20$ indicates that the rate of appearance of the smallest fragment is slower than in dilute polymer solutions for which a unique slope of -1.75 has been determined. This retardation of the fragmentation in concentrated polymer solutions may be attributed to the similar phenomenon, which leads to destabilization in a medium of concentration C^* [3]. If both aggregation and fragmentation processes coexist, the balance increases in favor of fragmentation with dilution.

Conclusion

These experiments demonstrated that the aggregate fragmentation strongly depends on the aggregation conditions. Obviously, our experiments showed that fragmentation currently lead to aggregate size distribution characterized by $S(t) = N(t) \cong 2$. Nothing was determined concerning the last step which is expected to give rise only to elementary particles. We expected that the scaling laws of fragmentation cannot describe this last step.

We tried to interpret our experimental results on the fragmentation kinetics on the basis of the theory of Cheng and Redner [6], and this tentatively led us to distinguish between the rate of fragmentation and the rate of aggregate break-up. When adsorption reached completion, it was *a priori* expected that the aggregate of size x would fragment instantaneously and give rise to x elementary particles, such as $a(x) \cong x$ as a result of the establishment of the stabilizing interfacial layer. Our results demonstrated that colloids did not fragment instantaneously as observed for aggregates when the density of the surface charges of the colloids increased [9]. In the present situation, the rate of break-up of the interparticle bridges between two or more particles depended on the hydrodynamic conditions leading to the aggregate formation. On the other hand, the rate of fragmentation was found to depend on the initial size and degree of coverage of the aggregate as well as on the concentration of the polymer solution. These considerations indicated that the polymer induced fragmentation involved complex phenomena which are known only poorly.

Acknowledgements This research was supported by the CNRS-PIR-SEM under the Project PR "Suspensions colloidales concentrées" and the following companies, Institut Français du Pétrole, Lafarge Coppée, Péchiney, Rhone-Poulenc and Total.

References

1. Napper DH (1983) Polymeric Stabilization of Colloidal Dispersions. Academic Press, New York
2. Gregory J (1978) In: Ives J (ed) The Scientific Basis of Flocculation. p. 101, Sijthoff & Noordhoff, Alphen aan de Rijn, The Netherlands, pp 101–130
3. Vincent B, Luckham PF, Waite FA (1980) J Colloid interface Sci 73:508–521
4. Pefferkorn E, Varoqui R (1989) J Chem Phys 91:5679–5686
5. Pefferkorn E, Elaissari A (1990) J Colloid Interface Sci 138:187–194
6. Cheng Z, Redner S (1988) Phys Rev Lett 60:24502453; (1990) J Phys A: Math Gen 23:1233–1258
7. Friedlander SK, Wang CS (1966) J Colloid Interface Sci 22:126–132
8. Pefferkorn E, Stoll S (1990) J Chem Phys 92:3112–3117
9. Stoll S, Pefferkorn E (1992) J Colloid Interface Sci 152:247–256; (1992) ibidem 152:257–264
10. Pefferkorn E, Widmaier J, Graillat C, Varoqui R (1990) Prog Colloid Polym Sci 81:169–173

Progr Colloid Polym Sci (1994) 97:71–74
© Steinkopff-Verlag 1994

SUSPENSION

V. Cabuil
N. Hochart
R. Perzynski
P. J. Lutz

Synthesis of cyclohexane magnetic fluids through adsorption of end-functionalized polymers on magnetic particles

Received: 16 September 1993
Accepted: 25 March 1994

Dr. V. Cabuil (✉)
N. Hochart
Laboratoire de Physicochimie Inorganique
URA CNRS Structure et Réactivité des
Systèmes Interfaciaux
Université Pierre et Marie Curie
4, Place Jussieu
75252 Paris Cedex 05, France

R. Perzynski
Laboratoire d'Acoustique et Optique de la
Matiére Condensée
URA CNRS 800
Université Pierre et Marie Curie
Paris, France

P. J. Lutz
Institut Charles Sadron
CNRS
Strasbourg, France

Abstract The synthesis of stable cyclohexane magnetic fluids constituted by maghemite nanoparticles, coated by polystyrene chains ($Mw = 13\,000$) is described. α-Lithium poly(styrene) sulfonated polymer chains are adsorbed on the cationic nanoparticles of an aqueous magnetic fluid through electrostatic interactions between the anionic sulfonate group and the cationic surface. The colloidal solution obtained by dispersion of the polymer coated particles in cyclohexane is characterized by magnetic, magneto-optic and viscosity measurements. The thickness of the adsorbed polymer layer is estimated to be equal to 8 nm.

Key words: Magnetic fluids
– adsorption – end-functionalized
polystyrene – solvation layer

Introduction

Synthesis of stable magnetic colloidal solutions is generally performed by adsorbing at the surface of magnetic nanoparticles tensioactive molecules, whose length is of the order of 2 nm [1]. Adsorption of polymer chains on magnetic particles has been described in some cases, but most often it leads to the coating by the polymeric molecules of particle aggregates [2].

We describe here the adsorption of end-functionalized macromolecules on maghemite particles. These particles are synthesized by a chemical method [3], and have an average diameter of 9 nm. They are obtained as ionic materials: they have surface charges, either positive or negative depending upon the pH value of the solution [4].

Such particles are dispersed in water (leading to aqueous ionic ferrofluids). They are able to adsorb various kinds of molecules, especially tensioactive ones [5]. Adsorption is due either to electrostatic interactions between the charges located on the surface of the particles and those of the tensioactive molecule, or to chelation of surface iron atoms by the polar head of the surfactant (for example, in the case of oleic acid [6]). The surfactant coated particles are then dispersable in several media leading to non-aqueous magnetic fluids. The usual surfactants are short chain molecules (about 2 or 3 nm) compared to polymeric chains. Adsorption of end-functionalized polymers on magnetic particles is studied here in order to produce particles surrounded by a layer whose thickness may be eventually monitored through the quality of the solvent towards the polymeric chain.

We have chosen to adsorb on maghemite particles α-lithium poly(styrene) sulfonated polymer chains ($Mw = 13\,000$ g/mol) for the following reasons: i) the lithium sulfonate function is well suited, as it may adsorb on charged magnetic particles using perhaps the same kind of process as that used for adsorption of surfactants; ii) the polystyrene chains can ensure dispersion of the particles in cyclohexane, and the polymeric nature of the chains may have a great influence on the stability of the sol.

Adsorption of the polymer on the magnetic particles and synthesis of cyclohexane magnetic fluids

Materials

i) Magnetic particles are obtained by making alkaline an aqueous mixture of ferric and ferrous chloride ($[Fe] = 0,1$ mol/l; $[FeCl_2]/[FeCl_3] = 0.33$) with ammonia to pH 9, at room temperature. Magnetite anionic particles are immediately obtained as spherical units with an average diameter of about 10 nm. The colloidal magnetite precipitate is acidified by nitric acid (2 mol/l), then oxidized by a boiling ferric nitrate solution (0.1 mol/l) until the precipitate becomes brick red, which means that the magnetite has been completely oxidized to maghemite. The particles obtained after this treatment are cationic and are dispersed in water to produce a stable acidic magnetic fluid, with a volume fraction of magnetic particles of the order of 0.1%. The surface charge density of the particles is about 0.2 C/m^2 and ionic strength, mainly nitric acid, is of the order of $4\ 10^{-3}$ mol/l.

ii) The polymer is an α-lithium polystyrene sulfonate, PSMS, of average mass $Mn = 13\,000$ g/mol.

Synthesis: It is obtained via anionic polymerization method according to a well established procedure [7]. Ionic polymerization methods are well suited for the synthesis of end-functionalized polymers because of the presence at the chain end of active sites. In addition, these methods yield linear polymers with polymerization degrees which can be chosen at will.

First, the polymerization of styrene is initiated by butyl lithium in a THF/Benzene mixture at $-30\,°C$, the polymerization degree being determined by the molar ratio of monomer converted to initiator. Once the polymerization of the monomer has been completed, induced deactivation of the active site with 1,3-propane sultone is achieved to fit the chain with the lithium sulfonate group. Just before addition, part of the solution is sampled for the purpose of

characterization. To decrease the probability of side reactions, it is necessary to reduce the nucleophilicity of the carbanion by an intermediate addition of 1,1-diphenylethylene. The reaction medium is kept over night to ensure completion of the reaction. The polymer solution is then poured into cold heptane, dried, recovered, diluted in THF, and precipitated again. This procedure has to be repeated to remove excess 1.3-propane sultone and the material is finely dried.

Characterization: The unfunctionalized sample was characterized by G.P.C. using a calibration with linear polystyrene samples. The molecular weight is compatible with the expected value:

$$M_w = 13\,000 \quad M_n = 12\,700 \quad M_w/M_n = 1.02$$

No characterization of the sulfonate sample was possible by G.P.C.: the α-lithium polystyrene sulfonate has a strong tendency to adsorb on the columns.

Adsorption of the polymer on the particles

The functionalized polystyrene is dispersed in acetone ($[PSMS]) = 1.7\ 10^{-3}$ mol/l). The dispersion (2 ml) is added to the acidic aqueous ferrofluid (volume fraction in particles 0.08%, 30 ml, $[HNO_3] = 0.2$ mol/l), under vigourous stirring, over 30 min, at room temperature. The precipitate containing the magnetic particles is isolated from the solution and washed with acetone three times, before being dispersed in cyclohexane (2 ml). Residual acetone is eliminated by heating at 56 °C.

Characterization of the magnetic fluids

The cyclohexane magnetic fluids thus synthesized have been characterized in order to determine the particle size and to estimate their stability especially in an homogeneous magnetic field.

Analysis of the magnetization curve

The magnetization curve of the cyclohexane magnetic fluid is characteristic of a superparamagnetic solution, with the same shape for the increase and the decrease of the applied magnetic field [8]. It has the same shape as that of the initial ionic aqueous magnetic fluid obtained by dispersion of the uncoated particles. That means that the coating with polymers has not induced any agglomeration phenomena

Progr Colloid Polym Sci (1994) 97: 71–74
© Steinkopff-Verlag 1994

that would have been detected by a destabilization of the fluid under magnetic field (i.e., the protection is efficient).

In the present case, the size distribution of particles deduced from the shape of the magnetization curve according to ref. [8] is fitted by a log-normal fit with parameters $D_0 = 8.2$ nm, $\sigma = 0.3$.

Relaxation of the birefringence signal

When submitted to a magnetic field, magnetic liquids become birefringent [9]. The analysis of the exponential decay of the birefringence signal when the field is cut-off (Fig. 1) is a measure of the rotational diffusion coefficient of the particle in the solvent [10], and leads to an hydrodynamic diameter of particles D_{bir}. The characteristic relaxation time τ_H gives R_H through the relation:

$$\tau_H = \frac{\pi D_{bir}^3 \eta}{6 k_B T},$$

where η is the viscosity of the solvent, T, the absolute temperature, and k_B, the Boltzmann constant.

In the case of aqueous magnetic fluids, the radius D_{bir} is related to the parameter D_0 through a relation taking into account the polydispersity [10]. For a σ equal to 0.3, the relation is:

$$D_{bir} \approx 3.1 \, D_0 \,. \tag{1}$$

The experimental device is described in ref. [9]. Samples were sufficiently dilute (volume fraction in particles less than 1%) in order to avoid interactions between particles.

The birefringence decay is plotted versus the time in Fig. 1 for the precursor aqueous magnetic field (1a) and for the magnetic fluid in cyclohexane constituted by the particles surrounded by polymer molecules (1b). In both cases, the semi-logarithmic plots are straight lines: that means that both magnetic fluids do not contain agglomerates which would relax more slowly than individual particles. The behavior of the cyclohexane magnetic fluid obtained by dispersion of the polymer coated particles has been compared to that of an "usual" cyclohexane magnetic fluid obtained by the dispersion of the same particles coated by surfactant molecules [5, 11, 12] (curve 1c).

The values obtained for the characteristic relaxation times lead to the following values of D_{bir}:

– for the aqueous precursor magnetic fluid: $D_{bir} = 27.2$ nm. Relation (1) is verified.

– for the cyclohexane magnetic fluid:

*surfactant coated particles: $D_{bir} = 30.3$ nm

*polymer coated particles: $D_{bir} = 43.6$ nm.

As the particles are the same in the three cases, and thus the polydispersity of the samples is identical, and as the birefringence decay is well described by a simple exponential law, it is reasonable to assume that the difference between the diameters of the ionic and surfactant coated particles on one hand, and the polymer coated particles on the other hand, may be attributed to the adsorption of the polymer.

Viscosity measurements

Magnetic fluids are Newtonian solutions as long as they are not agglomerated. The viscosity of the samples have been measured with a capillary vicosimeter at room temperature, for different values of the particle volume fraction ranging from 0 to 0.8%.

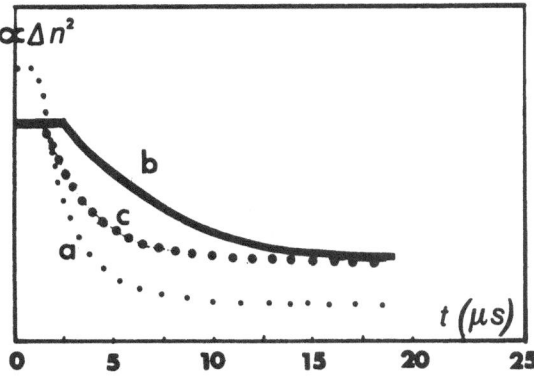

Fig. 1. Decay of the signal of birefringence Δn as a function of the time t. curve 1 a: for ionic particles dispersed in water; curve 1 b: for polymer coated particles dispersed in cyclohexane; curve 1 c: for surfactant coated particles dispersed in cyclohexane

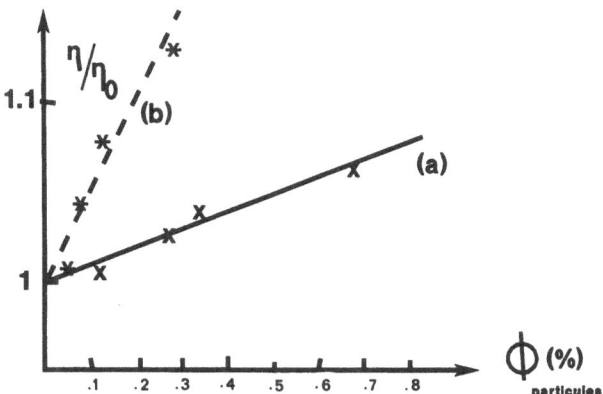

Fig. 2 Ratio of the viscosity of the magnetic fluids to the viscosity of the solvent, for the aqueous ferrofluid (a) and for the cyclohexane ferrofluid (b) constituted by polymer coated particles, as a function of the volume fraction in magnetic particles

74

V. Cabuil et al.
Adsorption of polymers of magnetic nanoparticles

In the case of the precursor aqueous magnetic fluid, as in the case of the cyclohexane one, the plot η vs $f(c)$, is a straight line (Fig 2) where c in the concentration in iron and proportional to the volume fraction in particles.

The comparison between the experimental law and the Einstein law for spherical particles allows us to get from the slope of the line, an average hydrodynamic diameter of the particle. For the aqueous magnetic fluid (uncoated particles), this diameter is found equal to 14.3 nm, although it is 25.2 nm for the polymer-coated particles.

Discussion and conclusion

Adsorption of end-functionalized polystyrene on magnetic nanoparticles allows the synthesis of stable cyclohexane magnetic fluids. Relaxation of birefringence and viscosity measurements are in good agreement: both of them lead to diameters about 1.7 times greater for polymer-coated particles than for ionic or surfactant coated particles. The difference between the diameters given for the same sample by both techniques is due to the specificity of the technique itself. Relaxation measurements investigate the tail of the distribution and the diameters given by this method are always much more greater than D_0 [9]. From viscosity measurements, one may estimate the thickness of the layer of adsorbed species on particles. For the studied sample, characterized by its mean diameter, $D_0 = 8.2$ nm, determined by magnetization measurements, which corresponds to a "physical" diameter slightly greater ($D_0 = 9$ nm) [8], and then to an average diameter $(\langle D^3 \rangle)^{1/3} = D_0 \exp(1.5\,\sigma^2) = 10.3$ nm. The thickness of the water layer surrounding the particles may be estimated from the diameter obtained from viscosity measurements (14.3 nm) of the order of 2 nm, and the thickness of the polymer "layer", of the order of 8 nm. This thickness is by about a factor of 2 greater than the hydrodynamic diameter of the corresponding unfunctionalized polystyrene chains free in solution (D_H of the order of Rg, where Rg is the radius of gyration of the polymer estimated to be 3.3 nm). This result is in agreement with the fact that adsorption of the functionalized polymer has induced an anisotropy in the molecule, which may no longer be described by a statistical distribution of segments arounds the center of gravity.

Acknowledgments The authors are greatly indebted to Dr. J. Bastide for having initiated the collaboration.

References

1. Bacri JC, Perzynski R, Salin D (1988) Endeavour 12 (2):76–83
2. Cabuil V (1994) Unesco Engineering and Technology division (ed), Magnetic Fluids and Applications, Handbook, to be published
3. Massart R (1981) IEEE Trans Magn MAG-17:1247
4. Jolivet J-P, Massart R, Fruchart J-M (1983) Nouv J Chim 7:325–331
5. Massart R, Cabuil V, Fruchart J-M, Roger J, Pons J-N, Carpentier M, Neveu S, Brossel R, Bouchami T, Bee-Debras A (1990) Eur Pat 9006484
6. Rocchiccioli-Deltcheff C, Franck R, Cabuil V, Massart R (1987) J Chem Research (S) 126:1209
7. Quirk RP, Kim J (1991) Macromolecules 24:45 15–4522
8. Bacri J-C, Perzynski R, Salin D, Cabuil V, Massart R (1986) J Magn Magn Mat 62:36–46
9. Bacri J-C, Perzynski R, Salin D, Cabuil V, Massart R (1987) J Magn Magn Mat 65:285–288
10. Bacri J-C, Perzynski R, Salin D, Servais J (1987) J Phys (Paris) 58:1385
11. Fabre P, Casagrande C, Veyssie M, Cabuil V, Massart R (1990) Phys Rev Lett 64:539–542
12. Cabuil V, Perzynski R, Bastide J (1994) Progr Colloid Polym Sci 97:75–79

Progr Colloid Polym Sci (1994) 97:75–79
© Steinkopff-Verlag 1994

SUSPENSION

V. Cabuil
R. Perzynski
J. Bastide

Phase separation induced in cyclohexane magnetic fluids by addition of polymers

Received: 16 September 1993
Accepted: 28 February 1994

Dr. V. Cabuil (✉)
Universite Pierre et Marie Curie
Laboratoire de Physicochimie
   Inorganique, Casier 63 "URA CNRS
Structure et Réactivité des Systemes
   Interfaciaux"
4, Place Jussieu
75252 Paris Cedex 05, France

R. Perzynski
Laboratoire d'Acoustique et Optique de la
Matiére Condensèe
URA CNRS 800
Université Pierre et Marie Curie
Paris, France

J. Bastide
Institut Charles Sandron
CNRS
Strasbourg, France

Abstract Magnetic colloidal solutions are prepared by dispersion of surfactant coated magnetic nanoparticles in cyclohexane. The particles are maghemite grains, with a size distribution described by a log-normal law of parameters $D_0 = 6.5$ nm, $\sigma = 0.35$. The surfactant used to ensure the dispersion of particles is a phosphate ester of a long chain alkylphenol. The colloidal solution is characterized by magnetic and optical measurements, in order to get either the size characteristics of the particles and the stability of the solution. Iron chemical titrations allow determination of the volume fraction in particles. Addition of non-adsorbing polymers dispersed in cyclohexane, in the present case two types of polydimethylsiloxane chains (PDMS) ($Mn = 17\,000$ and $Mn = 6500$), induces a phase separation which leads to a phase poor in magnetic particles coexisting with a phase rich in magnetic particles.

Key words Magnetic fluids – phase separation – depletion – magnetic colloids

Introduction

Colloidal stability of ionic ferrofluids has been investigated from the experimental point of view in ref. [1]. These ferrofluids are aqueous dispersions of maghemite particles (γ-Fe_2O_3), whose mean diameter may be monitored as being between 5 and 12 nm according to the experimental conditions of synthesis [2]. These particles have a surface charge, and stability of the dispersion is ensured by the screened electrostatic repulsions between particles. Thus, the experimental parameter used to improve the stability of the sol in ref. [1] was the ionic strength. By increasing this latter, a phase separation liquid-gas like is observed, leading to a liquid viscous phase rich in magnetic particles, coexisting with a liquid phase poor in magnetic particles. Such a phase separation may also be observed experi-mentally by decreasing the temperature or applying a magnetic field [3], and it is relevant to several theoretical works and experimental results on other colloidal systems [4–10].

In non-polar media, ferrofluids are obtained by adsorbing tensioactive molecules on magnetic particles. In the present work, magnetic fluids constituted by maghemite particles dispersed in cyclohexane are synthesized by adsorbing long-chain phosphorus esters on charged maghemite particles. The hydrophobic particles thus obtained are dispersed in cyclohexane leading to a mono-phasic magnetic fluid whose volume fraction in particles may range up to 8%; these have been used for synthesis of ferrosmectics [11]. These ferrofluids are characterized by their magnetization curve [12] and by relaxation of bire-fringence [13] in order to get the parameters of the size distribution and to estimate the stability of the colloidal solution.

A magnetic fluid is just a special case of colloidal solution. Thus, according to theoretical works [14–20] and to experimental results of several teams [21–27], it was expected to observe a destabilization of the sol if non adsorbing polymer was introduced into the solution, due to the induced depletion attractive forces between particles. As a matter of fact, when a polymer is added to a sterically stabilized magnetic fluid, a phase separation is observed, between a liquid phase concentrated in magnetic particles and another phase dilute in particles. We propose here a preliminary experimental study of this separation for one type of ferrofluids and for two non adsorbing polymers: a polydimethylsiloxane of average mass $Mn = 6500$ g/mol (PDMS 6500) and another one of mass $Mn = 17\,000$ g/mol (PDMS 1700).

Materials

Cyclohexane magnetic fluid

Magnetite particles are obtained through Massart's procedure [28] by alkalinisation of an aqueous mixture of ferric and ferrous chloride. The experimental conditions are chosen in order to get anionic particles with a mean diameter of about 7 nm [2]. Colloidal magnetite precipitate is then acidified by nitric acid and oxidized to maghemite by ferric nitrate. The cationic particles thus obtained are dispersed in water, then added to an anionic tensioactive (commercial product), which is a commercial mixture of long chain mono- and diesters of phosphoric acid neutralized by an amine. Adsorption of the surfactant is performed in aqueous medium, and the hydrophobic precipitate thus obtained is isolated, washed with methanol, and then dispersed in cyclohexane.

The volume fraction of the particles Φ may range up to 8%. The magnetization curve of a dilute sample ($\Phi < 1\%$) is characteristic of a superparamagnetic system, with a particle size distribution corresponding to a log-normal law of parameters $D_0 = 6.5$ nm, $\sigma = 0.35$.

When submitted to a magnetic field, such a solution becomes birefringent because of the alignment of magnetic particles along the direction of the magnetic field. The analysis of the birefringence decay when the field is cut off allows the determination of a characteristic diameter D_{bir} which is a hydrodynamic one (related to the rotational diffusion coefficient of particles), and which is strongly affected by the width of the size distribution. It has been found in ref. [1] that this diameter is the main parameter controlling the onset of the phase separation when it is induced by an increase of the ionic strength in ionic ferro-

fluids. In the case of the cyclohexane ferrofluid studied here, the decay is well described by a simple exponential, which indicates that the sample does not contain agglomerates that would relax with different characteristic times. The diameter D_{bir} is found equal to 28 nm which is in good agreement [29] with the parameters of the size distribution obtained from magnetic measurements.

Polymer

Two cyclohexane soluble polymers have been added to the magnetic fluid. Both were polydimethylsiloxane (PDMS) molecules obtained by precipitation and slight fractionation of a commercial product. The first one has an average mass $Mn = 17\,000$ g/mol, the second one an average mass of 6500 g/mol. These samples were rather polydisperse ($Mw/Mn = 1.8$). Their radius of gyration may be estimated from ref. [31]. For PDMS 17000, it is equal to 6 nm, although it is equal to 3 nm for PDMS 6500. Their hydrodynamic radius is thus of the order of 1.5 nm for the small polymer and 3 nm for the greater one.

Description of the phase separation

General features

Both polymers induce a phase separation in the initially monophasic magnetic fluid: a phase concentrated in magnetic particles appears in a more dilute one. Figure 1a shows what is observed with an optical microscope (enlargement × 225) when the phase separation has occurred. The concentrated phase is a liquid one and flows. Figure 1b shows a drop of concentrated phase elongated under an external magnetic field of 1600 A/m (20 Oe). After settling of the concentrated phase, a well-defined interface is observed between the phases.

Elementary analysis of Si in both phases, isolated and dried, shows that both of them contain PDMS.

In both cases, the phase separation is reversible and is followed by precipitation of the particles when the amount of PDMS is greater than a given threshold value.

Methods for characterization of the phases

Volume fraction of magnetic particles in the phases

The concentration of magnetic particles in both phases has been determined by chemical titration of iron. The procedure is the following one. To a given volume of monophasic cyclohexane magnetic fluid, is added a given volume of PDMS 6500 or PDMS 17000 (both polymers are

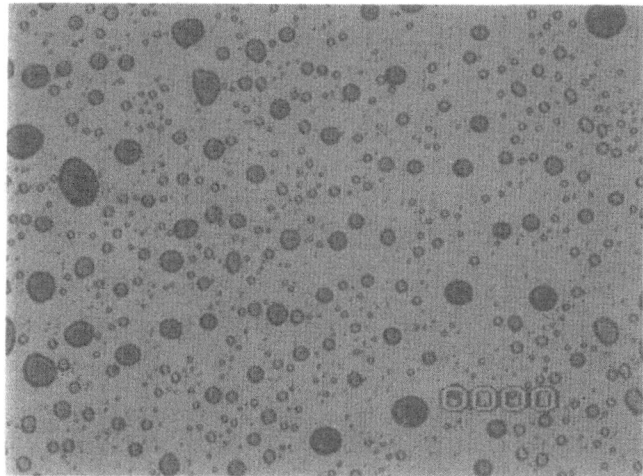

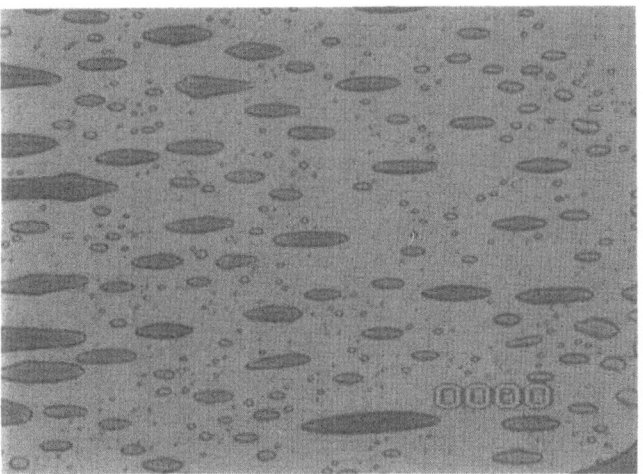

Fig. 1 Optical microscopy picture (enlargement × 225) of concentrated droplets in the dilute phase. a) without external magnetic field; b) with an external magnetic field of 1600 A/m (20 Oe)

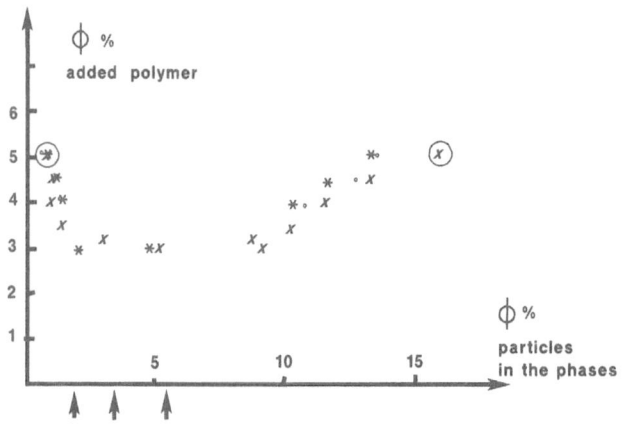

Fig. 2 Experimental phase diagram. Abscissa are volume fractions in magnetic particles; ordinates are volume fractions of added PDMS 17000. × corresponds to the ferrofluid of volume fraction in particles equal to 5.6%, · to the same ferrofluid dilute to 3.3%, *, to the same ferrofluid dilute to 1.8%. The encircled crosses correspond to the sample on which titration of Si has been performed

liquid although they are very viscous). The mixture is stirred for a few minutes and left for 2 days to let the concentrated phase settle at the bottom of the vessel. Then a given volume of the dilute phase and a given volume of the concentrated phase are sampled and iron titrations performed according to the procedure of ref. [30].

Volume fraction of PDMS in the phases

At the present time, we have not succeeded in making sufficiently precise measurements of the concentration of PDMS in the phases. Chemical titration of Si together with measurements of viscosity prove that PDMS is present in both phases with a concentration of the same order of magnitude. Si titrations made on the concentrated and dilute phases encircled on the diagram of Fig. 2, lead to a concentration in Si about four times smaller in the phase concentrated in particles than in the dilute one. But methods have to be improved to get more precise results and conclusions on the partition of the polymer in the phases.

Interfacial tension between the two phases

Measurement of the interfacial tension between the two phases has been performed by studying the peak instability at the interface between the phases. The procedure is the following one. Phases separation is performed and the separated sample is left at rest in a spectroscopic cell in order that the concentrated phase settles. When all the droplets of concentrated phase have settled, they form a phase at the bottom of the cell with a plane interface with the dilute phase. When a weak magnetic field is applied perpendicular to the interface and above a given intensity threshold of few A/m, a peak instability occurs [8], characterized by a critical wave length. This wave length at the threshold value of the field is related to the respective densities of the two phases and allows determination of the interfacial tension [32].

Preliminary experimental phase diagram

Figure 2 is a plot, for the present ferrofluid sample, of the concentration in magnetic particles in the dilute and concentrated phases, when the phase separation is induced by

78

V. Cabuil et al.
Phase separation in cyclohexane magnetic fluids

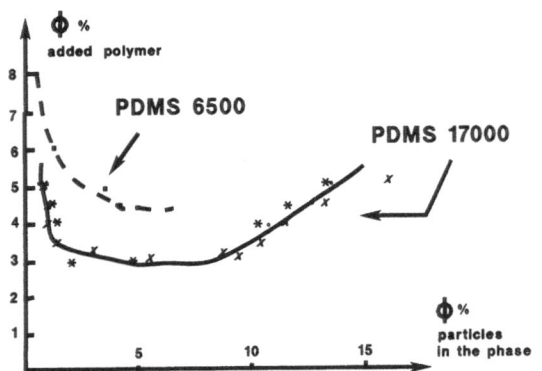

Fig. 3 Phase diagrams obtained when PDMS of different weight are added to the studied ferrofluid

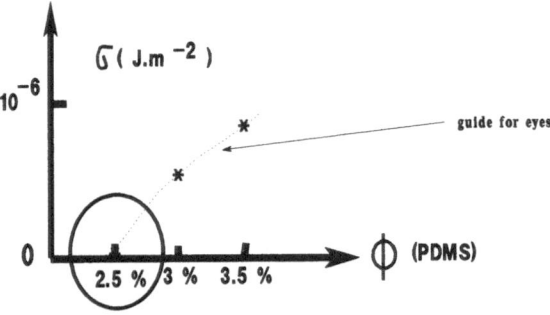

Fig. 4 Interfacial tension between concentrated and dilute phases as a funtion of the volume fraction of added polymers

PDMS 17 000. The abscissa are the volume fractions of magnetic particles in each of the phases, although ordinates are the volume fractions of added PDMS. As the system is not sufficiently described (concentrations of PDMS in both phases are not measured) we are not allowed to speak about a phase diagram. Nevertheless, the present diagram gives the PDMS concentration corresponding to the onset of the phase separation and the concentrations in magnetic particles of both phases. For the same magnetic fluid, the phase separation has been performed for three initial concentrations in particles, and all the points correspond to a unique master curve, unlike what was observed in the case of phase separation in aqueous ionic magnetic fluids [1].

Figure 3 illustrates the influence of the PDMS molar mass on the onset of phase separation and on the diagram in itself. PDMS of high mass induce phase separation for a lower threshold concentration value than PDMS of low mass: 2.5% for PDMS 17 000, to compare to 4% for PDMS 6500.

In Fig. 4 the interfacial tension σ between both phases is plotted in the case of a phase separation induced by PDMS 17 000, for some values of the volume fraction in polymer. The values of σ are very low and of the same order of magnitude as the ones observed in the case of phase separation in aqueous ionic magnetic fluids [32]. It decreases with the volume fraction of added polymer and seems to tend to zero near the threshold of the phase separation.

Discussion and conclusion

Extra polymeric chains, added to a monophasic cyclohexane magnetic fluid induce a phase separation between a liquid phase rich in magnetic particles and a phase poor in magnetic particles.

The former results show that the onset of the phase separation depends, as expected, on the molar mass of the polymer. The bigger the polymer, the more easily the phase separation is induced. It depends also on the particles size distribution in the magnetic fluid, especially on the width of this distribution: a large polydispersity lowers the polymer concentration threshold. For example, in the case of the present magnetic fluid sample ($D_0 = 6.5$ nm, $\sigma = 0.35$), the threshold concentration of PDMS 17 000 is $\Phi = 2.5\%$, although it is found equal to 1% for a sample characterized by the same D_0 but a greater value of σ ($\sigma = 0.4$). This point is similar to what is observed in the case of ionic magnetic fluids, for which the width of the size distribution, estimated through the value of D_{bir}, was a parameter controlling the onset of phase separation. It is also similar to the results of ref. [33–34] concerning depletion interactions used to produce monodisperse emulsions.

Measurements of interfacial tension between the two phases show the existence of a critical point at the onset of the phase separation. Nevertheless, these results are preliminary ones. The concentration of PDMS has to be determined in both phases and more experiments are necessary to build up a phase diagram and compare our results, concerning a rather polydisperse system of nanoparticles (diameter of the order of 10 nm), to the predictions of ref. [19], especially about the existence of a three phase region, which was found in the monodisperse system of polymethylmethacrylate spheres studied in ref. [27].

Acknowledgments The authors are greatly indebted to M. Carpentier and J. Fanton for technical assistance.

Progr Colloid Polym Sci (1994) 97:75–79
© Steinkopff-Verlag 1994

79

Reference

1. Bacri J-C, Perzynski R, Salin D, Cabuil V, Massart R (1989) J Colloid Interface Sci 132:43–53
2. Massart R, Cabuil V (1987) J Chim Phys 84:967–973
3. Bacri J-C, Perzynski R, Salin D, Cabuil V, Massart R (1990) J Magn Magn Mat 85:27–32
4. Victor J-M, Hansen J-P (1985) J Chem Soc Faraday Trans 281:43
5. Jansen JW, De Kruif CG, Vrij A (1986) J Colloid Interface Sci 116:681
6. Cowell C, Vincent B (1982) J Colloid Interface Sci 87:518
7. Vincent B, Edwards J, Emmett S, Croot R (1988) Coll and Surf 31:267
8. Rosensweig R, (1985) in Ferrohydrodynamics Cambridge Univ Press Cambridge
9. De Gennes PG, Pincus PA (1970) Phys Kond Mat 11:189–
10. Sano K, Doi M (1983) J Phys Soc Japan 52:2810–
11. Fabre P, Casagrande C, Veyssie M, Cabuil V, Massart R (1990) Phys Rev Lett 64:539–542
12. Bacri J-C, Perzynski R, Salin D, Cabuil V, Massart R (1986) J Magn Magn Mat 62:36–46
13. Bacri J-C, Perzynski R, Salin D, Cabuil V, Massart R (1987) J Magn Magn Mat 65:285–288
14. Asakura S, Oosawa F (1954) J Chem Phys 22:1255–1256
15. Vrij A (1976) Pure Appl Chem 48:471
16. Joanny J-F, Leibler L, De Gennes PG (1979) J Polymer Sci: Polymer Phys Ed 17:1085–1096
17. De Hek H, Vrij A (1981) J Colloid Interface Sci 84:409
18. Gast AP, Hall CK, Russel WB (1983) J Colloid Interface Sci 96:251–267
19. Lekkerkerker HNW, Poon WC-K, Pusey PN, Stroobants A, Warren PB (1992) Europhys Lett 20:559–564
20. Russier V, Douzi M (1994) J Colloid Interface Sci 162:356–371
21. De Hek H, Vrij A (1979) J Colloid Interface Sci 70:592–594
22. Vincent B, Luckham PF, Waite FA (1980) J Colloid Interface Sci 73:509–521
23. Pathmamanoharan C, De Hek H, Vrij A (1981) Colloid Polym Sci 769:769–771
24. Sperry PR (1984) J Colloid Interface Sci 99:97–108
25. Gast AP, Russel WB, Hall CK (1986) J Colloid Interface Sci 10:161–171
26. Patel PD, Russel WB (1988) J Colloid Interface Sci 131: 192–200
27. Poon WCK, Selfe JS, Robertson MB, Ilett SM, Pirie AD, Pusey PN (1993) J Phys II France 3:1075–1086
28. Massart R (1981) IEEE Trans Magn MAG-17:1274–1275
29. Bacri J-C, Perzynski R, Salin D, Servais J (1987) J Phys (Paris) 48:1385
30. Charlot G (1966) in "Les methodes de la Chimie Analytique" Masson et Cie Ed 737
31. Lapp A, Herz J, Strazielle C, (1985) Makromol Chem 186:1919–1934
32. Bacri JC, Salin D (1982) J Phys Lett 43:L-179–184
33. Bibette J, Roux D, Nallet F (1990) Phys Rev Lett 65:2470–2473
34. Bibette J (1991) J Colloid Interface Sci 147:474–478

Progr Colloid Polym Sci (1994) 97:80–84
© Steinkopff-Verlag 1994

SUSPENSION

An experimental study of a model colloid-polymer mixture exhibiting colloidal gas, liquid and crystal phases

S.M. Ilett
W.C.K. Poon
P.N. Pusey
A. Orrock
M.K. Semmler
S. Erbit

Received: 6 October 1993
Accepted: 30 October 1993

W.C.K. Poon (✉)
P.N. Pusey · A. Orrock
S.M. Ilett.
M.K. Semmler · S. Erbit
Department of Physics
The University of Edinburgh
Myfield Road
Edinburgh EH9 3JZ, United Kingdom

Abstract We report an experimental study of the phase behaviour of a model hard sphere colloid + non-adsorbing polymer mixture – sterically stabilised PMMA and polystyrene in cis-decalin. The addition of non-adsorbing polymer to an otherwise stable colloidal suspension can induce phase instability. This is due to an entropic effect whereby the polymer induces an effective "depletion" attraction between the particles. The topology of the resultant phase diagram (colloid volume fraction versus polymer concentration) depends on the ratio of polymer size (δ) to particle size (a), $\xi = \delta/a$. For $\xi < 0.20$, moderate concentrations of polymer cause the suspension to separate into coexisting colloidal fluid and colloidal crystalline phases whereas more polymer leads to "gel" states in which crystallization is supressed. For $\xi > 0.20$ three phase (colloidal gas, liquid, crystal) coexistence is observed for the first time, and is reported here for two systems of different polymer size ($\xi = 0.24$, and $\xi = 0.57$ at 23 °C). These results are in qualitative agreement with statistical mechanical predictions.

Key words Phase equilibria – dispersions – critical point – polymer – depletion potential

Introduction

It is known from experiment [1–7] that the addition of enough non-adsorbing polymer to a suspension of colloidal particles causes phase separation to occur. Understanding this phenomenon is of practical importance, as well as of fundamental interest, since many industrial products are, in essence, colloid + non-adsorbing polymer mixtures.

The earliest theoretical discussion of this subject is that of Asakura and Oosawa [1, 8], with a later, more comprehensive thermodynamic description proposed by Vrij [9]. Both theories assume that the phase separation of colloidal particles is a consequence of the "attractive depletion potential" resulting from the addition of the polymer. When the surfaces of two colloidal particles are separated by a distance less than the size of a polymer coil, polymer is excluded from a depletion region between the particles. The polymer therefore exerts a net osmotic force which pushes the two particles toward one another. This idea has formed the basis of other theoretical approaches [4, 10, 11, 12], and the qualitative features of this model have been verified by experiments with various colloid + non-adsorbing polymer mixtures [2–6, 13]. A common prediction in all these models is the dependence of the phase behaviour on the polymer size (δ) to colloid size ratio (a) (subsequently $\xi = \frac{\delta}{a}$), and the polymer concentration. For low values of ξ (< 0.3, [3–5]) a phase separation into a colloidal solid and a less dense colloidal fluid is predicted. For higher values of ξ then phase separation into two different colloidal fluid phases is also possible; hence the phase diagrams show a critical point analogous

Progr Colloid Polym Sci (1994) 97:80–84
© Steinkopff-Verlag 1994

to that seen for atomic and molecular systems. However the experimental observation of three coexisting colloidal phases was not reported. A recent statistical mechanical model by Lekkerkerker et al. [14] predicts that the triple point seen in previous theoretical phase diagrams should actually by an extended area of three-phase coexistence, (Figs. 2b, 3b in this paper), which should be easily observed experimentally.

A previous study has demonstrated that the collid + polymer system used here (sterically stabilised PMMA hard spheres + polystyrene in cis-decalin) provides a good model system to test the theoretical predictions of Lekkerkerker et al. for a value of $\xi = 0.08$ [7]. This present study aims to test the predictions for higher ξ, and in particular the existence of three-phase coexistence.

Materials and samples

These experiments used colloidal particles consisting of spherical poly (methyl methacrylate) (PMMA) cores stabilised by thin, 10–15 nm, chemically-grafted layers of poly-12-hydroxystearic acid [15]. The particles used had mean radii of 217 nm and 228 nm, and a size polydispersity of about 5% as measured by electron microscopy and dynamic light scattering. They were suspended in cis-decahydronaphthalene (cis-decalin). Studies using suspensions of this type have shown that the interparticle potential is steep and repulsive and is well approximated by that of hard spheres [16–18].

Polystyrene (PS) of molecular weight 390 000 (pressure Chemical Company, manufacturer's quoted value, $\frac{M_w}{M_n} < 1.10$), 2.85×10^6, and 14.4×10^6 (Polymer Laboratories, $\frac{M_w}{M_n} < 1.30$, and 1.21 respectively) was added, to give the phase behaviour observed. The size of the polymer coils (δ) was taken as the radius of gyration obtained using published experimental data of PS in cis-decalin [19].

Results

Figures 1, 2 and 3 show the phase behaviour for colloid-polymer mixtures with different values of ξ. Figures 1a, 2a, 3a, are obtained from the model [14], in which the y-axis represents the polymer concentration in a (hypothetical) reservoir of pure polymer in equilibrium with the system (c_{pr}). Figures 1b, 2b, 3b are the same results from the model plotted using a different y-axis, corresponding to the weight of polymer per unit volume of sample. This experimentally accessible representation is obtained from the former by multiplication by the "free volume fraction available to the polymer", α [14]. The sloping tie-lines of Fig. 1–3 clearly show that the model predicts partitioning

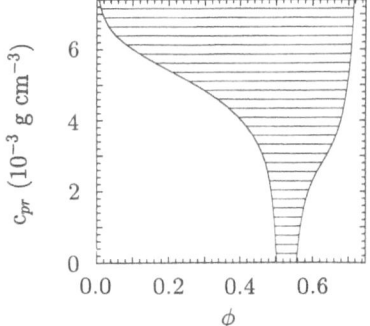

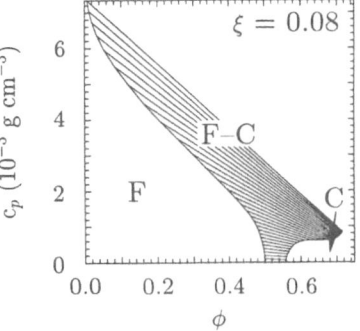

Fig. 1 Theoretical phase diagrams (polymer concentration versus colloid volume fraction (ϕ)). In each of Fig. 1–3, the y-axis in (a) is the polymer concentration in a reservoir of pure polymer solution in equilibrium with the colloid system (c_{pr}), whereas in (b) the y-axis is the polymer concentration in the actual sample (c_p). ξ is calculated using colloid radius = 217 nm, and polymer radius of gyration = 18 nm for molecular weight 390 000 at 19 ± 1°C. Symbols: [F] fluid; [F–C] fluid-crystal; [C] crystal

of the polymer between the different phases. The behaviour predicted depends on ξ, such that for $\xi < 0.32$ three regions are seen corresponding to colloidal fluid, colloidal fluid and colloidal crystal coexistence and colloidal crystal. At $\xi = 0.32$ the phase diagram qualitatively changes, with a region of colloidal gas-liquid crystal coexistence appearing. Regions of gas-liquid coexistence, gas-crystal coexistence and liquid-crystal coexistence are also seen. Hence the diagram now features a critical point analogous to that of pressure versus volume phase diagrams for molecular systems, which is seen clearly in the representations of Figs. 1a, 2a, and 3a.

Figures 4, 5, and 6 show the experimentally observed phase diagrams, obtained using polymer of three different molecular weights. In each case the topology of the phase diagram is qualitatively the same as that predicted by the model, except that gas-liquid-crystal coexistence is observed for a value of ξ lower than expected.

The diagram of Fig. 4 has already been discussed in detail by Poon et al. [7], but will be mentioned briefly again here. Samples with low colloid volume fractions

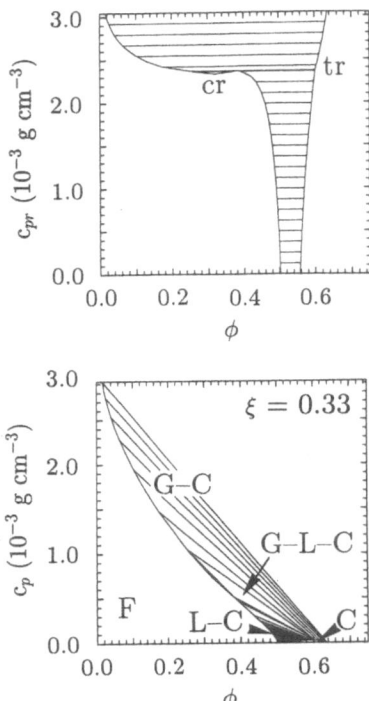

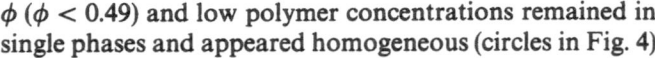

Fig. 2 Theoretical phase diagrams (polymer concentration versus colloid volume fraction (ϕ)). See note to Fig. 1. ξ is calculated using colloid radius = 228 nm, and polymer radius of gyration = 75 nm for molecular weight 5.31×10^6 at $23 \pm 1°$C. Symbols: [cr] critical point; [tr] triple point line; [F] fluid; [G–L–C] gas-liquid-crystal; [L–C] liquid-crystal; [G–C] gas-crystal; [C] crystal

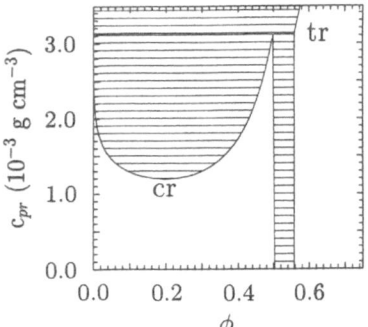

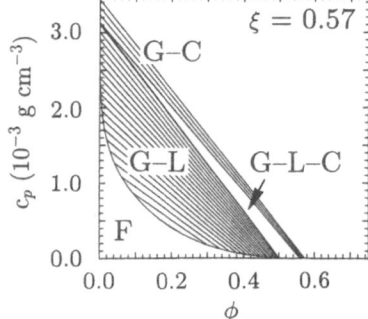

Fig. 3 Theoretical phase diagrams (polymer concentration versus colloid volume fraction(ϕ)). See note to Fig. 1. ξ is calculated using colloid radius = 228 nm, and polymer radius of gyration = 130 nm for molecular weight 14.4×10^6 at $23 \pm 1°$C. Symbols: [cr] critical point; [tr] triple point line; [F] fluid; [G–L–C] gas-liquid-crystal; [L–C] liquid-crystal; [G–C] gas-crystal; [C] crystal

ϕ ($\phi < 0.49$) and low polymer concentrations remained in single phases and appeared homogeneous (circles in Fig. 4)

In samples with higher polymer concentrations (squares in Fig. 4), colloidal crystallites, iridescent "specks" under white light illumination, began to be observed a few hours after mixing. Nucleation appeared to be homogeneous throughout the sample volume. within a day or so the crystallites settled under gravity, leaving supernant colloidal fluid separated from the polycrystalline phases by well-defined boundaries.

At still higher polymer concentrations (triangles in Fig. 4) crystallization was inhibited, and the resultant sample remained in a metastable "gel" state, as described elsewhere [7, 20]. The gel state is not predicted by the theory, although the formation of a tenuous, metastable, fractal-like arrangement under similar conditions has been previously observed in two completely different colloid-polymer systems [2, 13].

The behaviour of the samples in Fig. 5 is similar to that of those in Fig 4 for low colloidal and low polymer concentrations. However increasing amounts of added polymer gave samples exhibiting colloidal gas-liquid-crystal co-existence (crosses in Fig. 5). In these samples the crystal-

lites appeared a few hours after mixing, and began to fall to the bottom within a day. Simultaneously a boundary appeared close to the top of the sample between the lower turbid fluid-like region and an upper less turbid fluid-like region. This boundary moved down over a few days, at a rate much greater than that due to gravitational settling. The different turbidities of these two phases suggested that they had different colloid concentrations and were therefore termed "liquid" and "gas". No gas-liquid coexistence has yet been seen for this system, which is consistent with the infinitesimal gas-liquid coexistence region predicted by the model in Fig. 2b for ξ very close to the crossover value ($\xi_c^{\text{theory}} = 0.32$). Measurements at 12.5°C, for which $\xi = 0.20$, show a similar topology albeit with the phase boundaries shifted to higher polymer concentrations. We therefore conclude that $\xi = 0.20$ is very close to the crossover value (ξ_c^{expt}) required to give three-phase coexistence in an experimental system. Further added polymer gave gas-crystal coexistence (squares in Fig. 5). For even higher polymer concentrations crystallisation was again inhibited.

The colloid-polymer system shown in Fig. 6 features a polymer of even higher molecular weight (14.4×10^6), such that the value of ξ is much higher than the value

Progr Colloid Polym Sci (1994) 97:80–84
© Steinkopff-Verlag 1994

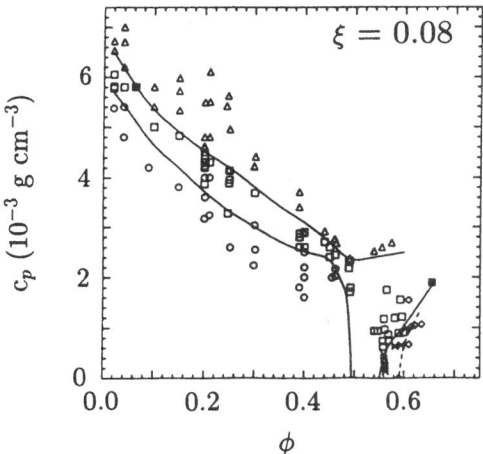

Fig. 4 Polymer concentration (c_p) versus colloid volume fraction (ϕ) phase diagram. ξ is calculated using colloid radius = 217 nm, and polymer radius of gyration = 18 nm for molecular weight 390 000 at 19 ± 1 °C. Symbols; [circle] fluid; [diamond] glass; [asterisk] crystal; [square] fluid-crystal; [triangle] gel

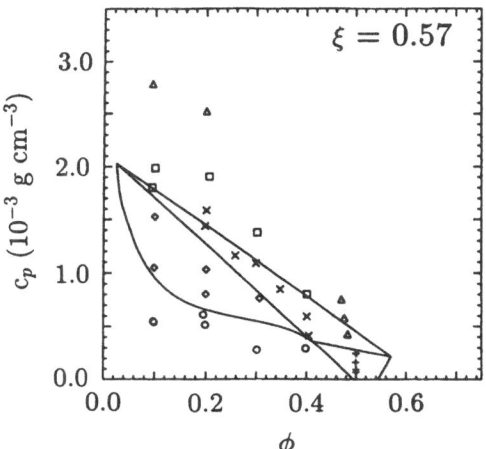

Fig. 6 Polymer concentration (c_p) versus colloid volume fraction (ϕ) phase diagram. ξ is calculated using colloid radius = 228 nm, and polymer radius of gyration = 130 nm for molecular weight 14.4 × 10⁶ at 23 ± 1 °C. Symbols; [circle] fluid; [diamond] gas-liquid' [cross] gas-liquid-crystal; [plus sign] liquid-crystal; [square] gas-crystal; [triangle] gel (?)

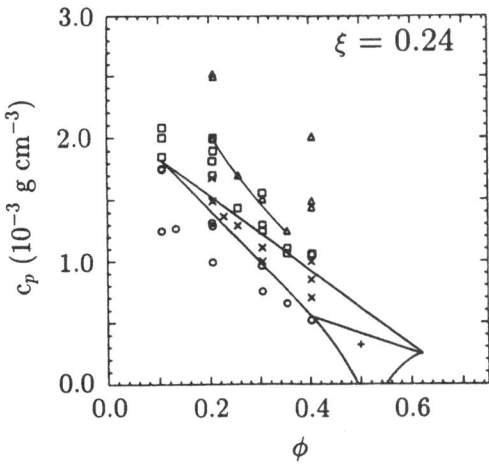

Fig. 5 Polymer concentration (c_p) versus colloid volume fraction (ϕ) phase diagram. ξ is calculated using colloid radius = 228 nm, and polymer radius of gyration = 54 nm for molecular weight 2.85 × 10⁶ at 23 ± 1 °C. Symbols [circle] fluid; [cross] gas-liquid-crystal; [plus sign] liquid-crystal; [square] gas-crystal; [triangle] gel (?)

For the three-phase coexistence samples of systems 5 and 6, the relative proportions of each phase was consistent with the phase diagram. For a given sample in the triangular region of gas-liquid-crystal coexistence the relative proportions of each phase can be calculated by comparing the areas enclosed by the three internal triangles formed by drawing a line from the sample coordinate to the vertices of the three-phase region [21].

The crystal phase of each of the three-phase coexistence samples of Fig. 5 were examined by static light scattering [17]. These results showed that the volume fraction of colloid in the crystal phase was constant ($\phi = 0.62$), and less than that of the crystal phase in the gas-crystal samples studied ($0.62 < \phi < 0.64$). These results are consistent with the phase diagram plotted.

Discussion

The experimental phase diagrams reported here qualitatively agree with the predictions of the model, and show the existence of a region of three-phase in a colloid system. However the results differ in detail from the predictions. One notable deviation is the appearance of three-phase coexistence for a significantly lower value of ξ (0.24). In fact measurements at 12.5°C, the theta temperature for PS in cis-decalin, also show similar behaviour, for which $\xi = 0.20$. The model [14] assumed that the polymer chains behave as hard spheres of radius δ with respect to the colloid particles, and so there is difficulty in choosing the

needed to give three-phase coexistence. One consequence of this is that the gas-liquid coexistence region is predicted by the model to be much larger than that for the system with the lower value of ξ, and accordingly, we found such a region in our experimental phase diagram (diamonds in Fig. 6). In these gas-liquid coexistence samples a boundary between the two fluid-like phases was observed within a day, with the turbidities of each phase being distinctly different.

.84

S.M. Ilett et al.
Experimental study of a model of colloid-polymer mixture

most appropriate value for δ. In this work we have taken the size of the polymer to be equal to the radius of gyration for PS in cis-decalin as reported by Berry [19] for the temperatures used here, which we believe gives a reasonable approximation with possible error of $\approx 10\%$ in ξ. For further detail see discussion in ref. [7].

A second potential source of disagreement between experiment and theory is the possible interpenetration of the particles by the polymer since the diameters of the PS polymer coils used in Figs. 4 and 5 (18 nm and 54 nm respectively) are comparable in magnitude to the thicknesses of the polymer coatings on the PMMA particles, $\approx 10–15$ nm.

An important limitation of the theory of [14] is that it is a mean field theory. The consequences of this, along with possible improvements are discussed in Poon et al. [7].

Note that these results suggest strongly that a potential of long enough range is needed for a thermodynamically stable critical point to appear on the phase diagram. This is expected to apply equally to atomic and molecular systems.

Finally, we mention that, since the presentation of the poster on which this paper is based, a paper by Leal Calderon et al. has appeared in which three-phase coexistence in a colloid-polymer mixture is also reported [22].

Note added in proof; a detailed report of the work presented here will be the subject of a forthcoming publication [23].

Acknowledgements Part of this work is funded by the Agriculture and Food Research Council, and attendance at this conference was made possible by the reciept of a Society of Chemical Industry "Sir Eric Rideal" Bursary. We are grateful to Prof R. H. Ottewill and Ms. F. Beach for providing the PMMA particles, and to Mr. T-T. Chui for characterizing these particles using light scattering.

References

1. Asakura S, Oosawa F (1954) J Chem Phys 22:1255–1256.
2. Sperry PR (1984) J Colloid Interface Sci 99:97–108.
3. Gast AP, Russel WB, Hall CK (1986) J Colloid Interface Sci 109:161–171
4. Vincent B, Edwards J, Emmett S, Croot R (1988) Colloids Surf 31:267–298
5. Patel PD, Russel WB (19899) J Colloid Interface Sci 131:192–200
6. Bibette J, Roux D, Pouligny B (1992) J Phys II France 2:401–424
7. Poon WCK, Selfe JS, Robertson MB, Ilett SM, Pirie AD, Pusey PN (1993) J Phys II France 3:1075–1086
8. Asakura S, Oosawa F (1958) J Polym Sci 33:183–192
9. Vrij A (1976) Pure Appl Chem 48:471–483
10. Sperry PR, Hopfenberg HB, Thomas NL (1981) J Colloid Interface Sci 82:62–76
11. de Hek H, Vrij A (1981) J Colloid Interface Sci 84:409–422
12. Gast AP, Hall CK, Russel WB (1983) J Colloid Interface Sci 96:251–267
13. Smits C, van der Most B, Dhont JKG, Lekkerkerker HNW (1992) Adv Colloid Interface Sci 42:33–40
14. Lekkerkerker HNW, Poon WCK, Pusey PN, Stroobants A, Warren PB (1992) Europhys Lett 20:559–564
15. Antl L, Goodwin JW, Hill Rd Ottewill RH, Owens SM, Papworth S, Waters JA (1986) Colloids Surf 17:67–78
16. Pusey PN, van Megen W (1986) Nature 320:340–342
17. Pusey PN, van Megen W, Bartlett P, Ackerson BJ, Rarity JG, Underwood SM (1989) Phys Rev Lett 63:2753–2756
18. van Megen W, Pusey PN (1991) Phys Rev A 43:5429–5441
19. Berry GC (1966) J Chem Phys 44:4550–4564
20. Pusey PN, Pirie AD, Poon WCK Physica A 201:322–331
21. Bartlett P (1990) J Phys Condense Matter 2:4979–4989
22. Leal Calderon F, Bibette J, Biais J (1993) Europhys Lett 23:653–659
23. Ilett SM, Orrock A, Poon WCK, Pusey PN Phys Rev E (accepted)

Progr Colloid Polym Sci (1994) 97:85–88
© Steinkopff-Verlag 1994

SUSPENSION

S.F. Schulz
H. Sticher

Surface charge densities and electrophoretic mobilities of aqueous colloidal suspensions of latex spheres with different ionizable groups

Received: 16 September 1993
Accepted: 30 November 1993

Dr. Susanne F. Schulz (✉)
Prof. Dr. H. Sticher
ETH-Institute of Terrestrial Ecology,
Bodenchemie
Grabenstrasse 3
8952 Schlieren, Switzerland

Abstract The total bare charges and electrophoretic mobilities of carboxylate latex and sulfate latex were measured as a function of pH and salt concentration. Titration curves fall together in a master curve, if the surface pH is calculated from the surface potential using a Gouy–Chapman model with small modifications. Carboxylate latex shows a spectrum of intrinsic dissociation constants with a linear increase of charge with pH, sulfate latex shows a smaller constant bare charge. The electrophoretic mobilities of the particles are strongly related to the electrolyte conductivities of counterions near the surface. Sulfate particles show a stronger decrease of mobility with ionic strength than particles with pH-dependent carboxylic groups where the increasing bare charge counterplays the screening effect.

Key words Charge densities – titration – electrophoretic mobilities – surface potential – zeta potential

The surface charge on suspended colloidal particles results from adsorption or desorption of small ions onto or from ionizable groups that are bound to the particle surface. Although the chemistry of the ionizable groups is believed to obey chemical equilibrium conditions, titration curves differ significantly from those of simple electrolytes. The composition of the supporting electrolyte is altered by the electrostatic potential of the surface charge in such a way that concentrations of possible counterions are increased while those of coions are decreased near the surface [1, 2]. The motion of the charged particles in external electric fields is expected to depend on the bare charge or the surface potential, but there also is a strong influence of the double layer that is not completely understood and verified from the side of experimental data [3–7]. Theoretical calculations [8–11], on the other hand, do not pay enough attention to the experimental situation to be applied easily.

On a charged surface the concentration of H^+ or other counterions is given by

$$[H^+]_{surface} = [H^+]_{bulk} \exp(-eV(a)/kT) \qquad (1)$$

from which the surface pH can be calculated by

$$pH_S = pH + eV(a)/kT/(\log 10), \qquad (2)$$

where e is the charge of the proton, $V(a)$ the electrostatic potential at the surface at $r = a$, and kT the thermal energy. $V(a)$ can be calculated from the measured surface charge density, the concentration of small ions, and the particle radius a using either the Gouy–Chapman (GC) model with modifications [1, 12, 13], the Debye–Hückel (DH) model, or the numerical solution of the Poisson–Boltzmann (PB) equation [11, 14]. If only a diffusive double layer has to be taken into account the choice of the model depends on the conditions of the ratio of radius to screening length (κa) and of electrostatic energy to kT. With l_B the Bjerrum length in water (0.174 nm at 25 °C) the GC expression for a symmetrical electrolyte of charge $z_i e$ reads

$$\sigma = e\kappa/(4\pi l_B)\, 2\sinh(z_i eV(a)/kT) \qquad (3)$$

The surface charge density σ, on the other hand, depends on the density of ionizable groups N_0, the intrinsic dissociation constants pK_A, and on the surface-pH pH_S. The surface charge can be studied either by titration methods or by the observation of the particle's reaction to an external electric field. If only H^+ counterions are present in the system, the total bare charge Z can be determined from the OH^- consumption in an acid base titration. The measurement of the electrophoretic velocity v in an electric field E leads to an electrophoretic mobility μ that is proportional to the surface charge at small surface potentials. At effective particle charge Q_{eff} can be defined by the balance of electric and friction forces on the moving particle:

$$\mu = v/E = Q_{eff}e/(6\pi\eta a) \tag{4}$$

According to model calculations [11] the mobility may exhibit a maximum and a decrease with large surface potentials corresponding to several kT and with small double layers compared to the particle radius.

We studied the charging behavior and corresponding electrophoretic mobilities of small latex particles with carboxylic and with sulfate groups. The samples were purchased from Polyscience (Carboxylate Latex size 100 nm) and from IDC (Sulfate Latex size 131 nm) and diluted to stock solutions of 1% in deionized water with mixed bed ion exchanger resin (Serva) to convert all the small ions to the H^+ or OH^- form. Acid base titrations were performed with a Schott Tr600 titrator and Metrohm components in salt solutions of $NaClO_4$ with $HClO_4$ and carbonate free NaOH. Electrophoretic mobilities were measured by electrophortic light scattering with a Malvern Zetasizer 3 in several types of electrolytes. Both experimental setups were stabilized to a temperature of 25 °C.

Figure 1 shows the results of acid base titrations and mobility measurements at different ionic strength of perchloric acid/sodium perchlorate/sodium hydroxide with latex volume fractions of about 0.5% for the potentiometric titrations and 0.005% for the electrophoresis. Since the stock suspensions were treated with mixed bed ion exchange resin, the measured charge density from the difference of latex titration curves and blank titration curves corresponds to the latex surface charge density, which is given in Fig. 1B. The LOW modification [12] of the GC expression (3) was used to calculate the surface potential. The surface pH was calculated using Eq. (2). If the charge densities are plotted versus surface pH the curves for different ionic strengths fall together within typical error bars for the measurement. For latex particles with carboxylic groups the surface charge densities versus surface pH resemble a dissociation behavior of a weak acid although the curves are considerably broader than described by a single pK_A. For latex particles with sulfate groups a smaller constant charge density is found on the particles in a pH range from 3 to 9, so that the pK_A values are probably below 2. PH dependent surface potentials for carboxylate latex range between 1 and $7\,kT$, for sulfate latex the values depend only on ionic strengths and amount to $2\,kT$ at 1 mM.

Figure 1A shows the corresponding electrophoretic mobilities of very dilute carboxylate latex particles in dependence of pH at three ionic strengths and in pure $HClO_4$ or pure NaOH. The mobilities mostly show a less pronounced pH dependence than the bare charges or potentials determined from acid base titration. Especially in the high pH region where titration data suggest a nearly linear

Fig. 1 pH dependence of surface charge densities (B) and electrophoretic mobilities (A) of carboxylate latex in $NaClO_4$ solutions of different ionic strength. The dashed lines (– – –) in (A) belong to measurements in pure $HClO_4$ (low pH) and in pure NaOH (high pH). The solid lines (———) in (B) indicate the charge densities of carboxylate latex plotted versus surface pH instead of bulk pH. Titration data of sulfate latex (*–*–*) are given for a 1 mM $NaClO_4$ solution

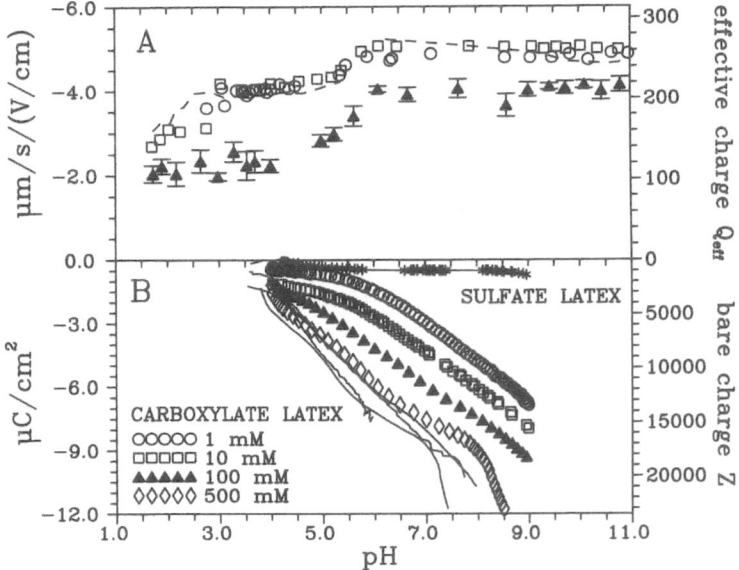

increase of charge the mobilities stay constant. The corresponding effective charges differ from the bare charges by orders of magnitude.

Figure 2 shows the mobilities of carboxylate latex and of sulfate latex in several types of electrolytes in dependence of the conductivity measured in the electrophoresis cell. There is a clear dependence on the type of counterion of the negatively charged particle. Sodium, potassium, and lithium lead to fairly high electrophoretic mobilities in salt solutions of pH about 5.5. The mobility of carboxylate latex in sodium hydroxide solutions is the same as in the neutral salt solutions at ionic strengths above 10^{-4} mol/l and slightly higher below.

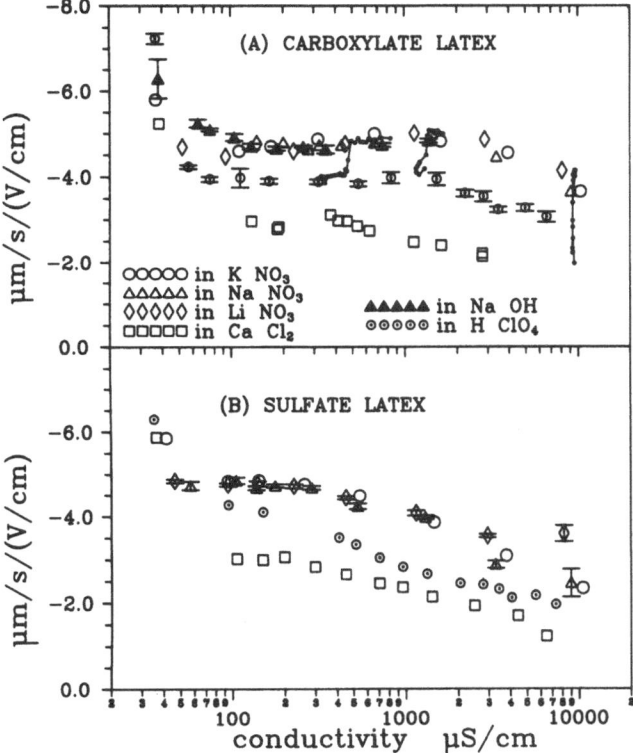

Fig. 2 Electrophoretic mobilities in various types of electrolytes in dependence of measured conductivities for (A) carboxylate and (B) sulfate latex. The leftmost points in each data set belong to zero salt concentration and are not very well defined for experimental reasons. The rightmost points in the 1-1 neutral salt data sets (pH 5.5–6) correspond to 91 mM solutions, the maximum $CaCl_2$ concentration is 29 mM, the maximum $HClO_4$ 20 mM. Data points on solid lines in (A) correspond to the measurements shown in Fig. 1(A) of 1, 10, and 100 mM $NaClO_4$

Hydrogen counterions lead to lower mobilities at the same conductivities in pure perchloric acid for both types of particles and calcium counterions to values that are roughly one-half of the values in 1-1 electrolytes. The data of Fig. 1A plotted into Fig. 2A show a crossover from the hydrogen counterion dominated curve to the sodium counterion dominated curve at data points which still belong to low pH values. For carboxylate latex a weak maximum in the mobilities appears at concentrations of 10 mM of K^+, Na^+, or Li^+ salts ($\kappa a = 16$) but at lower concentrations for perchloric acid (3 mM) and $CaCl_2$ (about 3×10^{-4} M).

The electrophoretic mobilities of sulfate latex spheres in Fig. 2B show a more pronounced decrease of mobility with ionic strength or conductivity than the particles with carboxylic groups. At counterion concentrations above 10^{-2} mol/l the specific conductance of the counterions clearly plays a role, since "slow" counterions like Li^+ lead to higher electrophoretic mobilities than "fast" counterions like K^+ and Na^+. At the two highest concentrations of $CaCl_2$ (20 to 30 mM) a slow aggregation of the sulfate latex particles can be observed by dynamic light scattering. Although the bare charge and surface potential of sulfate latex should be unaffected by moderate proton concentrations the mobility in pure perchloric acid is clearly smaller than for the neutral 1-1 electrolytes.

In conclusion these data indicate a strong dependence of the electrophoretic mobilities on the ionic composition of the double layer but only a weak dependence on the bare charge. The bare charge of these and similar latex colloids depends only on surface pH values and the types and densities of the ionizable groups [15] and are in accordance with diffuse double layers at the particle surface. The mobilities of particles with constant bare charge show a stronger dependence on electrolyte type and concentrations than particles with pH-dependent charge, where the effect of increasing bare charge is counterplayed by the increasing counterion density near the surface. The small values of mobility or effective charges compared to the bare charges on the particle can be understood qualitatively by the high electric conductivity near the particle surface that screens the external field. By this mechanism protons play a special role as highly conducting counterions even if they do not alter the bare charge. The influence of coions has not been studied in these experiments but other experiments [5] indicate an influence on the mobility from this side, too.

References

1. Hunter JR (1981) Zeta Potential in Colloid Science. Academic Press, London

2. Healy TW, White LR (1978) Adv Colloid Interface Sci 9:303–345

3. Harding IH, Healy TW (1985) J Colloid Interface Sci 107:382–397

88

S.F. Schulz et al.
Surface charge and mobilities of a latex spheres

4. Van der Linde AJ, Blijsterbosch BH (1990) Croatica Chemica Acta 63:455–465
5. Elimelech M, O'Melia CR (1990) Colloids and Surfaces 44:165–178
6. Shubin VE, Isakova IV, Sidorova MP, Men'shikova A Yu, Evseeva TG (1990) Translation from Kolloidnyi Zhurnal 52:935–941
7. Deggelmann M, Palberg T, Hagenbüchle M, Maier EE, Krause R, Graf C, Weber R (1991) J Colloid Interface Sci 143 318–326
8. Henry DC (1931) Proc Roy Soc London A133 106–129
9. Booth F (1950) Proc Roy Soc London A203 514–533
10. Wiersema PH, Loeb AL, Overbeek J Th G (1966) J Colloid Interface Sci 22 78–99
11. O'Brien RW, White LR (1978) J Chem Soc Faraday Trans II 74 1607–1626
12. Loeb AL, Overbeek J Th G, Wiersema PH (1961) The Electrical Double Layer Around a Spherical Colloid Particle, MIT Press
13. Ohshima H, Healy TW, White LR (1982) J Colloid Interface Sci 90 17–26
14. de Wit JCM, van Riemsdijk WH, Nederlof MM, Kinniburgh DG, Koopal LK (1990) Analytica Chimica Acta 232:189–207
15. Schulz SF, Gisler T, Borkovec M, Sticher H (1994) J Colloid Interface Sci 164:88–98

Progr Colloid Polym Sci (1994) 97:89–92
© Steinkopff-Verlag 1994

SUSPENSION

R. Despotović
Lj. A. Despotović
Z. Nemet
B. Biskup

On polycomponent colloid systems

Received: 16 September 1993
Accepted: 15 December 1993

Prof. Dr. R. Despotović (✉)
Lj. A. Despotović · Z. Nemet · B. Biskup
Colloid Chemistry Department
Ruder Boskovic Institute
P. O. Box 1016
Zagreb, Croatia

Abstract The influence of the surface active substances of the cationic (n-dodecyl amine sulphate S^+) + anionic type (sodium n-dodecyl sulphate S^-) on the characteristics of the negative silver iodide sols was investigated. Surfactants were present at the moment of silver iodide formation, and act markedly at very low concentrations. In order to obtain new data about interaction between $S^+ + S^-$ and AgI–I$^-$, a model of the formation of submicellar aggregates or associates is proposed, and measurements of surface tension, turbidity and microelectrophoresis were made. The results obtained indicate that $S^+ + S^-$ cause a change of the colloid properties of the stable silver iodide particles. It is supposed that the present inorganic sol particles cause the change of surfactant aggregates as well. Since $S^+ + S^-$ agglomerate could act as a charged ion of various electrophoretic characteristics, the stability of AgI–I$^-$ particles is dependent on the S^+/S^- ratio and the total surfactants concentration. All the experimental data show that the $S^+ + S^-$ produce mutual interactions in the polycomponent system cationic surfactant + anionic surfactant + AgI sol.

Key words Colloids – surfactants – electrophoresis – turbidimetry

Introduction

Polycomponent systems "surfactant + surfactant + inorganic sol" are important, both from a fundamental point of view and in several applications, such as industrial suspensions and detergents. Additives in the form of softeners, hardeners, accelerators, and retarders are, in fact, surfactants or surfactant mixtures which, under various conditions, bring about changes of the colloid state of given technological systems by their mutual interactions. These phenomena are known under different names, often differently described because of different visible interactions and end effects which can be explained by observing colloid process in simple and complex or polycomponent colloid systems. Systematic investigations have shown that by using the same surfactant and a given inorganic sol essentially different results are obtained when the same surfactant is present under different conditions or in different concentrations [1]. For mixed solutions of a cationic surfactant S^+ and an anionic surfactant S^-, colloid interaction is commonly observed over a wide range of mixing ratios and at total surfactant concentrations c_Σ far below critical micellar concentrations c_M [2]. In order to obtain new data about interactions between cationic surfactant + anionic surfactant with the negative diluted silver iodide sols, measurements of surface tension, turbidity and ultramicroscopic electrophoresis were made. It is essential for this study, based on the experimental results, that changes in colloid stability of the sols observed, can be attained by surfactant mixtures of $c_\Sigma = 0.00010 \ mol \ dm^{-3}$.

Experimental

Material

Water was twice distilled from the Zellner type apparatus (Duran 50, Jena, Schott und Gen.) and using Heraeus Bi 3 type quartz apparatus.

Analar grade B.D.H. and MERCK chemicals were used throughout experiments. Sodium dodecyl sulphate (S⁻) was of specially pure quality (B.D.H.) and was not further purified, and a solutions of S⁻ was prepared by weighing. n-Dodecylamine sulphate (S⁺) was prepared by dissolving n-dodecylamine (p.a. MERCK) in diluted sulphuric acid (p.a. MERCK). Using a water bath S⁺ was crystallized and recrystallized twice from ethanol containing a small amount of active carbon. The surfactant S⁺ was dried under vacuum and stored in a desiccator before weighing. A solution of NaI was standardized by a standard AgNO₃ solution using di-iodo-dimethylfluorescein as an adsorption indicator. A solutions of AgNO₃ were prepared using a standard AgNO₃ solutions (KEMIKA).

Preparation of polycomponent systems

The silver iodide sols were prepared by adding 0.002 M silver nitrate solution to an equal volume of 0.004 M sodium iodide solution containing S⁺ + S⁻ mixtures so that a stable silver iodide sol is obtained with AgI $0.0010 \, mol \, dm^{-3}$ with an excess of NaI $0.0010 \, mol \, dm^{-3}$. Total surfactant concentration of S⁺ + S⁻ was $c_\Sigma = 0.0001$ and/or $0.00005 \, mol \, dm^{-3}$. All the prepared systems were composed of 11 samples; at a constant total surfactant concentration c_Σ for each series the samples differ by 0.1 in molar fraction for a combination of the surfactant components present. All the systems prepared were thermostatted at 293 K using a HAAKE ultrathermostat. Before measuring all the polycomponent systems were aged throughout $t_a = 600$, 6000 and or 84600 s.

Tensometry

The surface tension of S⁺ + S⁻ mixtures σ (Nm⁻¹) was determined using a Lecomte de Ncuy semiautomatic torsion balance (Krüss) by means of the ring method. The average values of seven measurements are discussed using Fig. 1.

Turbidity

The measurements of tyndallometric values τ of polycomponent systems were made with a Pulfrich photometer in conjuction with a turbidimetric addition using red and green filters (Carl Zeiss, Jena). Results of turbidity are summarized in Figs 2 and 3.

Microelectrophoresis

The particle charge and the electrophoretic mobility of the system particles were determined by microelectrophoresis using the ultramicroscopic Smith–Lisse method [3]. Results of the zero point of particle charge, u_{+0-} (cm²s⁻¹V⁻¹) of the examined polycomponent systems are summarized in Table 1.

Results and discussion

According to the classical Hartley scheme, the surfactant ions at low concentrations are associated with their hydrophobic tail oriented toward the center, whereas the ionic groups are oriented toward the aqueous medium. By putting both the positively S⁺ (n-dodecylamine sulphate) and negatively S⁻ (sodium n-dodecyl sulphate) charged associates into contact, new colloid species $(S^+_{f_{s^+}} . S^-_{f_{s^-}})^{z+0-}$ are formed, where f_{S^+} and f_{S^-} correspond to molar fractions of the cationic S⁺ and anionic S⁻ surfactant present,

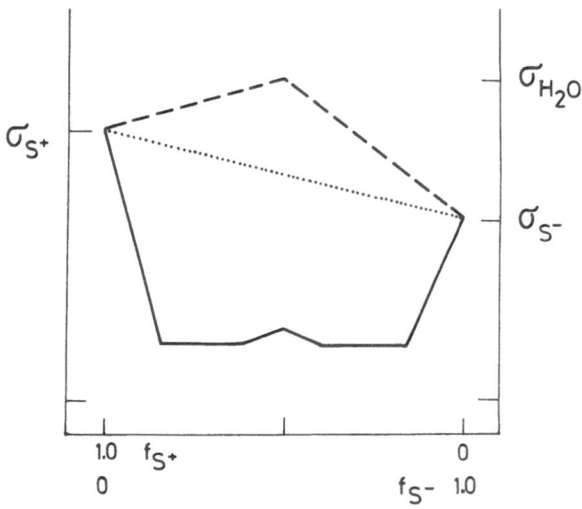

Fig. 1 Surface tension σ (Nm⁻¹) versus molar fraction f of cationic surfactant f_{S^+} and anionic surfactant f_{S^-} of S⁺ + S⁻ mixtures. σ_{H_2O} surface tension of pure water; σ_{S^+} surface tension of pure cationic surfactant solution; σ_{S^-} surface tension of pure anionic surfactant solution. Full line represents the average σ data obtained for the analyzed polycomponent system; dotted line corresponds to the ideal physical average $(\sigma \times f)_{S^+} + (\sigma \times f)_{S^-}$ and broken line corresponds to the σ value for systems in which S⁺ and S⁻ interact chemically

and z_{+0-} is the resulting electrostatic sign of the new colloid species. The new colloid acts very strongly on the interface layer causing a decrease of surface tension (Fig. 1, full line). Considering the system in which no new colloid is present, two possibilities could be thought of. One is the possibility of a simple physical mixture of two surfactants without colloid surfactant-surfactant interactions, and with the resulting surface tension as an average σ value corresponding to $\sigma = (f_{S+})(\sigma_{S+}) + (f_{S-})(\sigma_{S-})$ (Fig. 1, dotted line). The second possibility is of chemical interaction between S^+ and S^- with a resultant σ_{H_2O} value because of exclusion of both the surfactants from solution (Fig. 1, dashed line). Since neither one nor the other resultant curve was obtained, the mutual colloid interaction between S^+ and S^- appears as a reasonable possibility. The surfactant-surfactant aggregate is of a relative high molecular mass and volume with the electrostatic sign dependent on the molar fractions of cationic and anionic surfactant; by changing the molar fractions f_{S+} and f_{S-} it is possible to reach positively, negatively or zero charged $S^+ + S^-$ colloidal aggregates [2]. Using the negative stable silver iodide sol of $0.0010 \, \text{mol cm}^{-3}$ with an excess of NaI of $0.0010 \, \text{mol dm}^{-3}$, for different total surfactant + surfactant concentration c_Σ, the zero point of charge $u_0 \, (\text{cm}^2 \text{s}^{-1} \text{V}^{-1})$ of inorganic particle sols moves to a lower f_{S+} for higher c_Σ (Table 1.). In all cases f_{S+} is markedly higher as compared with f_{S-} because the inorganic particles are of negative electrostatic charge. The decrease of f_{S+} with an increase of c_Σ is reasonable since the total positive electrostatic capacity of surfactant aggregates is higher for higher c_Σ. At the same time, it is essential for this study, based on the results, that changes in colloid stability of negative silver iodide sols can be attained by colloid aggregates $(S_{f_{S+}}^+ S_{f_{S-}}^-)^{z+0-}$, and at c_Σ lower as compared with c_M of both the surfactants present (Table 1).

In order to obtain a general picture of the colloid interactions of the polycomponent system "cationic surfactant + anionic surfactant + inorganic stable sol," tyndallometric values τ were recorded as the function of the total $S^+ + S^-$ concentration c_Σ of different molar fractions f_{S+} and f_{S-} (Figs. 2 and 3). For the systems with $c_\Sigma = 0.00010 \, \text{mol dm}^{-3}$ two typical maxima and one minimum for the polycomponent system aged for $t_a = 600$ and $6000 \, \text{s}$ indicate Coulombic interactions between negative silver iodide particles and surfactant colloid aggregates (Fig. 2 triangles and open circle curves). The first curve slope corresponds to attractive forces causing electrostatic neutralization of the AgI crystal surface by surfactant colloid aggregates of positive sign, since the first Tyndall maxima are reached at $f_{S+} = 0.68$ at which conditions the zero point of charge reaches. The second Tyndall maximum corresponds to the formations of the polycomponent colloid system silver iodide particles surrounded by $(S_{f_{S+}}^+ S_{f_{S-}}^-)^{z+}$ aggregates. At the higher S^+ concentration for $f_{S+} = 0.68$ inorganic colloid particles are flocculated. Very similar results were obtained for both 600 and 6000 s aged systems. One day (86 400 s) aged colloids were mainly flocculated (Fig. 2, open square line). Between two Tyndall maxima a part of silver iodide particles was recharged indicating a deflocculation processes. It is of interest to note the regular time-dependent of the colloid stability (Fig. 2, $t_a = 600 \, \text{s}$ and $6000 \, \text{s}$ as compared with $t_a = 86 \, 400 \, \text{s}$). When increasing amounts of surfactant are added ($c_\Sigma = 0.00050 \, \text{mol dm}^{-3}$: Fig. 3.) to a solution, the tyndall value first goes through a maximum (at approx. $f_{S+} = f_{S-}$), then through a minimum (between $f_{S-} = 0.6$ and $f_{S+} = 0.9$) and then slightly increases. As a regular behavior for differently aged polycomponent systems the zero point of change is reached at $f_{S+} = 0.56$ because of

Table 1 The molar fraction f of n-dodecylamine sulphate f_{S+} and of sodium n-dodecyl sulphate f_{S-} at zero point of charge $u_0 (\text{cm}^2 \text{s}^{-1} \text{V}^{-1})$ of silver iodide particles in $S^+ + S^-$ mixtures of various total molar concentrations c_Σ

c_Σ (mol dm^{-3})	f_{S+}	f_{S-}
0.00010	0.68	0.32
0.00020	0.67	0.33
0.00030	0.62	0.38
0.00050	0.56	0.44

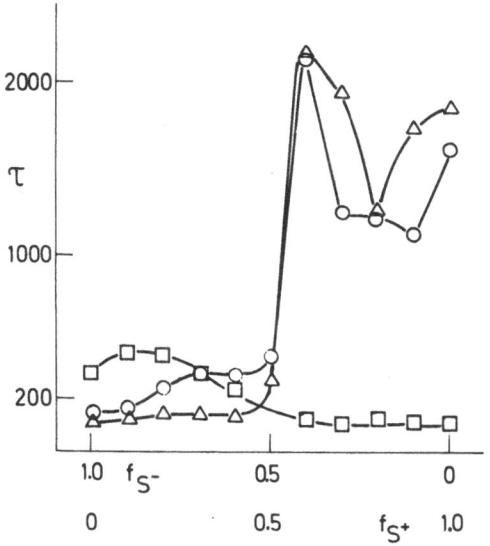

Fig. 2 Tyndall value τ as a function of molar fractions f of n-dodecyl-amine sulphate f_{S+} and sodium n-dodecylsulphate f_{S-}. System: The AgI sol $0.0010 \, \text{mol dm}^{-3}$ in $S^+ + S^-$ mixtures of total concentration $c_\Sigma = 0.00010 \, \text{mol dm}^{-3}$, at 293 K and aged for $t_a = 600 \, \text{s}$ (triangles), 6000 s (open circles), and 86400 s (open squares)

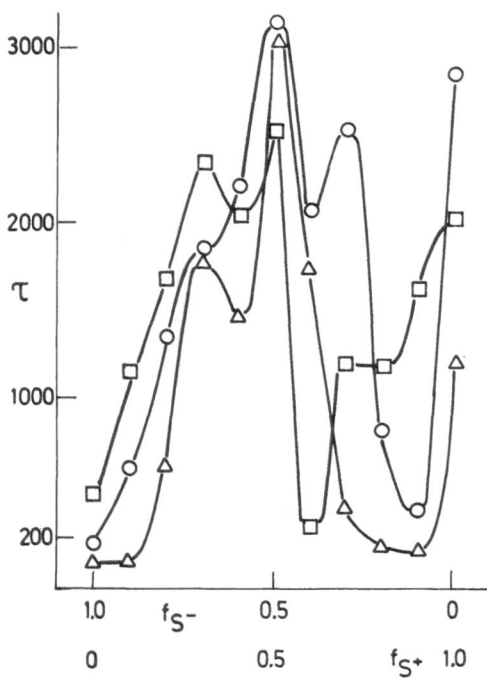

Fig. 3 Tyndall value τ as a function of molar fractions f of n-dodecyl-amine sulphate f_{s+} and sodium n-dodecyl sulphate f_{s-}. System: The AgI sol 0.0010 mol dm^{-3} in S$^+$ + S$^-$ mixtures of total concentration $c_\Sigma = 0.00050$ mol dm^{-3}, at 293 K and aged for $t_a = 600$ s (triangles), 6000 s (open circles), and 86400 s (open squares)

higher total positive electrostatic capacity for a surfactant mixture of $c_\Sigma = 0.00050$ mol dm^{-3} (Table 1). The resulting colloid stability, measured as a τ, is shifted to higher values. Following the Carpineti–Giglio model [4] by increasing c_Σ the number of $(S_{f_{s+}}^+ . S_{f_{s-}}^-)^{z+o-}$ colloidal aggregates decreases: big aggregates are widely spread and the time-dependence of the colloid stability appears to not be an important factor when comparing the systems with $c_\Sigma = 0.0001$ mol dm^{-3} of the present S$^+$ + S$^-$. It is very interesting that the system with the same inorganic sol, and the same concentration, shows such differences in colloid stability caused by an increase of c_Σ. supposing the silver iodide sol as a standard sol, a very strong shift of Tyndall maxima ($c_\Sigma = 0.00050$ as compared with $c_\Sigma = 0.00010$ mol dm^{-3} of S$^+$ + S$^-$: Figs. 2. and 3.) indicates a change of $(S_{f_{s+}}^+ . S_{f_{s-}}^-)^{z+o-}$ colloid structures confirming the supposed model of mutual interactions in polycomponent surfactant solutions [2]. We hope that a more detailed analysis of these systems will give an interesting picture of interactions between an inorganic sol and cationic + anionic surfactants mixture.

References

1. Despotović R (1989) Jorn Com Esp Deterg 20:331–340; Idem (1992) Ibid 23:295–306
2. Šarić A, Despotović R, Trikić S (1992) Progr Colloid Polym Sci 89:30–32
3. Smith ME, Lisse MV (1936) J Phys Chem 40:399–401
4. Carpineti M, Giglio M (1992) Lecture A4 at VIth ECIS Conference (1992) Graz

Progr Colloid Polym Sci (1994) 97:93–96
© Steinkopff-Verlag 1994

SUSPENSION

H.-J. Mögel
P. Brand
T. Angermann

Aggregation processes in solutions of basic aluminum chlorides

Received: 16 September 1993
Accepted: 25 March 1994

H.-J. Mögel (✉) · P. Brand
T. Angermann
Institut für Physikalische Chemie
Freiberg University of Mining and
Technology
Leipziger Straße 29
09599 Freiberg, FRG

Abstract A dynamic light scattering investigation of basic aluminium chloride solutions in alcohol/water mixtures confirms the existence of a wide range of particle sizes. There are three principally different kinds of particles: 2 nm particles which are probably chemically stable polycations, very stable particle in the size range 5 nm up to 8 nm, and very large agglomerates up to several hundred nm. It is assumed that these agglomerates consist of physically attracted small particles of the first and second type. The agglomerates can be destroyed by shearing into stable particles mainly of the second type.

Key word Aluminum chloride
– polycation agglomeration
– dynamic light scattering

Introduction

Basic aluminum chloride solutions are complex liquids which contain several kinds of polycations and aggregates in a wide range of particle size distribution. These substances have been used as antiperspirants and deodorants, as flocculants in waste water cleaning and as a chemical source for tailoring special ceramic materials by the sol-gel-process [1] and in heterogeneous catalysis. For this reason much work has been done to find correlations between the content of various species and the solution properties. However, it is difficult to analyze the structure, size, and electric charge of the oligomeric and polymeric cations and aggregates built up from the primary particles. Up to now only monomers, dimers, and tridecamers are precisely identified [2, 3]. Several experimental methods of analysis like gel filtration chromatography [2], viscosimetry [4], ultracentrifugation [5], and static light scattering [6] indicate the existence of highly complex particles in the colloid size range.

Our goal is to study the particle size distribution, the degree of reversibility of aggregation or polycondensation processes, and their time scale using the dynamic light scattering techniques.

Preparation of aluminum chloride solutions

From the large variety of methods for preparing aluminum chloride solutions, we choose the reaction of aluminum foil with 1.9 molar hydrochloric acid at boiling temperature. After reaching a molar ratio $n_{Al}/n_{Cl} = 1.9$ the remaining aluminum particles were separated by filtration. Several diluted solutions were prepared by addition of an alcoholic water mixture. Thereby, we obtained the following:

mixture 1: 2.3 mol Al/l mixture (= 40% ethanol + 60% water)

mixture 2: 2.0 mol Al/l mixture (= 50% ethanol + 50% water)

mixture 3: 1.6 mol Al/l mixture (= 60% ethanol + 40% water)

mixture 4: 1.3 mol Al/l mixture (= 70% ethanol + 30% water)

94

H.-J. Mögel et al.
Aggregation in basic aluminum chloride solution

mixture 5: 0.9 mol Al/l mixture (= 80% ethanol + 20% water)

Results and discussion

The size distribution of the solid content in the mixtures 1 to 5 was determined by a dynamic light scattering experiment using the Malvern equipment PCS 4700 with an Argon laser. At a scattering angle of 90° a laser light power of 200 mW was needed to get a good merit value for fitting the autocorrelation function for the small particles. In order to avoid artefacts from the influence of dust particles filtration by a 1.2 μm pore filter was carried out.

Instead of the number distribution in all figures the intensity ratio of the scattered and incident light is shown because this is the most sensitive measure to detect any change in the particle size distribution. The particle size here is the diameter of an effective sphere having the same translational hydrodynamic properties as the real anisometric particle. To be sure that the chemical reaction kinetics after preparation and mixture had finished, we studied 3-month old solutions at 25 °C.

We first investigated the influence of the ethanol content on the particle size distribution. Figure 1 shows a first peak at about 2–2.3 nm for each mixture. The higher the ethanol concentration, the larger the particle concentration in the small size range. Obviously, this feature indicates chemically stable primary particles which may be polycations. The second peak is shifted towards larger particle sizes with increasing water concentration. This effect can be explained by supposing that agglomerates build up from the small particles during the partial loss of the protective screening force arising from ethanol molecules.

Hence, there are two principally different kinds of colloid particles: chemically stable polycations and agglomerates.

A further topic is the question about the mechanical stability of the agglomerates. Figure 2 illustrates the influence of a shear field during the filtration operation by a 220 nm pore filter. The result only weakly depends on the ethanol/water concentration ratio. The agglomerates are broken into particles with diameters mainly in the interval from 5 nm to 8 nm. The instability against sheating is also observed for filter pore diameters of 440 nm or 800 nm. Generally, we found the trend to be the smaller

Fig. 1 Scattering intensity ratio of aluminum chloride solutions 3 months after preparation and filtration by 1.2 μm pore filter

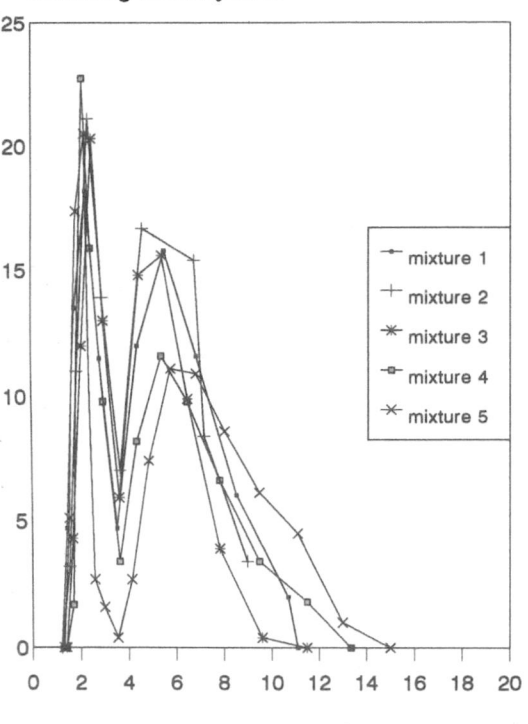

scattering intensity in %

particle diameter /nm

Fig. 2 Scattering intensity ratio of aluminum chloride solutions 3 months after preparation and filtration by 220 nm pore filter

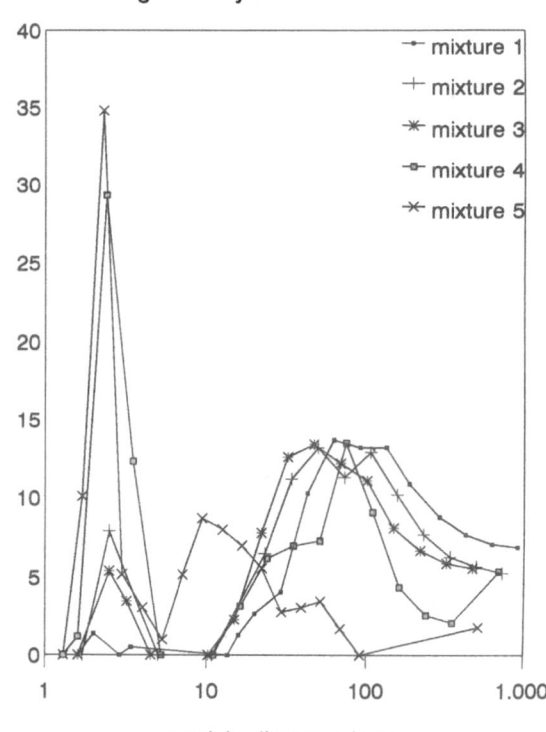

scattering intensity in %

particle diameter / nm

Progr Colloid Polym Sci (1994) 97:93–96
© Steinkopff-Verlag 1994

the pore diameter the more complete was the agglomerate destruction. The large particles are not the result of polycondensation reaction but rather of a physical attraction.

To study the problem of reversibility of particle growth and destruction we have carried out a diffusion experiment. A water drop was added at the top to the cylindrical sample cell containing the mixture 4 after filtration by 220 nm pore filter. A microamount of water passed through the solution onto the bottom of the cell. From here it was distributed upwards by diffusion increasing the local water concentration in the light scattering zone 1 cm from the cell bottom. After some time the diffusion through the whole cell was finished so that the local water concentration reached almost the same value as that with-

out water addition. In Figs 3, 4, and 5 the relative scattering intensity distributions are shown which were monitored over 1 day. The first peak remains at a constant particle size. The second peak is shifted towards higher particle sizes when the water concentration is increased. After rehomogenization by diffusion in the whole cell the protective force of ethanol is restored. The final picture shows a size distribution identical to the first one. The same measurement series was repeated at different sample cell heights. The results confirm the proposed interpretation. The consecutive measurements show the reversibility of growth and destruction processes. However, a time delay in dependence on the height was observed.

Fig. 3 Scattering intensity ratio of aluminum chloride solutions versus particle diameter 5 min, 10 min, 25 min, and 32 min after diffusion start

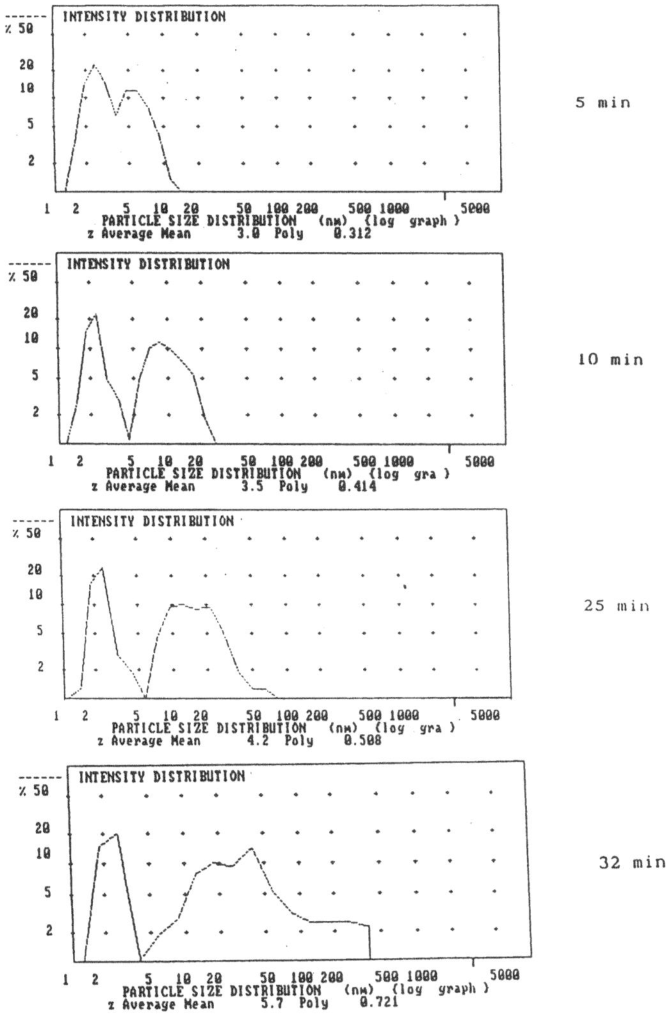

Fig. 4 Scattering intensity ratio of aluminum chloride solutions versus particle diameter 40 min, 80 min, 100 min, and 2 hours after diffusion start

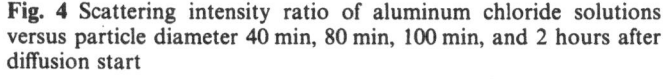

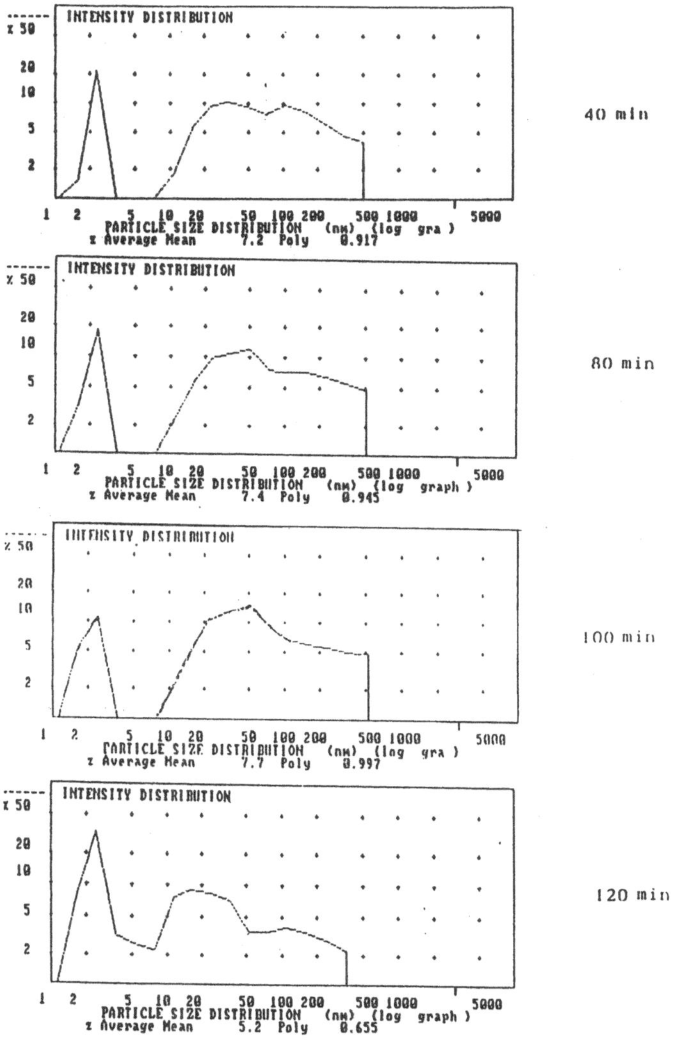

96

H.-J. Mögel et al.
Aggregation in basic aluminum chloride solution

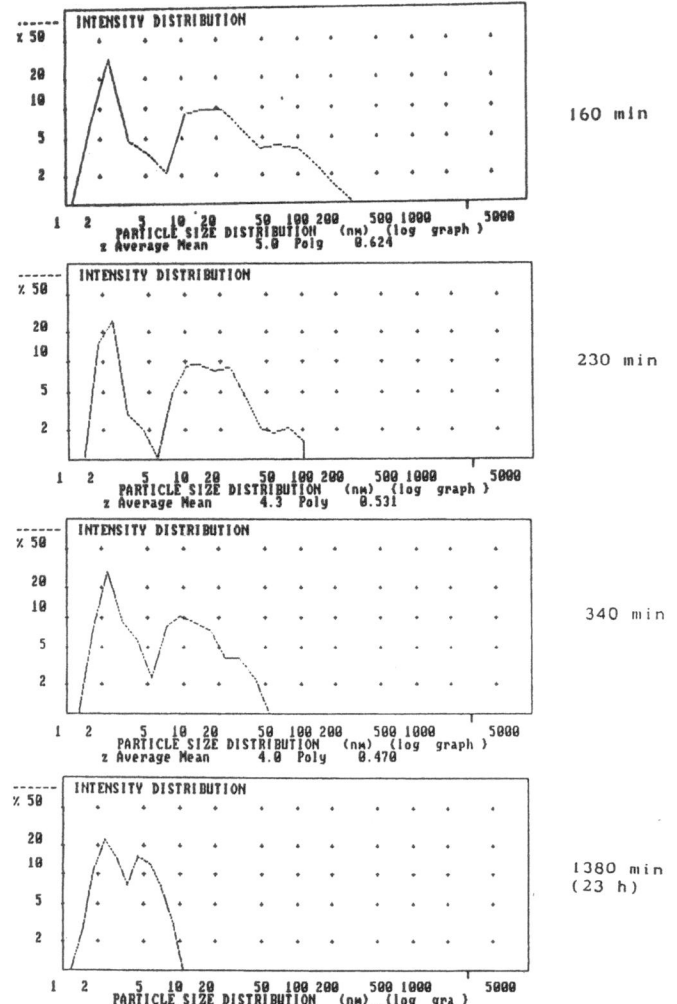

160 min

230 min

340 min

1380 min
(23 h)

Conclusions

The dynamic light scattering investigation of basic aluminum chloride solutions in alcohol/water mixtures confirms the existence of a widely range particle size spectrum. There are three different kinds of particles: 2 nm particles which are probably chemically stable polycations, very stable particles in the size range 5 nm up to 8 nm, and very large agglomerates up to several hundred nm. It is assumed that these agglomerates consist of physically attracted small particles of the first and second type. The agglomerates can be destroyed by shearing into stable particles mainly of the second type. The growth and destruction processes induced by changing the local water concentration are obviously reversible.

Fig. 5 Scattering intensity ratio of aluminum chloride solutions versus particle diameter 160 min, 230 min, 340 min, and 23 hours after diffusion start

References

1. Brand P, Dietzmann P (1992) Cryst Res Techn 27:529
2. Fitzgerald J (1988) In: Laden K, Felger CB (eds) Antiperspirants and Deodorants. Marcel Dekker New York, p 119
3. Akitt JW, Farthing A (1981) J Chem Soc Dalton Trans: 1617
4. Dobrev C, Trendafelov D, Doberva B (1981) Freiberger Forschungshefte A653:129
5. Aveston J (1960) J Chem Soc (London) 111
6. Petterson JH, Tyrell SY (1973) J Coll Interf Sci 43:389

Progr Colloid Polym Sci (1994) 97:97–102
© Steinkopff-Verlag 1994

SUSPENSION

K.R. Rogan
A.C. Bentham
G.W.A. Beard
I.A. George
D.R. Skuse

Sodium polyacrylate mediated dispersion of calcite

Received: 15 September 1993
Accepted: 20 December 1993

Dr. K.R. Rogan (✉) A.C. Bentham
G.W.A. Beard, I.A. George D.R. Skuse
Research Department
ECC International
John Keay House
St. Austell
Cornwall PL25 4DJ, United Kingdom

Abstract The stabilising action of sodium polyacrylate (NaPA) on colloidal dispersions of calcite has been investigated through measurement of viscosity, electrophoretic mobility and solution ion concentration. The dose of NaPA was in the range 0 to 28 mg per g of calcite and the dispersions were prepared at a solids content of 70% (by weight). The ionic strength of the dispersions increased with dose and was in the range ca. 5 to 500 mmol dm^{-3}.

The stabilising action of the NaPA was evident from the sharp fall in viscosity observed at low levels of addition, and the invariance of this low viscosity throughout the remainder of the dose range.

Electrophoretic mobility was converted into zeta potential using the mathematical procedures of O'Brien and White; in this conversion the colloid particles were treated as spheres with a number average radius of 193 nm.

The stability of the dispersion at low levels of NaPA addition was quantified by DLVO theory and readily attributed to electric double layer repulsion. However, at higher levels of addition, and with the encumbent double layer compression, the DLVO theory was found inadequate.

Recent calculations have shown that the acknowledgement of an interparticle steric repulsion can generate increasing colloidal stability at higher NaPA doses and so reconcile theory with experiment.

Key words Sodium-polyacrylate – calcite – dispersion-stability – slurry-viscosity – paper

Introduction

Aqueous slurries of finely ground calcite find wide applications in the paper industry. The ideal slurry is at a high solids content in order to minimise transportation and storage costs, and has a low viscosity to facilitate processing. In order to prepare such a slurry it is necessary to add a dispersant to reduce the viscosity. The industry standard dispersant for this application is sodium polyacrylate (NaPA).

This article describes the results of a fundamental study of the NaPA mediated dispersion of fine calcite. The re-sults include slurry viscosity, polymer uptake, zeta potential and solution ionic strength. The colloidal stability of dispersions was quantified using DLVO theory.

Materials

The dispersed phase. Fine calcite was prepared from an aqueous suspension of Carrara marble by dispersant-free grinding. The weight-averaged size distribution of the calcite particles was measured on a Micromeritics Sedigraph. The amount of material smaller than 2 μm was ca. 90%.

The weight-based distribution was translated into a number-averaged size distribution by the Micromeritics software. The number-averaged median radius and the geometric specific surface area of the particles were calculated to be 193 ± 23 nm and $5.7 \, m^2 \, g^{-1}$, respectively. This area was not significantly different from the measured nitrogen BET area of $6.3 \pm 1.0 \, m^2 \, g^{-1}$.

The supernatant obtained from a NaPA-free suspension of the carbonate was analysed for the more common elements found in aqueous environments by inductively coupled plasma atomic emission spectroscopy (ICP). This revealed the presence of Ca, Na and Mg and trace amounts of K; the Mg originates from small amounts of dolomite that are invariably found in Carrara marble.

The dispersing agent. The NaPA was obtained as a high concentration solution from Allied Colloids. The potentiometric titration of this solution against NaOH, gave a weight/weight percent NaPA concentration of $39.1 \pm 0.5\%$ and a Na^+ concentration of $4.168 \, mol \, kg^{-1}$. The weight-averaged molecular mass of the polymer was in the range 3500 to $9500 \, g \, mol^{-1}$ (low-angle laser light scattering).

Method

Contacting of dispersed phase and dispersing agent. The dispersed phase and the dispersing agent were contacted at a solids content of 70% for about 2 days as follows: Fine calcite equivalent to a dry weight of 140 g was weighed into a brass mixing pot. A known weight of NaPA was added such that the weight of NaPA per unit weight of calcite (i.e. the NaPA dose) was in the range 0 to 28 mg per g calcite. Finally, sufficient water was added such that the total weight of the contents of the pot was 200 g. The contents of the pot were homogenised.

Slurry viscosity measurement and slurry separation. The viscosity of the slurry was measured using a Brookfield viscometer at 100 rpm. About $2 \, cm^3$ of the slurry was retained while the remainder was separated into solid and supernatant phases by pressure filtration.

Analysis of supernatant. A part of the supernatant was analysed for the elements Na, Ca and Mg by ICP. In addition, a part of the supernatant was analysed for NaPA by gel permeation chromatography (GPC). The remainder of the supernatant was retained for electrophoretic mobility measurements.

Electrophoretic mobility measurements. An appropriately small amount of the slurry was redispersed in the super-

natant and the electrophoretic mobility of the suspended particles measured on a Malvern Zetasizer-4.

Repetition. This entire procedure was repeated 23 times. With each repeat the dose administered was a value in the range 0 to 28 mg per g calcite, such that the doseage difference between repeats was roughly 0.5 mg per g at low doses (i.e. < 10 mg per g) and roughly 5.0 mg per g at higher doses.

Results and discussion

Viscosity. The viscosity data of the 24 slurries are presented as a plot of viscosity against dose in Fig. 1. The sharp fall in viscosity at low levels of addition, and the invariance of this low viscosity throughout the remainder of the dose range demonstrated the stabilising action of NaPA on colloidal dispersions of calcite.

Ionic species. The principal ionic species present in each supernatant were Na^+, polyacrylate (PA^-), Ca^{2+} and Mg^{2+}. In general, above an addition level of ca. 1.0 mg per g, the solution concentrations of all these species increased with dose. At low dose levels

Fig. 1 Brookfield viscosity (at 100 rpm) as a function of Na-polyacrylate dose

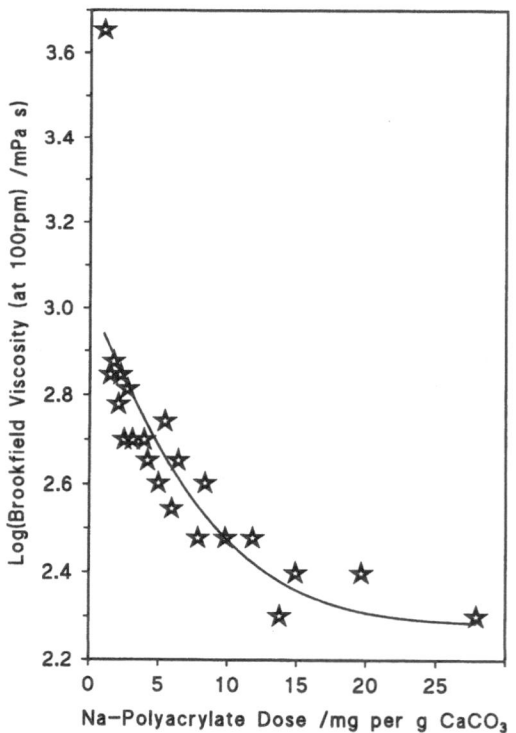

Progr Colloid Polym Sci (1994) 97:97–102
© Steinkopff-Verlag 1994

(ca. $\leq$ 1.0 mg per g), however, the concentration behaviour of the species was not uniform. Thus, no PA$^-$ was found in solution, while the concentrations of Ca^{2+} and Mg^{2+} decreased with increasing dose. In contrast, the concentration of Na$^+$ increased linearly, with an intercept close to the origin.

The behaviour of Ca^{2+} and Mg^{2+} concentration at low doses may be explained in terms of an initial reaction of these alkaline earth cations with NaPA to produce mixed cation polyacrylates (eg. sodium-calcium-polyacrylate) followed by the uptake of these polyacrylates by the calcite. This explanation implies that the polyacrylate acted as an ion exchanger and exchanged Na$^+$ ions in preference for Ca^{2+} and Mg^{2+} ions. From a thermodynamic point of view, such an exchange is predicted on the basis of the small solubility product values of Ca^{2+}/Mg^{2+} carboxylates and the vast difference between these values and those of Na$^+$ carboxylates [1, 2].

In a similar way, the increasing concentrations of Ca^{2+} and Mg^{2+} with dose above 1.0 mg per g may also be explained in terms of a polyacrylate ion-exchange reaction. However, this time Na$^+$ ions of the polyacrylate were exchanged for Ca^{2+} and Mg^{2+} ions from the solid phase, and the so-formed, water soluble mixed cation polyacrylate now remained in the solution phase.

Ionic strength. The molar ionic strength (I) of each supernatant was calculated in the conventional way. The I values of the supernatants are presented as a plot of I against dose in Fig. 2. The main contributors to I were Na$^+$ and PA$^-$.

Abstraction isotherm. The interaction of NaPA with calcite is better described as an abstraction rather than an adsorption process. This is because of the very likely occurrence of a chemical reaction between a portion (at least) of the interacting NaPA and the carbonate surface [3–6]. In this event, it is reasonable to expect that the NaPA/calcite system under study did not attain a state of equilibrium in the time scale of the experiment [7]. Thus, the concentration of NaPA found in solution ($\langle$NaPA$\rangle$) was considered as a residual concentration, at the time of sampling, rather than an equilibrium concentration.

The total amount of NaPA abstracted was evaluated, for each dose, from the difference between the amount of NaPA administered to the calcite and $\langle$NaPA$\rangle$. This abstracted amount was converted into an abstraction density (Γ) using the geometric specific surface area of the carbonate (5.7 m^2 g^{-1}).

For the sake of simplicity, all of the polyacrylate species detected and quantified by GPC were considered to be in the sodium form. That is, the small amount of mixed cation polyacrylate which may have been produced by ion

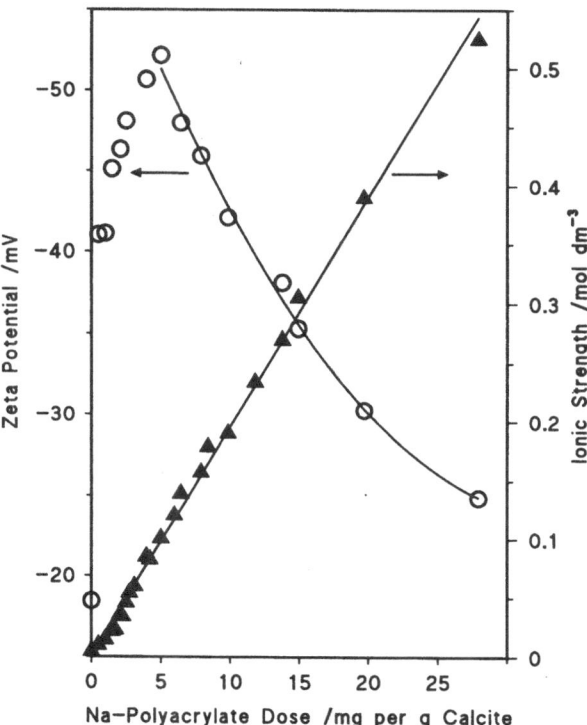

Fig. 2 Zeta potential and ionic strength as functions of Na-polyacrylate dose

exchange of the NaPA was considered negligible, to a first approximation.

An abstraction isotherm was constructed from a plot of Γ against $\langle$NaPA$\rangle$ (Fig. 3). This isotherm may be conveniently separated into the following three regions for the purposes of discussion:-

Region	$\langle$NaPA$\rangle$ range mg cm^{-3}	Dose range mg per g Calcite
1	0 to 0.9	0 to 2.2
2	2.1 to 10.2	2.5 to 6.4
3	> 10.2	> 6.4

Over the first two regions the isotherm has a rectangular hyperbolic shape, with an initial sharp rise in abstraction in Region 1 followed by a short plateau in Region 2. This shape of isotherm obeys the Langmuir model, and the abstraction data can be fitted to the linear equation

$$\langle NaPA \rangle / \Gamma = K_L / \Gamma_p + \langle NaPA \rangle / \Gamma_p, \qquad (1)$$

where K_L is a constant and Γ_p is the abstraction density at the plateau. Thus, a plot of $\langle$NaPA$\rangle/\Gamma$ against $\langle$NaPA$\rangle$ should yield a straight line having a slope of $1/\Gamma_p$ from which the value of Γ_p may be readily obtained. There was

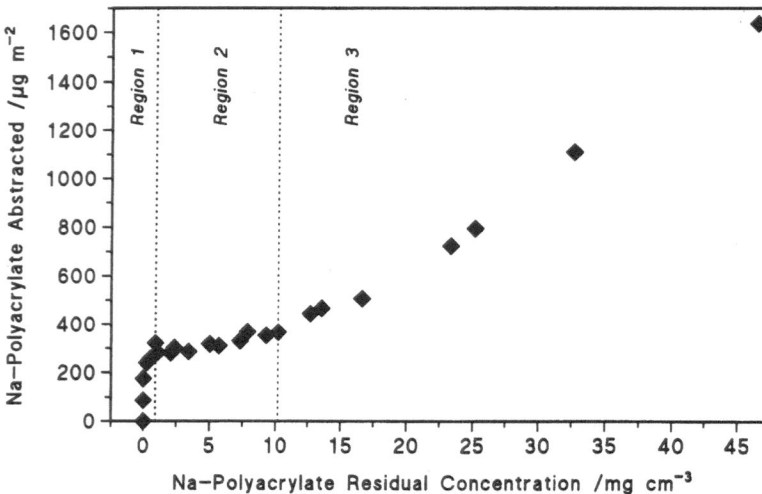

Fig. 3 Abstraction of Na-polyacrylate by calcite at a solids content of 70%

a good correlation between $\langle NaPA \rangle / \Gamma$ and $\langle NaPA \rangle$ up to the end of Region 2 (i.e. a dose of ca. 6 mg per g). The value of 370 $\mu g\,m^{-2}$ was obtained for Γ_p. The isotherm as a whole is of the type H3 in the Giles et al. [8] system of isotherm classification.

It may be inferred from the sharpness of the initial rise in abstraction (Region 1) that there was a high affinity of the NaPA molecules for the calcite surface [9], while the plateau at 370 $\mu g\,m^{-2}$ (Region 2) indicates the completion of an abstracted layer. The shortness of the plateau means that the abstracted layer exposed a surface to the bulk solution which had nearly the same affinity for more NaPA as that of the original surface.

There was a small, but discernable, increase in Γ (ca. 84 $\mu g\,m^{-2}$) across the plateau region. This may be attributed to the development of the monolayer with increase in dose.

It is known that the sizes of polyelectrolytes decrease appreciably with increase in ionic strength [10, 11]. This is because such an increase causes a shielding of charge centres and gives rise to a reduction in intra and inter segmental repulsions, resulting in the collapse of polymer chains [12]. It would appear reasonable to assume, therefore, that the increase in I across the plateau region brought about a collapse of the chains in the monolayer and the production of vacant sites on the calcite surface. Further NaPA was then able to bind to these vacant sites.

Thus, the Γ value at the start of the plateau (282 $\mu g\,m^{-2}$) refers to a monolayer of extended polymer chains, while the Γ value at the end of the plateau (366 $\mu g\,m^{-2}$) refers to a monolayer of collapsed polymer chains.

In the third region of the isotherm there is a rise in abstraction, and no correlation was found between $\langle NaPA \rangle / \Gamma$ and $\langle NaPA \rangle$. It may be inferred, therefore,

that further uptake of NaPA molecules occurred onto the layer of molecules initially abstracted, i.e. Region 3 describes multilayer formation. The reason for this multilayer formation may well lie with the increasing I of the slurry solution. The effect of I on the amount of polyelectrolyte abstracted from solution has been studied by several groups [13–16]. Irrespective of the relative signs of the surface charge and the polyelectrolyte, the abstracted amount generally increases with increasing I. Such behaviour arises from two sources: 1) the solvency of the solution for the polyelectrolyte decreases with increasing I, and 2) the shielding of the polyelectrolyte charge centres increases with I [12]. The combined effect of these two phenomena is to drive the polyelectrolyte into the abstracted state.

Zeta potential. The measured mobility was converted into zeta potential on the basis of the mathematical procedure developed by O'Brien and White [17], using the computer program containing this procedure written by White, Mangelsdorf and Chan [18]. The variation of zeta potential with dose is shown in Fig. 2.

Zeta potential increased with dose, reaching a maximum value of ca. -50 mV at a dose of ca. 6 mg per g, i.e. just prior to multilayer formation. At higher doses, the zeta potential decreased with increasing dose. The initial rise in zeta potential is readily attributable to the uptake of polyacrylate (anionic polyelectrolyte) onto the calcite surface. The fall in zeta potential at higher doses is reasonably explained in terms of the reduction in the decay distance of electrical potential incurred by the increase in ionic strength. Apparently, this collapse of particle electrical double layers (edls) outweighs the increase in electrical potential expected with polyelectrolyte uptake.

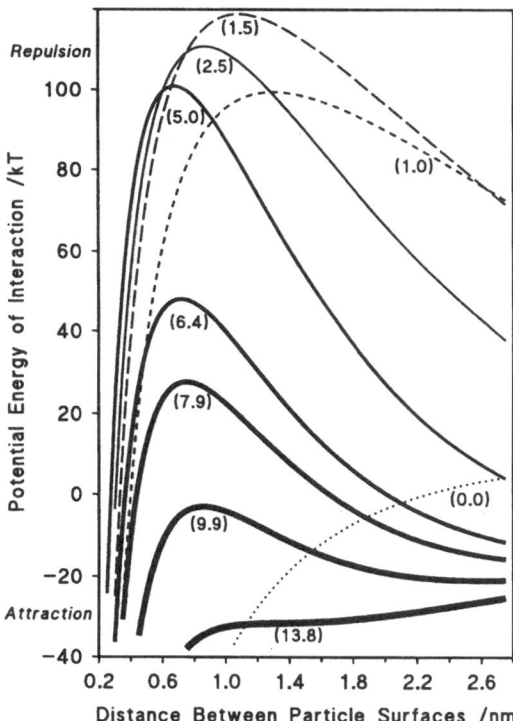

Fig. 4 Interaction energy curves: van der Waals attraction + electrostatic repulsion (no steric energy term). Values in parentheses are Na-polyacrylate dose in units of mg per g calcite

Particle-particle interactions. The potential energy of interaction (V_T) of NaPA-treated calcite particles, in a high solids slurry, was calculated from the sum of the potential energy of electrostatic repulsion (V_R) and the potential energy of attraction (V_A). The equations of Ottewill [19] and Verwey and Overbeek [20] were used to describe the variations of V_R and V_A, respectively, with distance.

The variation of V_T, scaled in units of kT, with distance between particle surfaces (interaction energy curve) is shown in Fig. 4, for a wide range of dose. The repulsive energy maximum of the interaction energy curve increases in height with increasing dose up to ca. 2 mg per g. Thereafter, the height of the energy maximum decreases with increasing dose. The increase at low dose is readily attributable to the increase in the zeta potential of these particles as they take up polyacrylate from low I solution.

At higher doses, the decrease in the height of the energy maximum may be attributed to the collapsing effect of I on edls, and the increasing importance of this effect at higher doses. Thus, although the zeta potential of the particles is observed to increase with dose above 2 mg per g, as poly-

acrylate is taken up, this is overshadowed by edl collapse. The net effect is a loss of electrostatic repulsive energy and the energy maximum diminishes with increasing dose.

Conclusions

NaPA has a high affinity for calcium carbonate surface and a monolayer is assimilated at quite low doses, ca. 2 mg per g. Hardly any polyacrylate remains in the continuous phase and the I of this medium is low and increases only slightly with dose. The polymer chains are in a relatively extended configuration. In this low I region the edls of particles are relatively extensive, and electrical potential decays only gradually with distance from particle surfaces. As a result, the initial uptake of polyacrylate (anionic polyelectrolyte) generates quite high, negative zeta potentials, ca. − 45 mV. In such a situation of extensive double layers and high zeta potentials, the electrostatic repulsion between approaching particles is strong and a slurry of the particles is colloidally stable.

With further addition, more and more polyacrylate appears in the continuous phase and a steady increase in I is observed with increase in dose. This leads to a collapse of surface polyacrylate chains and the production of vacant surface sites onto which further polyacrylate then binds. However, the rate of production of vacant surface cannot keep pace with the rate of deposition of polyacrylate chains and multilayer formation begins at a dose of ca. 6 mg per g. The increase in I also collapses edls and this effect begins to dominate over the observed increase in zeta potential arising from the uptake of polyacrylate. The net result is that interparticle electrostatic repulsion begins to diminish with increasing dose above 2 mg per g.

Inspite of the diminution in electrostatic repulsion above a dose of ca. 2 mg per g, the viscosity of slurries of polyacrylate covered particles continues to decay with increasing dose up to and beyond 6 mg per g. On the reasonable assumption that such a decay infers an increase in colloid stability, then such stability cannot be explained on the basis of electrostatics. Apparently, the notion that the colloidal stability of NaPA treated calcite slurries is due solely to the development of electrostatic repulsion between particles is an inadequate representation of the true picture above a dose of ca. 2 mg per g.

Recently calculations have shown that the acknowledgement of an interparticle steric repulsion, in NaPA treated calcium carbonate slurries, can generate increasing colloidal stability above ca. 2 mg per g and so reconcile theory with experiment.

References

1. Fuerstenau MC, Palmer BR (1976) In: Fuerstenau MC (ed) Flotation. AIME, New York, pp 151–152
2. Giesekke EW, Harris PJ (1984) Int Conf Miner Processing Johannesburg
3. Fuerstenau MC, Miller JD (1967) Trans AIME 238:153
4. Somasundaran P (1969) J Colloid Interface Sci 31:557
5. Han KN, Healy TW, Fuerstenau DW (1973) J Colloid Interface Sci 44:407
6. Rogan KR (1994) Colloid Polym Sci 272:82
7. Aplan FF, Fuerstenau DW (1962) In: Fuerstenau DW (ed) Forth Flotation. AIME, New York
8. Giles CH, MacEwan TH, Nakhwa SN, Smith D (1960) J Chem Soc p3973
9. Giles CH, MacEwan TH (1957) Proc 2nd Int Conf Surf Activity 2:339
10. Corner T (1983) In: Poehlein GW, Ottewill RH, Goodwin JW (eds) Science and technology of polymer colloids NATO ASI series E, Martinus Nijhoff, The Hague, pp 600–618
11. Munk P (1989) Introduction to macromolecular science, Wiley-Interscience, New York, pp 59–61
12. Cohen-Stuart MA, Cosgrove T, Vincent B (1986) Adv Colloid Interface Sci 24:143
13. Bonekamp BC, van der Schee HA, Lyklema J (1983) Croat Chem Acta 56:695
14. Cafe MC, Robb ID (1982) J Colloid Interface Sci 99:341
15. Takahashi A, Kawaguchi M, Kato T (1980) In: Lee L-H (ed) Adhesion and adsorption of polymers polymer science and technology volume 12b, Plenum, New York, pp 729–749
16. Takahashi A, Kawaguchi M, Hayashi K, Kato T (1984) In: Goddard ED, Vincent B (eds) Polymer adsorption and dispersion stability, ACS Symp Ser 240, pp 39–52
17. O'Brien RW, White LR (1978) J Chem Soc Faraday Trans II 74:1607
18. White LR, Mangelsdorf C, Chan YC (1989) University of Melbourne
19. Ottewill RH (1990) In: Candau F, Ottewill RH (eds) Scientific methods for the study of polymer colloids and their applications NATO ASI Series, Kluwer Academic, Dordrecht, pp 129–157
20. Verwey EJW, Overbeek JThG (1948) In: Theory of the stability of lyophobic colloids, Elsevier, Amsterdam, pp 160–163

Progr Colloid Polym Sci (1994) 97:103–109
© Steinkopff-Verlag 1994

SURFACTANTS

H. Hoffmann
S. Hofmann
J. C. Illner

Phase behavior and properties of micellar solutions of mixed zwitterionic and ionic surfactants

Received: 26 October 1993
Accepted: 31 January 1994

Prof. Dr. H. Hoffmann (✉) · S. Hofmann
J. C. Illner
Department of Physical Chemistry of the
University Bayreuth
Universitätsstr. 30
95440 Bayreuth, FRG

Abstract The charge density on micelles from alkyldimethyl-aminoxide was varied continuously by mixing the zwitterionic surfactant with ionic surfactants. The sign of the charge is of importance for the behavior of the micellar solutions. Large synergistic effects are observed for the combination of zwitterionic with anionic surfactants, but not with zwitterionic and cationic surfactants. The differences in the behavior are reflected in the values of the surface and interfacial tension, in the sphere-rod transition, in the general phase behavior and in the properties of mixed micellar solutions.

Key words Micellar solutions
– zwitterionic- ionic- surfactants

Introduction

Formulations of surfactants for various applications usually contain several different surfactants. Mixed surfactants often have superior properties in comparison to the properties of the individual components. A combination of two surfactants can be more surface active, have a better washing power or a higher foamability effect than the components [1]. Chemical formulations consist of ionic surfactants which are mixed with nonionic surfactants which can be alkylpolyglycolether, sugar surfactants or zwitterionic surfactants. Such systems have been studied in detail [2]. Synergistic effects are usually readily revealed in results of surface and interfacial tension measurements. Often, one finds that the surface tension or the interfacial tension of micellar solutions against a hydrocarbon has a minimum for a certain mixing ratio [3]. The synergism can also be revealed in the phase diagram. Tiddy et al. observed a hexagonal phase in mixtures for hexadecyl-dimethylammoniumpropanesulfonate with SDS, which was reached for much lower surfactant concentrations than any hexagonal phase which has been observed in a purely binary system [4]. Typical signs of synergistic effects in mixed systems of ionic and anionic surfactant's

are maxima of the viscosity as a function of the mixing ratio. Large synergistic effects can always be expected when two surfactants can pack to a denser film at the bulk or micellar interface than the components on their own. This is often the case with the combination of charged and uncharged surfactants because the components require a fairly large area for different reasons [5]. Charged headgroups require a large area because of mutual electrostatic repulsions and their steric area requirement is low. For uncharged headgroups it is usually the other way around. In mixed systems the uncharged headgroups can therefore approach much closer to the charged headgroups than to the uncharged headgroup. In formulations for the stabilization of emulsions it is obvious that mixed charged/uncharged surfactants can thus be more effective than the two components alone because both short range steric and long-range electrostatic repulsive interactions can contribute to the stabilization of the systems. The electrostatic interaction is usually fully developed with about 20% of ionic surfactant. Large mole fractions of ionic surfactant only lead to counterion condensation and shield some of the charge density of the interphase.

In the present work experimental results will be presented for mixtures of tetradecyldimethylaminoxide with SDS, tetradecyltrimethylammoniumbromide and tet-

104

H. Hoffmann et al.
Phase behavior and properties of micellar solution

radecyldimethyloxoniumchloride. We also present some experimental results for micellar solutions with the mixed surfactants in the presence of different amounts of the cosurfactant hexonal. Some properties for the $C_{14}DMAO/SDS$ system have already been published by Weers et al. [6] For this combination, we also have shown previously that supramolecular structures are formed under shear for mixed solutions of these surfactants. Mixed solutions show the phenomenon of shear thickening [7]. In this publication, we are more interested in the micellar structures in the solutions at rest and in the different phases of the systems. We are studying the influence of the charge density on the phases and, in particular, how the sign of the charge is of importance for the properties.

In the classic zwitterionic systems the cationic charge is usually separated by at least one or more CH_2–groups from the negative charge. For the aminoxides the positive and negative charge are directly connected to each other. We would still consider it is a zwitterionic surfactant. Actually, this seems to be justified because the data which are presented are very similar to the results that have reported for mixtures of alkyldimethylbetain + SDS by Iwasaki et al. [8]

Experimental results and discussion

In Fig. 1 the phase diagram for the ternary system of $C_{14}DMAO/SDS/H_2O$ is shown. The phase diagrams for these two components has already been reported by Kekicheff and by Oetter et al. [9, 10] Both surfactants have a hexagonal phase in the range of 40 to 60% by weight of surfactant. The system $C_{14}DMAO$ also has,

within a narrow region before the hexagonal phase, a nematic N_c–phase. Pure samples of SDS show no nematic phase. However in samples of commercially available SDS one usually finds an N_c phase. The most prominent feature of the mixed system is the extension of the hexagonal phase in the mixing ratio between 2:8 to 8:2 to lower concentrations and in particular the nematic phase which is already observed for 10% by weight. To our knowledge, this is the only nematic N_c–phase which has been observed for such a low surfactant concentration. The nematic phase can be recognized on the texture with a polarization microscope. It can be aligned in a magnetic field. Due to the high viscosity of the samples the alignment takes a long time. Typical polarization micrographs for the phases are shown in Fig. 2 a b c.

In Fig. 3, we show the various phases which are observed when hexanol is added to mixtures of $C_{14}DMAO/SDS$ of a constant surfactant concentration of 100 mM. The phase diagram for mixtures of $C_{14}DMAO$ with $C_{14}TMABr$ and hexanol has previously been reported [11]. It was found that a L_α–phase was observed for the whole mixing ratio. With increasing molar ratio of the ionic surfactant both, the phase boundaries of the L_1 and the L_α–phases were shifted to a higher cosurfactant surfactant ratio. For the $C_{14}DMAO/SDS$ system we observe the L_α–phase only for mole fractions of SDS of less than 30%. The reason for the different behavior for the two systems is not clear at present. We note that the phase boundary of the L_1–phase for small charge densities de-

Fig. 2a Polarization micrograph for the nematic phase before alignment in a magnetic field in the ratio of $C_{14}MDAO:SDS = 5:5$ and with the overall surfactant concentration of 15% by weight

C_{14} DMAO:SDS = 5:5
15 wt%
nematic Schlierentexture
before alignment

Fig. 1 The ternary phase diagram $C_{14}DMAO/SDS/H_2O$

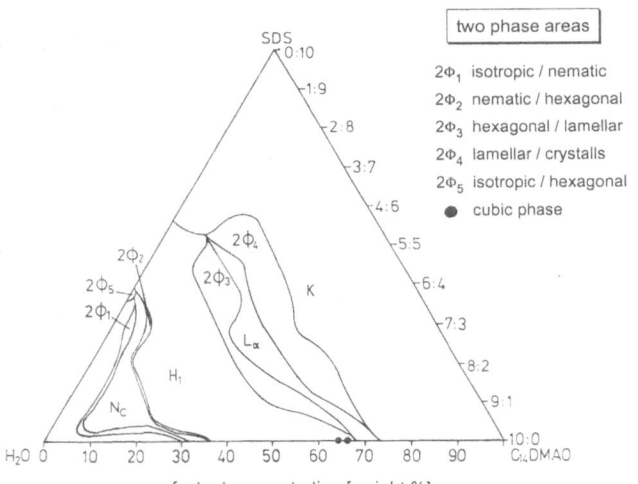

surfactant concentration [weight %]

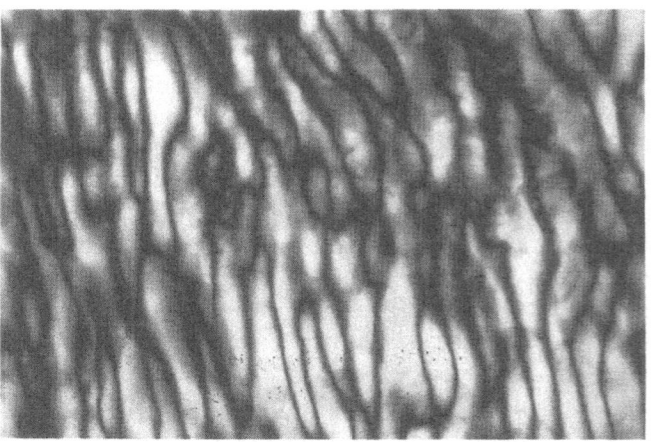

Progr Colloid Polym Sci (1994) 97:103–109
© Steinkopff-Verlag 1994

C_{14} DMAO:SDS = 5:5
15 %
after alignment in a
magnetic field

Fig. 2b Polarization micrograph for the nematic phase of the same composition and concentration after alignment in a magnetic field

C_{14} DMAO:SDS = 5:5
55 wt%
network texture
lamellar phase

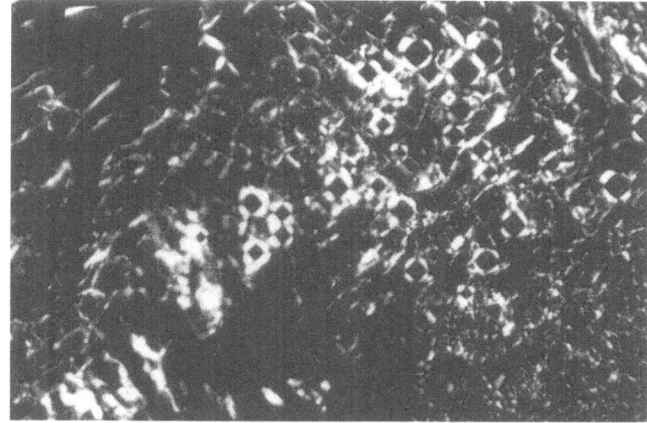

Fig. 2c Polarization micrograph for the lammellar phase in the ratio of C_{14}DMAO:SDS = 5:5 and with the overall surfactant concentration of 55% by weight

Fig. 3 The phase diagram for increasing hexanol concentrations in mixtures of C_{14}DMAO and SDS. The surfactant concentration was cnstant 100 mM and the temperature 25 °C

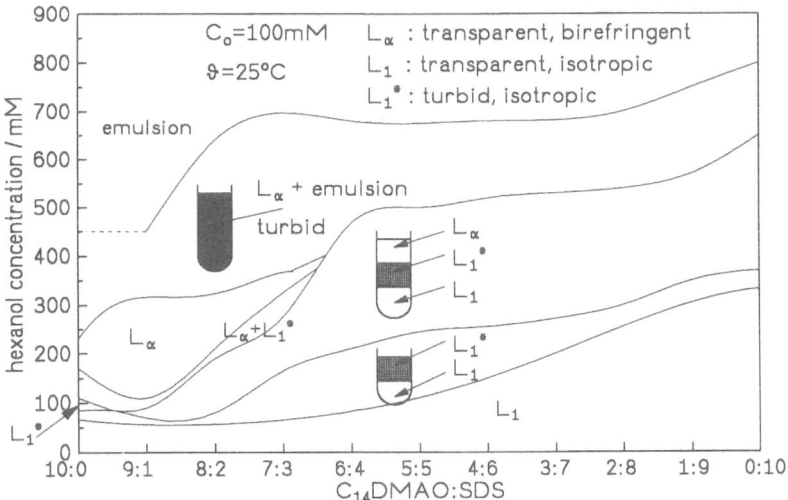

pends little on the X_s–ratio. This is probably an indication that the hydrophilicity of the surfactant in these mixing ratios is about the same. This feature correlates with the extension of the hexagonal phase in the ternary diagram to low surfactant concentrations!

Surface and interfacial tension measurements

In Fig. 4a and b surface and interfacial tension measurements for the three combinations C_{14}DMAO/SDS,

C_{14}DMAO/C_{14}TMABr and C_{14}DMAO/HCl are presented. For the first combination we observe a large synergism. The surface and interfacial tensions are much lower for the mixtures than for the two components. The minimum is at a mixing ratio of about 7:3. The main lowering occurs when a small fraction of one component is added to the other components, while in the mixing ratio between 8:2 and 2:8 the changes are not as large. The large synergism is also present when the surfactant mixtures contain cosurfactant and this cosurfactant is incorporated in the micells. It should be noted that the synergism in this

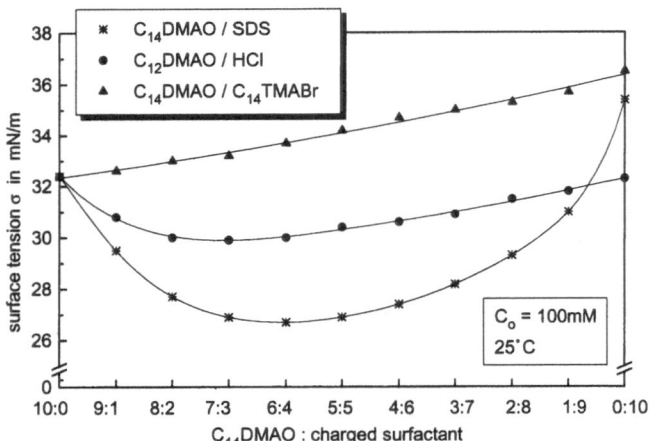

Fig. 4a Surface tension measurements for mixtures of the three surfactant combinations $C_{14}DMAO/SDS$, $C_{14}DMAO/C_{14}TMABr$ and $C_{12}DMAO/HCl$ are presented for 25 °C and 100 mM surfactant concentration

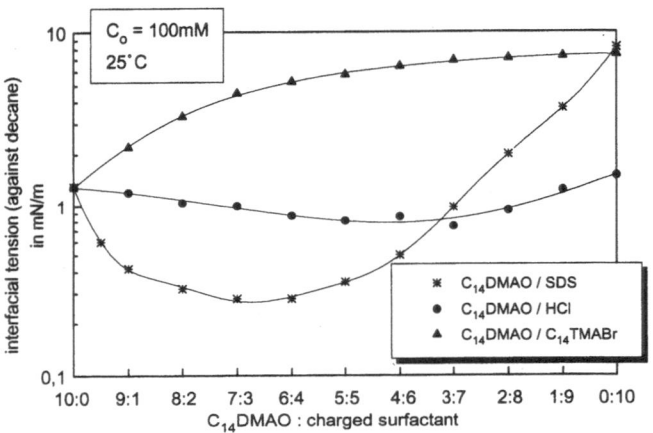

Fig. 4b Interfacial tension measurements against decane are presented for the three surfactant combinations $C_{14}DMAO/SDS$, $C_{14}DMAO/C_{14}TMABr$ and $C_{14}DMAO/HCl$ for 25 °C and the overall surfactant concentration of 100 mM

combination is effected by the protonation equilibrium of the aminoxides [12]. Mixing SDS to the micelles of $C_{14}DMAO$ increases the basicity of the aminoxide and leads therefore to a larger fraction of protonated groups than without the SDS and hence to an increase of pH [13]. The fraction of protonated aminoxides is however still small and around a few percent. The synergism is somewhat less if the protonation is suppressed by the addition of NaOH to the solutions. It is however still strong. The main point of the synergism is thus a result of the interaction of SDS with the unprotonated aminoxide. The synergism between the aminoxide and SDS is probably similar

in nature to the strong synergism that is observed between cationic and anionic surfactants [14]. It is an indication of the preferred binding and mixing between the two surfactants. We thus can expect that the two surfactants have a strong tendency to mix and, consequently, there is probably no tendency for demixing on a local scale in a micelle which is formed from the two components. This is in agreement with SANS measurements on mixtures of the components. No demixing was observed with the contrast variation technique [15].

The surface and interfacial tension measurements of mixtures of $C_{14}DMAO/C_{14}TMABr$ reveal a completely different situation. For this combination we find a strong increase of the surface and interfacial tensions of $C_{14}DMAO$ with small additions of the cationic surfactant. The data seem to indicate a tendency for demixing of the two surfactants even though the two surfactants are so similar in their structure. Contrast variation of SANS for micelles of these combinations have not yet been made. It is thus not clear whether demixing occurs. The surface tension values makes it clear however that in combinations with small mole fractions of cationic surfactant the $C_{14}DMAO$ becomes much more hydrophilic.

Combinations of $C_{14}DMAO/C_{14}DMAOH^{+}Cl^{-}$ show a behavior which is intermediate between the two previous ones. The surface and interfacial tension values do not vary much with the mixing ratio. There is a small tendency for synergism. In this combination the charge can simply be placed on the micelles by mixing an $C_{14}DMAO$ solution with an equimolar $C_{14}DMAO$ and HCl solution. pH measurements indicate that up to a mixing ratio 5:5 the H^{+} ions are more or less all located on the aminoxide and the pH is in the neutral range.

Viscosity measurements

Some viscosity data are given in Fig. 5a, b. Figure 5a gives the zero shear viscosity for 6:4 mixing ratio of $C_{14}DMAO/SDS$ against the total concentration of surfactant. The viscosity increases abruptly at C^{*} which is the overlap concentration and increases up to and into the l.c. phase region. In the double log plot the slope of the viscosity against the concentration is about 8. Such high scaling exponents have also been observed for other charged surfactant systems [16]. A high exponent is probably a result of increased growth of the micelles for $C > C^{*}$ [17]. In Fig. 5b the viscosities of mixed surfactant systems with 20% charge are plotted against the cosurfactant concentration. For each mixed systems we observe an increase of the viscosity with the cosurfactant concentrations which is almost linear in the semilog plot. We observe however that the increase of η^{0} is much larger for the SDS system

Progr Colloid Polym Sci (1994) 97:103–109
© Steinkopff-Verlag 1994

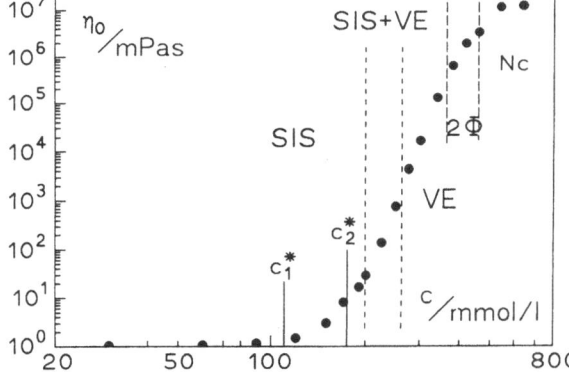

Fig. 5a The zero shear viscosities for mixtures of $C_{14}DMAO:SDS = 6:4$ in different phases are printed against the surfactant total concentration. Data taken at 25 °C

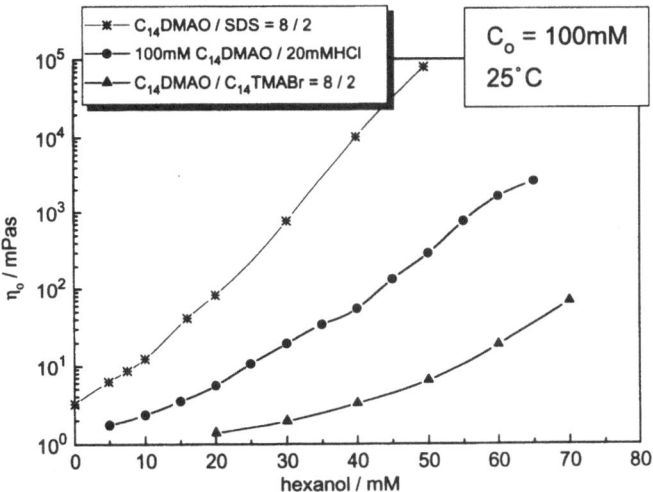

Fig. 5b The zero shear viscosities of the three 20% charged systems $C_{14}DMAO/SDS$, $C_{14}DMAO/C_{14}TMABr$ and $C_{14}DMAO/HCl$ for 25 °C and 100 mM surfactant concentration measured for increasing hexanol concentrations

than for the $C_{14}DMAOH^+Cl^-$ system which is larger again than for the $C_{14}TMABr$ system. For the last system the micelles in the solution are of globular shape and a sphere rod transition occurs with increasing hexanol concentrations while in the other two systems rodlike micelles exist already in the surfactant solutions and the rods become longer with increasing cosurfactant concentration. From the viscosities it seems that the difference of the three systems only indicates that the rodlike micelles in the three different systems have different lengths and thus different rotation times that control the viscosities. We tried to determine the length of the rods by electric biref-

ringence measurements and observed qualitatively completely different results for the three systems.

Electric birefringence measurements

Electric birefringence results for the $C_{14}DMAO/$ $C_{14}TMABr$ systems have already been reported [7]. With the pulse method, one observes a simple signal with a rise and a decay time. Both time constants increase with the hexanol concentration up to the L_1–phase boundary. The time constants of the birefrigence measurements show the same behavior as the viscosities, which is an indication that the viscosities are controlled by the rotation of the rodlike micelles. For the negatively charged mixed micelles the situation is completely different as is shown in Fig. 6 for several solutions with a mixing ratio of the surfactants of 6, 2:3, 8 and increasing cosurfactant concentration. Without hexanol, one observes a small signal with a negative sign. The decay time is less than 0, 5 µs, which corresponds to rods which are shorter than 200A°. With increasing cosurfactant concentration the birefringence amplitude is increasing and becomes longer. At 20 mM hexanol the fast process is still present, but it is overcompensated by a second process of opposite sign. Finally, at 70 mM hexanol concentration we observe in addition to the two processes a third process which has the same sign as the first one. Similar complicated electric birefringence signals have been observed on other binary and ternary surfactant systems with charged rodlike micelles and on other colloidal systems and on polyelectrolytes [18]. The phenomenon is usually referred to as electric birefringence anomaly. Several explanations have been proposed for the phenomenon. The explanations differ somewhat for the second process which is only observed in the small concentration region in which the charged rodlike micelles begin to overlap. An experimental result shows that during this process some of the rodlike micelles are oriented perpendicular to the electric field. Cates has proposed that this process originates from domains of correlated rods [19]. These domains would than have a large electric polarisability perpendicular to their main axis. Theoretical calculation's based on this model seem to support this model. A somewhat different explanation was proposed by Hoffmann et al. It was argued that in the concentration region where the second process occurs the electric double layers overlap in the direction of the rods, but not in the perpendicular direction. It is for this reason that it becomes possible to align the rods perpendicular to the electric field [20]. The third process is usually explained as a hindered rotation of the entangled rods. These electric birefringence results show that the rods, when they are charged with SDS or with $C_{14}TMABr$ respond differently

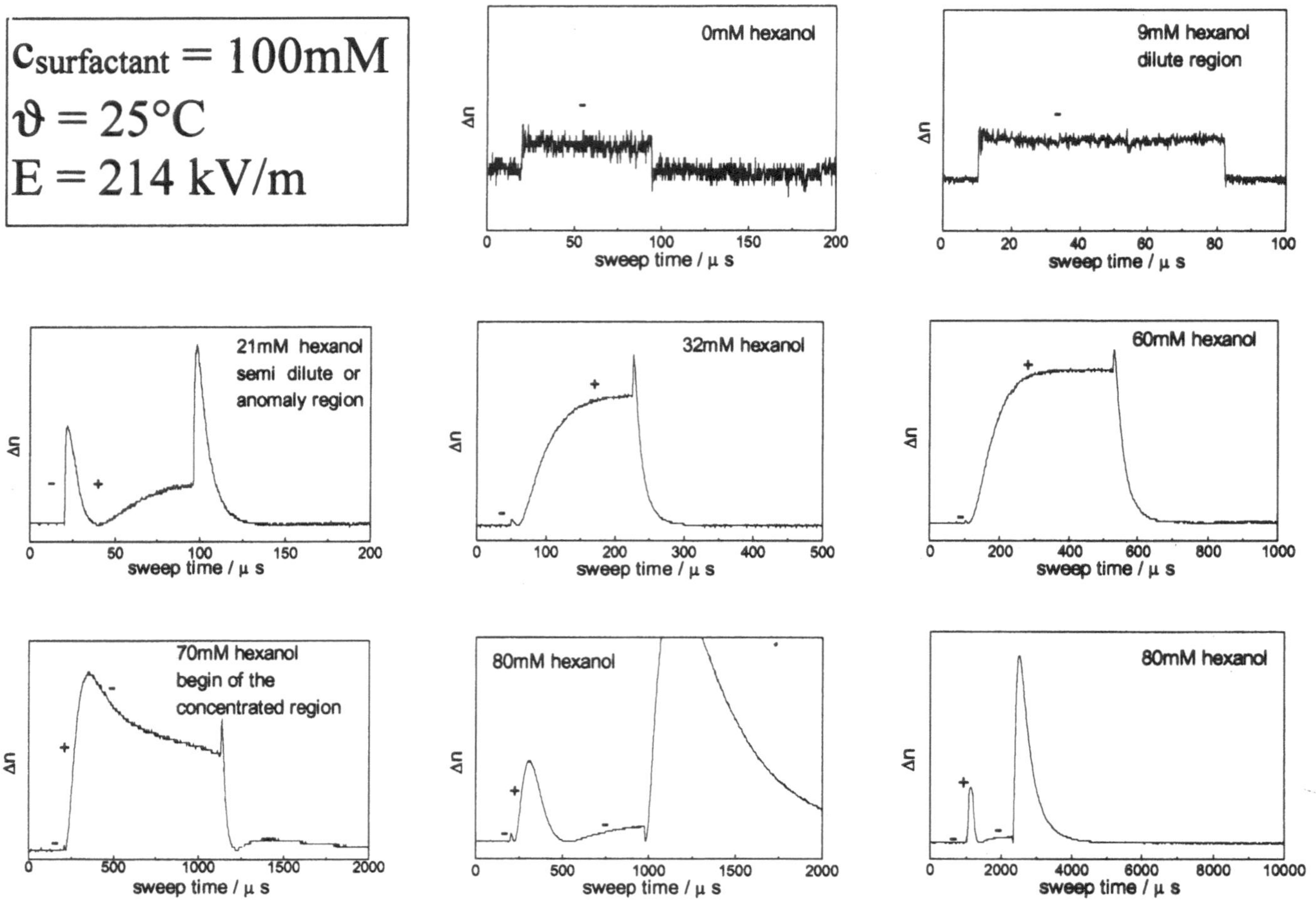

$$c_{surfactant} = 100mM$$
$$\vartheta = 25°C$$
$$E = 214\ kV/m$$

0mM hexanol

Δn

sweep time / μ s

9mM hexanol
dilute region

Δn

sweep time / μ s

21mM hexanol
semi dilute or
anomaly region

Δn

sweep time / μ s

32mM hexanol

Δn

sweep time / μ s

60mM hexanol

Δn

sweep time / μ s

70mM hexanol
begin of the
concentrated region

Δn

sweep time / μ s

80mM hexanol

Δn

sweep time / μ s

80mM hexanol

Δn

sweep time / μ s

Fig. 6 Electric birefringence signals of the system C_{14}DMAO:SDS = 6, 2:3, 8 with increasing hexanol concentration. The temperature was 25 °C, the surfactant concentration 100 mM and the electric field strength 214 kV/m

to an electric field. The rods of C_{14}DMAO when charged with HCl shows the anomaly and behaves again like the rods which are charged with SDS. The sign of the charge is thus not the reason for the different behavior. We have to find therefore a different explanation of this subtle phenomenon. An answer to the problem may be in the local distribution of the charges on the mixed micelles. The interfacial tension results indicate a strong mixing for SDS and alkylaminoxide and a tendency to demix for the C_{14}DMAO/C_{14}TMABr combination. It is thus conceivable that there is local charge separation on the mixed micelles for the latter system and the intermicellar interactions between the charged rods could thus be different for the two systems.

The time constants that have been evaluated for the different process are given in Fig. 7. They show the same pattern as the systems when the surfactant concentration was varied.

Fig. 7 The relaxation times of the four different relaxation processes which where observed in the 100 mM system C_{14}DMAO:SDS = 6, 2:3, 8 with increasing hexanol concentration

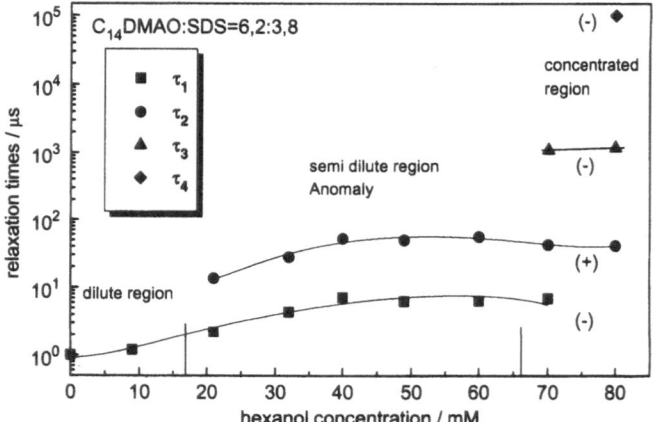

Progr Colloid Polym Sci (1994) 97:103–109
© Steinkopff-Verlag 1994

Conclusions

The surface activity and the phase behavior of micellar solutions of the zwitterionic surfactant $C_{14}DMAO$ with SDS and with the cationic surfactants $C_{14}TMABr$ and $C_{14}DMAOH^+Cl^-$ was studied. A large synergistic behavior is observed in the properties of the $C_{14}DMAO/SDS$ combination. Mixtures of the two surfactants are more surface active than the two components on their own. With increasing mole fractions of the anionic surfactant X_a a deep minimum is observed for the surface tension and the interfacial tension. Mixtures of the two surfactants form a nematic N_c–phase over a large mixing ratio. The N_c–phase extends far to the water corner and begins already at a concentration of 10% by weight of surfactant. The hexagonal phase also begins at smaller concentrations than in the binary systems. Mixtures of the surfactants are highly viscoelastic in the L_1–phase. The viscosity passes over a maximum with increasing X_a. Micellar solutions with small rodlike micelles show the electric birefringence anomaly at the overlap concentration. The synergism of the two surfactants is still strong in the presence of the cosurfactant hexanol.

Mixtures of $C_{14}DMAO$ with the cationic surfactant $C_{14}TMABr$ show an antisynergistic behavior. The surface and interface tension increase rapidly with increasing mole fractions of the cationic surfactant X_c. In the bulk solutions the formation of rodlike micelles is suppressed with X_c. As for mixtures with SDS, the rods grow with the addition of the cosurfactant. At the overlap concentration such micellar solutions do not show the electric birefringence anomaly.

Micelles which are charged by adding HCl to the $C_{14}DMAO$ solutions behave similar as when the micelles are charged with SDS. The electric birefringence anomaly can be observed again. It is likely that the synergistic and antisynergistic behavior has its cause in the local distribution of the electric charges on the micelles.

References

1. Clint JH (1990) The Structure, Dynamics and Equilibr. Properties of Coll Systems 71–84
2. a. Bucci S, Fagotti C, Degeorgio V, Piazza R (1991) langmuir 7; b. Chang CH, Wang N-HL, Franses EI (1984) Colloids and Surfaces 62:321–332; c. Holland PM (1984) ACS Symp 253; d. Guering P, Nilsson PG, Lindman B (1985) J Colloid Interf Sci 105; e. Esumi J, Sakamoto Y, Meguro K (1990) Colloid Interf Sci 134, No. 1
3. Rosen MJ (1991) Langmuir 7
4. Saul D, Tiddy GJT, Wheeler BA, Wheeler PA, Willis E (1974) J Chem Soc Faraday Trans I 70:169
5. Tamori K, Esumi K, Meguro K, Hoffmann H (1991) J Colloid Interf Sci 147:33
6. Weers JG, Rathman JF, Scheuing DR (1990) Colloid Polym Sci 268, 832
7. Hofmann S, Rauscher A, Hoffmann H (1991) Ber Bunsenges Phys Chem 95, 135
8. Iwasaki T, Ogawa M, Esumi K, Meguro K (1991) Langmuir 7
9. Kekicheff P, Gabrielle G-Mandelmont (1989) J Colloid Interf Sci 131 No. 1
10. Hoffmann H, Oetter G, Schwandner B (1987) Progr Colloid Polym Sci 73, 95–106
11. Hoffmann H, Thunig C, Valiente M (1992) Colloids and Surfaces 67, 223–237
12. Imae T, Konishi H, Ikeda S (1986) J Phys Chem 90, 1417
13. Haegel FH, Dissertation, Bayreuth (1987)
14. Kästner U, Hoffmann H, Dönges R, Ehrler R (1993) given lecture Lund
15. Pilsl H, Hoffmann H, Hofmann S, Kalus J, Kencono AW, Lindner P, Ulbricht W (1993) J Phys Chem 97, 2745–2754
16. Hoffmann H (1993) given lecture ACS Chicago
17. MacKintosh FC, Safran SA (1990) Europhys Lett
18. Angel M, Hoffmann H, Krämer U, Thurn H (1989) Ber Bungsenges Phys Chem 93, 184–191
19. Cates ME, Marques CM, Bouchaud J.-P (1991) J Chem Phys 94, 8529–8536
20. Hoffmann H, Krämer U (1990) The Structure, Dynamics and Equilibr. Properties of Coll Systems, 385–396

Progr Colloid Polym Sci (1994) 97:110–115
© Steinkopff-Verlag 1994

SURFACTANTS

Mixtures of gelling agarose with non-ionic surfactants or block-copolymers: Clouding and diffusion properties

M.H.G.M. Penders
S. Nilsson
L. Piculell
B. Lindman

Received: 16 September 1993
Accepted: 21 March 1994

M.H.G.M. Penders (✉) · L. Piculell
B. Lindman
Physical Chemistry 1
Chemical Center
University of Lund
Box 124
22100 Lund, Sweden

S. Nilsson
Rogaland Research
Prof. Olav Hanssensväg 15
Box 2503
Ullandhaug
4004
Stavanger, Norway

Abstract The clouding and diffusion behavior of nonionic micellar systems of dodecyl hexaoxyethylene ($C_{12}E_6$), dodecyl octaoxyethylene ($C_{12}E_8$) glycol monoethers and a triblock copolymer of composition $E_{13}PO_{30}EO_{13}$ (PE6400) have been investigated in agarose gels and solutions with and without sodium thiocyanate. In the presence of agarose the clouding temperature of the nonionic surfactant decreased on cooling and a hysteresis behavior was observed. However, the gelation temperature of agarose remained practically unchanged on the addition of surfactant; also, the diffusion of the surfactant was reduced due to obstruction caused by the polymer.

Key words Surfactant – agarose – diffusion – block-copolymer – micelles

Introduction

Recently, studies of micellar diffusion in physical (polymer) gels have been of interest [1–3]. In this work the study of *non-ionic clouding* surfactants in gels and solutions of agarose, a *non-ionic gelling* polysaccharide, without and with sodium thiocyanate (NaSCN), is reported. We have focused on the clouding and diffusion properties.

Agarose in water forms a gel, composed of aggregated double-helices, at temperatures below 41 °C, on cooling. On heating, however, the gel melts at about 80 °C, presumably because of stabilization of the helices through aggregation [4–6]. The large thermal hysteresis in the gel-solution transition facilitates comparison of mixtures in the solution with mixtures in the gel at the same temperature. As non-ionic *clouding* surfactants, we have used hexaoxyethylene ($C_{12}E_6$) and octaoxyethylene ($C_{12}E_8$) glycol mono (n-dodecyl) ethers and a poly[ethylene oxide]-poly[propylene oxide]-poly[ethylene oxide] (PEO-PPO-PEO) block-copolymer, PE6400 ($EO_{13}PO_{30}EO_{13}$). Triblock copolymers of the PEO-PPO-PEO type, commercially known as Pluronics®, are important non-ionic surface active agents (see, e.g., [7–24]). At higher temperatures the *micellar* state is predominant, whereas at lower temperature the *monomeric* state prevails. The critical micellization temperature (CMT) for PE6400 decreases with increasing concentration. At elevated temperatures non-ionic surfactants containing ethylene oxide (EO) segments (e.g., $C_{12}E_6$, $C_{12}E_8$ and PE6400), display a clouding behavior in water. The temperature at which the clouding takes place (or cloud-point) is affected by the presence of salts [16, 18, 22]. Addition of NaSCN increases the solubility of non-ionic surfactants [2, 3, 18, 22] and also destabilizes agarose gels [6, 25] (salting-in effect). Addition of NaCl has the opposite effect (salting-out behavior). The influence of NaSCN (1.0 M) on the properties of agarose/water and non-ionic surfactant/water systems is discussed in more detail in Results and Discussion section.

By choosing the combination of a *gelling* polymer and a *clouding* surfactant in water, it is possible to probe both the influence of agarose on the clouding behavior of the surfactant and the effect of surfactant on the gelation behavior of agarose with light transmittance measurements. With the FT-PGSE ^{1}H NMR technique the transport properties (self-diffusion) of micellar systems in agarose gels and solutions vs. temperature were measured, which also provides information about the polymer network (discussed later).

Experimental

Materials

Agarose (type VIII, for isoelectric focusing, No. A-4905) was obtained from Sigma (St Louis, Missouri, USA) and used without further purification. NaSCN was of analytical grade. The agarose solutions were prepared by dissolving agarose in the appropriate solvent (water or salt solution) in sealed glass tubes, which were heated in boiling water with occasional shaking.

$C_{12}E_6$ and $C_{12}E_8$ were purchased from Nikko Chemicals, Tokyo, Japan, and Pluronic® PE6400 (molecular weight 3000, wt% PEO = 40) was obtained from BASF Aktiengesellschaft. All surfactants were used without further purification.

For the preparation of the samples, Millipore water was used in the case of the light transmittance measurements and D_2O (99.8% purity, supplied by Merck or Dr. Glaser AG Basel) in the case of the NMR self-diffusion studies. All solutions were prepared by weight.

Methods

Light transmittance measurements versus temperature, using a cooling (or heating) rate of 0.3 °C/min, were performed with a 5 cm path length cell in a Hitachi Perkin-Elmer (Model 124) double-beam spectrophotometer. The temperature was controlled by the circulation of thermostatically regulated water through the jacketed cell. From the transmittance vs. temperature curves the cloud-points of the surfactant on heating and cooling as well as the gel and melting temperatures of agarose were determined [2, 3].

^{1}*H NMR self-diffusion* measurements were carried out on a JEOL FX-60 spectrometer, operating at 60 MHz, using the FT-PGSE technique, as described in more detail by Stilbs [26]. With this technique one uses a 90°-τ-180°-τ-echo pulse sequence, with two added rectangular magnetic field gradient pulses of magnitude G, separation time

Δ and duration time δ. The echo amplitude at time 2τ is given by [27]

$$A(2\tau) = A(0)\exp[-2\tau/T_2 - \gamma^2 G^2 D\delta^2(\Delta - \delta/3)], \qquad (1)$$

where T_2 is the transverse relaxation time, and γ the magnetogyric ratio for the proton. The self-diffusion coefficients D were determined by measuring the echo amplitude A as a function of δ, keeping G and Δ fixed. For all the experiments $\Delta = 140$ ms and $G = 16.7$ mT/m or 40.0 mT/m, depending on the size of the diffusion coefficient. The temperature control during the experiments was within 0.5 °C.

Results and discussion

Influence of NaSCN (1.0 M) on the properties of agarose/water and non-ionic surfactant/water systems

In Table 1 and Figs. 1 and 2 the effect of NaSCN (1.0 M) on the behavior of agarose/water and (non-ionic surfactant)/water systems is presented. Results concerning the gelation of agarose and the clouding of surfactants following from light-transmittance measurements are shown in Table 1. In Figs. 1 and 2 the results of FT-PGSE ^{1}H NMR measurements concerning the self-diffusion of non-ionic surfactants vs. temperature are given. From Table 1 it follows that the gel point (T_g) of agarose on cooling is lowered drastically in the presence of 1.0 M NaSCN from 41.0 to 20.5 °C (On addition of non-ionic surfactants ($C_{12}E_6$, $C_{12}E_8$ and PE6400) T_g remains practically unchanged [2, 3]). The decrease in T_g is in accordance with results published before [6, 25] where it was shown that the gelling ability of agarose is weakened by the addition of NaSCN caused by the adsorption of SCN^- ions to the polymer [6]. Also the gel melting point (T_m) of agarose is lowered on addition of NaSCN.

Table 1 Influence of NaSCN (1.0 M) on the gelation of agarose (1 wt %) and on clouding of non-ionic surfactants (1 wt %). The symbols are explained in the text.

System		T_g/°C	T_m/°C	T_{cl}/°C
Agarose	in water	41.0	76.8	
	in 1.0 M NaSCN	20.5	65.3	
$C_{12}E_8$	in water			78.9
	in 1.0 M NaSCN			> 100
$C_{12}E_6$	in water			51.8
	in 1.0 M NaSCN			69.9
PE6400	in water			55–57
	in 1.0 M NaSCN			70–71

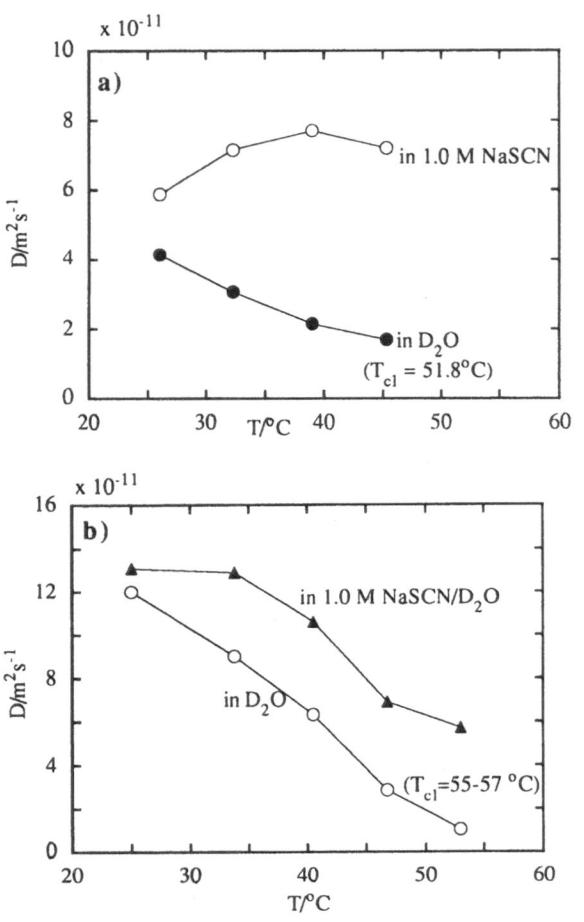

Fig. 1 Self-diffusion coefficients vs. temperature of $C_{12}E_6$ (1 wt %) and PE6400 (3 wt %) in D_2O without and with salt (1.0 M NaSCN). T_{cl} represents the cloud-point of the surfactant/water system. a) $C_{12}E_6$ b) PE6400

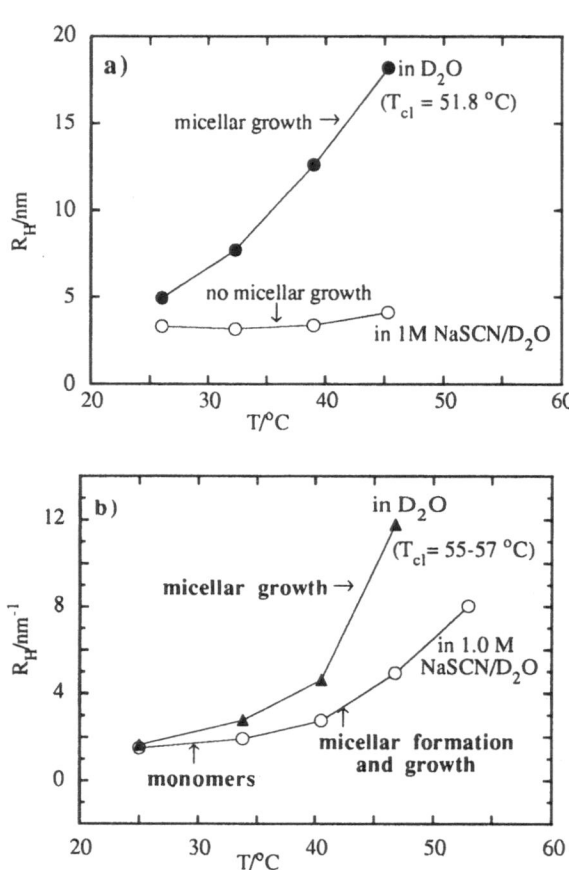

Fig. 2 Hydrodynamic radius R_H versus temperature for $C_{12}E_6$ and PE6400 in D_2O without and with added salt (1.0 M NaSCN). T_{cl} represents the cloud-point of the surfactant/water system. a) $C_{12}E_6$ b) PE6400

The presence of NaSCN (1.0 M) raises the clouding temperature (T_{cl}) of $C_{12}E_8$, $C_{12}E_6$ and Pluronic PE6400 ($EO_{13}PO_{30}EO_{13}$) and thus increases the hydrophilicity of the surfactant (see Table 1). The rise in cloud-point of the PE6400 system from 55°–57° to 70°–71 °C is in accordance with the results shown by Pandya et al. [18]. The "salting-in" behavior of NaSCN has also been observed in aqueous ethyl hydroxyethyl cellulose (EHEC)/NaSCN systems [28]. Addition of NaSCN (1.0 M) to EHEC (1 wt %) gives an increase in cloud-point from circa 65° to 75 °C.

From Fig. 1 it can be seen that on addition of NaSCN (1.0 M) the self-diffusion coefficients of $C_{12}E_6$ (1 wt %) and PE6400 (3 wt %) are increased compared to the salt-free case. This effect is enhanced at higher temperatures. In the case of $C_{12}E_8$, however, there is no significant increase in self-diffusion coefficient on addition of NaSCN [2]. It is

seen in Fig. 2 that on addition of NaSCN (1.0 M) the hydrodynamic radii R_H, using the Stokes–Einstein relation for spheres, for $C_{12}E_6$ and PE6400 are decreased compared to the salt-free case at temperatures between 30° and 45 °C. In the salt-free case a strong increase in R_H with increasing temperature is observed in this temperature range. Apparently, due to the presence of NaSCN (1.0 M) the growth of the $C_{12}E_6$ micelles is suppressed on increasing the temperature (Fig. 2a), which is caused by a slight enrichment of SCN^- ions at the micellar surface. The addition of NaSCN also reduces the tendency of formation and growth of the PE6400 micelles on increasing the temperature from 30° to 45 °C (Fig. 2b). In the case of $C_{12}E_8$ no marked increase in R_H of the micelles (no micellar growth) at increasing temperature below 50 °C is found, as expected [2, 29, 30]. The R_H values for $C_{12}E_8$ micelles stay practically unchanged in the presence of NaSCN (1.0 M).

Progr Colloid Polym Sci (1994) 97:110–115
© Steinkopff-Verlag 1994

Clouding and diffusion of non-ionic surfactants in the presence of agarose

In the presence of agarose coils the cloud-point of the non-ionic surfactant on cooling (T_{cc}) is decreased due to an "incompatibility" effect (repulsive *coil*-micelle interactions). This is demonstrated in Table 2 for $C_{12}E_8$, $C_{12}E_6$ and PE6400. The decrease in T_{cc} is enhanced at higher agarose/and or surfactant concentrations [2, 3]. A similar result has been found by Sjöberg et al. [31–34] who observed a depression of the cloud-point for poly(ethylene glycol) solutions on the addition of low molecular weight saccharides or dextran.

The observed decrease in cloud-point for non-ionic surfactants on cooling in the presence of agarose coils may be interpreted in terms of energy and/or entropic contributions. In the former case, short-range pair interactions between sugar units of agarose, surfactant headgroup, and water molecules play an important role. In the latter case, the decrease in cloud-point can be explained by the fact that the polymer segment density decreases near the surface of the micelle, due to the loss in configurational entropy experienced by a polymer close to a surface. This "depletion" gives rise to a net attraction between the micellar particles.

From Table 2 it follows that in the presence of agarose (1 wt %), T_{cc} is lowered to 48–49 °C in the case of PE6400 (1 wt %) and to a value < 42 °C for $C_{12}E_6$ (1 wt %). The agarose/PE6400 system in water is difficult to handle and for the agarose/$C_{12}E_6$ system there is no temperature region where it is homogeneous, since at temperatures below 42 °C the agarose starts to gel. The miscibility of both the agarose/PE6400 and the agarose/$C_{12}E_6$ couple, however, can be improved on addition of NaSCN which according to the findings of Hofmeister [35, 36] (lyotropic series) displays a "salting in" effect.

The cloud-point of $C_{12}E_8$, $C_{12}E_6$ and PE6400 in agarose *gels* on heating (T_{ch}) are situated at a higher temperature than T_{cc} for the surfactants in agarose *solutions* [2, 3]. The difference between T_{ch} and T_{cc} for the surfactants is enhanced at higher surfactant concentrations. Evidently, the interaction between agarose *coil* and micelles is more repulsive than the *gel*-micelles interaction,

since in the latter case the agarose *gel* network leaves more space available for the micelles than in the *coil* state. In the case of $C_{12}E_8$ and $C_{12}E_6$/1.0 M NaSCN [2] there is a slight difference between T_{ch} and the cloud-point of the surfactant in absence of agarose, whereas in the case of PE6400/1.0 M NaSCN these two temperatures are approximately the same [3].

Fig. 3 D/D_0 vs. temperature of non-ionic surfactants in agarose (1 wt %)/1.0 M NaSCN gels and solutions. a) $C_{12}E_8$ (1 wt %) b) $C_{12}E_6$ (1 wt %) c) PE6400 (3 wt %)

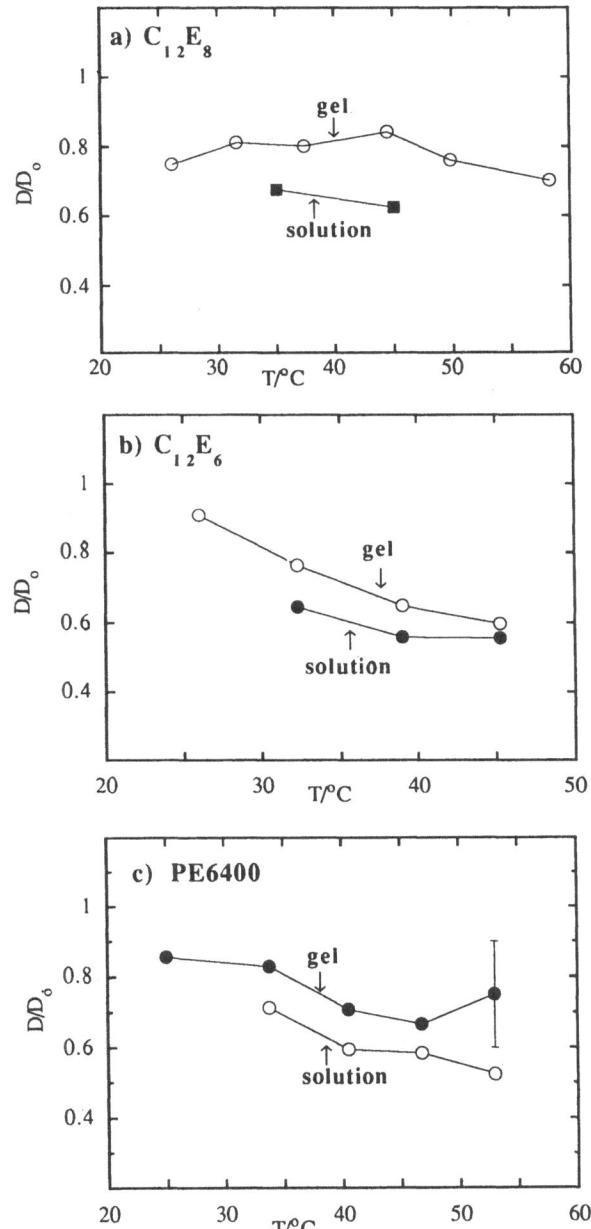

Table 2 Depression of the cloudpoint of the surfactant (1 wt %) on cooling in the presence of agarose (1 wt %)

Surfactant	T_{cc}/°C without agarose	T_{cc}/°C in the presence of agarose
$C_{12}E_8$	78.9	69.7–73.8
$C_{12}E_6$	51.8	< 42
PE6400	55–57	48–49

In Fig. 3 the results of the FT-PGSE NMR measurements concerning the self-diffusion of $C_{12}E_8$, $C_{12}E_6$ and PE6400 in aqueous agarose/1.0 M NaSCN gels and solutions vs. temperature are given. It can be seen that the self-diffusion coefficient D of the non-ionic surfactant in the presence of agarose is smaller than D_0, the self-diffusion coefficient of surfactant in 1.0 M NaSCN/D_2O in the absence of agarose. This obstruction is stronger in the *solution* than in the *gel* state (see Fig. 3). The faster diffusion of the surfactant in the *gel* system may be explained by the increase of distances between the agarose chains in the network and the available volume fraction for the surfactant due to the formation of double helices.

A similar results has been found by Johansson et al. [1, 37–40] in their self-diffusion study of $C_{12}E_8$, PEG, glucose and sucrose in K^+-κ-carrageenan gels and in Na^+-κ-carrageenan solutions at 25 °C.

From Fig. 3a it follows that D/D_0 for $C_{12}E_8$ in agarose gels is approximately constant in the temperature region between 25° and 60 °C. The D/D_0-value (0.75–0.80) turns out to be higher for agarose gels than for K^+-κ-carrageenan gels [1]. Johansson et al. found a D/D_0-value for $C_{12}E_8$ of about 0.60 at 1 wt% K^+-κ-carrageenan gel (mainly coils present) and 0.45 at 3 wt% (mainly helices present). The higher D/D_0-value in the case of agarose might be explained by the fact that the network structure of the aggregated agarose gel is more open (larger mesh-size) in comparison with the less aggregated K^+-κ-carrageenan gel.

In contrast to $C_{12}E_8$, D/D_0 for both $C_{12}E_6$ and PE6400 in agarose gels and solutions decreases at increasing temperature (see Figs. 3b–c). The decrease in D/D_0 is probably due to the fact that at higher temperatures, apart from the obstruction effect, also formation and growth of $C_{12}E_6$ and PE6400 micelles plays an important role.

Concluding remarks

In this paper we have reported clouding and diffusion properties of non-ionic surfactants ($C_{12}E_8$, $C_{12}E_6$ and PE6400) in agarose gels and solutions with and without NaSCN (1.0 M).

It follows that on addition of agarose the clouding temperature of non-ionic surfactants on cooling is decreased due to the incompatibility of the agarose/surfactant couple. In other words, the interaction between agarose and surfactant is repulsive. There is also a hysteresis in clouding of the surfactant in the presence of agarose on cooling and heating indicating that the interaction between agarose *coil* and micelles is more repulsive than the *gel*-micelles interaction. The gelation temperature of agarose on cooling stays practically unchanged on addition of non-ionic surfactants.

The miscibility of the agarose/(non-ionic surfactant) couple in water can be increased on addition of NaSCN. In the latter case, the gelation and melting temperature of agarose is decreased and the clouding-point of the non-ionic surfactants is increased.

The self-diffusion of non-ionic surfactants is retarded due to the presence of agarose. The obstruction effect of agarose is stronger in gels (more open structure) than in solutions. The presence of NaSCN impedes micellar growth on increasing the temperature.

Acknowledgment This work was financially supported by grants from the Swedish Institute and the Wenner-Gren Center Foundation.

References

1. Johansson L, Hedberg P, Löfroth J-E (1993) J Phys Chem 97:747–755
2. Penders MHGM, Nilsson S, Piculell L, Lindman B (1993) J Phys Chem 97:11332–11338
3. Penders MHGM, Nilsson S, Piculell L, Lindman B (1994) J Phys Chem 98:5508–5513
   (1972) J Mol Biol 68:153–172
5. Arnott S, Fulmer A, Scott WE, Dea ICM, Moorhouse R, Rees DA (1974) J Mol Biol 90:269–284
6. Piculell L, Nilsson S (1989) J Phys Chem 93:5596–5601
7. Al-Saden AA, Whateley TL, Florence AT (1982) J Colloid Interface Sci 90:303–309
8. Zhou Z, Chu B (1988) Macromolecules 21:2548–2554
9. Zhou Z, Chu B (1988) J Colloid Interface Sci 126:171–180
10. Tontisakis A, Hilfiker R, Chu B (1990) J Colloid Interface Sci 135:427–434
11. Wu G, Zhou Z, Chu B (1993) Macromolecules 26:2117–2125
12. Wanka G, Hoffmann H, Ulbricht W (1990) Colloid Polym Sci 268:101–117
13. Brown W, Schillén K, Almgren M, Hvidt S, Bahadur P (1991) J Phys Chem 95:1850–1858
14. Almgren M, van Stam J, Lindblad C, Li P, Stilbs P, Bahadur P (1991) J Phys Chem 95:5677–5684
15. Almgren M, Bahadur P, Jansson M, Li P, Brown W, Bahadur A (1992) J Colloid Interface Sci 151:157–165
16. Bahadur P, Li P, Almgren M, Brown W (1992) Langmuir 8:1903–1907
17. Bahadur P, Pandya K (1992) Langmuir 8:2666–2670
18. Pandya K, Lad K, Bahadur P (1993) J Macromol Sci – Pure Appl Chem A 30:1–18
19. Pandya K, Bahadur P, Nagar TN, Bahadur A (1993) Colloids and Surfaces A 70:219–227
20. Tiberg F, Malmsten M, Linse P, Lindman B (1991) Langmuir 7:2723–2730
21. Linse P, Malmsten M (1992) Macromolecules 25:5434–5439

22. Malmsten M, Lindman B (1992) Macromolecules 25:5440–5445
23. Malmsten M, Lindman B (1993) Macromolecules 26:1282–1286
24. Mortensen K, Pedersen JS (1993) Macromolecules 26:805–812
25. Watase M, Nishinari K (1989) Carbohydr Pol 11:55–66
26. Stilbs P (1987) Prog NMR Spectrosc 19:1–45
27. Stejskal EO, Tanner JE (1965) J Chem Phys 42:288–292
28. Karlström G, Carlsson A, Lindman B (1990) J Phys Chem 94:5005–5015
29. Nilsson P-G, Wennerström H, Lindman B (1983) J Phys Chem 87:1377–1385
30. Lindman B, Wennerström H (1991) J Phys Chem 95:6053–6054
31. Gustafsson Å, Wennerström H, Tjerneld F (1986) Polymer 27:1768–1770
32. Gustafsson Å, Wennerström H, Tjerneld F (1986) Fluid Phase Eq 29:365–371
33. Sjöberg Å, Karlström G (1989) Macromolecules 22:1325–1330
34. Sjöberg Å, Karlström G, Tjerneld F (1989) Macromolecules 22:4512–4516
35. Hofmeister F (1888) in: Naunyn-Schmiedebergs Archiv für Experimentelle Pathologie und Pharmakologie (Leipzig) 24:247–260
36. Collins KD, Washabaugh MW (1985) Quart Rev Biophys 18:323–422
37. Johansson L, Löfroth J-E (1991) J Colloid Interface Sci 142:116–120
38. Johansson L, Skantze U, Löfroth J-E (1991) Macromolecules 24:6019–6023
39. Johansson L, Elvingson C, Löfroth J-E (1991) Macromolecules 24:6024–6029
40. Johansson L, Löfroth J-E (1993) J Chem Phys 98:7471–7479

Progr Colloid Polym Sci (1994) 97:116–120
© Steinkopff-Verlag 1994

SURFACTANTS

K. Lunkenheimer
H.-R. Holzbauer
R. Hirte

Novel results on adsorption properties of definite n-alkyl oxypropylene oligomers at the air/water interface

Received: 1 October 1993
Accepted: 31 March 1994

Dr. K. Lunkenheimer (✉)
Max-Planck-Institut für
Kolloid- und Grenzflächenforschung
Rudower Chaussee 5
12489 Berlin-Aldershof, FRG

Dr. H. Holzbauer
Zentrum für Makromolekulare Chemie
Rudower Chaussee 5
12489 Berlin-Adlershof, FRG

Prof. R. Hirte
Technische Fachhochschule Wildau
Friedrich-Engels Straße 63
15742 Wildau, FRG

Abstract Adsorption properties at the air/water interface obtained with definite *n*-octyl-oligo oxypropylene ethers were investigated. Equilibrium surface tension-concentration isotherms at 295 K reveal a transition region, separating two surface regions with different molecular arrangement. Consequences for theoretical considerations on adsorption phenomena as well as for practical applications of nonionic surfactants are discussed.

Key words *n*-octyl oligo-oxypropylene ethers – air/water interface – adsorption properties – surface equation of state – alternation (even/odd) effects

Introduction

Certain nonionic manufacturers have found that a combination of propyleneoxide with ethyleneoxide gives enhanced detergency [1, 2]. A new and interesting series of materials, the Pluronics, have been developed. By varying the amounts of propylene and ethylene oxide, the hydrophobe/hydrophile balance can be adjusted to any particular requirement.

Nonionic oxypropylene surfactants are now advantageously used in various fields of application because of low toxicity and easy biodegradability. Adsorption phenomena at the solid and liquid interfaces and dependence on the oxypropylene chain length with respect to applications properties have been investigated [3–6]. However, there are only very few quantitative results on this type of nonionic amphiphile [7, 8] as they are usually available as oxyethylene/oxypropylene copolymers with a statistically distributed number of EO/PO units per molecule.

We have synthesized definite *n*-octyl oligooxypropylene ethers possessing a purity of 98–99%. We report on novel results of their adsorption properties at the air/water interface.

Experimental

Substances

The oxypropylene oligomers were obtained directly from the reaction of *n*-octanol with propylene oxide using 1% NaOH as catalyst at 283–403 K until an average degree of propoxylation of six was reached. The individual oligomers were obtained from the reaction mixture by fractional distillation. Repeatedly implemented fractionations using an automatically operating high performance distillation apparatus from Fischer (FRG), were required to obtain the individual *n*-octyl oligooxypropylenes with

Progr Colloid Polym Sci (1994) 97:116–120
© Steinkopff-Verlag 1994

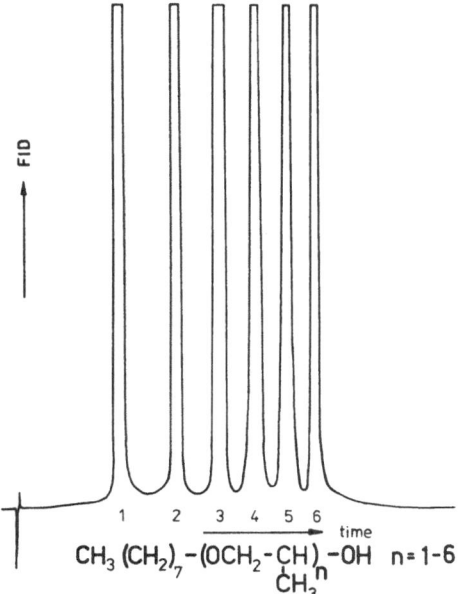

Fig. 1 Gas chromatogram of a mixture of six definite oligomers ($n = 1 - 6$) of n-octyl oligooxypropylene ethers $C_8(PO)_n$ (FID-signal of flame ionization detector)

a purity of $\geq 98\%$ according to gas chromatografic analyses (cf. Fig. 1).

Methods

Surface-active trace impurities were removed from aqueous stock solutions until the necessary grade of surface-chemical purity was reached by applying a peculiar purification device (see below [9]).

Surface tension of the aqueous solutions was determined by using a Lauda ring tensiometer taking into consideration modifications necessary for the measurement of surfactant solutions [10, 11].

Philosophy of Investigations

Today, scientists dealing with amhiphile properties are fond of phenomena such as molecular architecture, self-organization, fractals, supramolecular chemistry, more generally speaking, by terms like "structure" and "interaction" [12, 13]. Our contribution deals with the very basic thermodynamic properties of adsorption. As one can see, new quantitative information on the molecular behavior in the adsorption layer could be gained. However, in the long run it was obtained from surface tension measurements "only".

Certain requirements have to be followed strictly in order to obtain reliable information of the amphiphiles' adsorption properties at fluid interfaces. These can be summarized as follows:

i) When applying the ring method to solutions of amphiphiles three interfacial phenomena caused by frictional effects at the (hydrophilic) vessel wall, by straining effects in the adsorption layer [10], and by wetting defects at the ring surface [11] may give rise to serious errors. These errors can be avoided by applying appropriate measures [10, 11].

ii) Investigations on interfacial properties of amphiphilic compounds necessitate a special grade of purity to escape the false effects of stronger surface-active contaminants. Due to their physico-chemical peculiarity amphiphiles usually still contain traces of the stronger surface-active parent compounds, which, being present in the bulk phase in negligible concentrations, are extraordinarily enriched in the interface. Thus, they have an enormous falsifying influence on the properties of the adsorption layer only. Conventional chemical purification techniques are generally not suitable for removing such impurities.

Hence, in addition to the above-mentioned chemical purification procedure, we applied an automatically operating purification system, developed by us recently [9] in connection with a convenient criterion [14] to guarantee absence of impurities at the surface also. The principle of this purification procedure consists of repeatedly implemented operations of compressing, sucking off and dilating the adsorption layer.

iii) According to [15] equilibrium surface tension (σ_e)-concentration (c) isotherms of soluble amphiphiles can be described by surface equations of state for "ideal" and/or "regular" surface behavior. Ideal behavior is described by Langmuir's (Eq. (1)) and/or Szyszkowski's (Eq. (2)) equation, regular behavior by Frumkin's equation (3), according to

$$\Gamma = \Gamma_\infty \frac{c}{(a + c)}, \tag{1}$$

$$\sigma_0 - \sigma_e = RT\Gamma_\infty \ln(1 + c/a), \tag{2}$$

$$\sigma_0 - \sigma_e = -RT\Gamma_\infty \ln(1 - \Gamma/\Gamma_\infty) - a'(\Gamma/\Gamma_\infty)^2. \tag{3}$$

Here, σ_0 and σ_e stand for the surface tension of the pure solvent and the equilibrium surface tension of surfactant solution, respectively, Γ and Γ_∞ for the surface concentration and saturation surface concentration, a for the surface activity parameter or the bulk-surface distribution coefficient, and a' for the surface interaction parameter which, according to [15] can be correlated with the partial molar free energy of surface mixing of surfactant and solvent at infinite dilution H^s by

$$a' = \Gamma_\infty H^s. \tag{4}$$

At high dilution's Eqs. (1) to (3) result in Henry's equation:

$$\sigma_0 - \sigma_e = RT\Gamma. \tag{5}$$

Recently, we have shown [16] that the σ_e; c-isotherms of surfactants can be well described within part of the adsorption interval by the above equations. However, they usually fail to satisfactorily describe the entire concentration interval. Therefore, we have put forward the concept of considering the adsorption layer as consisting of two energetically discriminatable species of the same chemical individual. These two species were found to occur separately at lower and/or higher bulk concentrations, but occur as a surface mixture across a rather small transition interval at medium concentrations. The transition is described by a transition function α in the following way

$$\sigma_e = \alpha(\sigma_e^{I}) + (1 - \alpha)(\sigma_e^{II}). \tag{6}$$

σ_e^{I} and σ_e^{II} represent the two different state functions. Hitherto, we succeeded in describing various homologous series of amphiphiles thermodynamically reasonably with high precision in agreement with the molecules' geometry.

Doing so, we have always found that the region of lowest concentrations had to be described by a Henry equation which differs from that resulting from Eqs. (2) and (3) at lower concentrations. Thus, we could quantitatively discriminate the two adsorption states by a difference in their standard enthalpy of adsorption.

Results

Figure 2 represents the equilibrium surface tension–concentration isotherms of the *n*-octyl oxypropylene oligomers at 295 K. As may be suggested at first sight,

certain irregularities in the adsorption properties are obviously to be expected. Thus, for example, the shift of the isotherms as dependent on the number of the oxypropylene units $(PO)_i$ seems to occur rather arbitrarily in comparison with that observed with homologous series of surfactants.

The theoretical evaluation of these isomers revealed another peculiarity unknown so far. Attempts to fit the isotherms by our above approach to a surface equation of state were satisfactory for the species $(PO)_1$ only. The higher the number of PO units, the less satisfactory was the best fit. Hence, we tentatively matched the isotherms by a modification of the surface equation suggesting that the first state of adsorption layer may extend well beyond the Henry region. In terms of the surface parameters, this then resulted in the following preconditions:

$$c^{I} \leq c^{II}, \tag{7}$$

$$\Gamma_{\infty}^{I} \leq \Gamma_{\infty}^{II}, \tag{8}$$

$$a^{I} < a^{II}. \tag{9}$$

Equations (7) to (9) imply that there are two discriminatable states of the adsorbed species in the surface, each of which can be described by a separate Langmuir-Szyszkowski isotherm. However, the results of the best-fits revealed that these adsorption states exist separately at the margins of the isotherm only, i.e., at either very low or very high concentrations.

This is illustrated by Fig. 3 for $C_8(PO)_4$. The dotted lines denote the courses of the isotherm which were obtained by extrapolation from the adsorption parameters of

Fig. 3 Best-fit of the surface pressure-concentration isotherm of aqueous solutions of *n*-octyl tetraoxypropylene ether as obtained from the model of surface mixture. Dotted lines refer to the curves extrapolated from the single surface states at either very low (I) or very high (II) concentrations. $\alpha(---)$ represents the curve of the transition function

Fig. 2 Equilibrium surface tension-concentration isotherms of aqueous solutions of *n*-octyl oligooxypropylene ethers at 295 K

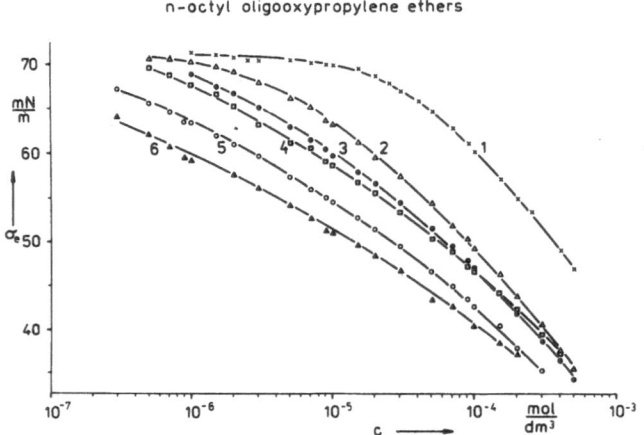

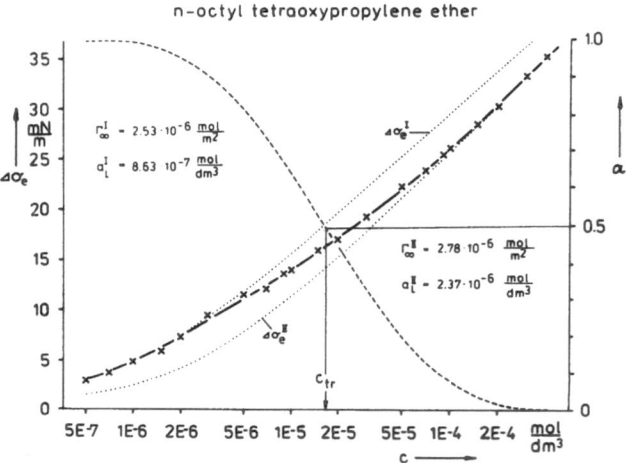

each single adsorption state concerned. It becomes quite evident that most of the adsorption has to be described by a surface mixture of two energetically discriminitable species of the same chemical individual. α denotes the relative portion of species I of the surface excess. The concentration referring to $\alpha = 0.5$ is called concentration of transition (C_{tr}).

Opposite to typical surfactant systems where c_{tr} occurs at comparatively low surface pressures of a few mN/m only, the transition concentrations of the $C_8(PO)_n$

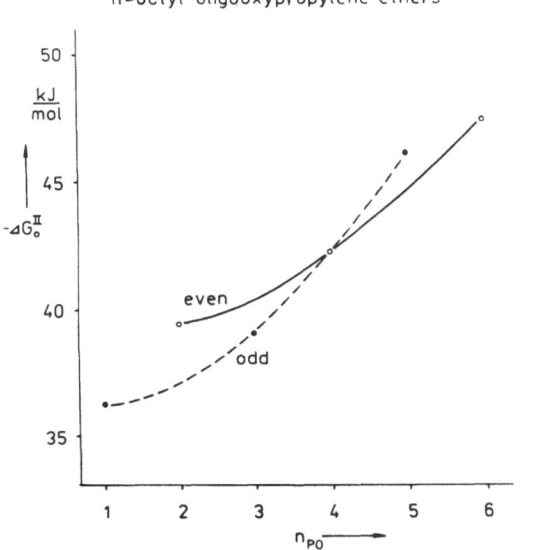

Fig. 4 Standard free energy of adsorption of ΔG_0^{II} of surface state II in dependence on the number n of oxypropylene units (PO)

Fig. 5 Limiting surface area per molecule adsorbed A_{min}, belonging to surface state II in dependence on the number n of oxypropylene units

oligomers are related to rather high surface pressures of 15 to 25 mN N/m. With increasing number of PO units c_{tr} occurs at increasing surface pressures.

The following figures illustrate another interesting feature of the adsorption parameters. Figure 4 shows the standard free energy of adsorption ΔG_0^{II} which is obtained from the surface activity parameter a^{II} by

$$\Delta G_0^{II} = RT \ln a, \tag{10}$$

according to [15]. ΔG_0^{II} increases with increasing n (PO). However, there is a distinct difference between the odd and the even numbered oligomers. This phenomenon is also observed for the limiting surface area demand per molecule A_{min}, as obtained from $A_{min} = (\Gamma_\infty^{II} \times N_L)^{-1}$ (Fig. 5). It is even more pronounced when the difference between the standard free energy of adsorption for the two surface states concerned (Fig. 6) and c_{tr} (Fig. 7) are plotted against the oxypropylene chain number.

Discussion

As one can see, two novel features in the adsorption properties of soluble amphiphiles are found. Firstly, there is the adsorption of a single amphiphilic compound appearing as a surface mixture within almost the entire concentration interval. As shown in [16] the corresponding differences in the adsorption enthalpies can reasonably be attributed to different molecular arrangements in the adsorption layer.

Secondly, there is a distinct effect of alternation with respect to the number of oxypropylene units per molecule in all adsorption parameters. This effect seems to increase with increasing n.

Fig. 6 Differences in standard free enthalpies of adsorption of surface states I and II in dependence on PO-number. (The arrow refers to the value obtained for n-octanol.)

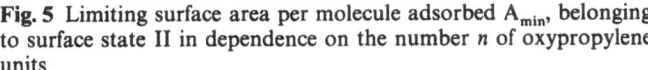

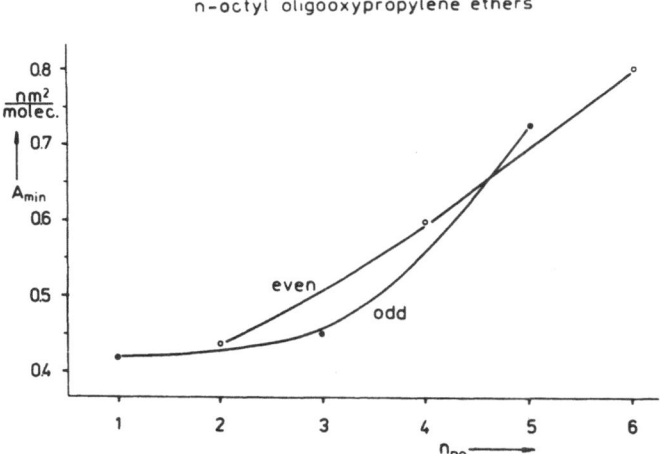

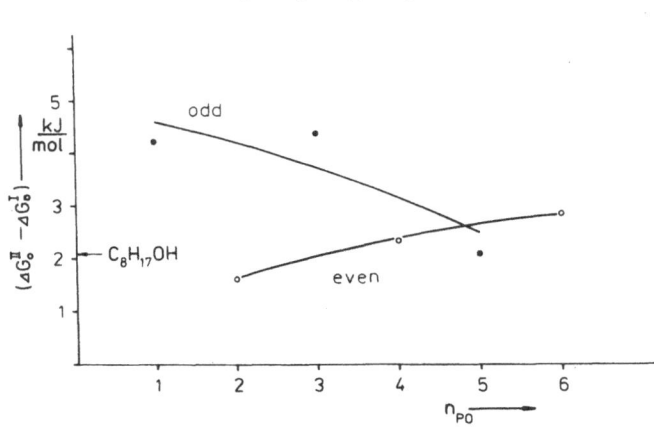

120

K. Lunkenheimer et al.
Adsorption properties of definite *n*-alkyl oxypropylene oligomers

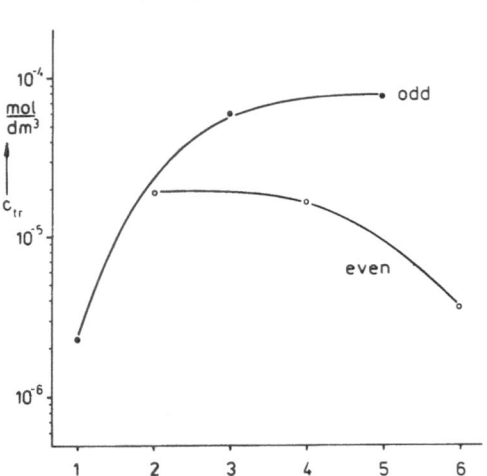

Fig. 7 Concentration of transition c_{tr} in dependence on PO-number

Introducing this concept to the oxypropylene oligomers of this investigation, we have to take into consideration the possibility of hydrogen bonding at the terminal hydroxyl group and at the oxygen atoms of the oxypropylene ether groups. Doing so, one can imagine complex alternating effects across the molecules such that one segment of it may experience an amplification, whereas another one may undergo a quenching of this effect depending on the PO-number. With increasing PO-number molecular segments undergoing effects of complete and/or partial quenching (amplification) can formally be established by this concept. If such an effect is true it will necessarily be reflected in the adsorption properties of the compounds concerned. Thus, the unusual phenomena of alternation can be understood at least qualitatively.

Phenomena of alternation in the adsorption parameters of various homologous series of soluble amphiphiles were detected only recently by us [17–19]. So far, there is no theory by which these phenomena could satisfactorily be explained.

However, Gutmann's first rule of donor-acceptor interaction may serve as a suitable hypothesis for understanding [20].

According to this rule there will always be a lengthening of the bond between the donor (acceptor) atom and the neighboring one, provided that a donor acceptor bond is formed. Alternatively lengthening and shortening of bonds can be induced by it across the entire molecule, resulting in different net charges at the terminal atoms for the even- and odd-membered chains.

Conclusions

The results on the adsorption properties of definite *n*-octyl oxypropylene oligomers give new insight into the mechanism of interaction within the adsorption layer. This is of interest for practical and theoretical applications. With respect to industrial application one can conclude that the surface properties of oxyethylene-oxypropylene copolymers do not only depend on the definite number of the oligomer units, but also on their position and their sequence along the amphiphilic copolymer.

Moreover, these results may be a useful contribution to understanding adsorption properties of polymeric amphiphiles in terms of surface thermodynamics.

Finally, the results represent another challenge in searching for an adequate theoretical description of adsorption phenomena at fluid interfaces, which so far has still been missing.

References

1. Schick MJ (1967) Nonionic Surfactants. Marcel Dekker, Inc, New York
2. Davidsohn A, Milwidsky BM (1972) Synthetic Detergents. Leonhard Hill, London p 28
3. Kronberg B, Stenius P, Thorssell Y (1984) Colloids Surf 12:113
4. Cho CS, Song SC, Kunou M, Akaide T (1990) J Colloid Interface Sci. 137:292
5. Miano F, Bailey A, Luckham PF, Tadros TF (1992) Colloids Surf 62:111
6. Dubyaga EG, Konoplev EG, Zakharova TA (1992) Vysokomol Soedin (Plastics Fabrication and Uses) Ser A, 34:113
7. Holzbauer, H-R, Herbst M (1988) Tenside Detergents 25:308

8. Sokolowski A, Burczyk B, Holzbauer H-R, Herbst M (1991) Colloids Surf., 57:307
9. Lunkenheimer K. Pergande H-J, Krüger H (1987) Rev Sci Instrum 58:2313
10. Lunkenheimer K, Wantke K-D (1981) Colloid Polymer Sci. 259:354
11. Lunkenheimer K (1989) J Colloid Interface Sci 131:580
12. Evans DF, Ninham BW (1986) J Phys Chem 90:226
13. Lehn J-M (1988) Angew Chem 100:91
14. Lunkenheimer K, Miller R (1987) J Colloid Interface Sci. 120:176
15. Lucassen-Reynders EH (1976) Progr Surf Membr Sci 10:253

16. Lunkenheimer K, Hirte R (1992), J Phys Chem 96:8683
17. Lunkenheimer K, Hirte R Lectures held at VIIth Intern Conf Surf Act Subst, Bad Stuer, Germany, 1988; 10th European Conference "Chemistry of Interfaces", San Benedetto, Italy, 1988; 11th European Conference "Chemistry of Interfaces", Strausberg (Berlin), Germany, 1990
18. Lunkenheimer K, Burczyk B, Piasecki A, Hirte R (1991) Langmuir 7:1765
19. Lunkenheimer K, Laschewsky A (1992) Progr Colloid Polym Sci 89:239
20. Gutmann V (1978) The Donor-Acceptor Approach to Molecular Interactions. Plenum Press, New York and London, 1978

Progr Colloid Polym Sci (1994) 97:121–127
© Steinkopff-Verlag 1994

SURFACTANTS

F. Mallamace
N. Micali
C. Vasi
S. Trusso
M. Corti
V. Degiorgio

Raman, depolarized and Brillouin scattering studies on nonionic micellar solutions

Received: 1 November 1993
Accepted: 10 December 1993

Prof. F. Mallamace (✉)
Dipartimento di Fisica
Universita' di Messina
98166 Vill. S. Agata C.P. 55, Messina, Italy

N. Micali·C. Vasi·S. Trusso
Istituto di Tecniche Spettroscopiche
del C.N.R.
98166 Vill. S. Agata, Messina, Italy

M. Corti· V. Degiorgio
Dipartimento di Elettronica
Universita' di Pavia
Via Abbiategrasso
27100 Pavia, Italy

Abstract The structure of water in aqueous solutions of polyoxyethylene nonionic amphiphiles $C_{10}E_5$, is studied by Raman, depolarized Rayleigh, and Brillouin scattering along an isothermal path crossing the isotropic one-phase region from 0 to 1 amphiphile volume fraction ϕ.

Key words Water – amphiphile – Raman and Brillouin scattering

Introduction

Aqueous solutions of non-ionic polyoxyethylene amphiphiles ($C_mH_{2m+1}(OCH_2CH_2)_nOH$ or C_mE_n for short) have been the subject of many studies [1–2] in order to clarify the structural and dynamical properties of the supramolecular aggregates formed above the critical micelle concentration (CMC). We consider the $C_{10}E_5$ solution, above the mesophase regions and below the cloud-point curve, where it is possible to follow a continuous isothermal path which crosses the isotropic one-phase region from 0 to 100% of the amphiphile. In order to explain its structural properties along this path (it is well-known that above the CMC $C_{10}E_5$ forms globular micelles, whereas it is not known up to what amphiphile concentration the solution can still be described as a water continuous dipersion of amphiphile aggregates, and how the system evolves towards the pure liquid amphiphile phase) the system has recently been the subject of careful studies of x-ray and neutron scattering [3]. In particular,

from such studies it has been observed that along this isothermal path ($T = 35\,°C$) the solution is structured for all concentrations ranging from the micellar region to the pure liquid amphiphile. As the volume fraction ϕ increases, the micellar structure becomes less and less sharp, but some orientational correlations between neighboring amphiphile molecules are preserved even at high concentrations. From the small-angle neutron data (SANS) [3] a structure peak is clearly observable up to $\phi = 0.95$, but the pure $C_{10}E_5$ neutron spectra were absolutely flat. This is due to the fact that the neutron scattered intensity is mainly determined by the large contrast between the deuterated water and the hydrogenated amphiphile. More precisely, the interpretation of scattering data (SANS and SAXS) leads to the following conclusions: large orientational correlations exist among neighboring amphiphile molecules, and at high surfactant concentrations ($\phi > 0.7$) the system behaves essentially as a block-copolymer melt. The existence of a structure peak ($\phi = 0.95$ for neutrons, and $\phi = 1$ (pure amphiphile) for x-ray) is a direct result of the block structure of the surfactant monomer with attractive head-head and tail-tail interactions and repulsive head-tail interactions. This phenomenon, proposed by P.G. de Gennes [4], is known as the correlation-hole effect. The use of deuterated water enhance via the hydration the contrast between hydrophilic and hydrophobic groups of the amphiphile molecule allowing for the observation of this latter phenomenon.

Therefore, from the picture proposed by SANS and SAXS in these non-ionic amphiphile solutions, it turns out that water molecules play a significant role. A part of water is hydrogen-bounded (HB) to the polyoxyethylene head groups of the amphiphile. In particular, the analysis of SANS data reveals that: the oxyethylene groups are hydrated, the average number of bound water molecules per group is $n_w \simeq 1.5$ for $\phi < 0.7$, and $n_w \simeq 2$ for $\phi > 0.7$; furthermore, for $\phi > 0.7$ all water molecules are bound via HB to the oxyethylene chains. The fact that, at high concentrations, there is no free water in the system is consistent with the low values of the electric conductivity measured in the solution.

On this basis, we consider that Raman, depolarized and Brillouin light scattering can give confirmation of this structural model proposed for these non-ionic amphiphile solutions by the analysis of SANS and SAXS data. Considering that the Raman Scattering constitutes a powerful tool in order to study vibrational dynamics, we try to investigate the structural properties of water in non-ionic amphyphile solutions through the analysis of the spectral region of O–H stretching vibrations. It is well known that O–H stretching is very sensitive to the molecular organization of water; In particular, for pure water it can provide, as a function of the thermodynamic variables (T, P, etc.), detailed information on the structure corresponding to a particular state (solid, liquid, vapor, supercooled) [5]. Here, we present results obtained from Raman scattering measurements, at 35 °C with $C_{10}E_5$ solutions. The resulting data agree with the structural model proposed by SANS and SAXS experiments, giving new detailed information about the vibrational dynamics of water molecules hydrogen-bounded with the polyoxyethylene head groups. In particular, we show that the O–H stretching vibrations of such bounded water are analogous to the glassy water [6]. Additional interesting information about such characteristic water-amphiphiles structures can be obtained from the analysis of depolarized or Rayleigh–Wing data (central frequency contribution of the scattered light). As is well known this spectral contribution are sensitive to the molecular rotational motion and can probe the water rotational dynamics in the system. The obtained linewidth (or the corresponding relaxation time) can be related, in the present case, to the hydrogen bond (HB) rotational dynamics [7]. In summary, we can have,

by means of this latter technique, direct information on the "local" properties of water in the neighborhood of amphiphilic molecules.

Considering that the system maintains a well-defined micellar structure for $\phi \leq 0.75$, a study of the acoustic propagation in the high frequency regime can be interesting and comparison made with the ultrasound data [8,9]. Sound absorption and, in particular, sound velocity are directly related to the bulk compressibility of the medium, thus we can probe the collective properties (ensamble of spherical micelles) of the system as modulated by interactions [9]. In particular, the attractive interaction between micelles can originate, as pointed out from current theories on the liquid state [10, 11], from extended aggregates. The kinetics and the dynamics of such clusters are dominated by the percolation phenomenon. In this respect, we perform, in our water-amphiphile system, additional and extensive Brillouin scattering measurements that give new insight into the structure and dynamics of very dense micellar systems.

Experimental results and discussion

The phase diagram of $C_{10}E_5$ in H_2O [1] shows a cloud-point curve with a minimum at about 45 °C and liquid crystalline regions in the range of amphiphile volume fraction between 0.5 and 0.85 with temperature between 0° and 20 °C; therefore, we worked along an isothermal path (T = 35°C), that did not hit any mesophase regions; for depolarized Rayleigh and the Brillouin, we performed a measurement also at 25°C. The studied concentrations were: $\phi = 0.3$, 0.5, 0.6, 0.7, 0.75, 0.87, 0.91 and the pure amphiphile. For all the scattering measurements the exciting source was the 5145 Å line from an Ar^+ laser, the scattering geometry was the usual 90° arrangement. The scattered light was collected through a Glan-Thompson polarizer with an extinction coefficient better than 10^{-7}. The sample was thermostatted in an optical cell to within 0.02°C.

Raman scattering measurements were performed using a triple monochromator (Spex Ramalog V). The measured spectra were taken in the range 2900-3800 cm^{-1}, with a resolution of 4 cm^{-1}, both in the parallel (VV) and orthogonal (VH) polarizations. The depolarized Rayleigh and the Brillouin scattering were performed using a high resolution double pass double monochromator (DMDP), SOPRA model DMDP 2000, with a half-width at half maximum (HWHM) resolution of 700 MHz. The resolution of the DMDP is comparable to that of a Fabry–Perot interferometer working at a free spectral range of ~ 50 GHz. For the study of the Brillouin contribution

Progr Colloid Polym Sci (1994) 97:121–127
© Steinkopff-Verlag 1994

this instrument has several advantages over the Fabry-Perot interferometer: an optical stability of several days, it eliminate the problems arising from the periodic boundary conditions and, in addition, the instrument has an exceptionally high stray-light rejection. All depolarized spectra are measured in the frequency range $-100 - +100$ cm^{-1}, and Brillouin data in the range $-10 - +10$ GHz.

For Brillouin spectra, we performed the data analysis using a well established procedure [12]. In particular, we used a convolution method between the hydrodynamic triplet and the instrumental response to obtain the frequency shift value $\Delta\omega(k)$ (i.e. the sound velocity).

Raman scattering

Starting from the obtained spectra of OH stretching vibration in both the polarization geometries, we calculate the isotropic part of the Raman intensity $I_{iso}(\omega)$. We calculated the OH stretching spectrum of water in the amphiphile solution using the following relation:

$$I_{is}^{\phi}(\omega) = I_{is}^{\phi,wat}(\omega)(1 - \phi) + I_{is}^{amph}(\omega)\phi \; ,$$

where $I_{is}^{\phi}(\omega)$, was considered as the sum of two contributions: one due to the water $I_{is}^{\phi,wat}(\omega)$, and the second to the pure amphiphile $I_{is}^{amph}(\omega)$, both weighted for the corresponding concentrations. Figure 1 shows $I_{is}^{\phi,wat}(\omega)$ for $\phi = 0.3$, 0.6, 0.7 and 0.75; in the same figure is also reported, for comparison the OH stretching of pure water at the same temperature as the amphiphile solutions ($T = 35°C$).

We use, for interpretation of the resulting OH stretching data, a recent theoretical model for water that can be considered intermediate between the two classes of models historically considered: i.e., continuous models and discrete models [13, 14]. In particular, both of them have in common the fact that a local four-coordinated environment with low density is preferred for the water structural arrangement. In such a model [14], each water molecule is assigned to one to five species, according to the number (from zero to four) of HB. Then, using percolation concepts it is shown that tetrabounded molecules tend to cluster, giving rise to finite regions, *patches*, whose structural properties are different from those of the remainder. In these terms, water molecules can be divided into two classes: "open" water in which a regular tetrahedral structure exists, and "closed" water that behaves like a continuum, being the mixing of all the remainder molecules. This model, stressing that "open" water is related to low density structures, explains quite well the thermodynamic properties of water and also the measured spectra of OH stretching in supercooled [15] and amorphous water [6].

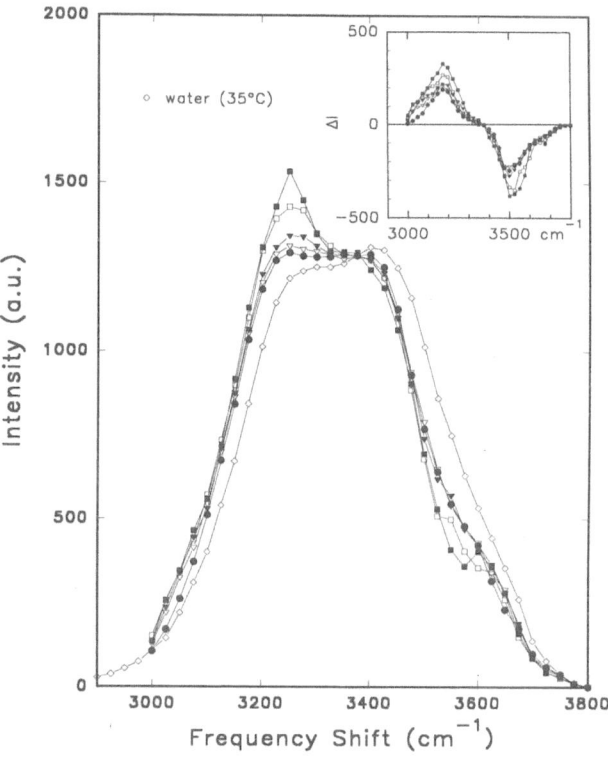

Fig. 1 The isotropic OH stretching $I_{is}^{\phi,wat}(\omega)$ for $\phi = 0.3, 0.5, 0.6, 0.7, 0.75$; for comparison, the OH stretching of pure water at the same temperature of the amphiphile solutions ($T = 35°C$). In the inset are reported the differences between the OH stretching spectra of water in the amphiphile solutions and the corresponding spectra of pure water

"Open" and "closed" contributions fall in two different frequency regions of the OH stretching Raman spectra [15]; while the "open" contribution has a mean peak centered at about 3150 cm^{-1}, the corresponding peak for "closed" structures is centered at about 3500 cm^{-1} [15].

As can be observed from Fig. 1, the spectrum of pure water differs from the spectrum of water in the mixture also for the less concentrated solution; the comparison of pure water spectrum with the spectrum for $\phi = 0.3$ shows that in the amphiphile solution a larger amount of water molecules is bonded in structures of low density. From the inset of Fig. 1, where are reported the differences between the OH stretching spectra of water in the amphiphile solutions and the corresponding spectra of pure water, it can be observed that an increase in the amphiphile content results in an increase in the open water contribution to the spectrum. This behavior can be explained considering the suggestions of SANS and SAXS [3], where it is proposed that for all values of ϕ below a saturation value ϕ_S ($\phi_S \sim 0.75$), we can have a certain quantity of water bounded to the polyoxyethylene head groups. This

ϕ range is the concentration interval for which well-defined micellar structures are present in the system. More specifically, in the concentration region where stable micelles are present, SANS and SAXS [3] data can be well described considering a three-component model: the hydrocarbon region (hydrophobic chain of the amphiphile), the hydrophilic region (polyoxyethylene head groups together with bound water, with an average number of water molecules bounded per oxyethylene group $n_w = 1.5$) and a region of free water. The value of n_w can be roughly calculated evaluating the "open" water contribution to the area of the OH stretching spectrum, once this latter has been normalized so as to cover a unit area. The obtained value ranges, within the experimental uncertainty between 1.4 and 1.7, and agrees with the SANS value.

With regard to the molecular organization of the solution at high surfactant concentration, $\phi > \phi_S$, we can only give qualitative confirmation that all the water present in the system is bound to oxyethylene groups. From the related spectra shown in Fig. 2 ($\phi > 0.75$), it is evident there is a significant difference between the spectra with ϕ above and below the saturation value. There is a further suggestion that for $\phi < \phi_S$ water molecules are arranged in

Fig. 2 Isotropic OH stretching contributions of water for samples with $\phi > 0.75$. In the inset is reported the isotropic OH stretching contribution for glassy water [6]

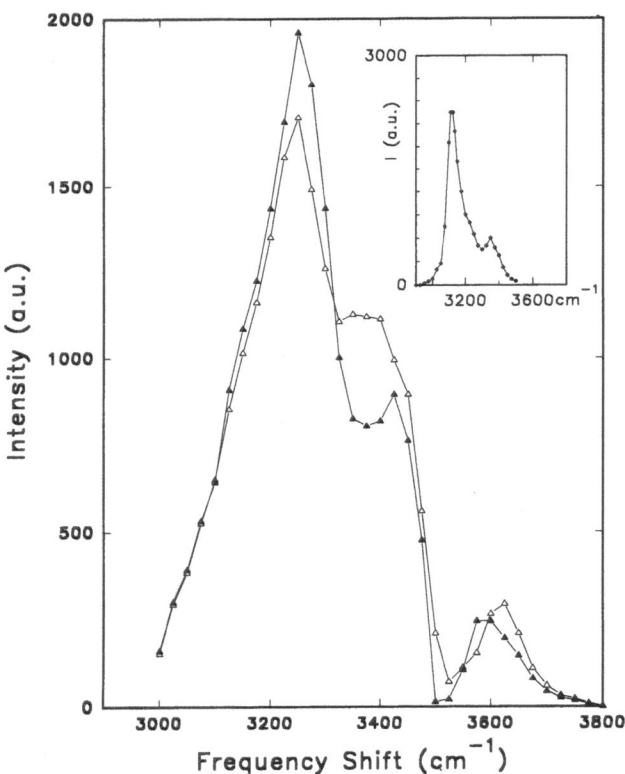

the amphiphile solution in a different way in comparison with the concentrations where the surfactant molecules are aggregated in micellar structures. In particular, the OH stretching spectra for solutions with $\phi = 0.87$ and 0.91 are entirely located in the region of "open" water; the percentage of "closed" water, in comparison with solutions of lower amphiphile volume fractions, is irrelevant. The dominant spectral contribution is located at the frequency of about 3200 cm^{-1}. Such a result gives an indication that a very large amount of water molecules are bonded to the amphiphile, and the corresponding structure reflects a local environment with a low density in comparison to the bulk water. As a proof of this result, we show in the inset of Fig. 2 the isotropic OH stretching spectrum of amorphous solid water in a film with a thickness of ~ 1 μm, prepared by vapor deposition, at $T = 100$°K [6]. As can be observed, the spectra of solutions at high volume fractions are similar to those of glassy water, the relevant difference is in the frequency value of the mean peak, but this is a temperature effect (in glassy water this frequency is temperature dependent: increasing with increasing T [6]).

Depolarized scattering

It is well known that the nonshifted depolarized light scattering is caused by the fluctuations of the traceless part of the polizability tensor [8, 16]. The corresponding time correlation function $G^{anis}(t)$ and its Fourier transform $I_{VH}(\omega)$ can be characterized by various contributions, which depend on the different mechanisms involved in the scattering processes. The information which can be extracted from the measured spectra is mainly related to the translational motion, as reflected by the density correlation function; more precisely, to the rotational motion of the molecules. In this paper, we show that depolarized Rayleigh scattering experiments can be used to probe the dynamics of water in the present complex liquid system, giving, therefore, a "local" information on the water properties in the neighborhood of amphiphilic molecules. A comparison with pure water spectra, at the same temperatures, reveals the same linewidths, so we can assume that the observed spectral contributions are related with the hydrogen bond (HB) dynamics. The free amphiphile molecular rotation linewidth or the contribution related to the rotational dynamics of polyoxyethylene terminal groups falls in a spectral region smaller than the present instrumental resolution. In Fig. 3, we show, in the frequency range -30 to $+30$ cm^{-1}, the spectrum for the suspension with $\phi = 0.6$.

Because the rotational contributions have a Lorentzian shape, and the instrumental resolution is a Gauss–Lorentzian, we fit our spectra with these two

Progr Colloid Polym Sci (1994) 97:121–127
© Steinkopff-Verlag 1994

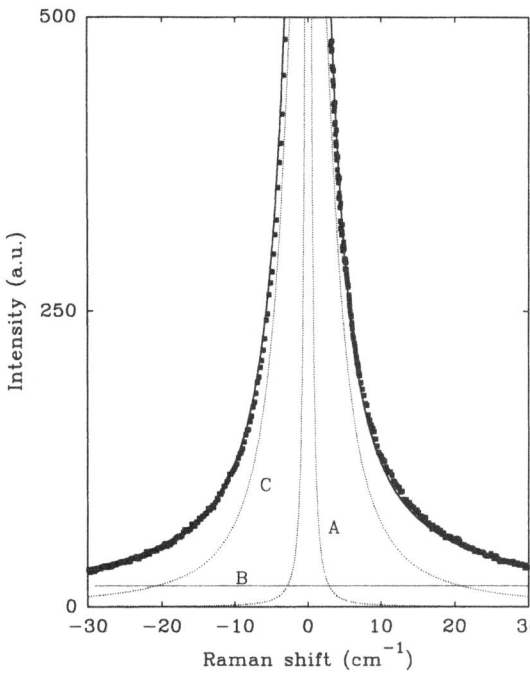

Fig. 3 Anisotropic spectrum for suspension with $\phi = 0.6$. The continuous line represents the best fit of the data; C is the contribution caused by molecular rotational motions, A the instrumental resolution, and B the constant background

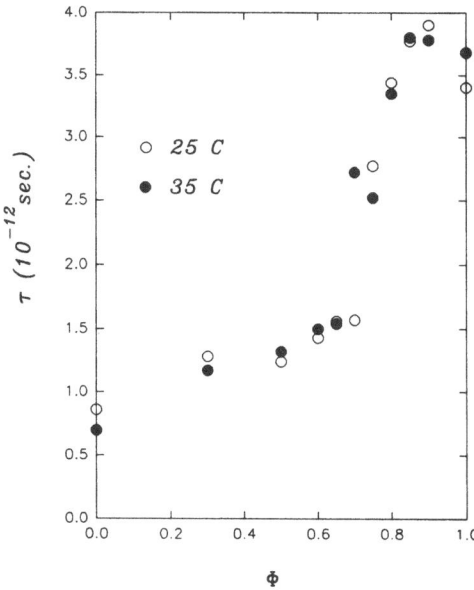

Fig. 4 The relaxation time τ of the water molecular rotational contribution versus the volume fraction ϕ

functions plus a constant. The continuous line in Fig. 3 represents the best fit of the data; in the same figure the contribution caused by molecular rotational motions is indicated as C, the instrumental resolution as A, and the constant background as B. We consider, in order to discuss the obtained results, the relaxation time τ related to the measured linewidth Γ (the HWHM of the rotational contribution). The results as a function of ϕ (measured at two temperatures, 25 and 35 °C) are reported in Fig. 4, where for comparison are also reported the pure water and pure amphiphile values. From this figure, we observe two different behaviors above and below ϕ_S (the saturation value $\phi_S \sim 0.75$). Namely, for $\phi < \phi_S$, we observe a small increase in this rotational relaxation time; the comparison of the suspension data with the water data ($\phi = 0$) gives the definitive confirmation that with such measurements we observe rotational dynamics due to the water molecule, that as it is well known, is related to the HB dynamics [17]. For $\phi > \phi_S$ such contributions give evidence of a dramatic slowing-down; note that the final τ value (for $\phi = 0.91$) is about the same as that of deep, supercooled water (T = − 27 °C). Such results agree very well with the other scattering results (RAMAN, SAXS and SANS). The slowing-down in the HB dynamics is related to the process of hydration of the amphiphile oxyethylene chains. In addition, the measured pure amphiphile with a OH head

group) τ values agree with the observed correlation-hole effect.

Brillouin scattering

The fully polarized scattered intensity vs. the frequency ω, at a given k, gives information on the dynamical structure factor $S(k, \omega)$ which is the Fourier transform of the k-th component of the density-density correlation function $G_\rho(k, t) = \langle \delta\rho(k, 0)\, \delta\rho(k, t) \rangle$. The dynamic structure factor depends on the viscoelastic behavior of the system under examination and reflects its collective properties modulated by the interactions. Therefore, through a direct measurement of the hypersound propagation it is possible to give detailed information of the structural and dynamical properties of the system.

In particular, the elastic properties of a medium are characterized in terms of the complex longitudinal modulus $M = M' + iM''$ directly connected with the quantities measured in a Brillouin experiment, i.e., the velocity V and the absorption coefficient α. The velocity is associated with the real part of M by $V^2 = M'/\rho$ (ρ is the average density). Furthermore, since M is related to the compressional modulus K and the shear modulus G ($M = K + 4G/3$), the Brillouin data are also sensitive to the shear rigidity of the system, although only longitudinal properties are probed directly. Therefore, higher sound velocity is associated with a behavior typical of solid-like or associated structures, and slower sound velocity with a liquid-like behavior [9].

As pointed out by the current theoretical models for the liquid state an assembly of hard spheres, because of interparticles attractive interactions, will tend to form extended structures by means of percolation phenomena. More generally, such a system, depending on the temperature and concentration, can also perform a transition from the liquid to a crystalline or a glassy state; so that it can be used to explore the physical properties of these phases of condensed matter. Colloidal systems, as it is well known, constitute a class of models for which it is possible to study these properties typical of a dense system of hard spheres. Also, micelles, because of their weak attractive interaction, will tend to form extended clusters whose size and concentration will increase with ϕ. On this basis, we can use the present water-nonionic amphiphile system (as shown by SANS is easy to change the volume fraction maintaining the micellar structure) in a Brillouin experiment in order to study the properties of such intermicellar aggregates. Figure 5 shows the obtained hypersonic velocities (~ 6 GHz) for the $C_{10}E_5$ water system as a function of ϕ at the temperature of 35°C. In the same figure are reported, with the values corresponding to pure water and pure amphiphile, the ultrasound velocity data (5 MHz) that refer to some concentrations studied here. As can be observed, in the range $0 \leq \phi \leq 0.75$, a strong dispersion is present, that increases with ϕ.

Such a dispersion (different values in the sound velocities at the same concentration for different frequencies, 5MH and 6 GHz) can be easily connected with the presence of an intermicellar structure. Due to weakness of the attractive interaction between micelles [18], existing clusters could live only on short time scales. Therefore, at high frequencies (hypersound) the time scale of the measurement will be faster than the dynamics of the interactions, and the system is viewed as an instaneous connected network. By contrast, at low frequencies (long-times) the dynamic of the system reflects a collection of non-interacting particles in which shear stresses are relaxed. Consequently, for short times the solid-like network is able to support shear stresses and exhibits a finite elastic modulus. The data of Fig. 5 confirm the presence at high ϕ of a connected intermicellar network.

For micellar systems, as showed by Weitz and coworkers [18], the observed elastic behavior can be entirely connected with the percolation processes. In such a case the increase in the elastic modulus must scale as:

$$\Delta M' \sim (\phi - \phi_P)^t ,$$

where ϕ_P is the volume fraction corresponding to the percolation threshold, t is the percolation exponent and $\Delta M'$ is the difference between the measured M' and the corresponding value calculated from an effective medium model M'_e, i.e.

$$\Delta M' = M' - M'_e ; M'_e = (1 - \phi) M'_{water} + \phi M'_{C_{10}E_5} .$$

Fig. 5 The measured hypersonic velocities V as a function of ϕ at the temperature of 35°C. Also reported for some concentrations are the corresponding ultrasound velocity data measured at 5 MHz

Fig. 6 Scaling behavior of $\Delta M'$ (for T = 25°C and 35°C) as a function of $(\phi - \phi_P)$

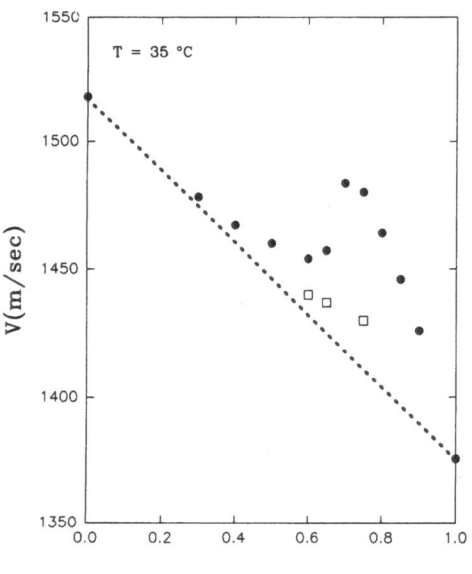

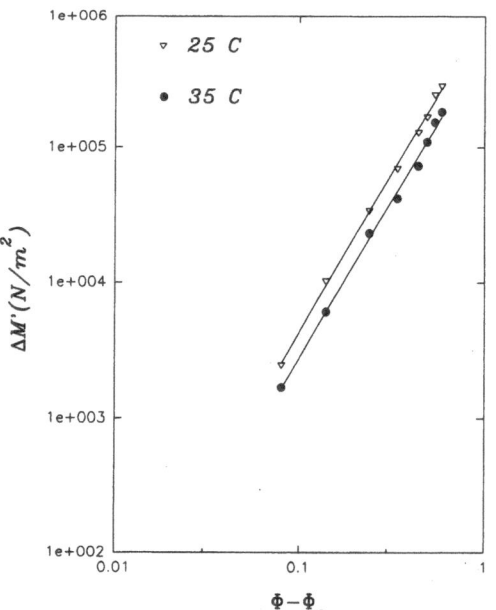

Progr Colloid Polym Sci (1994) 97:121–127
© Steinkopff-Verlag 1994

In Fig. 6, we report, for the two temperatures 25° and 35 °C, on a log-log scale $\Delta M'$ as a function of $(\phi - \phi_P)$ having put $\phi_P = 0.16$ as a result of the best fit. The obtained value for t is 2.2 ± 0.2. Such a value for ϕ_P is in agreement with the value obtained for a different micellar system in ref. [19]. The t value is in complete agreement with the value is in complete agreement with the value proposed from theoretical studies of percolation [19].

The strong decrease in V observed for $\phi > \phi_S$ can be related to a change in the structure of the system. In a concentration region the micellar structure is gradually destroyed and the system evolves a more disordered phase. As shown by the observation of the correlation hole effect (SAXS and SANS data) the system behaves, in such high ϕ region, as a block-copolymer melt. A more detailed study by means of Brillouin scattering is in progress in a concentrated region in order to have a deeper understanding of the physical phenomena related to the transition from the micellar to the block-copolymer melt phase.

Conclusions

We have performed different scattering experiments in a water amphiphile solution in the isotropic single-phase region, where for a volume fraction lower than $\phi = 0.75$ the system is arranged in a well defined micellar structure. From the Raman and the depolarized Rayleigh wing scattering, we have studied the properties of water in such a complex system. As a conclusion, both the reported studies agree with the structural picture proposed for this suspensions by SAXS and SANS measurements, i.e., water molecules are partially bounded to the oxyethylene head groups of the sufactant for amphiphile volume fractions ϕ lower than 0.75; above such a value all the water present in the system is bound to the oxyethylene groups. In addition, the water structure around the surfactant has a local structure corresponding to an environment with low density.

Considering that Brillouin scattering is related to the viscoelastic properties of a system and reflects its collective properties modulated by the interactions, we give information about the presence, in this water amphiphile suspension, of extended percolating structures originated by the intermicellar interaction. This is a result that, in particular suggests that the present system can be successfully used to test the findings of hard spheres theoretical models for liquid state.

Finally, all the different scattering techniques give the information that above a saturation value, $\phi_S \sim 0.75$, the system behaves like a block-copolymer melt.

References

1. Degiorgio V (1985) In: Degiorgio V, Corti M (eds) Physics of Amphiphiles, Micelles, Vesicles and Microemulsions. North-Holland, Amsterdam, p 303; and refs. cited therein
2. Magid LJ (1987) In: Schick MJ (ed) Nonionic surfactants: physical chemistry. Dekker, New York
3. Degiorgio V, Corti M, Piazza R, Cantu'L, Rennie AR (1991) Colloid and Polym Sci 269:501; Barnes IS, Corti M, Degiorgio V, Zemb T (1992) to be published
4. de Gennes PG (1969) Scaling concepts in polymer physics. Cornell University Press, Ithaca, p. 65
5. Walrafen G (1972) In Franks F (ed) Water a comprehensive treatise. Plenum Press, New York 1:161
6. Li PC, Devlin JP (1973) J Chem Phys 59:547; Sivakumar TC, Rice AS, Sceats MG (1978) J Chem Phys 69:3468
7. Montrose CJ, Bucaro JA, Marchall-Coakley J, Litovitz TA, (1974) J Chem Phys 60:5025; Conde O, Teixeira J Mol Phys 44:525
8. Fabelinkii IL (1968) Molecular Scattering of Light. Plenum, New York.
9. Litovitz TA, Davis CM (1965) In Mason P (ed) Physical Acoustic. Academic New York II, pt A, Chap. 5.
10. van Megen W, Snook I (1984) Adv. Colloid Interface Sci 21:119; Hess W, Klein R (1983) Adv Phys 32:173; Tough RJA, Pusey PN, Lekkerkerker HNW, van den Broeck C (1986) Mol Phys 59:595; Pusey PN, van Megen W (1986) Nature (London) 320:340
11. See e.g. Lekkerkerker HNW, in this issue
12. Evans AB, Powels JC (1974) J Phys A 7:1944
13. See for example: Eisemberg DE, Kauzmann W (1969) The structure and properties of water. Oxford University Press, Oxford
14. Stanley HE, Teixeira J (1980) J Chem Phys 73:3034
15. D' Arrigo G, Maisano G. Mallamace F, Migliardo M, Wanderlingh F (1981) J Chem Phys 75:4264
16. Berne BJ, Pecora R (1976) Dynamic light scattering. Wiley, New York
17. Aliotta F, Vasi C, Maisano G, Majolino D, Mallamace F, Migliardo P (1986) J Chem Phys 84,:4731; Mazzacurati V, Nucara A, Ricci MA, Ruocco G, Signorelli G (1990) J Chem Phys 93:7767; Scicrtino F, Geiger A Stanley HE (1992) J Chem Phys 96:3857
18. Ye L, Weitz DA, Scheng P, Bhattachrya, Huang JS, Higgins MJ (1989) Phys Rev Lett 63:263; Ye L, Liu J, Scheng P, Huang JS, Weitz DA (1993) J de Physique IV 3:C1 183.
19. Stauffer D (1985) Introduction to Percolation Theory. Taylor and Francis, London

Progr Colloid Polym Sci (1994) 97:128–129
© Steinkopff-Verlag 1994

SURFACTANTS

Studies of 1-C$_{16}$-2-C$_6$-PC and 1-C$_6$-2-C$_{16}$-PC rodlike micelles by small-angle neutron scattering

T.-L. Lin
Y. Hu
S.-H. Chen
M.F. Roberts
J. Samseth
K. Mortensen

Received: 16 September 1993
Accepted: 1 March 1994

Prof. Tsang-Lang Lin (✉) · Y. Hu
Department of Nuclear Engineering
National Tsing-Hua University
Hsin-Chu, Taiwan 30043, ROC

S.-H. Chen
Department of Nuclear Engineering
Massachusetts Institute of Technology
Cambridge, Massachusetts MA02139, USA

M.F. Robert
Department of Chemistry
Boston College
Chestnut Hill, Massachusetts MA02167,
USA

J. Samseth
Institutt for Energiteknikk
PO BOX 40, N-2007, Kjeller, Norway

K. Mortensen
Risø National Laboratory
Roskilde, DK-4000, Denmark

Abstract Small-angle neutron scattering measurements have been made to determine the structure of the 1-hexadecanoyl-2-hexanoyl-phosphatidylcholine (1-C$_{16}$-2-C$_6$-PC) and 1-hexanoyl-2-hexadecanoyl-phosphatidylcholine (1-C$_6$-2-C$_{16}$-PC) lecithin micelles in aqueous solutions. Both these two isometric lecithins were found to form polydispersed rodlike micelles. The parameters of the rodlike structure were determined from the measured scattering intensity distributions. The lecithin molecules forming the straight section of the rodlike micelle were found to occupy a surface area of only 53 Å^2 per molecule. This low surface area per molecule indicates that the wedge shaped hydrophobic parts of the these lecithin molecules conform very well in the rodlike structure.

Key words Small-angle neutron scattering – lecithin – micelle – rodlike micelle

Small-angle neutron scattering (SANS) measurements have been made to determine the structure of the 1-hexa-decanoyl-2-hexanoyl-phosphatidylcholine (1-C$_{16}$-2-C$_6$-PC) and 1-hexanoyl-2-hexadecanoyl-phosphatidylcholine (1-C$_6$-2-C$_{16}$-PC) lecithin micelles in aqueous solutions. Both the two isomeric synthetic lecithins have one long hydrocarbon chain and one short hydrocarbon chain. The investigated concentrations are from 1 mM to 30 mM for 1-C$_{16}$-2-C$_6$-PC, and from 0.2 mM to 3 mM for 1-C$_6$-2-C$_{16}$-PC. The samples are prepared in D$_2$O solutions. SANS measurements were done at the Risø National Laboratory for 1-C$_{16}$-2-C$_6$-PC micellar solutions, and at the High Flux Beam Reactor of the Biology Department of Brookhaven National Laboratory for 1-C$_6$-2-C$_{16}$-PC

micellar solutions. Figure 1 shows some of the measured scattering spectra plotted in log scales, ln($I(Q)$) versus ln(Q). The profiles of these scattering spectra for different concentrations are similar to each other. For high concentration samples, one can see clearly from Fig. 1 that the scattering intensity increases rapidly with decreasing Q in the low-Q region ($Q < 0.03$ Å^{-1}). This indicates either there are strong attractive interactions between these large rodlike micelles [1], or the lecithins might form giant flexible rods at high concentrations and these giant flexible rodlike micelles might entangle with each other [2]. The ln($I(Q)Q$) versus Q^2 plots show that both lecithins indeed form very large rodlike micelles. The critical micellar concentrations (CMC) of 1-C$_{16}$-2-C$_6$-PC and 1-C$_6$-2-C$_{16}$-PC

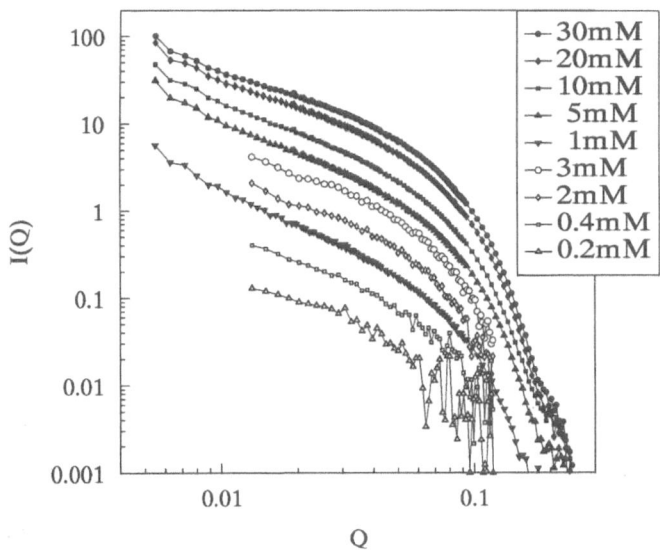

Fig. 1 The $\ln(I(Q))$ versus $\ln(Q)$ plots of the scattering intensity distributions for $1\text{-}C_{16}\text{-}2\text{-}C_6\text{-}PC$ samples (solid dots), and for some $1\text{-}C_6\text{-}2\text{-}C_{16}\text{-}PC$ samples (open dots), where $I(Q)$ is in units of cm^{-1} and Q is in units of $Å^{-1}$

PC, and 19.9 ± 0.8 Å for $1\text{-}C_6\text{-}2\text{-}C_{16}\text{-}PC$. These two cross-sectional radii of gyration are much larger than that for short-chain (with 6−8 carbons in each hydrocarbon chain) lecithin rodlike micelles. The aggregation number per unit rod length are determined to be 2.7 ± 0.3 Å^{-1} for $1\text{-}C_{16}\text{-}2\text{-}C_6\text{-}PC$, and 2.6 ± 0.5 Å^{-1} for $1\text{-}C_6\text{-}2\text{-}C_{16}\text{-}PC$. Each lecithin molecule in the straight section of these rodlike micelles occupies about 53 Å^2 at the surface of the hydrocarbon core. The low value of the surface area occupied by each lecithin molecule in the cylindrical section of the rodlike micelles indicates that the wedge shaped hydrophobic parts of the $1\text{-}C_{16}\text{-}2\text{-}C_6\text{-}PC$ and $1\text{-}C_6\text{-}2\text{-}C_{16}\text{-}PC$ molecules conform very well in the rodlike structure. The size distribution of the rodlike micelles at very dilute concentrations (which are supposed to have negligible inter-particle interactions) can be obtained by using the indirect Fourier transform method [4]. The obtained length distribution of the polydispersed rodlike micelles has the same typical profile as that for short-chain rodlike micelles [3]. The weight averaged length of the polydispersed rodlike micelles formed by $1\text{-}C_6\text{-}2\text{-}C_{16}\text{-}PC$ at dilute concentrations of 0.2 mM and 0.4 mM are respectively determined to be 317 and 383 Å.

are much lower than the CMC of the short-chain lecithins and they can form large rodlike micelles at relatively low concentrations. From $\ln(I(Q)Q)$ versus Q^2 plots, the radius of gyration across the cross-section of these rodlike micelles, R_c, is found to be 18.5 ± 0.2 Å for $1\text{-}C_{16}\text{-}2\text{-}C_6\text{-}$

Acknowledgements The authors would like to thank the Risø National Laboratory and the Brookhaven National Laboratory for providing the neutron beam time. T.-L. Lin acknowledges the support of the National Science Council, ROC, grant NSC 82-0208-M-007-082.

References

1. Lin T-L, Chen S-H, Gabriel NE, Roberts MF (1986) J Am Chem Soc 108:3499−3507

2. Schurtenberger P, Magid LJ, King SM, Lindner P (1991) J Phys Chem 95:4173−4176

3. Lin T-L, Chen S-H, Gabriel NE, Roberts MF (1987) J Phys Chem 91:406−413

4. Glatter O (1980) J Appl Cryst 13:7−11

Progr Colloid Polym Sci (1994) 97:130–133
© Steinkopff-Verlag 1994

SURFACTANTS

E.J. Staples
L. Thompson
I. Tucker
J. Penfold

Adsorption from mixed surfactant solutions containing dodecanol

Received: 16 September 1993
Accepted: 24 July 1994

L. Thompson (✉) · E.J. Staples
I. Tucker
Unilever Research
Port Sunlight
Quarry Road East
Bebington, Wirral L63 2JW,
United Kingdom

J. Penfold
ISIS Science Division
Rutherford Appleton Laboratory
Chilton
Didcot, Oxon, United Kingdom

Abstract Specular neutron reflection has been used to measure adsorption at the air/water interface from mixed solutions of ethoxylated nonionic surfactants ($C_{12}E_n$) and of these surfactants with sodium dodecyl sulphate (SDS), in the presence and absence of dodecanol. Measurements were made over a wide range of concentrations, in excess of the critical micellar concentration. In the absence of dodecanol our observations are consistent with the predictions of Regular Solution Theory. This is not the case with dodecanol present.

Key words Mixed surfactant adsorption – Regular Solution Theory – on neutron reflection

Introduction

The introduction of Regular Solution Theory (RST), initially attributed to Corkill [1] and subsequently developed by others [2–4], has provided a convenient framework for the prediction of the solution and surface properties of surfactant mixtures, including critical micellar concentrations (CMC), monomer and micelle compositions, adsorbed layer composition and surface tension, using readily obtainable surface chemical data for the pure components and for a single mixture. Although semi-empirical, this approach provides a useful insight into surfactant adsorption at concentrations relevant to most practical applications, that is, well above the CMC. The objective of the present work is to establish the value of the Neutron Reflection technique as a means of investigating the composition of adsorbed layers at concentrations well in excess of the CMC, and to assess the ability of RST to cope with surfactant solutions containing dodecanol. Dodecanol is a common impurity, which is insufficiently soluble to form micelles itself, but which is readily incorporated into the micelles of other surfactants.

Experimental details

Protonated nonionic surfactants $C_{12}E_3$, $C_{12}E_8$ were obtained from Nikkol. SDS and n-dodecanol were obtained from BDH. Deuterated nonionic surfactants for the neutron reflection measurements, $C_{12}E_3$, $C_{12}E_5$, $C_{12}E_8$ and $C_{12}E_{12}$ were synthesised at Unilever, Port Sunlight, by a procedure involving addition of the appropriate ethoxylate oligomer to 1-bromo-dodecane with a fully deuterated alkyl chain. Protonated $C_{12}E_{12}$ was synthesised by the same procedure, using non-deuterated 1 bromo dodecane. Deuterated 1 bromo dodecane, deuterated SDS and deuterated dodecanol were obtained from MSD Isotopes Ltd and used without further purification. Deuterium oxide (D_2O) was supplied by Sigma. High purity water (Elga Ultrapure) was used throughout, and

Progr Colloid Polym Sci (1994) 97:130–133
© Steinkopff-Verlag 1994

the glassware and Teflon troughs for the neutron reflection measurements were cleaned by soaking in 1% Decon solution followed by extensive rinsing. The neutron reflection measurements were carried out on the reflectometer CRISP [5] at the ISIS pulsed neutron source, where the measurements have been made using the fixed geometry (angle of incidence of 1.5°) white beam time of flight method (using wavelengths from 0.5 to 6.5 Å) in the Q range 0.05 to 0.65 Å^{-1}. The experimental procedures are now well established and are described in detail elsewhere [6].

Results and discussion

The adsorption of SDS/$C_{12}E_3$ mixtures in 0.1 M NaCl has been measured by neutron reflection over a range of concentrations in order to compare it with the predictions of RST. These predictions were based on the CMC and area/molecule measurement obtained from surface tension data. The composition (65 mol% $C_{12}E_3$) was selected because of its association with optimum detergency conditions [7, 8]. In Table 1 the parameters derived from surface tension analysis and used in the subsequent Regular Solution Theory analysis using the treatments of Rubingh [2] and of Holland [3] are presented. The CMC data at various compositions are adequately fitted using a value for the interaction parameter, β, of -2.4. Table 1 also shows that the adsorption data obtained from neutron reflectivity are entirely consistent with the surface tension data. Substitution of the neutron results into the Regular Solution analysis makes no appreciable difference to the predictions.

Figure 1 shows the adsorption of SDS and $C_{12}E_3$ separately as well as the total surface excess (determined independently) as a function of concentration. In Fig. 2 the data are presented as the proportion of SDS in the adsorbed layer, and are compared to the theoretical results obtained from Regular Solution Theory. Good agreement

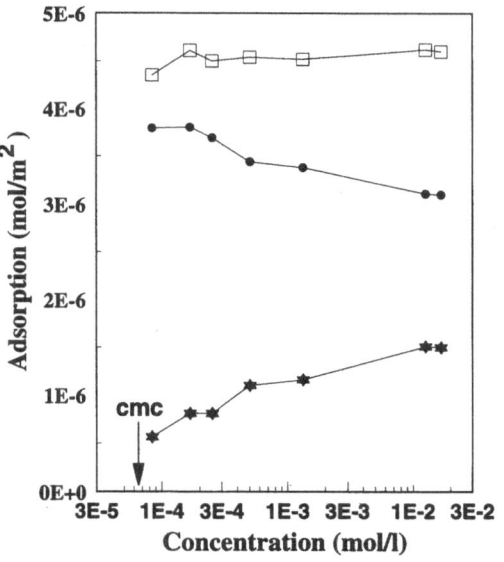

Fig. 1 Variation of adsorption with total surfactant concentration for 35/65 mol ratio SDS/$C_{12}E_3$ mixtures. □, total adsorption; ●, $C_{12}E_3$ adsorption; *, SDS adsorption

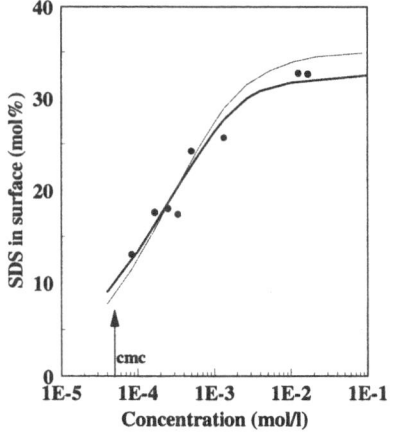

Fig. 2 Effect of total surfactant concentration on adsorbed layer composition for 35/65 mol ratio SDS/$C_{12}E_3$. ●, neutron reflection data; ——— Regular Solution Theory prediction using $\beta_s = -3.40$; ----, RST prediction for micelle composition, using $\beta = -2.4$

is obtained. Moreover, it is noted that the RST prediction for the micellar composition is very close to the composition of the monolayer over the whole range of concentration.

In subsequent measurements we have investigated adsorption from solutions of three component mixtures comprising anionic/nonionic and nonionic/nonionic surfactants together with dodecanol in the presence of 0.1 M NaCl. Figure 3 shows adsorption at three different concentrations of an "equivalent" 26/39/35 mol ratio mixture of

Table 1. Characterisation of SDS/$C_{12}E_3$ in 0.1 mol dm^{-3} NaCl by surface tension and by neutron reflectivity.

	SDS	$C_{12}E_3$	SDS/$C_{12}E_3$ (35/65 mol ratio)
CMC (mol dm^{-3})	1.5×10^{-3}	3.0×10^{-5}	4.2×10^{-5}
a_{hg}/nm^2 (surface tension)	0.36	0.39	0.40
a_{hg}/nm^2 (neutron)	0.37	0.36	0.37
Surface tension at CMC/mN m^{-1}	33.8	27.2	27.5
β	–	–	-2.4
β_s	–	–	-3.4

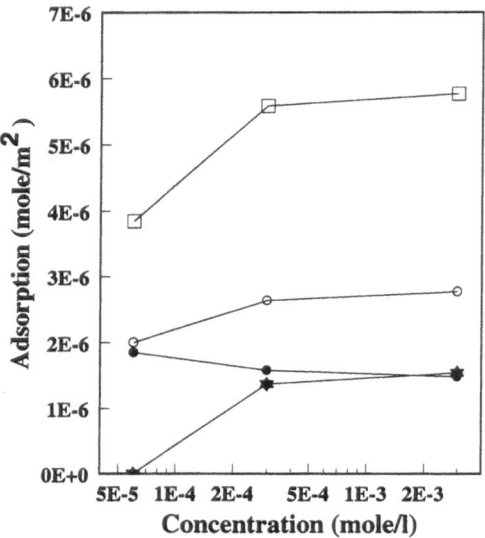

Fig. 3 Variation of adsorption with total surfactant concentration for 35/39/26 mol ratio mixtures of SDS/$C_{12}E_5$/docecanol. ○, dodecanol; ●, $C_{12}E_5$; *, SDS; □, total adsorption

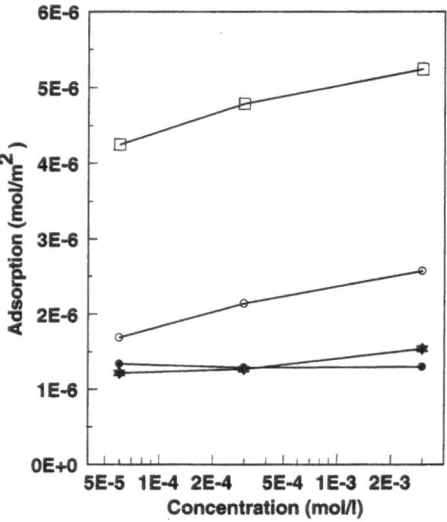

Fig. 4 Variation of adsorption with total surfactant concentration for 50/30/20 mol ratio mixtures of $C_{12}E_8$/$C_{12}E_5$/dodecanol. ○, dodecanol; ●, $C_{12}E_5$; * $C_{12}E_8$; □, total adsorption

dodecanol/$C_{12}E_5$/SDS where the $C_{12}E_3$ component of the system used in Figs. 1 and 2 was replaced by a 3/2 $C_{12}E_5$/dodecanol combination. The average ethoxylate chain length of this solution is 3.0 and at a solution concentration of 1% it gives cloud points in a range of mixtures with SDS which are within 2 °C of those obtained with $C_{12}E_3$ itself. Comparison of Figs. 1 and 3 shows that total adsorption of the system containing dodecanol is much higher because of high levels of dodecanol adsorption. Although the bulk properties of the $C_{12}E_5$/dodecanol based mixture are similar to those of $C_{12}E_3$ in the limit of high concentration, the surface adsorption indicates that complex compositional changes occur upon dilution.

It might be expected, from earlier work [9] and from the data in Fig. 2, that the surface concentration of the most surface active component (dodecanol), would decrease with increasing total surfactant concentration. For the systems investigated here, however, we find that the opposite occurs: the dodecanol content of the monolayer increasing with concentration, giving rise to an overall increase in total adsorption. This is shown in Fig. 3 for the dodecanol/$C_{12}E_5$/SDS system and in Fig. 4 for the dodecanol/$C_{12}E_5$/$C_{12}E_8$ system. The observed changes in adsorption with concentration must reflect the evolution of the monomer concentrations of the components. These in turn are related to the composition and energy of formation of the associated micelles. For systems such as the $C_{12}E_3$/SDS mixtures described earlier, near the CMC the most surface-active component is preferentially adsorbed at the air/water interface and into micelles whilst in

Table 2. Critical micelle concentrations

Surfactant*	CMC/mol dm^{-3}
$C_{12}E_5$	6.4×10^{-5}
$C_{12}E_8$	10^{-4}
SDS	1.5×10^{-3}
Dodecanol/$C_{12}E_5$ (10/90)	5.2×10^{-5}
Dodecanol/$C_{12}E_5$ (20/80)	4.0×10^{-5}
Dodecanol/$C_{12}E_5$ (40/60)	5.6×10^{-5}
Dodecanol/$C_{12}E_5$/$C_{12}E_8$ (20/39/50)	4.0×10^{-5}
Dodecanol/$C_{12}E_5$/SDS (26/39/35)	5.0×10^{-5}

* All measurements were carried out in the presence of 10^{-1} mol dm^{-3} NaCl.

the limit of very high concentration the micelle composition must be that of the bulk. This means that the monomer concentration of the most surface-active species and hence its concentration in the adsorbed layer decline with increasing concentration. In Figs. 3 and 4 we observe the opposite trend. Here, the increase in dodecanol adsorption with concentration must be matched by an increase in both the dodecanol monomer concentration and the dodecanol content of the micelle. This can occur in systems which exhibit a minimum in the CMC versus composition curve which is below that of the pure components. In this case the composition with the minimum CMC can obviously be identified with the micelle composition with the lowest free energy. Other surfactant compositions will, at their CMC, result in micelle compositions that are biased toward the minimum energy composition. Table 2 shows that the $C_{12}E_5$/dodecanol system has a CMC minimum at

Progr Colloid Polym Sci (1994) 97:130–133
© Steinkopff-Verlag 1994

about 20 mol % dodecanol. Therefore at bulk compositions which contain more than 20% dodecanol, the monomer concentration of the dodecanol will increase with total surfactant concentration in response to the shift to a more dodecanol-rich micelle. Extending this argument to the three component systems of Figs. 3 and 4 is more difficult because accurate CMC determinations are difficult owing to curvature of the $\gamma/\log C$ curves. Nevertheless, the approximate CMC's shown in Table 2 for the ternary mixtures used in Figs. 3 and 4 again suggest the presence of minima in the CMC versus composition curves.

Acknowledgement The authors gratefully acknowledge the contribution of Mr MP Nicholls in synthesising the deuterated nonionic surfactants.

References

1. Clint JH (1990) Bloor DM, Wyn-Jones E (eds) In: The Structure, Dynamics and Equilibrium Properties of Colloidal Systems. Kluwer Academic Publishers, Amsterdam, 76
2. Rubingh DN (1979) In: Mittal KL (ed) Solution Chemistry of Surfactants, Vol 1. Plenum Press NY, p 337
3. Holland PM (1986) Colloids and Surfaces, 19:171
4. Scamehorn JF (1986) ACS Symposium Ser. 311:1
5. Penfold J, Ward RC, Williams WG (1987) J. Phys. E. Sci. Inst., 20:1411
6. Lee EM, Thomas RK, Penfold J, Ward RC, (1989) J Phys Chem 93:381
7. Thompson L, (1992) in "Surfactants in Lipid Chemistry", JHP Tyman (Ed), Royal Society of Chemistry Special Publication, 118:56.
8. Thompson L (1994) J Colloid Interface Sci 163:61
9. Penfold J, Thomas RK, Simister EA, Lee EM and Rennie AR (1990) J. Phys. Condens Matt. 21:SA411

Progr Colloid Polym Sci (1994) 97:134–140
© Steinkopff-Verlag 1994

SURFACTANTS

H. Edlund
A. Lindholm
I. Carlsson
B. Lindström
E. Hedenström
A. Khan

Phase equilibria in dodecyl pyridinium bromide – water surfactant systems

Received: 16 September 1993
Accepted: 15 January 1994

Dr. Lindström (✉)
H. Edlund · A. Lindholm
I. Carlsson · E. Hedenström
Chemistry Department
Mid Sweden University
851 70 Sundsvall, Sweden

A. Khan
Physical Chemistry 1
Chemical Center
Box 124
University of Lund
22100 Lund, Sweden

Abstract The phase equilibria in four binary (temperature vs. composition) surfactant systems, each containing one of the four isomers of dodecyl pyridinium bromide (DPB) have been studied experimentally by water deuteron NMR and polarizing microscopy methods and theoretically by the Poisson–Boltzmann cell model.

Key words Dodecyl pyridinium bromide – phase diagram – crystal – deuteron NMR

Introduction

Long-range electrostatic effect [1, 2] and surfactant molecular constraints [3] are dominating factors that determine aggregation phenomena (e.g., aggregate shape) and the existence of stability regions of homogeneous phases in a phase diagram for an ionic surfactant system. Different monovalent counterions are expected to produce small differences in electrostatic interactions thus leading to similar phase stabilities [4–10]. Moreover, the valency of the headgroup [11] and the counterion [4, 12] plays an important role in the aggregation. Apart from the electrostatic effects of the counterions and the headgroups, counterion hydration, especially dealing with strongly hydrated ions, may have significant influence on surfactant self-association processes [10]. One could expect that also the charge distribution in a large headgroup will influence the aggregation processes [13]. In order to develop a description of the role of the charge distribution of the polar headgroup in the self assembly process, we have undertaken this study of the phase equilibria in the dodecyl pyridinium-water systems. The surfactant molecules are isomers with the hydrocarbon chain attached to different positions in the pyridine ring. Binary phase diagrams have been determined for the four different isomers. The results are discussed in terms of electrostatic effects and molecular packing constraints. Identification of phases and equilibrium phase boundaries in surfactant systems are obtained by non-evasive ^2H NMR and polarizing microscopy methods. In the theoretical calculations, the possibility of counterions to reside in the headgroup region has been taken into consideration. The hydration effect may lead to different stability regions of phases or in some cases new phases may be formed. We have therefore chosen to use the same counterion, Br^-, in all systems to avoid the effects caused by different counterions.

Experimental section

Materials

Unless otherwise stated, starting materials and solvents were used as received from commercial suppliers. GC analysis were carried out using a capillary column (Hewlett Packard, crosslinked 5% phenyl methyl silicone, SE54-type, 22 m, 0.31 m, 0.31 mm I.D., $d_f = 0.52$ μm, carier gas N_2 (10 psi), split ratio 1/20). Thin-layer chromatography (TLC) was performed on silica plates (Merck, 60, pre-coated aluminum foil) using ethyl acetate in hexane, and developed by means of ultraviolet irradiation. IR spectra were recorded as pellets (2 mg of substance and 200 mg of KBr) using a Perkin Elmer 782 infrared spectrometer. Boiling and melting points are uncorrected. NMR spectra were recorded in $CDCl_3$ and with chloroform as internal standard using a Jeol EX270 (270 MHz ^{1}H, 67.8 MHz ^{13}C) spectrometer. Elemental analyses were carried out by Mikrokemi, Uppsala, Sweden.

Preparation of surfactants

1-Dodecylpyridinium bromide monohydrate, see Fig. 1. a) The title compound was prepared by a method described previously by Jacobs et al. (14, 15). For physical data see Table 1. *2-Dodecylpyridinium hydrobromide*, see Fig. 1. (b), *3-Dodecylpyridinium hydrobromide*, Fig. 1. (c) and *4-Dedecylpyridinium hydrobromide* (16), Fig. 1. (d) The dodecylpyridinium hydrobromides were prepared using a method described previously by Jacobs et al. for the preparation of decylpyridinium hydrobromides. (13) to a suspension of 0.15 mol sodium amide (prepared from sodium (3.5 g, 0.15 mol), $Fe(NO_3)_3 \times 9 H_2O$ (0.43 g, 1.1 mmol) and 170 mmol and 170 ml of liquid ammonia) [14, 17] was added the appropriate methylpyridine (0.15 mol). After stirring for 0.5 h 1-bromoundecane (0.15 mol) was added during 0.1 h, the suspension was stirred and allowed to reach ambient temperature over night. The residue was treated dropwise with ethanol (6 ml) followed by water (125 ml). The diethyl ether extract (2 × 100 ml) of the solution was dried (MgSO$_4$), the solvent was evaporated off and the remaining methylpyridine was removed by distillation leaving an oil as residue. This oil was treated with 0.15 mol of hydrobromic acid in an ethanol-water solution. The precipitated dodecylpyridinium hydrobromide was purified by crystallization from acetone or ethanol-diethyl ether solutions. For physical data of the 2-,3- and 4-dodecylpyridinium hydrobromides

Table 1 Physical data of 1-, 2-, 3- and 4-dodecyl pyridinium bromide

Surfactant a–d	Yield %	m.p. °C	Chemical purity%	IR cm^{-1}	^{1}H NMR ppm.	^{13}C NMR ppm
a	54	74–76 lit.[ref.14] 74–75	>99[a]	3386, 2915, 2846 1635, 1485, 1470 1175, 777, 682, 455.	d 0.83 (3H, t, J = 6.6 Hz), 1.00–1.40(18H, m), 2.00 (2H, apparent quintet, J = 7.4 Hz), 4.95 (2H, t, J = 7.4 Hz, 8.13(2H, t, J = 7.1 Hz), 8.51(1H, t, J = 7.8 Hz), 9.45(2H, d, J = 5.9 Hz).	11.8, 20.3, 23.8, 26.8 27.0, 27.1, 27.2, 27.3, 27.3, 29.6, 29.7, 59.8, 126.3, 126.3, 126.3, 142.9, 142.9
b	47	71–72	>99.7[b]	2911, 2844, 2549 1612, 1461, 1152 783, 718, 628, 510.	d 0.82(3H, t, J = 6.8 Hz), 1.15–1.45(18H, m), 1.85(2H, apparent quintet, J = 7.4 Hz), 3.23(2H, t, J = 7.8 Hz), 7.71(1H, d, J = 7.9 Hz), 7.83 (1H, t, H = 6.8 Hz), 8.35–8.45(1H, m), 8.70(1H, d, J = 5.6 Hz), > 10(1H, bs).	11.8, 20.0, 26.7, 26.9, 27.0 ,27.1, 27.2, 27.2, 27.3, 27.3, 29.6, 30.9, 122.2, 124.6, 138.3, 143.6, 155.7.
c	58	97–98	>99.7[b]	2916, 2844, 2673 1603, 1535, 1462 1247, 1123, 810, 688.	d 0.83(3H, t, J = 6.6 Hz), 1.15–1.35(18H, m), 1.66(2H, apparent quintet, J = 7.4 Hz), 2.85(2H, t, J = 7.8 Hz), 7.94(1H, dd, J = 7.9 Hz), 8.27(1H, t, J = 8.3 Hz), 8.68(1H, s), 8.77(1H, d, J = 5.3 Hz), > 10(1H, bs).	13.9, 22.5, 28.8, 29.1, 29.2, 29.3, 29.4, 29.5, 29.5, 30.3, 31.7, 32.7, 126.7, 138.2, 139.9, 143.2, 145.9.
d	36	137–139 lit.[ref.16] 137.6– 138.9	>99.7[b]	2911, 2840, 2670, 1627, 1603, 1510 1462, 1177, 810, 512.	d 0.83(3H, t, J = 6.4 Hz), 1.15–1.40(18H, m), 1.69(2H, apparent quintet, J = 7.4 Hz), 2.87(2H, t, J = 7.8 Hz), 7.78(2H, d, J = 6.3 Hz), 8.82(2H, d, J = 6.3 Hz), > 10(1H, bs). lit.[ref.16]	13.9, 22.5, 29.0, 29.1, 29.2, 29.3, 29.4, 29.5, 29.5, 29.6, 31.7, 36.2, 126.8, 126.8, 140.1, 140.1, 164.5

[a] Chemical purity (^{1}H NMR) as 1-dodecylpyridinium hydrobromide hydrate.

[b] Chemical purity (GC) as 2-, 3-, and 4-dodecyl pyridine respectively.

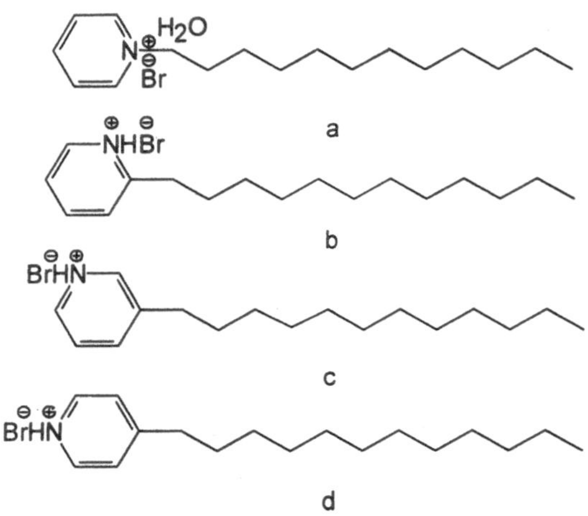

Fig. 1 The four dodecyl pyridinium bromide isomers studied in this work, the a) 1-dodecyl pyridinium bromide, b) 2-dodecyl pyridinium bromide, c) 3-dodecyl pyridinium bromide, and d) 4-dodecyl pyridinium bromide molecule

see Table 1. Anal. Calcd for $C_{17}H_{30}BrN$: C, 62.19; H, 9.21; N, 4.27. Found: 2-dodecylpyridinium hydrobromide C, 62.0; H, 9.40; N, 4.25, and 3-dodecylpyridinium hydrobromide C, 62.0; H, 9.35; N, 4.25.

Sample preparation

The samples were prepared by weighing appropriate amounts of substances into glass tubes which were flame sealed. The samples with liquid crystalline materials were centrifuged at regular intervals for 2 weeks until they attained equilibrium.

Methods

The phase diagrams of the surfactant systems were determined by a combination of polarizing microscopy and NMR techniques.

Polarizing microscopy

The samples were first examined between crossed polaroids for sample homogeneity and occurrence of birefringency as described elsewhere [18]. Liquid solutions and cubic liquid crystalline phases have isotropic structures, as a result, they produce dark background in the polarizing microscope. The textures of anisotropic liquid crystalline samples were studied at room temperature using a microscope equipped with a hotstage. The changes in the texture of samples were then examined as a function of

temperature, and the temperature was increased at a rate of 2 K per minute. The heating rate was slowed down near the phase transition temperature and at the phase transition, the samples were thermostated for several minutes.

²H NMR

²H NMR of deuterated water is a well established technique for studying the phase equilibria and phase boundaries of two- and multicomponent surfactant systems [5, 19]. The deuteron nucleus has a spin quantum number of unity and it possesses an electric quadrupole moment. For an anisotropic medium, e.g., a liquid crystalline phase, the interaction of the quadrupole moment with the electric field gradients at the nucleus generates a powder spectrum with two peaks. This observed quadrupole splitting, Δ, measured in Hz, as the peak-to-peak distance in a spectrum in hexagonal and lamellar liquid crystalline phases depends on the fraction of deuterons in one or more anisotropic sites, the quadrupole coupling constant and the average molecular ordering of water in the sites. It has been shown that $\Delta_{lam} = 2\Delta_{hex}$ if the local conditions are the same [20]. Moreover, the Δ-value is dependent on surfactant concentration. In an isotropic phase, such as a micellar solution, this interaction is averaged to zero as a result of rapid and isotropic molecular motion. In this case the spectrum consists of a sharp singlet. For the same reasons one obtains a singlet in a cubic liquid crystalline phase.

For a heterogeneous system consisting of two or more phases, a superposition of the ²H NMR spectra characterizing the phases is obtained. Thus, for a heterogeneous system containing a mixture with one anisotropic and one isotropic phases, a doublet and a central singlet will be observed. Therefore, the analysis of ²H NMR spectra provides a direct determination of the phase diagrams of surfactant systems.

²H NMR experiments were performed at a resonance frequency of 41.47 MHz on a Jeol EX 270 pulsed FT spectrometer equipped with a super-conducting magnet of 6.34 T. A variable temperature control unit was used to control the temperature at the airflow in which the 10 mm (i.d) NMR tube containing the samples tube was placed in the NMR probe. The accuracy of the probe temperature was better than ± 0.5 °C.

Results and Discussion

The phase diagrams of the four binary surfactant-water (composition versus. temperature) systems determined by

Progr Colloid Polym Sci (1994) 97:134–140
© Steinkopff-Verlag 1994

combined ^{2}H NMR and polarizing microscopy methods are shown in Fig. 2. They were constructed on the basis of wt% of each component. However, the molecular weight of the surfactant molecules ($M_r = 328.34$ g/mole) are equal and therefore the different systems can be compared directly. In other cases it would be preferable to use molar concentration units.

The samples in the pure E- and D-phases in the 1-,2-and 3-DPB systems produce single splittings in their ^{2}H NMR spectra and the splitting values (Δ) increases with increased surfactant concentration, whereas in the two-phase regions, ($L_1 + E$), ($E + I$) and ($I + D$), the spectra consists of a quadrupolar splitting and a central isotropic singlet. However, in the 4-DPB system no splittings were obtained. The exact location of the phase boundaries for this system could not be determined accurately by NMR, so approximate phase boundaries were obtained only by polarizing microscopy. This is indicated by dashed lines in Fig. 2.

The Krafft temperature, T_k, was above the room temperature for all studied systems in dilute aqueous solutions, except for the 1-dodecyl pyridinium bromide system. The electrostatic interactions are stronger when the charge in the surfactant ion is localized and sterically accessible. In the 1-DPB molecule, a great deal of the charge at the nitrogen atom is sterically inaccessible, thus preventing strong interactions in the solid. The other isomers have the nitrogen atom located further away from the hydrocarbon chain and the electrostatic interactions are stronger in the crystal than the solvation energy in a solution would be. The Krafft point of the systems studied increased with the surfactant concentration.

The surfactants were easily soluble in water. Critical micellar concentration measured by the surface tension method at 40 °C was observed to be about 10 mM for the 1-and 4-DPB systems. In the 2-and 3-DPB systems the CMC was higher, about 15 mM. Initially, they formed small spherical micelles and these may undergo spherical-to-rod shape transformation at higher surfactant concentrations. Preliminary NMR self-diffusion measurements showed that in the 3-DPB system this transformation is facilitated. In a study of decyl pyridinium bromides, similar results were found [13]. At even higher surfactant concentrations the first liquid crystalline phase appeared [8, 21]. We found that these systems form normal hexagonal [8] but not discrete cubic [8, 21] phases as the first liquid crystalline phase.

The aggregation of ionic surfactants is dominated by the electrostatic effect and surfactant geometrical constraints. The geometrical constraints may be described by a packing parameter [22, 23], $P = v/al$, where v is the alkyl chain volume, l is the optimal chain length, and a is the headgroup area. For $P < 1/3$, one expects spherical micelles; $1/3 < P < 1/2$, rod-shaped micelles; $1/2 < P < 1$, vesicles or lamellae; and $P > 1$, inverted structures, e.g., reversed micelles or reversed hexagonal liquid crystals.

Fig. 2 Binary phase diagram for the a) 1-dodecyl pyridinium bromide–water system, b) 2-dodecyl pyridinium bromide–water system, c) 3-dodecyl pyridinium bromide-water system, and d) 4-dodecyl pyridinium bromide-water system

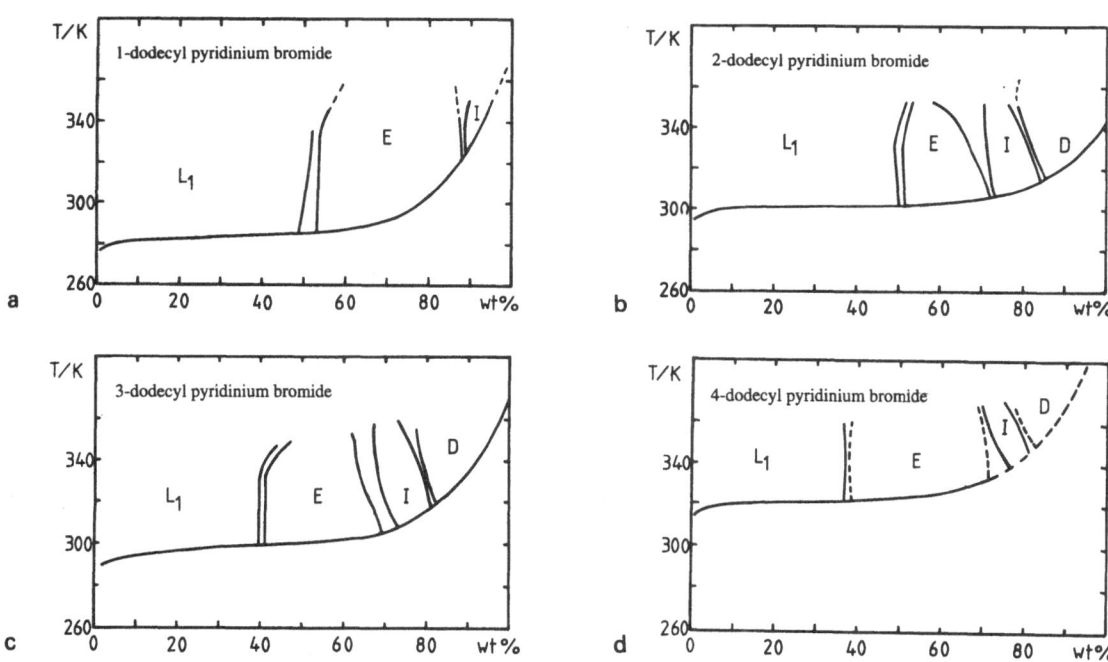

Since the size of the headgroups in the systems is approximately equal, the differences in the aggregation behavior is therefore expected to depend mainly on the differences in the charge distribution of the headgroup. Thus, the 4-DPB molecule, for example, could simply be considered as a surfactant molecule with a longer chain (hydrocarbon chain plus diameter of the pyridine ring) than the 1-DPB molecule. In addition, tilting of the headgroup can be expected especially in the 1-DPB system since the electrostatic charge resides mainly at the nitrogen atom and the rest of the pyridine ring tends to avoid water contact. The result should be a larger headgroup area leading to unfavorable packing in a lamellar phase. It was also observed that no lamellar liquid crystalline phase exists in the 1-DPB surfactant/water system.

In the 1-DPB system, the counterions would be expected to be located at a longer distance from the headgroup charge because this is very close to the hydrocarbon chain. However, if the headgroup is tilted, most of the charge, which is located at the nitrogen atom, becomes more accessible. This would lead to a small micellar region. On the other hand, the headgroup area is large, probably because of the tilting, which is favorable for micelle formation. It can be seen in Fig. 2 that the micellar region for the 1-DPB system extend to over 50 wt% of the surfactant. In the 2- and 3-DPB systems, the tilting can be expected to be smaller. The headgroup charge will thus be more or less inaccessible and the counterions cannot approach very close, leading to the same effect as for systems with strongly hydrated counterions. In the 4-DPB surfactant molecule most of the pyridinium ring can be considered to belong to the hydrocarbon chain. Thus, the behavior should be the same as for a surfactant with a chain length that is the sum of the lengths of the hydrocarbon chain and the diameter of the pyridine ring. The experimental results agree with this explanation since the micellar phase extends to only 36 wt% of surfactant concentration compared to 48 wt% and 40 wt% for the 2- and 3-DPB systems, respectively.

In all DPB systems studied, a hexagonal liquid crystalline phase is formed at higher surfactant concentration. Some representative Δ-values obtained in the hexagonal and lamellar liquid crystalline phases are shown in Table 2.

In order to rationalize the Δ-values in more detail, we have used the conventional two site model with a division into "free" and "bound" water molecules. Assuming the ordering of free water molecules to be negligible, we may express the splitting as a function of molar ratio between surfactant and water as [24]:

$$\Delta = nv_Q \frac{S \cdot X_{surf}}{X_w} = nv_Q S \frac{(1 - X_w)}{X_w} , \tag{1}$$

Table 2 Representative deuterium splittings in hexagonal and lamellar phases at 50 °C. See also Fig. 3.

Surf. Conc.	1-DPB	2-DPB	3-DPB
wt%	Δ/Hz	Δ/Hz	Δ/Hz
Hexagonal phase			
50	450	700	320
55	510	800	460
60	610	990	540
65	690	1230	590
70	850		710
75	990		
Lamellar phase			
85		2100	830

where X_{surf} is the mole fraction of surfactant and X_w is the mole fraction of water in the system, n is the average number of hydrating water molecules per surfactant molecule, S is the order parameter and v_Q is the deuteron quadrupole coupling constant, about 220 kHz [25]. Assuming n and S to be constants for the compositions discussed here, we may write Eq. (1) as

$$\Delta = k \frac{(1 - X_w)}{X_w} \tag{2}$$

The following observations can be made from the measured ^{2}H NMR splitting values. A plot of Δ vs. $(1 - X_w)/X_w$ yields straight lines passing through the origin when the hydration properties of surfactants are independent of water content. The only effect of increasing the water concentration in such a system is to increase the amount of free water, the so-called ideal swelling behavior.

The plots of Δ vs. $(1 - X_w)/X_w$ show the swelling of the hexagonal phases in Fig. 3. However, the plots in the water-poor regions of the hexagonal and lamellar phases do not pass through the origin. These results are expected since the two-site model is unlikely to be applicable in very concentrated systems.

The isotropic cubic liquid crystalline phases are formed after the hexagonal phases and they appear to have a bicontinuous type structure [26]. Thermal stability of the liquid crystalline phases was rather high and the difference in thermal stability between systems was also small.

Calculated phase boundary micellar/hexagonal phases. A thermodynamic model developed by Jönsson et al. [1, 2] has been used to calculate the phase boundary between the micellar and hexagonal phases. The system is divided into a number, N, of identical cells. The cells and aggregates are approximated as spheres in the micellar phase and cylinders in the hexagonal phase. In every cell, the aggregate is centered. The size and number of cells

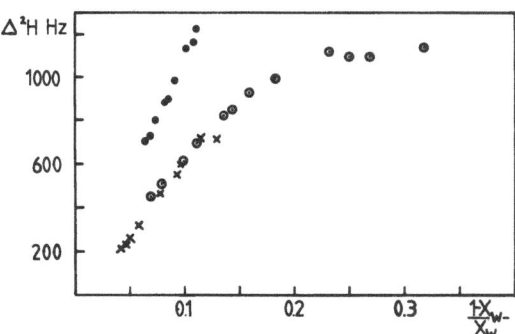

Fig. 3 Water deuteron quadrupolar splittings plotted against the ratio of the mole fractions of amphiphile and water in the hexagonal phases of 1-,2- and 3-dodecyl pyridinium bromide-water system, ⊙ 1-OPD, ● 2-DPB and × 3-DPB

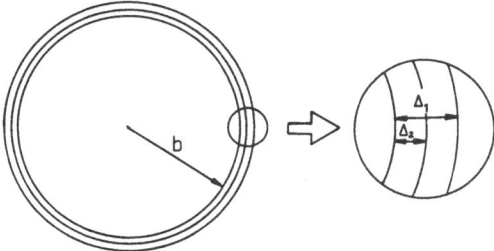

Fig. 4 Model of an aggregate (cylinder or sphere) used in the theoretical calculations. Δ_1 is the distance from the hydrocarbon chain to the surface of charge and b is the chain length. Δ_2 is the distance from the hydrocarbon chain to the water region

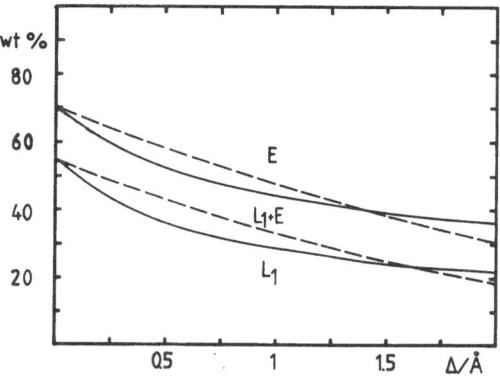

Fig. 5 Phase boundary between the micellar phase (L_1) and the two phase region ($L_1 + E$) and this two phase region and the hexagonal phase region (E) as a function of Δ_1, calculated at 328 K; full lines when $\Delta_2 = 0$ and dashed lines when $\Delta_1 = \Delta_2$

allowed to enter into the hydrocarbon core. The results of the calculations are shown in Fig. 5.

depend on surfactant concentration and aggregate dimension. In the calculations Tanford's formulas [27] have been used to estimate the extended chain length and in the calculations this value has been used as in earlier works. The Poisson–Boltzmann differential equation in the appropriate symmetry for the electrostatic field outside the positively charged aggregate is solved numerically. The electrostatic potential as a function of the distance from the center of the cell and hence the distribution for all ions is obtained. It is then possible to calculate the chemical potential of amphiphile and water. When, for example, micelles and the hexagonal aggregates are in equilibrium, the calculated amphiphile chemical potentials for the spheres and cylinders are equal. Water is considered as a continuous medium with the dielectric constant ε_w.

Two cases were considered in our calculations. In the first case, the charges of the head groups were smeared out into a continuous surface charge density at the micellar surface. In the other case, the distance from the hydrocarbon core to the charged surface (spherical or cylindrical) was varied as shown in Fig. 4. The volume charge density, ρ, of the volume between this surface and the hydrocarbon core, was kept constant. In our calculations, it was possible for counterions to reside in this volume, but they were not

Concluding remark

The phase diagrams of the four dodecyl pyridinium bromide isomers in water resemble each other. The order of appearance of the different phases with increased surfactant concentration is the same in all cases, micellar, hexagonal, cubic and in the 2-, 3-, and 4-DPB systems a lamellar phase. However, some significant differences between them can be observed. In the 1-DPB system there is a large micellar and hexagonal, but no lamellar phase, which indicates a large headgroup area possibly due to the tilting of the pyridine ring. In the aggregation, the 2-DPB surfactant molecule behaves as if its hydrocarbon chain length is slightly longer than the chain of the 1-DPB molecule, the 3-DPB molecule seems longer than the 2-DPB and the 4-DPB molecule seems longer than the 3-DPB molecule. From both the experiments and the theoretical calculations a trend can be found. The phase boundary between the micellar phase and the micellar and hexagonal two-phase region changes gradually to lower concentrations for the 1-, 2-, 3- and 4-DPB systems respectively as the phase boundary between the two phase region and the hexagonal phase region does. This decrease is consistent with the first case in the theoretical calculations where the hydrocarbon core extends to the charged surface and thus the calculations suggest that there is tilting of the headgroups to some extent in all the systems studied.

Acknowledgements We are grateful to University College of Sundsvall/Härnösand for financial support and to Kai Kangassalo at Swedish Institute for Materials Technology for building the hostage used in the microscopy studies. Valuable comments on the manuscript by Prof. Hakan Wennerström and Dr. Bengt Jönsson at Physical Chemistry 1, Lund University are gratefully acknowledged.

References

1. Jönsson B, Wennerström H (1981) J Colloid Interface Sci 80:482
2. Jönsson B, Wennerström H (1987) J Phys Chem 91:338
3. Israelachvili JN (1985) Intermolecular and Surface Forces. Academic Press, New York and London
4. Wennerström H, Khan A, Lindman B (1991) Adv in Colloid and interface Sci 34:433
5. Khan A, Fontell K, Lindblom G, Lindman B (1992) J Phys Chem 86:4266
6. Khan A, Fontell K, Lindman B (1982) Colloids and Surfaces 11:401
7. Khan A, Fontell K, Lindman B (1985) Prog Colloid Polym Sci 70:30
8. Maciejewska D, Khan A, Lindman B (1987) Prog Colloid Polym Sci 73:174
9. Fontell K, Khan A, Lindström B, Maciejewska D, Puang-Ngern S (1991) Colloid Polym Sci. 269:727
10. Wennerström H, Lindman B (1979) Phys. Rep. 52:1
11. Hagslätt H, Söderman O, Jönsson B, Johansson B-Å (1991) J Phys Chem 95:1703
12. Lindström B, Khan A, Söderman O, Kamenka N, Lindman B (1985) J Phys Chem 89:5313
13. Jacobs P, Anacker EW (1973) J Colloid Interface Sci 44:505
14. Ames DE, Bowman RE (1952) J Chem Soc 44:1057
15. Jacobs, PT Geer, RD Anacker EW (1972) J Colloid Interface Sci 39:611
16. Sudhölter EJR, Engberts JBFN, de Jeu WH (1982) J Phys Chem 86:1908
17. Furniss BS, Hannaford AJ, Smith PWG, Tatchell AR, Vogels Textbook of Practical Organic Chemistry. Fifth edition, Longman Scientific & Technical, John Wiley and Sons, New York 1989, pp. 1034–1035
18. Rosevear FD (1968) J Soc Cosmet Chem 19:581
19. Ulmius J, Wennerström H, Lindblom G, Arvidsson (1977) Biochemistry 16:5742
20. Wennerström H, Lindblom G, Lindman (1974) Chem Sci 6:97
21. Balmbra R, Clunie (1969) Nature Lond 222:1159
22. Israelachvili JN, Michell DJ, Ninham BW (1976) J Chem Soc Faraday Trans. 2 72:1525
23. Mitchell DJ, Ninham BW (1981) J Chem Soc Faraday Trans 2 77:609
24. Wennerström H, Persson NO, Lindman B (1975) ACS Symp Ser 9:253
25. Glasel JA (1972) in "Water, A Comprehensive Treatise" (F. Franks, ed.), Vol. 1, p. 215, Plenum Press, New York and London
26. Kang C, Khan A (1993) J Colloid Interface Sci. 156:218
27. Tanford C (1972) J Phys Chem 76:3020

Progr Colloid Polym Sci (1994) 97:141–145
© Steinkopff-Verlag 1994

SURFACTANTS

AOT, influence of impurities on the phase behavior

W. Sager
R. Strey
W. Kühnle
M. Kahlweit

Received 24 September 1993
Accepted 24 January 1994

Dr. W. Sager (✉)
Department for Physical and
Macromolecular Chemistry
Gorlaeus Laboratories
University of Leiden
Postbus 9502
Einsteinweg 55
2300 RA Leiden, The Netherlands

R. Strey · W. Kühnle · M. Kahlweit
Max-Planck-Institut für
Biophysikalische Chemie
Postfach 28 41
37018 Göttingen, FRG

Abstract A peculiar feature often observed within the sodium diethylhexylsulfosuccinate (AOT)–water–oil phase diagram is an isolated two-phase island, when water in oil (w/o) microemulsions are formed at low temperatures. In this work, the influence of the hydrolysis products of AOT, namely monoester and alcohol, on the phase behavior has been modeled by adding SDS and octanol to $(H_2O/NaCl)$–decane–AOT systems. Addition of small amounts of either compound, frequently present as impurities in AOT, leads to drastic shifts in the mean temperature of the three-phase regions in opposite directions. In particular, addition of octanol changes the shape of the one-phase region adjacent to the three-phase body.

Key words AOT – sodium diethylhexylsulfosuccinate – phase behavior – microemulsions – phase diagrams

Introduction

AOT (sodium di-(2-ethylhexyl)sulfosuccinate) is one of the classic examples of an ionic surfactant that forms microemulsions without adding a cosurfactant. However, the phase diagrams published for AOT samples purified using a variety of methods, (see, e.g., [1, 2]), differ considerably with respect to the extensions and shapes of the microemulsion regions, indicating that small amounts of residual impurities have a large effect. Hydrolysis of this diester [3] leads to the formation of more hydrophilic compounds such as the monoester and the dicarboxylate ion as well as the more hydrophobic alcohol.

An often observed peculiar feature of the water–oil–AOT phase diagram (Fig. 1 top) is the occurrence of an isolated two-phase island at low temperatures when w/o microemulsions are formed (see e.g. [1, 4]). Within this island, which shrinks with increasing temperature towards the oil-corner of the Gibbs' triangle, a mixture separates into an upper w/o microemulsion-phase and an almost pure water-phase. The tielines connecting both phases in equilibrium therefore do not lie in the plane of the paper, which indicates that according to the phase rule the system contains more than three components. In comparison, the phase diagram of an "ordinary" Winsor II system (three-component system) shows a central miscibility gap with tielines leading from the water-corner of the Gibbs' triangle to the oil-rich-side of the binodal (Fig. 1 bottom).

In order to understand the origin of this isolated two-phase island, we modeled the effects of the hydrolysis products of AOT, frequently present as impurities in low concentration, by adding sodium dodecyl sulfate (SDS) and octanol respectively. To follow trends in the phase behavior of systems containing more than three components systematically, a reference point is required. We used in this investigation the three-phase body as reference. Mixtures of water, oil and surfactant can separate into three distinct phases by varying, e.g., temperature or salt

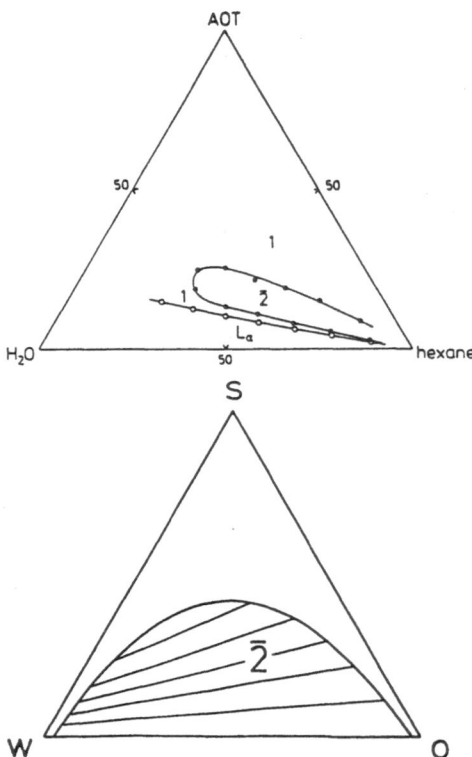

Fig. 1 *Top:* Gibbs phase triangle obtained for a system of water–hexane–AOT at 25 °C [4]. The phase diagram shows the often observed isolated two-phase island ($\bar{2}$), in which a w/o microemulsion phase is in equilibrium with an almost pure water phase. The island is surrounded by an one-phase w/o microemulsion (1); at lower surfactant concentration a lamellar phase (L_α) is observed *Bottom:* Schematic phase diagram for an ordinary Winsor II system obtained in a ternary system. The tielines in the central miscibility gap ($\bar{2}$) lead from the water-corner of the Gibbs triangle to the oil-rich side of the binodal

Fig. 2 Vertical section at $\alpha = 50$ wt% through the phase prism with the Gibbs triangle as base and the temperature as ordinate (top) to determine the position and the shape of the three phase body (3) (bottom). $\bar{2}$ indicates a two-phase region, in which a o/w microemulsion phase is in equilibrium with an excess oil phase. The "fish tail end point", $\tilde{X}$, indicating the minimum amount of surfactant needed to solubilize equal amounts of water and oil, was used as reference point

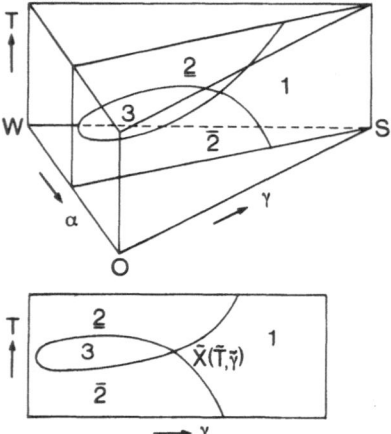

concentration [5]. The "fish tail end point", $\tilde{X}$ (see Fig. 2), which can be detected by a vertical cut through the phase prism at a water to oil ratio of 1 ($\alpha = 50$ wt %) is present as a point, line or plane in three, four and five component systems respectively and can therefore be used as a reference point in multicomponent systems. The water–oil–AOT system forms three phases only in the presence of salt [6].

Experimental section

Throughout the whole investigation we used AOT Mikroselect purchased from Fluka, whose purity was checked using thin layer chromatography. Samples were prepared by mixing appropriate masses of aqueous sodium chloride solutions (w), decane (o) and surfactant (s). The concentration of salt in the aqueous phase was given by $\varepsilon = \text{NaCl}/(\text{NaCl} + \text{water})$. Phase diagrams were performed erecting a vertical section through the phase triangle at $\alpha = 50$ wt % (Fig. 2), where $\alpha = o/(o + w)$. The temperatures of the phase boundaries were detected as a function of the surfactant concentration γ ($\gamma = s/(s + o + w)$) by equilibrating the samples in a thermostatted water bath using a magnetic stirrer. The concentration of the additive (SDS or octanol) in the surfactant mixture was given by $\delta = \text{additive}/(\text{additive} + \text{AOT})$. To obtain the phase diagrams at a given δ, a sample at high surfactant concentration was consecutively diluted with equal amounts of NaCl solution and decane. Phase separation took place within a few minutes up to several hours. The position of the phase boundaries was reproducible for both raising and lowering the temperature.

Results and discussion

The effect of SDS on the phase behavior is shown in Fig. 3. Upon addition of SDS the one-phase region shifts to lower temperatures. The fish tail end point, $\tilde{X}$, of the system (H$_2$O/NaCl)–decane–AOT (filled squares in Fig. 3) lies for $\varepsilon = 0.6$ wt % at 40 °C. At $\delta = 7$ wt %, $\tilde{T}$ is shifted down to 14 °C. The shape of the one-phase region adjacent to the three-phase body, however, does not change significantly in the δ-range of 1–7 wt %. Figure 4 reveals that the addition of octanol leads to a shift of the one-phase region to higher temperatures. The fish tail end point, $\tilde{X}$, of the system (H$_2$O/NaCl)–decane–AOT (filled squares in Fig. 4) lies for $\varepsilon = 0.4$ wt % at 24.5 °C. At $\delta = 7$ wt %, $\tilde{T}$ is shifted up to 30 °C. The addition of octanol also leads to a drastic change in the shape of the one-phase region adjacent to the three-phase body. The lower phase bound-

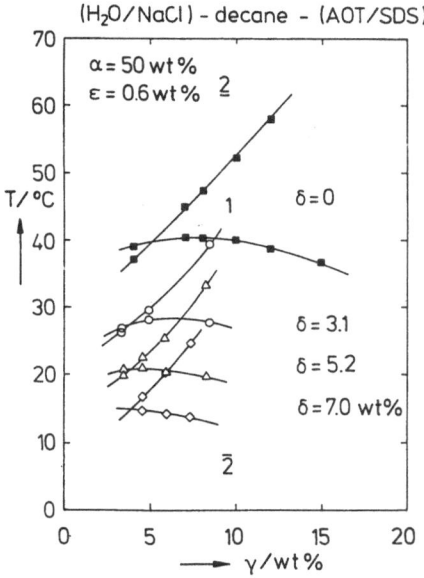

Fig. 3 Phase diagram of (H$_2$O/NaCl)–decane–(AOT/SDS) systems for different SDS to AOT ratios δ as a function of total surfactant concentration at $\alpha = 50$ wt % and $\varepsilon = 0.6$ wt %. The quaternary system ($\delta = 0$) is represented by filled squares

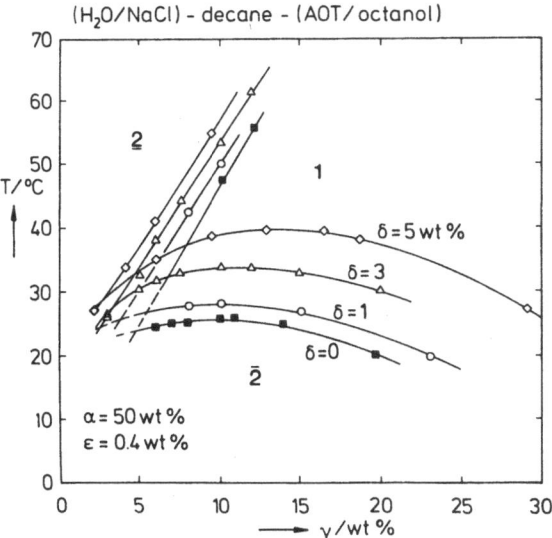

Fig. 4 Phase diagram of (H$_2$O/NaCl)–decane–(AOT/octanol) systems for different octanol to AOT ratios δ as a function of total surfactant concentration at $\alpha = 50$ wt % and $\varepsilon = 0.4$ wt %. The quaternary system ($\delta = 0$) is represented by filled squares

ary bends increasingly upon addition of octanol. An isothermal cut through this hump reveals the phase sequence of 1, $\bar{2}$, 1, $\underline{2}$ with decreasing total surfactant concentration, thereby indicating that the one-phase region is reentered at low surfactant concentration.

The dependence of $\tilde{T}$ on δ is shown for both additives in Fig. 5. The addition of SDS and octanol lead to drastic changes in $\tilde{T}$, acting in different directions. For both additives, $\tilde{T}$ depends almost linearly on δ, independent of the salt concentration (ε) necessary to obtain a three-phase body. Upon adding SDS the "surfactant mixture" becomes more hydrophilic which causes $\tilde{T}$ (and the HLB-temperature) to decrease. Octanol partitions between the interface and the bulk-oil-phase. Upon addition of octanol the "surfactant mixture" becomes more hydrophobic, while the hydrophobicity of the oil decreases. Both changes cause $\tilde{T}$ (and the HLB-temperature) to rise. Long chain alcohols (e.g., octanol) cannot be treated as a pseudo-component owing to their partitioning between the interface and the bulk oil-phase. If the alcohol is added to a ternary system either with the surfactant or the oil, its interfacial-concentration will change upon dilution depending on the equilibrium concentration of the alcohol in the oil. Dilution of a sample, prepared with a fixed ratio of alcohol to surfactant (δ), by adding equal amounts of water and oil, (Case A), would therefore decrease the concentration of the alcohol in the interface. Dilution of a sample, prepared with a fixed ratio of alcohol to oil (β), with water and a solution of the alcohol in the oil of the same β, (Case B), would increase the interfacial-concentration of the alcohol.

To demonstrate how the addition of alcohol can generally affect the shape and position of the three-phase body and its adjacent one-phase region, Fig. 6 displays the phase diagram of the nonionic system: water–octane–pentaethylene monodecylether (C$_{10}$E$_5$)–octanol. Addition of octanol leads in Case A (const. δ) to a decrease of the three- and one-phase temperature intervals with increasing γ [7]. In Case B (const. β) the three- and one-phase temperature

Fig. 5 Dependence of $\tilde{T}$ on δ for octanol and SDS at $\varepsilon = 0.4$ wt % and $\varepsilon = 0.6$ wt %

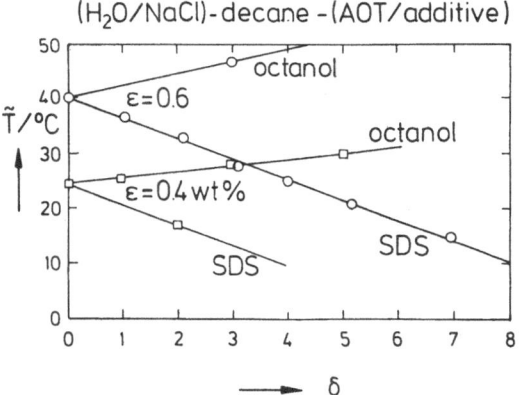

144
W. Sager et al.
AOT influence of impurities on the phase behavior

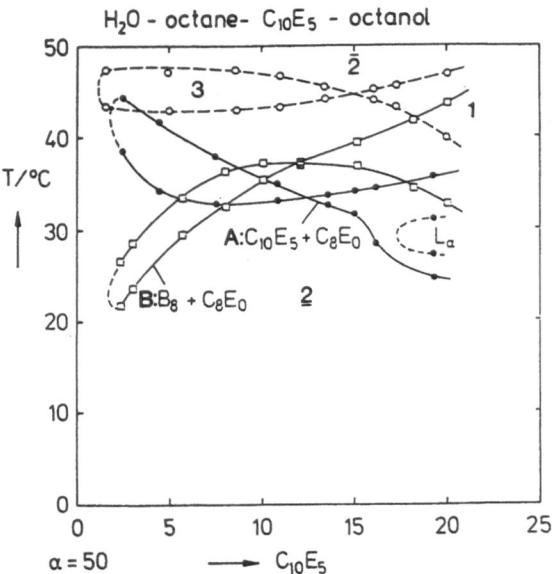

Fig. 6 Phase diagram of H_2O–octane–$C_{10}E_5$–octanol (C_8E_0) systems. The ternary system is represented by a dotted line. The quaternary systems are shown for: (A) constant octanol to $C_{10}E_5$ ratio δ of 10 wt % (filled circles) and (B) constant octanol to octane ratio β of 2 wt % (open squares)

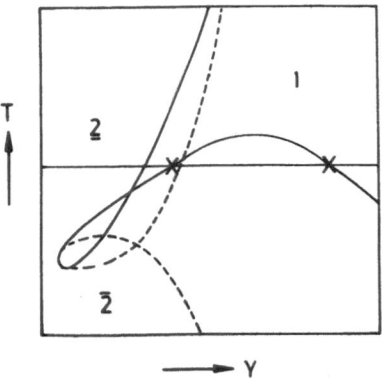

Fig. 7 Schematic diagram illustrating the influence of octanol on the shape of the one-phase region adjacent to the three phase body. The dotted line represents the pure ternary system, whereas the full line shows the system in the presence of octanol. The two crosses mark the phase boundaries of the two-phase island at $\alpha = 50$ wt % for that particular temperature

intervals increase with increasing γ. The "fish" of the ternary system (dotted line) will be therefore approached at low γ for A and at high γ for B. The drift of the "fish head"

(A and B) causes the upper (A) and lower (B) phase boundary of the one-phase region adjacent to the three-phase body to bend. Since nonionic and ionic surfactants reveal an opposite phase behavior with respect to temperature, the mirror image of the "fish" in case A, reflected at a plane parallel to the γ-axis, corresponds to the AOT–octanol case, discussed in Fig. 4. The schematic diagram illustrating the influence of octanol on the shape of the one-phase region adjacent to the three-phase body is shown in Fig. 7. The dotted line represents the pure ternary system. The increase of the octanol concentration with γ causes a shift of the one-phase region to higher temperatures. The ternary system will therefore be reached at low total surfactant concentration. The two crosses (Fig. 7) mark the phase boundaries of the two-phase island at $\alpha = 50$ wt % for that particular temperature. The hump-like shape indicates that the extensions of the isolated island in the Gibbs triangle shrink with rising temperature.

Conclusions

The peculiarity in the frequently observed phase behavior of AOT can be modeled by the controlled addition of SDS and octanol. The addition of SDS and octanol, respectively, leads to a systematic shift of the fish tail end point temperature ($\tilde{T}$). SDS as the more hydrophilic component decreases $\tilde{T}$, whereas octanol as the more hydrophobic component increases $\tilde{T}$. The position of $\tilde{T}$ (or the HLB-temperature) is therefore no purity criterion. This position depends on the different hydrolysis product residues, since the presence of both compounds leads to a drastic shift in $\tilde{T}$, but acts in different directions. Partitioning of octanol between the interface and the bulk oil phase bends the lower one-phase phase boundary and causes the formation of the isolated two-phase island in the Gibbs' triangle. The extensions of the isolated island depend for a given temperature on the amount of octanol present as impurity. Since the concentration of octanol in the interface decreases with decreasing surfactant concentration in the "ternary" system, the ternary "fish" will be approached at low total surfactant concentration.

Acknowledgement We thank T. Lieu for assistance with phase diagram measurements and D. Luckmann for drawing the figures.

Progr Colloid Polym Sci (1994) 97:141–145
© Steinkopff-Verlag 1994

References

1. Kunieda H, Shinoda K (1979) J Colloid Interface Sci 70:577
2. Tamamushi B, Watanabe N (1980) Colloid Polym Sci 258:174
3. Fletcher PDI, Perrins NM, Robinson BH, Toprakcioglu C (1985) In: Luisi PL, Staub B (eds) Biological and technological relevance of reversed micelles and other amphiphilic structures in apolar media. Plenum, New York, p. 69
4. Sager W, Sun W, Eicke H-F (1992) Progr Colloid Polym Sci 89:284
5. Kahlweit M, Strey R, Busse G (1990) J Phys Chem 94:3881
6. Kahlweit M, Strey M, Schomäker R, Haase D (1989) Langmuir 5:305
7. Kahlweit M (1993) Tenside Surf Det 30:1

Progr Colloid Polym Sci (1994) 97:146–150
© Steinkopff-Verlag 1994

SURFACTANTS

A. Khan
O. Regev
A. Dumitrescu
A. Caria

Mixed surfactants: Sodium bis(2-ethyl-hexyl)sulpho-succinate- didodecyldimethyl-ammonium bromide- water system

Received: 24 September 1993
Accepted: 24 January 1994

Ali Khan (✉) Oren Regev[†]
Adina Dumitrescu Annalisa Caria
Division of Physical Chemistry 1
Chemical Centre
Box 124
University of Lund
22100 Lund, Sweden

[†] Present address: Department of Chemical
Engineering, Ben-Gurion University, Box 653,
84105 Beer-Sheva, Israel

Abstract Phase equilibria for the mixed surfactant system didodecyl-dimethylammonium bromide (DDAB)-Sodium di-(2-ethylhexyl) sulphosuccinate (AOT)-water are studied by ^{2}H NMR, optical- and cryo-transmission electron microscopy methods at 303 K. The phase behavior displayed by the mixed system as well as the stability region of various single phases are discussed in terms of surfactant molecular packing considerations and electrostatic effects.

Key words Mixed surfactants – didodecyldimethylammonium bromide – sodium di-(2-ethylhexyl)sulphosuccinate – phase equilibria – vesicles – liquid crystals – L_3 phase NMR

Introduction

Aqueous equimolar mixtures of oppositely charged ionic surfactants exhibit novel features that are often absent in the parents' surfactant–water systems. Studies which are concerned mostly with the very dilute region [1–3] indicate that the mixed system a) is more surface active than their parents' components; b) often precipitates forming surfactant crystals; c) has Krafft point higher than that of individual surfactant system; and d) forms stable, single-walled vesicles especially with excess of one of the parent's components. However, there are only few reports on phase equilibria for the mixed systems in concentrated regions. Phase equilibria for the mixed system with one single alkyl chain cationic - one single chain anionic [4], and one double chain cationic – one single chain anionic [3] surfactants are published. Here, we report a preliminary investigation of phase equilibria for the system didodecyldimethylammonium bromide (DDAB)-Sodium di-(2-ethylhexyl)sulphosuccinate (AOT)-water at 303 K. Like many other double-tailed ionic surfactants, DDAB (Fig. 1) forms lamellar liquid crystals (D) with water, but what is unique with this system is that it forms two lamellar liquid crystalline phases -D_1 and D_2, at low (4–30 wt %), and high (83–91 wt %) surfactant concentrations, respectively, and the two lamellar phases coexist through an extensive two-phase, $D_1 + D_2$, region [5, 6]. Cryo-TEM micrographs show the presence of single- and multi-walled vesicles in aqueous dispersions below 3 wt% of surfactant [7].

AOT (Fig. 1) is an asymmetric surfactant with two branched alkyl chains and it displays a rich polymorphism with water. Multi-lamellar vesicles are detected by cryo-TEM in lamellar dispersions at low surfactant concentrations [7] prior to the formation of a single lamellar phase which extends to high surfactant contents (10–70 wt%). With increasing concentration, a bicontinuous cubic phase (77–80 wt% surfactant) and a reversed hexagonal liquid crystalline phase ($\geq$ 83 wt% surfactant) are formed [8].

Experimental section

Sodium di-(2-ethylhexyl)sulphosuccinate (Aerosol OT) and didodecyldimethylammonium bromide were obtained from Fluka Chemie AG, Switzerland, and were used as

Progr Colloid Polym Sci (1994) 97:146–150
© Steinkopff-Verlag 1994

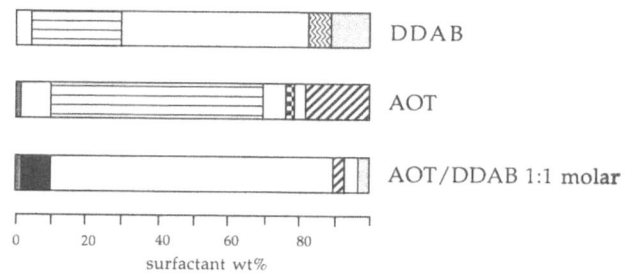

Fig. 1 Phase diagrams for the aqueous systems of a) DDAB, b) AOT and c) DDAB-AOT (equimolar) at 303 K.
Phase notations:
■ – isotropic solution phase (L); ▨ – reversed hexagonal phase (F);
⊠ – cubic phase (I); ▤, ▥ – lamellar liquid crystalline phases (D).
▢ – hydrated crystals (G); □, ■ – appropriate two-phase regions.

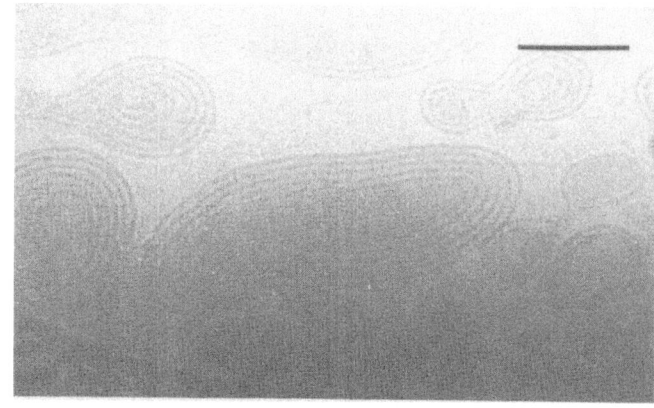

Fig. 2 Cryo-TEM micrographs of 1 wt% of equimolar DDAB-AOT aqueous solution at 303 K (bar = 200 nm)

received. ^2H_2O (99.7 atom % ^2H) was obtained from Norsk Hydro, Norway.

Samples were prepared by weighing appropriate amounts of each component into 10 mm glass ampoules which were then flame-sealed. The samples were mixed by centrifugation and equilibrated as described previously [3].

Identification of phases and equilibrium phase boundaries were obtained by combined methods of water ^2H NMR and polarizing microscopic studies [3].

Anisotropic liquid crystals are identified by microscopic textures and the isotropic phases are non-birefringent and do not show any texture in the polarizing microscope.

A ^2H NMR spectrum yields a doublet in a single anisotropic liquid crystalline phase, a singlet in an isotropic phase and a superposition of the spectra of the independent phases in mixtures containing isotropic and anisotropic phase. Thus, the analysis of the ^2H NMR lineshapes and their intensity vs. sample compositions yields the phase boundary of different phases in a phase diagram. ^2H NMR spectra were recorded at 15.371 MHz on the Brukar MSL 100 spectrometer. Aggregate structures in the very dilute part of the phase diagram were studied by the cryo-transmission electron microscopic method as described previously [9].

Results and discussion

AOT/DDAB (equimolar)-water phase behavior.

The Krafft point of a dilute aqueous solution of an equimolar composition of AOT and DDAB is below room temperature (25 °C). With less than 1 wt% of surfactants in water, a bluish colored solution is obtained and cryo-TEM micrographs taken in the solution show the presence of

multi-walled vesicles with high polydispersity (Fig. 2). A precipitation is obtained between 1–10 wt% of surfactants (Fig. 1). Few crystals are separated from the bulk liquid and the wet crystals when warmed to about 50 °C show a focal conic microscopic texture typical of hexagonal liquid crystals. A two-phase region consisting of anisotropic liquid crystals and an isotropic liquid phase is obtained between 10 and 90 wt% surfactants, above which only a single anisotropic liquid crystalline phase is stable. The liquid crystals have hexagonal microscopic texture and produce a powder-type single quadrupolar splitting in the ^2H NMR spectra. From the position of the liquid crystalline phase in the phase diagram (water poor region), it can be concluded that the phase consists of reversed-type hexagonal rods (F phase). The liquid of the two-phase mixture formed between 10–90% of surfactants is separated and the ^1H NMR spectra do not show the presence of any detectable amounts of surfactants in the liquid phase. The F phase has a very limited stability range (90–93 wt%) and above 93 wt% of surfactants, a two-phase mixture of a liquid and F phase is formed. The liquid phase was not analyzed. However, the phase may consist of reversed-type micelles. Addition of a little excess DDAB to the mixture leads to the formation of a viscous isotropic solution phase. Above 97 wt% of surfactants, the system consists of hydrated surfactant crystals.

AOT-DDAB-water phase behavior

A partial isothermal pseudo-ternary phase diagram for the system AOT-DDAB-water is presented at 303 K in Fig. 3. Single phases formed in the system are identified and most of the two-phase and three-phase mixtures are detected. As no attempt has been made to obtain the area of occurrence

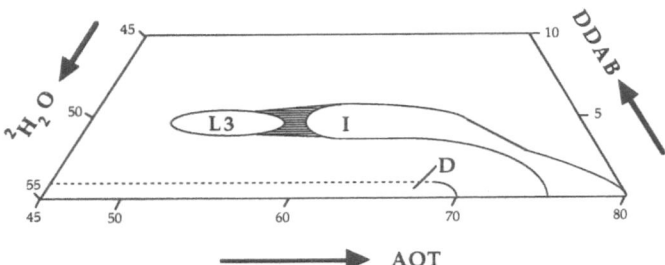

Fig. 3 Phase diagram in the DDAB-poor part of the DDAB-AOT-water system at 303 K. L3; isotropic solution and I; cubic liquid crystalline phase

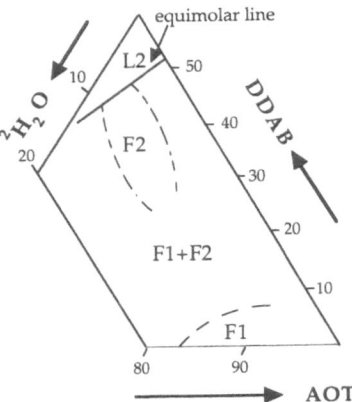

Fig. 4 Phase diagram in the water-poor side of the DDAB-AOT-water system at 303 K. F1 and F2; reversed hexagonal liquid crystalline phases

of the heterogeneous regions, these are not shown in the figure. The solution phase formed with AOT in water (solubility of AOT = 1.3%) cannot solubilize any detectable amounts of DDAB (not shown in Fig. 3). The lamellar phase of the binary AOT system is easily destabilized and forms multi-phase mixtures by adding small amounts of DDAB (maximum 2%). On further addition of DDAB (3–5 wt%), a low viscous isotropic solution phase, L_3, is detected between 50–57 wt% of AOT. The bicontinuous cubic phase, I, formed in the binary AOT-water system extends considerably to the water axis with small amounts of DDAB and the I phase coexists with L_3 phase by a two-phase region. Below 45 wt% of AOT and above the stability region of the D phase the triangular phase diagram is dominated by two- and three-phase regions.

L_3 and I phase obtained in this study may be compared with those reported [10, 11] for the AOT-water-NaCl system. For both systems, the stability range of the L_3 phase is very sensitive to the concentration of both DDAB and NaCl. However, the phase with DDAB has very limited capability to swell with water ($\approx$ 50% H_2O), whereas with NaCl, the L_3 phase can swell as much as with 95% of H_2O. On the other hand, the cubic phase with DDAB can extend to about 62% of water against less than 40% of water with NaCl. Microstructures of L_3 and I phases with DDAB (under study) are expected to parallel, respectively, those reported for the system with NaCl [11].

There exist two reversed hexagonal liquid crystalline phases- F_1 phase with AOT in water contents less than 18 wt% and F_2 phase with equimolar mixtures of AOT and DDAB at a water concentration of about 10 wt% (Fig. 4). The F_1 phase cannot solubilize more than 5 wt% of DDAB, but F_2 phase has the capability to solubilize rather large amounts of AOT within their respective stability limits. However, the two liquid crystalline phases do not seem to merge into one homogeneous reversed liquid crystalline phase. Instead, they coexist through a two-phase, $F_1 + F_2$, region as revealed by two doublets in the water ^2H NMR spectra. The magnitude of the inner split-

ting is comparable with that of single splitting measured in the homogenous F_2 phase and that of the outer splitting with the one obtained from the single splitting recorded in the homogeneous F_1 phase. To the best of our knowledge, this finding is the first example showing the coexistence of two hexagonal liquid crystalline phases in a surfactant system. However, we have not yet succeeded in determining the equilibrium boundary lines of single and two-phase liquid crystalline regions.

Molecular geometry and aggregate structure.

Phase microstructure and the stability of a phase in the phase diagram for the ionic surfactant systems are determined primarily by the long-range electrostatic effects [12] and surfactant molecular packing considerations [13]. The geometrical considerations are based on the critical packing parameter (CPP) of the surfactant which is defined as $CPP = v_c/(a_0 l_c)$, where v_c and l_c are the alkyl chain volume and length, respectively, and a_0 is the polar headgroup area. For ionic surfactants, the effective area of the headgroup is largely determined by electrostatics. This has the consequence that the CPP depends on both surfactant concentration and salt content. The surfactant geometry dictates the allowable geometries of the aggregates that will form in solution. Different values of CPP are compatible with different geometric shapes of the aggregates. Cylindrical micelles are formed when CPP lies between 1/3 and 1/2 and when CPP > 1/2, first, the system forms highly curved bilayer vesicles and then, the lamellar phase (with flat bilayers) which has the highest stability at CPP $\approx$ 1. Reversed micelles and hexagonal liquid crystals with high reversed curvature are preferred structures for CPP > 1.

Progr Colloid Polym Sci (1994) 97:146–150
© Steinkopff-Verlag 1994

For the systems studied here, equimolar mixtures of two lamellar-biased double-tailed surfactants, AOT and DDAB, apparently form a pseudo four-tailed zwitterionic surfactant which results in a smaller effective headgroup and larger hydrophobic region than in the individual surfactants. The effective size of the head group, especially at high concentration of surfactant mixtures decreases, resulting CPP > 1. This in turn will lead to the formation of aggregates with reversed curvature as found experimentally. For highly dilute mixtures, the mixed surfactants are known to form vesicles and/or precipitate (surfactant crystals). A crystalline precipitate is expected when the two surfactants are linear and symmetric in alkyl chain length since the surfactant pairs can pack efficiently into a crystalline lattice. If the surfactants are branched and/or contain a bulky substituent in the tail group, the precipitate phase stability is reduced relative to that of vesicular phase. Here, DDAB consists of two symmetrical alkyl chains, but AOT is a branched asymmetric double-chained surfactant with a bulky headgroup. Moreover, at very high dilution, the CPP value for the mixture is not expected to vary greatly from that of the parents' surfactants. Consequently, the system is able to retain lamellar-type structure (multi-walled vesicles) at highly dilute mixtures before the formation of surfactant crystals at relatively high surfactant content.

Existence of various single phases in the AOT-rich-side of the ternary AOT-DDAB-water system can also be understood from the consideration of surfactant packing parameter. As mentioned earlier, the lamellar phase has its highest stability at CPP ≈ 1. Addition of small amounts of DDAB to the lamellar phase formed in the AOT-water system will cause partial neutralization of the lamellar surface (smaller head group area) and produce a larger hydrophobic volume, and the combined effects will lead to a higher CPP value. Hence the lamellar phase is destabilized and new phases, like the bicontinuous L_3 solution phase or the bicontinuous cubic liquid crystalline phase are preferred. The inability of the L_3 phase to swell extensively to the water axis compared to that with NaCl is due to the constraints imposed to the packing by the DDAB molecules in surfactant aggregation.

Reversed hexagonal liquid crystalline phases, F_1 and F_2, exist at very high surfactant concentrations where CPP > 1. On addition of DDAB to the F_1 phase (AOT-water system) and AOT to the F_2 phase (AOT/DDAB (equimolar)-water system), the liquid crystalline phases are stabilized since the CPP value by this addition is not expected to alter to any significant extent. However, the coexistence of two hexagonal liquid crystalline phases is unexpected. More data are necessary to establish that the two liquid crystals are in thermodynamic equilibrium.

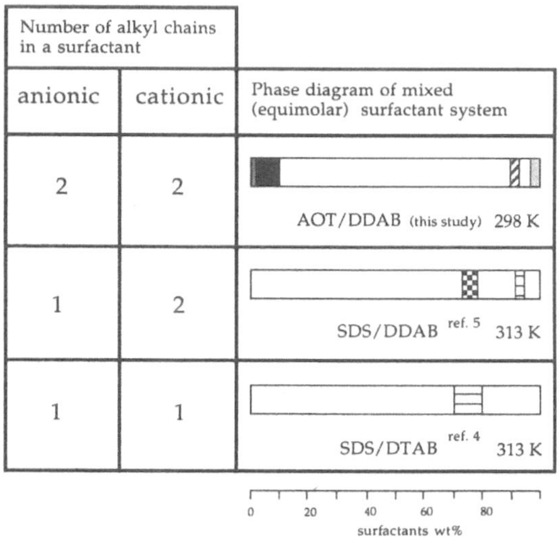

Number of alkyl chains in a surfactant		Phase diagram of mixed (equimolar) surfactant system
anionic	cationic	
2	2	AOT/DDAB (this study) 298 K
1	2	SDS/DDAB ref. 5 313 K
1	1	SDS/DTAB ref. 4 313 K

(scale: 0   20   40   60   80   surfactants wt%)

Fig. 5 Phase diagrams of equimolar mixtures of anionic/cationic surfactants with water. For phase notations, see Fig. 1

Conclusion

There are few studies of complete phase diagrams of mixed oppositely charged ionic surfactant systems [3, 4]. The phases formed by equimolar mixtures of surfactants in water for these systems are summarized in Fig. 5. Addition of excess amounts of one of the parents' surfactants brings the new complexity to the phase behavior, with the formation of various new phases with variable stability ranges. The interplay of both surfactant geometric packing parameter and electrostatic effects can qualitatively explain the phase behavior displayed by the mixed systems. The results indicate that the desired shape of surfactant aggregates as well as the stability region of single phases can be formulated by a suitable blending of the surfactants in mixed systems. The study is important for both theoretical modeling and microstructure engineering.

Acknowledgement Eduardo Marques is thanked for useful comments on the paper. The project is financed partly by the Swedish Research Council for Engineering Science. The stay of one of us (O. R.) was made possible by a grant from the Swedish Institute.

References

1. Kaler EW, Herrington KL, Miller DD, Zasadzinski JAN (1992) In: Chen SH, Huang JS, Tartaglia P (eds) Structure and Dynamics of Strongly Interacting Colloids and Supramolecular Aggregates in Solution. Kluwer Academic Publishers, Dordrecht, The Netherlands, pp 571
2. Kamenka N, Chorro M, Talmon Y, Zana R (1992) Colloids and Surfaces 67:213
3. Marques E, Khan A, Miguel MG, Lindman B (1993) J Phys Chem 97:4729
4. Jokela P, Jönsson B, Khan A (1987) J Phys Chem 91:3291
5. Fontell K, Ceglie A, Lindman B, Ninham BW (1986) Acta Chem Scand A40:247
6. Warr GG, Sen R, Evans FD, Trend JE (1988) J Phys Chem 92:774
7. Regev O, Khan A: this volume
8. Rogers J, Winsor PA (1967) Nature (London) 216:477.9
9. Bellare JR, Davis HT, Scriven LE, Talmon Y (1988) J Electr Microsc Tech 10:87
10. Fontell K (1975) ACS Symp Ser 9:270
11. Balinov B, Olsson U, Söderman O (1991) J Phys Chem 95:5931
12. Jönsson B, Wennerström W (1987) J Phys Chem 91:338
13. Israelachvili JN (1985) Intermolecular and Surface Forces, Academic Press, New York

Progr Colloid Polym Sci (1994) 97:151–153
© Steinkopff-Verlag 1994

SURFACTANTS

Surfactant aggregation in organic solvents:
physical gels and "living polymers"

P. Terech
V. Rodriguez

Received: 30 September 1993
Accepted: 31 March 1994

P. Terech (✉)
CEA-Département de Recherche
Fondamentale sur la Matiére Condensée
SESAM/PCM
17, rue des Martyrs
38054 Grenoble Cedex 9, France

V. Rodriguez
Laboratoire des Spectroscopie Moléculaire
 et Cristalline
Université de Bordeaux I
351, cours de la Libération
33405 Talence Cedex, France

Abstract Aggregation of some non-ionic surfactants in organic solvents gives long chains which can overlap in three-dimensional networks to give viscoelastic materials. Depending on the chemical functionality of the amphiphilic molecules, the related networks can be either permanent or transient in a range of temperatures. Two classes of thermoreversible gelifying materials are distinguished: the "strong" physical gels or crystalline gels and the so-called "living polymers" or "weak" physical gels, respectively. Fiber rigidity and crystallinity are typical structural features of the former type, while chain flexibility and statistical disorder characterize the latter type. With "living polymers" a dynamical process of scission/recombination of the chains competes with their dynamics of motion. Structural and rheology investigations are used to distinguish the two classes. A subtle hydrophobicity/polarity balance of the surfactant/solvent couple determines the aggregation processes and the related structures and variations.

Key words Gels – living polymers – organic solvents – structures – dynamics

Aggregation of some non-ionic amphiphilic molecules in organic solvents can give fiber-like structures in given thermodynamic conditions defined by the solvent and surfactant types, the concentration and the temperature. The kinetics of the aggregation process occurs in a few seconds or hours (depending on the binary system) when a concentration threshold is passed: the solution evolves suddenly from monomeric species (or dimers) towards very long aggregates subsequently entangled to give gel-like materials. Some systems behave like solids while some others are viscoelastic fluids. The concentration threshold is less than 1% and the rheological properties of the stationary state distinguish the crystalline gels (or "strong" physical gels) and the "weak" gels. In both cases an elastic modulus G' can be probed. For the first class, G' is measured in the whole frequency range including towards the zero limit (long times of the order of some tenths of seconds). The related numerical G' values are relatively high (in the context of the so-called "soft matter"): i.e., G' ca. 5.10^4 dyn. cm^{-2} for a fatty acid in benzene [1]. For the second class, G' is measured only in a restricted frequency range, excluding the lower frequencies and is significantly lower (i.e. G' ca. 10^3 dyn. cm^{-2} at 1000 rad. s^{-1} for the organometallic complex of Fig. 2 [2] at a comparable concentration of 1%). Both classes of materials are thermoreversible: it is sufficient to heat the sample to recover a warmed liquid phase where the "monomeric" amphiphilic species are randomly tumbling. A new decrease of temperature makes the binary system unstable and produce a gel which can be in turn metastable and evolves through phase separation. The variety of macroscopic behaviors (mechanical, stability, optical, thermal properties) is built upon structural specificities characterizing the couple surfactant/organic solvent. A very brief overview of the physical chemistry underlying these surfactant-made organogels is given below.

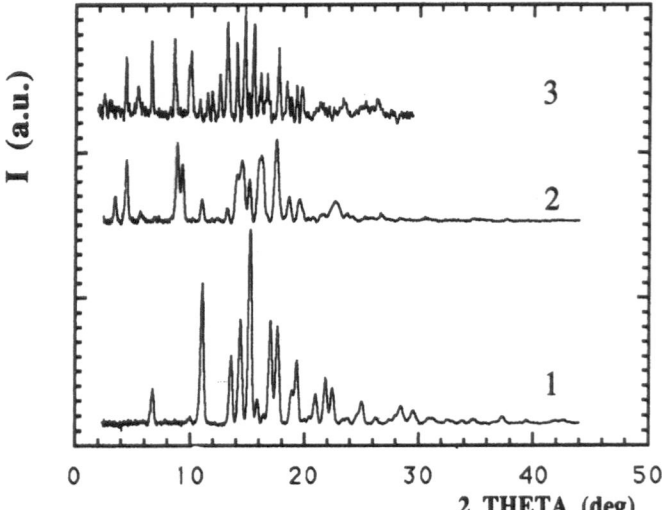

Fig. 1 Wide angle x-ray diffraction patterns of a homo androstanyl steroid (di-n-propyl-17,17 aza-17 a D-homo 5α androstanol-3β) gelator [3] : 1) crystalline powder, 2) aerogel from cyclohexane 3) gel in cis-decalin (scattering from the solvent and air have been subtracted). A typical example of a "crystalline gel" where the structures are different in the various states 1,2,3.

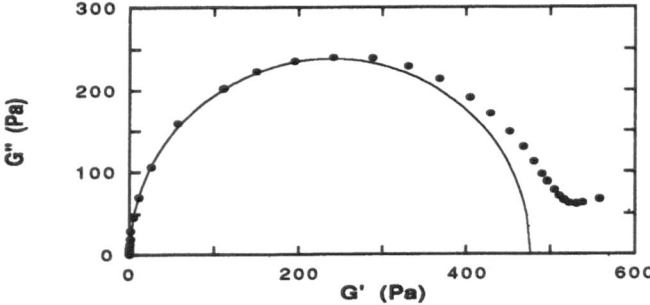

Fig. 2 Dynamical rheology of a bicopper tetracarboxylate complex [3] (bicopper tetra ethyl-2 hexonate) in tert-butyl cyclohexane at $C = 1.1\%$wt [5]. The graphical representation (Cole–Cole plot), G'' (loss modulus) versus G' emphasizes the monoexponential character of the stress relaxation

Basics of the physical chemistry of surfactant organogels

The aggregation step creates a three-dimensional solid-like network which can be either permanent or transient in a range of temperatures:

1) When the junction zones of the network are crystalline microdomains where the fiber-like aggregates are merging, the material is solid-like and can exhibit a yield stress. Usually, the fibers are infinitely long (several microns) and their rigidity is very high. The number of aggregated surfactant molecules per unit length of fiber n_L is large (40 mol. Å^{-1} for the 12-hydroxy octadecanoic fatty acid of ref. [1]). Bending processes of the crystalline fibers would require high energies so that the interactions between them are developed by lateral growth of crystalline super-aggregates of variable shapes and symmetries (cylindrical, helical, lamellar). The molecular ordering in the fibers of the gel or in the collapsed network of the aerogel, is usually different from the one in the single crystal (the symmetry is usually lower in the various gel states organizations). The shape of the cross-section of the fibers can vary from square to rectangular (from fibers to ribbons) or helical structures. Correlations can be drawn between the solvent type and the shapes of the aggregates (determined from small angle neutron or x-ray scattering experiments, SANS or SAXS), their short-range structures (form x-ray diffraction experiments, WAXS see Fig. 1 and [4]), the optical activity and the "crystallinity" of the samples. The chemical functionality of the surfactant determines the aggregation mechanism and the internal molecular organization in the fibers and their aggregates and accounts for the large variety of surfactant gelators available (steroids, fatty acids, organometallic complexes, alcohols, aromatic derivatives etc···). Hydrogen bondings, coordination bondings and various electron transfer processes are the most common "driving forces" for the 1D aggregation (fiber axis growth) and for the cross-sectional extension (junction zone growth).

2) when the chain-like aggregates are integrating in dynamically disordered zones (called in the following "entanglements" by analogy with polymers), the mechanical cohesion on long time scales cannot be insured, the network is transient and the sample is a viscoelastic fluid. The chain length or the molecular weight distribution is also a thermal equilibrium and according to the nature and the strength of the monomer-monomer interaction within the chain, a kinetical breaking/recombination process can exist. This process can be in competition with the dynamics of motion of the individual chains and defines the "life" of the chains by contrast to the "ordinary" polymers (unbreakable chains). For instance, if the motion is described by the reptation model, a chain is moving by curvilinear diffusion out of the tube formed by the topological obstacles along with its contour length (reptation) when a break occurs at a given distance from its extremity which rules the efficiency of the breaking/recombination steps for the stress relaxation. As a result, in some given thermodynamic conditions, a single exponential stress relaxation can be obtained (Maxwell fluid). In addition, n_L is much smaller to allow a greater flexibility of the chain-like aggregate. An extreme case of a "living" molecular thread [2] in apolar hydrocarbons is presented in Fig. 2. When exposed to a small strain at high frequency,

Progr Colloid Polym Sci (1994) 97:151–153
© Steinkopff-Verlag 1994

the stored energy (elastic modulus) is larger than the dissipated energy (loss modulus) and the deformed sample can return to its original shape while at low frequencies the flowing properties of a liquid are recovered.

In all these systems, the structures of the aggregates at various scales (shapes, sizes, flexibility, internal molecular arrangement) are directly affected by a slight modification of the chemical constitution of the solvent or of the surfactant gelators thus providing a way to modify the macroscopic behaviours of these materials (i.e., stability, mechanical, optical and thermal properties).

Acknowledgements. The authors are most grateful to ILL and EMBL (Grenoble, France) where the WAXS and SANS experiments were made and to NIST (Maryland, USA) where the rheological data were obtained. P. Maldivi is acknowledged for providing the organometallic complex.

References

1. Terech P (1992) J Phys II France 2: 2181–2195 and references cited therein
2. Terech P, Schaffhauser V, Maldivi P, Guenet JM (1992) Langmuir 8: 2104–2106 and references cited therein
3. Terech P (1989) J Phys France 50: 1967–1982 and references cited therein
4. Terech P, Rodriguez V, McKenna GB, Barnes J (Langmuir (October 1994)
5. Terech P, Maldivi P, Dammer, J phys II France, (October 1994)

Progr Colloid Polym Sci (1994) 97:154–157
© Steinkopff-Verlag 1994

SURFACTANTS

M. D'Angelo
G. Onori
A. Santucci

Study of micelle formation in aqueous sodium *n*-octanoate solutions

Received: 16 September 1993
Accepted: 25 February 1994

Prof. Dr. G. Onori (⊠)
M. D'Angelo A. Santucci
Dipartimento di Fisica
Universita' di Perugia
V.A. Pascoli
06100 Perugia, Italy

Abstract The critical micelle concentration (CMC) for sodium *n*-octanoate was determined from ultrasound velocity in the 15–90 °C temperature range at 2.5 °C intervals. From the temperature effect on the CMC, free energy, entropy and enthalpy of micellization are determined on the basis of pseudo-phase separation model of micellization. Large changes in the enthalpy and entropy were observed on increasing the temperature. These changes nearly compensate and thus make only a small contribution to the free energy of micellization.

Key words: Surfactants – *n*-octanoate – micelles – thermodynamic properties – ultrasound velocity

Introduction

Hydrophobic interactions are believed to be closely related to the micellization process and to play an important role in stabilizing the native structure of proteins. It is usually accepted that hydrophobic processes are driven by positive entropy changes resulting from the release of structured water when non-polar groups interact with one another.

This traditionally held view of hydrophobic processes seems incorrect. As first observed by Shinoda [1], the mole ordered hydration structure formation around the solute molecules is, indeed, accompanied by a large decrease in entropy; this is, however, more than compensated by an even greater enthalpic effect. Thus, the net consequence of the effect of hydrophobic hydration is to enhance the solubility of non polar species and to disfavor their aggregation.

These divergent views have recently lead to questions about the precise origin of the hydrophobic interactions and have stimulated studies on hydrophobic process under conditions where enthalpy-entropy effects associated with the unique three-dimensional structure of water are diminished [2–7].

Micelle formation is a typical "hydrophobic process"

in water that constitutes an attractive model because the entropy and enthalpy of micellization can be obtained relatively easily through measurement of the temperature effect on the critical micelle concentration (CMC). New insight into hydrophobicity has been recently gained by an analysis of thermodynamics of micelle formation for tetradecyltrimethylammonium bromide in water across an extensive temperature range [2]. Large, but compensating changes in the enthalpy and entropy were observed on increasing temperature while the free energy of micellization was found almost independent of the temperature. In recent works [7, 8], it has been suggested that the extent of enthalpy and entropy contribution associated with structural reorganization of water can be modulated by addition of short chain alcohols. The study of the effect of ethanol on the critical micelle concentration of aqueous sodium dodecylsulphate reveals changes in the entropy and enthalpy that nearly compensate each other, giving a net free energy which is almost independent of alcohol concentration [7]. From both studies [2, 7] it results, according to Shinoda's suggestion [1], that the hydrophobic effect is exceedingly insensitive to changes in the structure of water. Water structuring effects almost cancel out and play only a small part in the free energetic of aggregation.

Progr Colloid Polym Sci (1994) 97:154–157
© Steinkopff-Verlag 1994

In the present paper we report preliminary results on the temperature dependence of the CMC in aqueous solutions of sodium *n*-octanoate obtained from ultrasound velocity in the 15–90 °C temperature range at 2.5 °C intervals. Thermodynamics properties of *n*-octanoate solutions are discussed in terms of the temperature dependence of the free energy, enthalpy and entropy of micellization.

The main aim of this investigation is to provide further insight into the origin of the "hydrophobic effect" via studies of micellar aggregation in aqueous solutions.

Experimental section

Ultrasonic velocities were measured by means of a sing-around velocimeter, model 6080, available from Nusonic Corp., New Jersey, USA. Some measurements were also performed using a variable-path interferometer working at 2 MHz; the values thus obtained are coincident with those of the sing-around velocimeter within the experimental error limits. The temperature control was better than 0.1 °C and the measurements of sound velocity wave reproducible within ± 0.1 m s^{-1}.

From the values of sound velocity and density ρ, the adiabatic compressibility β_s was calculated by means of the relation $\beta_s = l/v^2\rho$. The densities used to calculate β_s were measured with an Anton Paar DMA 512 device. Sodium *n*-octanoate (Fluka, analytical grade) was used without any further purification. The water used was bidistilled and all mixtures studied were prepared by weighing the components.

Results and discussion

Figure 1 shows the sound velocity (v) and compressibility (β_s) vs. surfactant mole fraction (x_2) diagrams for aqueous solutions of sodium *n*-octanoate at 30 °C. As may be seen in the figure the diagrams consist of two linear portions with a well defined break at the CMC in agreement with the previously reported CMC values for these systems [9]. The two linear portions can be assigned to the monomeric and micellar forms. The linear decrease in β_s (or increase in v) below the CMC can be ascribed to hydration effects at both ionic groups and hydrocarbon moiety of surfactant molecules monodispersed in the solvent. It is attended a remarkable lowering in this effect as $x_2 > \text{CMC}$ because the micelle formation decreases hydrophobic hydration.

Recently, infrared radiation has been used to spectroscopically examine the formation of micellar aggregates in aqueous solutions and a large shift to a lower frequency was observed for the C–H stretching modes of some ionic

surfactants upon micelle formation [10–12]. Figure 2 shows the concentration dependence of the antisymmetric methyl stretching frequency for aqueous solutions of sodium *n*-octanoate at 30 °C. As may be seen in the figure at lower concentration the frequency has a constant value indicative of a monomeric phase. This is followed by a range of concentration within which the frequency changes abruptly as a function of concentration. This decrease in frequency is observed in the same concentration range where a discontinuity in β_s or v is observed and can be assumed indicative of the transition to micellar phase. Both constant value of frequency (Fig. 1) and the

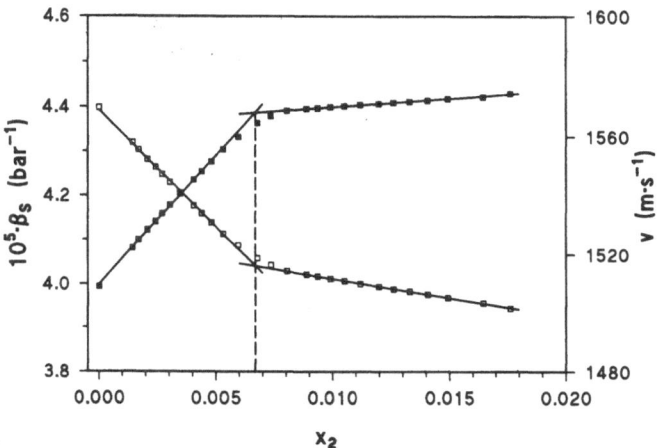

Fig. 1 Experimental sound velocity (v) (■) and compressibility (β_s) (□) values vs. surfactant mole fraction (x_2) for aqueous solutions of sodium *n*-octanoate at 30 °C. (——): obtained from linear fitting

Fig. 2 IR frequency of the antisymmetric methyl stretching as a function of sodium *n*-octanoate concentration ($T = 30$ °C)

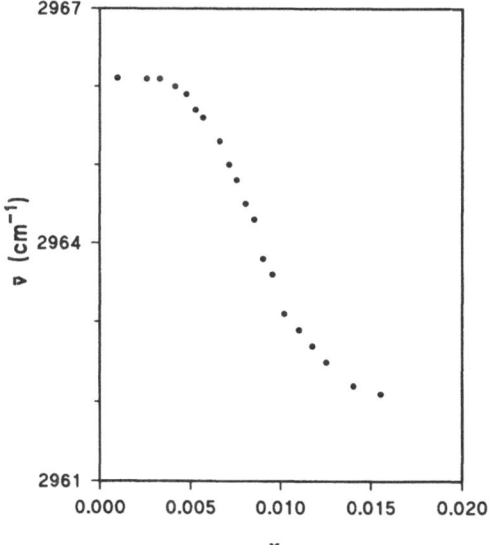

very high linearity in the v (or β_s) vs. x_2 (Fig. 2) observed for $x_2 < $ CMC indicate negligible premicellar aggregation.

At all temperatures in the 15–19 °C range a well defined break in the v vs. x_2 plot was observed. The CMC was determined by fitting the data points above and below the break to two equations of the form $v = ax_2 + b$ and solving simultaneously for the point of intersection. The linearity is very high and the intersection is determined to an accuracy better than ± 0.00004 in CMC.

These results are collected in Table 1 and shown in Fig. 3. The plot follows a concave, upward curve having a minimum at a certain temperature T^*. The occurrence of minima in CMC vs. temperature plots is common to most ionic surfactants. It has been shown [13] that irrespective of alkyl chain length, polar head group, and counterions, CMC vs. temperature data can be fitted by reduced equations having the form,

$$\frac{CMC(T)}{CMC^*} - 1 = K \left| 1 - \frac{T}{T^*} \right|^{\gamma}, \tag{1}$$

where CMC* and T^* are the minimum CMC and T values, respectively, and γ an exponent whose numerical value is 1.74. Our CMC vs. temperature data can be reproduced with an accuracy close to the experimental uncertainty by Eq. (1) with CMC* = 0.00610, $T^* = 332.6$°K, $K = 6.0$ and $\gamma = 1.74$ (continuous line in Fig. 3).

According to the monodispersed phase separation model of micellization, the standard free energy of micelle formation per mole of surfactant is given by

$$\Delta G_m^0 = RT \ln(CMC), \tag{2}$$

with CMC expressed as a mole fraction.

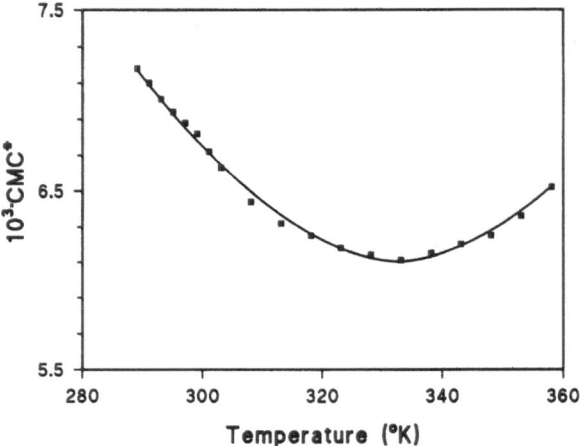

Fig. 3 Critical micelle concentration of sodium n-octanoate as a function of temperature. (■): experimental points. (——): calculated according to Eq. (1)

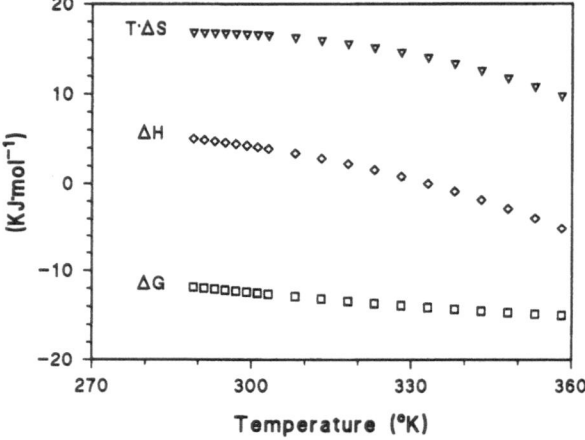

Fig. 4 Thermodynamics quantities calculated according to Eqs. (2) to (6) as a function of temperature

The standard enthalpy change per mole of monomer can then calculated by applying the Gibbs–Helmotz equation

$$\Delta H_m^0 = - RT^2 \left\{ \frac{\partial \ln(CMC)}{\partial T} \right\}, \tag{3}$$

and the standard entropy change from the equation

$$\Delta S_m^0 = \frac{(\Delta H_m^0 - \Delta G_m^0)}{T} \tag{4}$$

The factor $\partial \ln(CMC)/\partial T$ in Eq. (3) has been obtained by calculating the polynomial

$$\ln(CMC) = A + BT + CT^2 + DT^3 + \dots. \tag{5}$$

Table 1 Critical micelle concentration as a function of temperature

Temperature (K)	x_2^*
289.1	0.00718
291.1	0.00710
293.1	0.00701
295.1	0.00694
297.1	0.00688
299.1	0.00682
301.1	0.00672
303.1	0.00663
308.1	0.00644
313.1	0.00632
318.1	0.00625
323.1	0.00618
328.1	0.00614
333.1	0.00611
338.1	0.00615
343.1	0.00620
348.1	0.00625
353.1	0.00636
358.1	0.00652

Progr Colloid Polym Sci (1994) 97:154–157
© Steinkopff-Verlag 1994

which best fit the data. Since a three-order equation fits the data with an accuracy close to the experimental uncertainty, higher terms in the series were not considered. By differentiating Eq. (5), one obtain

$$\Delta H_m^0 = -RT^2(B + 2CT + 3DT^2) \, . \qquad (6)$$

The thermodynamic quantities so calculated are shown in Fig. 4 as function of temperature.

Large changes are observed in the enthalpy $[\Delta(\Delta H_m^0) = 10.2 \, \text{KJ} \cdot \text{mol}^{-1}]$ and entropy $[\Delta(\Delta S_m^0) = 31 \, \text{J} \cdot \text{mol}^{-1} \cdot \text{deg}^{-1}]$ in the temperature range studied. These changes nearly compensate and thus make only a small contribution to the free energy of micellization

$[\Delta(\Delta G_m^0) = 3.1 \, \text{KJ} \cdot \text{mol}^{-1}]$. At the lowest temperature ΔG_m^0 is primarily entropic but becomes increasingly enthalpic as the temperature is raised. The observed behavior is in line with the analysis performed by Shinoda [1] of the hydrophobic effect and with the idea that water becomes less and less anomalous as it is heated.

Obviously, due to the hypothesis present in the phase separation model of micellization [14], the values obtained for thermodynamic quantities should be considered approximate. Direct measurements of micellar enthalpies using the microcalorimetry technique are now in progress in our laboratory and the results will be reported in due course.

References

1. Shinoda K (1977) J Phys Chem 81:1300–1302
2. Evans DF, Wightman PJ (1982) J Colloid Interface Sci 86:515–524
3. Ramadan MS, Evans DF, Lumry R (1983) J Phys Chem 87:4538–4543
4. Ramadan MS, Evans DF, Lumry R, Philsons S (1985) J Phys Chem 89:3405–3408
5. Evans DF, Ninham BW (1986) J Phys Chem 90:226–234
6. Shinoda K, Kobayashi M, Yamaguchi N (1987) J Phys Chem 91:5292–5294
7. Onori G, Santucci A (1991) Chem Phys Letters 189:598–602
8. Onori G, Passeri S, Cipiciani (1989) J Phys Chem 93:4306–4310
9. Mukerjee P, Mysels KJ (1971) in "Critical micelle concentrations of aqueous surfactant systems" NSRDS-NBS-36 (U.S. Government Printing Office, Washington)
10. Umemura JH, Cameron DG, Mantsch HH (1980) J Am Chem Soc 84:2272–2277
11. Umemura JH, Mantsch HH, Cameron DG (1981) J Colloid Interface Sci 83:558–568
12. Yang PW, Mantsch HH (1986) J Colloid Interface Sci 113:218–224
13. La Mesa C (1990) J Phys Chem 94:323–326
14. Van Os NM, Daane GJ, Haandrikman G (1990) J Colloid Interface Sci 141:199–217

Progr Colloid Polym Sci (1994) 97:158–162
© Steinkopff-Verlag 1994

SURFACTANTS

M. D'Angelo
G. Onori
A. Santucci

Structure and state of water in reversed aerosol OT micelles: an infrared study

Received: 16 September 1993
Accepted: 4 March 1994

Prof. Dr. G. Onori (✉)
M. D'Angelo·A. Santucci
Dipartimento di Fisica
Universita' di Perugia
V.A. Pascoli
06100 Perugia, Italy

Abstract The structure of water in bis(2-ethylhexyl)sodium sulfosuccinate (AOT) micelles has been studied as a function of the $[H_2O]/[AOT]$ ratio (W) by using the absorption IR due to O–H stretching modes in the 3800–3000 cm^{-1} range. Two systems have been studied: water/AOT/carbon tetrachloride and water/AOT/n-heptane. The results show that IR spectra can be expressed as sum of contributions from interfacial and bulklike water. The fraction of water in the two "regions" within the water pool was evaluated as a function of W. From the data it appears that a continuous variation in the water properties inside micellar cores occurs rather than a two step hydration mechanism. The solubilization of water is described in terms of hydration of the AOT head group and Na$^+$ counterions. The maximum hydration number of AOT was found to be 3 both in CCl$_4$ and n-heptane.

Key words Surfactants – bis(2-ethylhexyl)sodium sulfosuccinate – IR spectroscopy –micelles

Introduction

A reversed micelle is an aggregate of surfactant formed in a non polar solvent. The main interest in the reversed micelles is based on their ability of dissolving water in their core. This provides a unique opportunity to study the properties of water aggregates close to the ionic centers (head polar groups and the corresponding counter-ions of the surfactant) without the interference due to large quantities of bulk water. The water solubilized in reversed micelles, in many respects, is similar to the interfacial water present near the biological membranes or at protein surfaces; so, it is of interest to study the state of water in reversed micelles as a model of specific water in biological systems.

Reversed micelles formed by the surfactant bis(2-ethylhexyl)sodium sulfosuccinate (AOT) in alkanes (e.g., heptane or iso-octane) have been the most widely investigated [1–3]. Reversed micellar systems formed by AOT can solubilize relatively large amounts of water: the maximum amount of cosolubilized water is a sensitive function of the temperature and, for a specified temperature, of the chemical nature of the dispersion medium. Several features of these systems remain to be explained, especially on a molecular level. One of these concerns the characterization of the different local structures of the water in the micellar core and the determination of their relative amounts as a function of the molar ratio $[H_2O]/[AOT]$ (W), temperature and dispersion medium.

Although most of the properties of water in reversed AOT micelles have been closely investigated, little attention has been paid to their spectra of vibrational bands [4–8]. Infrared spectroscopy is a technique particularly suitable to detect hydrogen bonds and it has been often employed to study solvent modification in aqueous solutions. In a previous paper [8], we reported an IR investigation of water–AOT–carbon tetrachloride reversed micelles in the 4000–3000 cm^{-1} range, where absorptions due to O–H stretching modes of H$_2$O are present. The results show that the IR spectra can be expressed as the sum of contributions from interfacial and bulklike water. The fraction of water in the two "regions" within the water pool was evaluated as a function of W and it has been possible to explain the data in terms of a continuous equilibrium between water molecules present in the two regions.

In the present paper IR spectroscopy in the O–H stretching region is utilized to study the reverse micelle system n-heptane–AOT–water as a function of the parameter W. The water binding capacity of AOT in n-heptane is higher than in CCl_4, so differences in the filling mechanism of AOT reversed micelle are expected in the two solvents.

Experimental section

AOT 99% (Alfa Product), purified by recrystallization from methanol and drying in vacuum, was stored in vacuum over P_2O_5. Bidistilled water and n-heptane (purity > 99.5%) were used without additional purification. Some residual water molecules remain bound to the AOT molecules after the drying process of the surfactant. Analysis of the water content of purified AOT and n-heptane mixtures with Karl Fisher titrator revealed the presence of 0.2 moles of water residual per mole of AOT. Such a small residual of water was considered as a part of the total water in the mixtures under study.

The $AOT/H_2O/n$-heptane mixtures for IR measurements were prepared by weight. IR spectra were recorded by means of a Shimadzu Mod. 470 infrared spectrophotometer equipped with a variable path-length cell and CaF_2 windows. Typical path-lengths employed were 50 to 800 μm for $AOT/H_2O/n$-heptane mixtures. Pure water spectra were taken with shorter path-lengths.

The molar-extinction coefficient of water was calculated by using the expression $\varepsilon = A/(c \cdot d)$, where A is the absorbance, c the water concentration in $mol \cdot l^{-1}$ and d the cell depth in centimeters.

Gaussian curve fitting was achieved with a computer application of the Marquardt algorithm

Fig. 1 (——): O–H stretching band for pure water (d) and water in $H_2O/AOT/n$-heptane system at selected values of W (a), (b), (c); (- - -): Gaussian components from least squares fitting

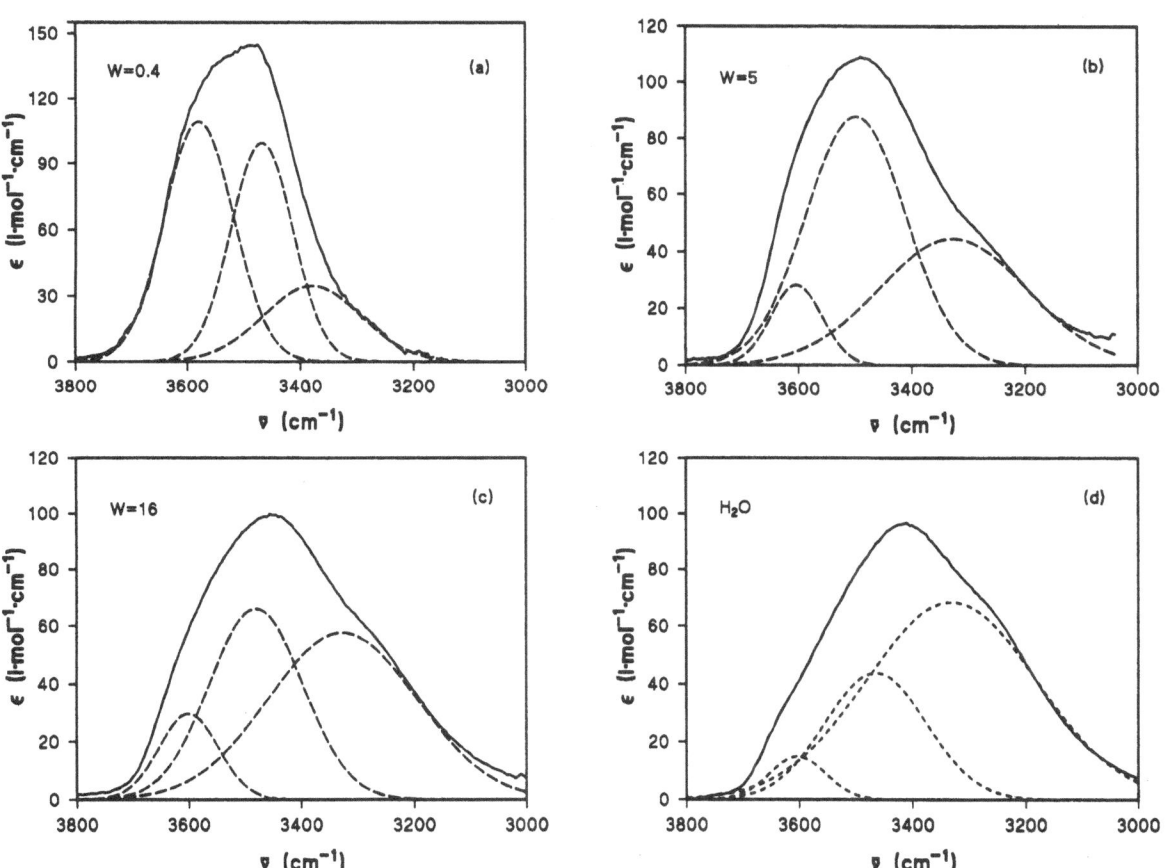

160
M. D'Angelo et al.
IR study of water in reversed micelles

Results and discussion

Parts a–d of Fig. 1 show the O–H stretching region for pure water and water in $H_2O/AOT/n$-heptane system at selected values of W. The spectrum of pure water (Fig. 1d) can be fitted very well in terms of three Gaussian components centered at $3603 \pm 6 \, cm^{-1}$ (bandwidth = $74 \pm 7 \, cm^{-1}$), at $3465 \pm 5 \, cm^{-1}$ (bandwidth = $130 \pm 10 \, cm^{-1}$) and $3330 \pm 20 \, cm^{-1}$ (bandwidth = $206 \pm 8 \, cm^{-1}$). These values are in good agreement with those reported in literature [6, 7].

The hydrogen bonding in water has been extensively studied [9, 10] and there are recent numerous theoretical and experimental reports [4–8, 11–18]. The highest frequency component at $3603 \, cm^{-1}$ represents just a small part of the total ($\sim 7\%$) and is usually assigned to the non-H-bonded or weakly H-bonded O–H groups [11, 16, 18]. The components at $3330 \, cm^{-1}$ shifted down by $300 \, cm^{-1}$ with respect to frequency absorptions due to free O–H groups has been related to molecules in more regular structure with unstrained H bonds, and the component at $3465 \, cm^{-1}$ to molecules in more distorted structures with energetically unfavored H bonds [6, 7, 13, 17].

It should be noted that the interpretation of the broad band associated with the O–H stretching vibrations of water is still a matter of controversy in the literature [12]. Such a situation arises in part from the fact that the problem of interpretation of the H_2O spectrum exceeds the limits of pure spectroscopic problems and involves a model for the structure of the water. However, water is so far not satisfactorily described.

The IR spectrum of surfactant entrapped water is significantly different from that of pure water, indicating that the water solubilized in the reversed micelles lacks the normal hydrogen-bonded structure present in the bulk water. However, the total peak area, A, of the O–H stretching band of water has been found to increase linearly with water content, as it is predicted by Beer's law with $\varepsilon = 3.7 \cdot 10^4 \, 1 \cdot mol^{-1} \cdot cm^{-2}$. This value for the molar extinction coefficient is equal within the experimental errors to that of pure water. The same result has been obtained by using both CCl_4 [8] or n-heptane as dispersion medium.

To quantify the changes in the O–H stretching region, we fitted each spectrum as a sum of three Gaussian components (see part a–c of Fig. 1) The fitted curves are practically indistinguishable for the measured ones. The parameters characterizing the Gaussian components (peak frequency and bandwidth) differ just a little from those of pure water. They depend on the water-to-surfactant ratio, gradually changing with W toward the values characteristics of pure water (Fig. 2).

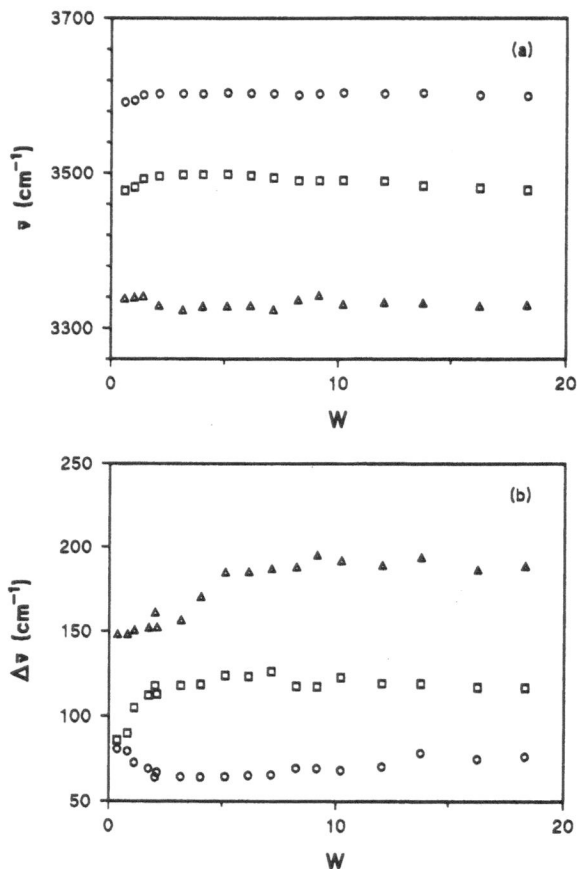

Fig. 2 Peak frequency (a) and bandwidth (b) of the Gaussian components of the water spectrum in $H_2O/AOT/n$-heptane system as a function of W. Component centered at $\sim 3603 \, cm^{-1}$ ($\bigcirc$), at $\sim 3465 \, cm^{-1}$ ($\square$), at $3330 \, cm^{-1}$ ($\triangle$)

Figure 3 shows the variation of the ratio between the area of any Gaussian component (A_i) to the total area (A) as a function of W. Due to the independence of ε on W, it is reasonable to assume the curves in Fig. 3 are representative of the variations of the different fractions of –OH groups assigned to each Gaussian component. At low W values the solubilized water exists mainly as "bound" water molecules whose static and dynamic properties are determined by local interactions with Na^+ counterions and the strong dipole of AOT polar groups. From the figure it appears that the fraction of peak area A_3/A, assigned to water in regions characterized by a regular tetrahedral connectivity, gives just a little contribution at low W ratios. This finding is consistent with the interpretation proposed for this quantity and with the expected breakdown of tetrahedral H-bonded water structure in the hydration region. As W increases, A_3/A gradually increases towards to the A_3^0/A^0 value characteristic of pure water.

As previously pointed out on the $H_2O/AOT/CCl_4$ system [8], we can assume two types of water ("bound"

Progr Colloid Polym Sci (1994) 97:158–162
© Steinkopff-Verlag 1994

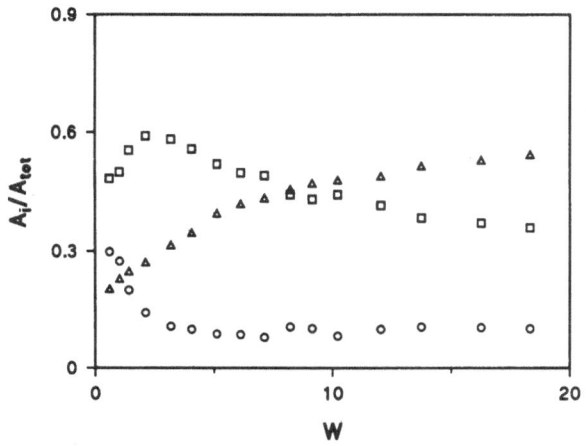

Fig. 3 Ratio of the area of the i-th Gaussian component (A_i) to the total peak area (A) vs. W. Symbols are the same as in the Fig. 2

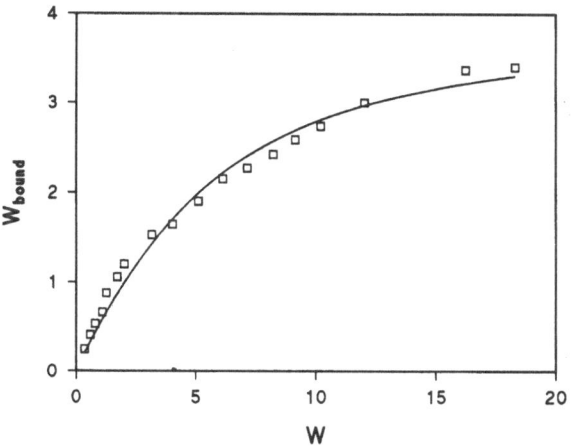

Fig. 5 Plot of W_{bound} vs. W for the H$_2$O/AOT/n-heptane system

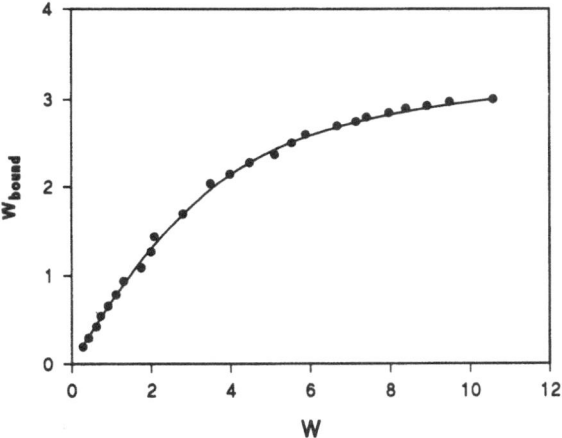

Fig. 4 Plot of W_{bound} vs. W for the H$_2$O/AOT/CCl$_4$ system

and "bulk") coexisting and exchanging quickly. If one assumes $A_3/A = 0$ for the "bound" and $A_3/A = A_3^0/A^0$ for the "bulk" water spectrum respectively, the mole fraction of bulklike water, P_{bulk}, in the micellar core can be evaluated as

$$P_{bulk} = \frac{A_3/A}{A_3^0/A^0}$$

and the concentration of bonded water per AOT molecules as

$$W_{bound} = (1 - P_{bulk}) \cdot W.$$

Both for the H$_2$O/AOT/CCl$_4$ (Fig. 4) and for H$_2$O/AOT/n-heptane (Fig. 5) systems on increasing W, W_{bound} gradually increases reaching a value of ~ 3 and $W \sim 6$, above which it remains constant. So, the bound

water region seems to hold three water molecules per AOT molecule and its formation is nearly complete at $W > 6$. The observed behavior is qualitatively in agreement with the data in the literature referring to several physico-chemical properties of water solubilized in reversed AOT micelles [1–3].

One can tentatively attribute the three binding sites on AOT molecules to the three oxygen atoms of the SO$_3^-$ head group to which water molecules could be hydrogen bonded. The environments of these molecules could be different from that of bulk water and would to be complete for $W \sim 6$.

It is of note that from the data it appears that there is a continuous variation in the water properties inside micellar cores rather than a two-step hydration mechanism. It was previously shown [8] that the water pool formation in the AOT reversed micelle is well interpreted in terms of a continuous equilibrium between bonded and bulk water

$$N + H_2O \rightleftharpoons N \cdot H_2O,$$

where N is not an occupied site in the hydration zone and N·H$_2$O an occupied one. The fitting result is shown by the continuous curve in Figs. 4 and 5.

From these preliminary results the IR technique appears to be promising when applied to the study of structural and dynamical properties of water in these or similar systems where the H-bonded network of bulk liquid water is significantly changed by the interaction with the interface molecules. It should be noted that the time scale of stretching vibrations of the water molecules makes possible the detection of species with lifetimes as short as 10^{-13} s, which is shorter than most rearrangement processes occurring in solution.

162

M. D'Angelo et al.
IR study of water in reversed micelles

References

1. Eicke HF (1980) Top Curr Chem 87:86–145
2. Luisi PL, Giomini M, Pileni MP, Robinson BH (1988) Biochim Biophys Acta 947L209–246
3. Chevalier Y, Zemb T (1990) Rep Prog Phys 53:279–371
4. Seno M, Sawada K, Araki K, Iwamoto K, Kise H (1980) J Colloid Interface Sci 78:57–64
5. Sunamoto J, Hamada T, Seto T, Yamamoto S (1980) J Colloid Interface Sci 78:57–64
6. Mac Donald H, Bedwell B, Gulari E (1986) Langmuir 2:704–708
7. Jain TK, Varshney M, Maitra A (1989) J Phys Chem 93:7409–7416
8. Onori G, Santucci A (1993) J Phys Chem 97:5430–5434
9. Luck WAP Ed, (1974) Structure of Water and Aqueous Solutions, Verlag: FRG
10. Schuster P, Zundel G, Sandorfy C, Eds (1976) The Hydrogen Bond, North-Holland Publishing Co, Amsterdam
11. Luck, WAP (1973) in "Water: a Comprehensive Treatise" Franks, F, Ed Plenum, New York 2:235–321
12. Maréchal Y (1991) J Chem Phys 95:5565–5573
13. Luck WAP (1980) Angew Chem Int, Ed Engl 19:28–41
14. Stillinger FH (1980) Science 209:451–457
15. Zilles BA, Person WB (1983) J Chem Phys 79:65–77
16. Tso TL, Lee EKC (1985) J. Phys Chem 89: 1612–1618
17. D'Arrigo G, Maisano G, Mallamace F, Migliardo P, Wanderling F, (1981) J Chem Phys 75:4264–4270
18. Giguère Pa, Pingeon-Grosselin M (1986) J Raman Spectroscopy 17:341–344

Progr Colloid Polym Sci (1994) 97:163–165
© Steinkopff-Verlag 1994

SURFACTANTS

D. Tsiourvas
C. M. Paleos
A. Malliaris

Monomeric and polymeric bola-amphiphiles based on the succinic and maleic anhydrides

Received: 16 September 1993
Accepted: 31 January 1994

Dr. A. Malliaris (✉)
D. Tsiourvas · C. M. Paleos
N. R. C. "Demokritos"
Agia Paraskevi
Athens 153 10, Greece

Abstract Monomeric and polymeric bola-amphiphiles based on the anhydrides of succinic and maleic acid were synthesized. Their aggregational properties were studied by physicochemical methods including electrical conductivity, fluorescence probing and video-enhanced microscopy. The results show that in aqueous solutions these amphiphiles form aggregates of varying sizes, from ordinary spherical micelles to large conglomerates.

Key words Bola-amphiphiles
– surfactants – polymers
– aggregation – micelles

Introduction

The succinic and maleic anhydrides, as well as the poly-maleic acid anhydride, have been reported as the starting materials for the synthesis of thermotropic, amphiphilic-type liquid crystals [1–3] and also of monomeric and polymeric amphiphiles capable of forming molecular organizations in aqueous media [4, 5]. In those reports the interaction of the anhydrides with long chain primary or secondary amines or alcohols in a molar ratio of 1:1 was employed. According to this reaction scheme for each reacting lipophilic moiety, one carboxylic group is generated, which is susceptible to further functionalization, thus producing ordinary surfactants with only one polar group. In the present study the succinic, maleic and polymaleic anhydrides interacted with either 11-aminoundecanoic or 12-hydroxydodecanoic acid, and two polar groups were generated, leading to the formation of bola-amphiphiles as shown in the reaction scheme below. These bola-amphiphiles were studied from the point of view of their organizational behavior in aqueous media.

Experimental

Synthesis

Polymaleic anhydride was obtained by γ-irradiation (43 Mrads) of maleic anhydride [6] in a Co60 source. The degree of polymerization was determined by vapor pressure osmometry (Knauer vapor pressure osmometer) and found equal to ca. 70 monomeric units.

All anhydride derivatives, monomeric and polymeric, were prepared by refluxing for several hours 0.013 M of the corresponding anhydride dissolved in acetone, with 0.01 M of 11-aminoundecanoic or 12-hydroxydodecanoic acid. Because of the low solubility of 11-aminoundecanoic acid in acetone, refluxing in this case continued for ca. 24 h, while for the other acid the reaction was considered complete after ca. 6 h. Subsequently, the solvent was distilled off while the excess of the monomeric and polymeric anhydrides was removed with water. The products were dried and recrystallized from acetone. The monomeric and polymeric carboxylic amphiphiles were dissolved in ethanol and neutralized with equimolar quantities of sodium ethoxide to produce the final Na salts.

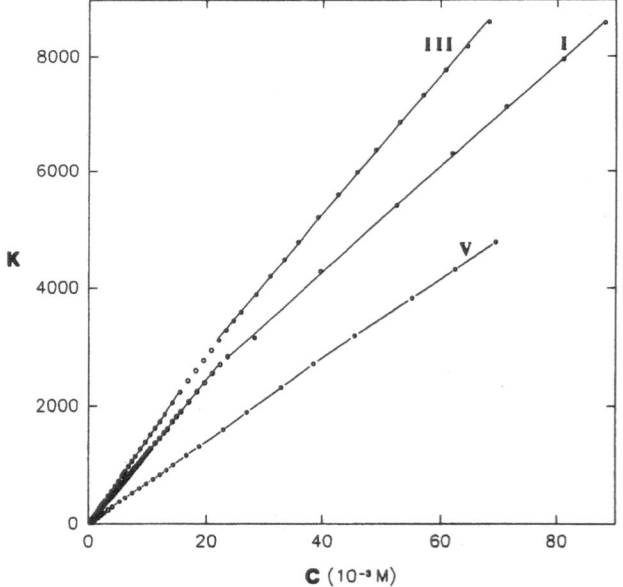

Analysis

Calculated for $C_{15}H_{25}O_5NNa_2$ (I): C = 52.17%, H = 7.30%, N = 4.06%. Found: C = 52.57%, H = 7.63%, N = 3.96 %. Calcd. for $C_{16}H_{28}O_6$ (II): C = 60.74%, H = 8.92%. Found: C = 60.30%, H = 9.08%. Calcd. for $C_{15}H_{23}O_5NNa_2 \cdot H_2O$ (III): C = 49.86%, H = 6.97%, N = 3.88%. Found: C = 49.17%, H = 7.04%, N = 3.53%. Calcd. for $C_{16}H_{26}O_6$ (IV): C = 61.13%, H = 8.34%. Found: C = 60.65%, H = 8.62%. Calcd. for $C_{15}H_{25}O_5N$ (V): C = 60.18%, H = 8.42%, N = 4.68%. Found: C = 60.11%, H = 8.58%, N = 4.50%. Calcd. for $C_{16}H_{26}O_6 \cdot H_2O$ (VI): C = 57.80%, H = 8.49%. Found: C = 58.77%, N = 8.72%.

Electrical conductivity measurements were performed using the E512 Metrohm-Herisau conductometer in conjunction with thermostatted conductivity cells capable of stabilizing the temperature at $\pm 0.1\,°C$. Triply distilled water having electrical conductivity of only few μS was used for the preparation of all solutions.

Absorption spectra were obtained using a Varian-Cary 210 spectrophotometer. Fluorescence spectra of extensively zone refined pyrene (Aldrich 99 + %) solubilized in the aggregates, were recorded on a Jasco FP-777 spectrophotometer. In all solutions the pyrene concentration was kept constant and very low viz. $5 \times 10^{-6}M$, to prevent excimer formation.

For the video-enhanced microscopy (VEM) we employed a previously described system [5].

Results and discussion

From the synthetic point of view the interest focuses on the fact that the equimolar reaction of a cyclic anhydride (monomeric or polymeric) with ω-amino or ω-hydroxy acid opens the ring and results in the formation of an additional carboxylic group (see the reaction scheme). Thus, in a single step, two carboxylic groups appear affording bola-amphiphilic monomeric and polymeric structures.

All aqueous solutions of the monomeric salts show a pronounced inflection point in their corresponding electrical conductivity vs concentration plots in the region $2.2–2.6 \times 10^{-2}M$ studied here (Fig. 1). This proves the formation of micellar aggregates with a well-defined critical micelle concentration (CMC). The polymeric surfac-

Fig. 1 Typical conductivity vs. concentration plots of monomeric (I, III) and polymeric (V) bolaamphiphiles, K in μS

Progr Colloid Polym Sci (1994) 97:163–165
© Steinkopff-Verlag 1994

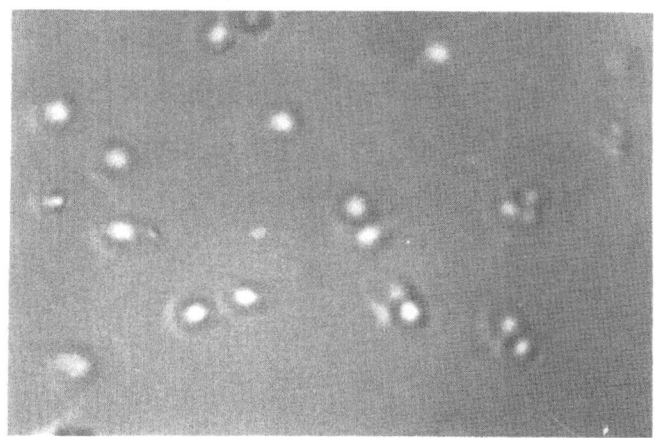

Fig. 2. Micrograph of aggregates of polymer VI (5×10^{-2}M) obtained by VEM

tants on the contrary, show a continuous curvature in their plots without a clear break-point. This fact indicates that aggregation probably starts at very low concentration, while as the concentration of the solution increases the large aggregates formed do not follow ordinary micellar behaviour.

The findings of the electrical conductivity measurements were further supported by the fluorescence probing technique. It is well known that the details of the emission spectra of pyrene solubilized in these aggregates provide reliable information concerning the general structural characteristics of the molecular organizates [7, 8]. Thus, regarding the monomeric surfactants it was found that the intensity ratio of the first to the third fluroescence peaks of pyrene (I_1/I_3) revealed that below the CMC (ca. $2.2 - 2.6 \times 10^{-2}$M) pyrene was dissolved in water (I_1/I_3

for water was found to be 1.75) suggesting that micelles were not formed. Above the CMC pyrene sensed a non-aqueous environment similar to that of acetone in terms of polarity (I_1/I_3 for acetone was found to be 1.75), therefore molecular organization was present. For the polymeric surfactants the results of fluorescence indicated that even at the lowest concentration studied (1×10^{-3}M) there was evidence of encapsulation of pyrene in a non-aqueous environment, attributed to the formation of intramolecular micelles. With increasing concentrations the I_1/I_3 ratio decreases continuously reaching values similar to those for SDS aqueous solution above its CMC. In addition, it was found that when the concentrated surfactant solution (above CMC) was filtered through a 3μ Millipore filter most of the pyrene was removed from the solution. This shows that a great percentage of these surfactants is present in the solution in the form of large aggregates. However, a distribution of the aggregate size is implied by the fact that if the solution is filtered before the addition of pyrene, through even a 0.1 μm filter, and then pyrene is added, a micelle-like environment is detected by the fluorescence of the probe.

Visual confirmation of the above results was obtained by means of VEM. While aggregates were not observed below the CMC of the monomeric surfactants, a large number of them having sizes between 1 and 4 μm was discernible at concentrations above CMC. For the polymeric surfactants at the low concentration range only few small aggregates (up to 1.5 μm) were observed, while at higher concentrations (Fig. 2) the number of the aggregates increased as well as their sizes (up to 4 μm).

In conclusion, our results indicate that both, the monomeric and the polymeric bolaamphiphiles studied here undergo spontaneous organization in water. They form molecular conglomerates which range in size from ordinary spherical micelles to very large aggregates.

References

1. Paleos CM, Margomenou-Leonido-poulou G, Margaritis LH, Terzis A (1985) Mol Cryst Liq Cryst 129:127
2. Tsiourvas D, Paleos CM, Dais P (1989) J Polym Sci, Polym Chem Ed, 38:257
3. Tsiourvas D, Paleos CM, Dais P (1990) J Polym Sci, Polym Chem Ed 39:1263
4. Malliaris A, Paleos CM, Dais P (1987) J Phys Chem 91:1149
5. Tsiourvas D, Paleos CM, Malliaris A (1993) J Polym Sci, Polym Chem Ed 31:387
6. Lang JL, Pavelich WA, Clarey HD (1963) J Polym Sci 1:1123
7. Malliaris A (1987) Adv Colloid Interface Sci 27:153
8. Malliaris A (1988) Intern Revs Phys Chem 7:95

Progr Colloid Polym Sci (1994) 97:166–170
© Steinkopff-Verlag 1994

SURFACTANTS

N. Micali
S. Trusso
C. Vasi
F. Mallamace
D. Lombardo
G. Onori
A. Santucci

Aggregation properties of a short chain nonionic amphiphile (C₄E₁) in water solutions

Received: 1 November 1993
Accepted: 15 February 1994

Dr. Norberto Micali (✉)
S. Trusso · C. Vasi
Istituto di Tecniche Spettroscopiche
del CNR 98166 Vill. S. Agata,
Salita Sperone, Messina,
Italy

F. Mallamace · D. Lombardo
Dipartimento di Fisica
dell' Universita' di Messina
98166 Vill. S. Agata, C.P. 55 Messina,
Italy

G. Onori · A. Santucci
Dipartimento di Fisica dell' Universita'
di Perugia
Via A. Pascoli
06100 Perugia, Italy

Abstract We have studied the depolarized Rayleigh–Wing light scattering of aqueous solutions of 2-butoxyethanol (C_4E_1) as a function of the temperature and concentration. The measured spectra give information on the rotational dynamics of water molecules. The data analysis, with the results of recent surface tension measurements, confirm the amphiphilic character of the ethoxylated alcohol molecules.

Key words Water – amphiphile – aggregation

Introduction

Interest in the properties of water solutions of nonionic amphiphile molecules is well known; in particular for long-chain polyoxyethylene monoalkyl esters (with chemical formula $C_mH_{2m+1}(OCH_2CH_2)_nOH$ or C_mE_n for short) which can form well defined supramolecular aggregates as micelles, or microemulsions (on adding a mineral oil). In addition, such systems exhibit a tendency to demix with increasing temperature, and separate at a lower critical solution temperature (LCST) for concentrations lying in the very water-rich region of composition [1]. The LCST is usually located above the room temperature [1].

Normal alcohols (C_mE_0), and alkoxyethanols (C_mE_1) can be considered low n and m members of the C_mE_n series. Depending on the length (hydrophobicity) of the alkyl chain, they exhibit different phase behaviors. However, both systems, display similar trends: for $m \leq 3$ they are completely miscible with water, while for $m > 3$ they become immiscible, displaying a closed loop of solubility. In fact, C_4E_1 (2-butoxyethanol or BE) aqueous solutions demix at a LCST $T_c = 42.9\,°C$ and $X_c = 0.052$ (X represents the BE mole fraction) [2], while all the others alkoxyethanols with $m < 3$ are completely miscible in water at all temperatures and concentrations. The phase diagram of BE mixture is quite similar to that observed in long-chain C_mE_n amphiphile solutions [1]. These latter features and the presence of a definite maxima in the partial molar heat capacity [3] at a characteristic concentration, make the BE solution quite similar to C_mE_n, suggesting that among the C_mE_1, BE has the minimal length to form micellar aggregates. Although the conclusive confirmation of such structures is far from being completely established, the investigation of these systems is very useful in order to clarify the effect of amphiphilic molecules length (and in particular the requirement of a minimal length) on the formation of surfactant structures.

Much experimental data coming from different techniques [3–9] give an indication that some kind of micellar structure is present in the mixture above an amphiphile molar fraction X of about 0.018. Recent viscosity, small-angle neutron scattering (SANS) [10] and light scattering data, [11] on water–butoxyethanol (C_4E_1) mixtures as a function of temperature, $-10 \leq T \leq 45\,°C$ in the concentration range $0.015 \leq X \leq 0.09$, indicate that these amphiphilic alcohols remain monomolecularly dispersed in water at very low concentrations (lower than 0.018), while at higher concentrations the amphiphilic molecules form micellar-like structures that become more effective when the temperature decreases.

At present no definitive structural model exists for the nature of such aggregates: for example, it is not clear if they are composed of molecules of the same species, or are to be considered a mixture. Two very recent studies seem to confirm that micelles originate in the water-BE system: one is the measure of the surface tension [12] and the another is a recent and accurate SANS experiment [13], both performed at different concentrations and temperatures. In this latter case, the corresponding spectra clearly show the existence of micellar aggregates in BE solutions above the critical micelle concentration (CMC). The size of micelles is constant; its shape is spherical and the radius corresponds to the length of the surfactant. In addition, the presence of concentration fluctuations is observed at all the temperatures that strongly contribute to the scattered intensity. Such fluctuations increase with temperature and are dominant near the demixing line. This suggests that the stability of the micelles decreases with increasing temperature and can be related to the exchange of alcohol molecules between micelles.

More conclusive are the surface tension results. As can be observed, from Fig. 1 in which is reported the mixture surface tension γ (measured at $T = 4$ and $40\,°C$), the behavior of γ as a function of the BE concentration C is typical of aqueous surfactant solutions forming micelles [14]. In particular, on increasing C (the surfactant in the present case) γ decreases; when the CMC is reached, γ remains constant and does not change with further C-increases. The obtained CMC values are in good agreement with others determined by the use of different techniques such as partial molar specific heat and sound velocity [4, 9].

In order to analyze, from a molecular point of view, the properties of such structures, we have performed on the water-BE system a study of anisotropic light scattering. Light scattering is a powerful tool for the study of the structural and dynamical properties of materials. In particular, it can give information on the rotational dynamics

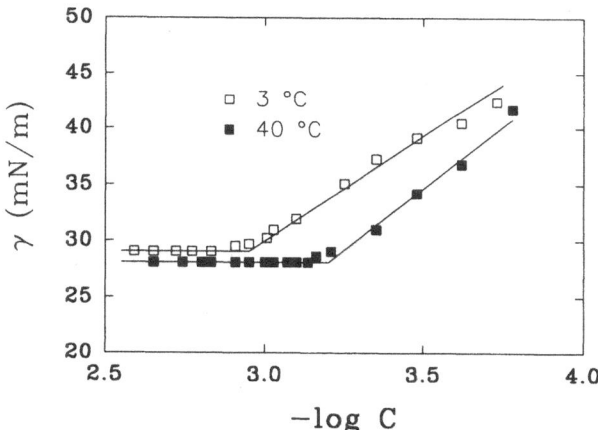

Fig. 1 Surface tension γ as function of C (mol·l^{-1}), of BE-water, at $T = 4$ and $40\,°C$

as reflected by the density correlation function; the information which can be extracted from the measured spectra is mainly related to the rotational motion of the molecules as influenced by the presence of structural arrangements. More precisely, we show that depolarized scattering can be used to probe the dynamics of water in our complex liquids giving "local" information on the water properties in the neighborhood of micellar aggregates. In addition, we hope that, by studying the rotational dynamics of water molecules in the presence of well defined aggregates, we can obtain further information on the structural properties of the solution.

Experimental results and discussion

The solutions studied, together with the pure BE, have the following concentrations: $X = 0.015, 0.035, 0.048, 0.052, 0.07$ and 0.09. The depolarized Rayleigh scattering was performed using a double monochromator with a half width at half maximum (HWHM) resolution of 700 MHz in the usual 90° scattering geometry with an Ar^+ laser operating at 5145 Å as the exciting source. It is well known that the nonshifted depolarized light scattering is caused by the fluctuations of the traceless part of the polarizability tensor [15]. The measured scattered intensity $I_{VH}(\omega)$ is the Fourier transform of corresponding time correlation function $G^{anis}(t)$ and is directly related to the molecular rotational motion. This is reflected in the measured spectra as Lorentzian contributions related to the exponential time decay of the local order. On this basis, our data were fitted to either a single or a double Lorentzian plus the instrumental response function. As a result of such a procedure we conclude that the pure BE spectrum is well described by a single Lorentzian line, while all other spectra can be fitted only in terms of two significant contributions. The HWHM and the relative intensity of each Lorentzian line are obtained directly by the fitting procedure. The pure alcohol exhibits a HWHM of about 1 cm^{-1} with a weak temperature dependence, while the solutions show two contributions which are strongly dependent on T. The results of many studies performed on bulk water, in normal and supercooled regions, show two Lorentzian lines: a *fast* one, centered at about 40 cm^{-1} (nearly independent of T), and a *slow* one ranging from about 1.7 cm^{-1} ($T = 20 °C$) to about 8 cm^{-1} ($T = 50 °C$); we relate the two spectral contributions observed in the solutions to the water dynamics. In Figs. 2 and 3 we report, in a time representation, as a function of T for the different concentrations studied, the results of the *slow* and *fast* modes respectively; for comparison the results for pure water (coming from different experiments) [16, 17] and for pure BE are also shown. It can be observed that the alcohol rotational dynamics is slower in comparison with that corresponding to water. Another measured quantity is the integrated area of the two different Lorentzian contributions that represent the number of scatterers for the respective modes. In Fig. 4 is reported (for several concentrations versus the temperature) the ratio of the area of the *fast* contribution with the total area; as a result we conclude that the relative number of scatterers of the two distinct modes is independent of the concentration and the temperature.

In Figs. 2 and 3, we have confirmation that the two contributions, *fast* and *slow*, observed from the measured spectra are due to the dynamics of water within the solutions. We stress that the well known *slow* contribution ($1.7–8 \text{ cm}^{-1}$) is related to the hydrogen bond dynamics of water molecules. In fact, the corresponding time follows an Arrhenius temperature dependence with an activation energy that corresponds to the hydrogen-bond energy [18]. The origin of the *fast* contribution ($\sim 40 \text{ cm}^{-1}$), observed very recently [17], is unclear but seems to be related to the correlations of orientational modes of water molecules [19], and the results of Fig. 4 agree with such an interpretation. The behavior of this mode, within the experimental error, is independent of the temperature in the range $-10–20 °C$. (Fig. 3) and is about the same as pure water; on the contrary, for $T > 20 °C$, τ_f increases with T while the corresponding time for pure water remains constant.

Also the relaxation time τ_s (*slow* contribution) shows two different behaviors·for the two temperature regimes, $T < 20 °C$ and $T > 20 °C$, respectively. Also in this case, we observe that *slow* relaxation time behaves, with temperature, in the same way as pure water; in particular, for $T < 20 °C$, τ_s, measured at the different water-BE concentrations, shows the same Arrhenius temperature depend-

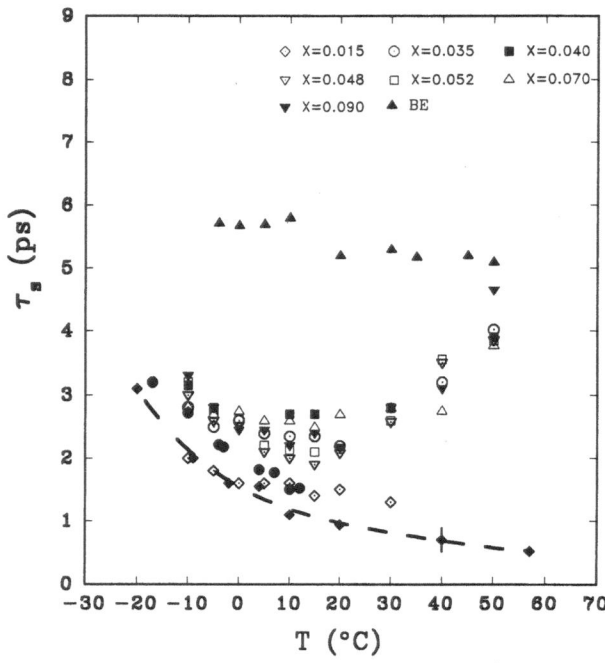

Fig. 2 Temperature behavior of the *slow* relaxation time τ_s for the different studied solutions. Data for water refers to ref. [16] (full dots) and ref. [17] (full rhombs)

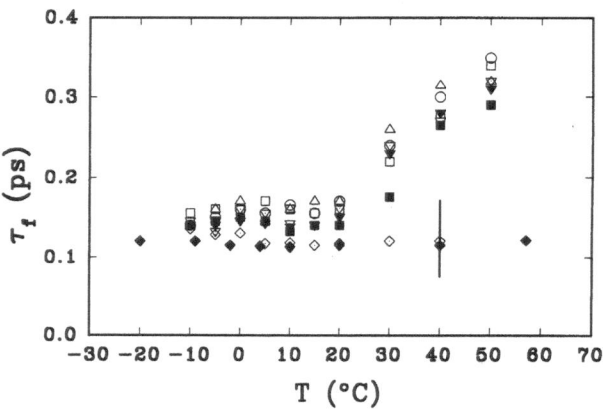

Fig. 3 Temperature behavior of the *fast* relaxation time τ_f for the different studied solutions. Data for water refer to ref. [17] (full rhombs)

ence as pure water (dotted lines in Fig. 2). At the same temperature the relaxation time of the BE aqueous solution is higher ($\sim 30\%$) than that corresponding to the bulk water. For high temperatures, $T > 20 °C$, also for this mode we observe a remarkable difference with water: a sharper increase in τ_s with increasing T. However, we can rationalize the results of the present measurements.

In the low temperature region the two relaxation times behave, on changing T, in a way similar to the corresponding times for pure water. More precisely, while τ_f is nearly

Progr Colloid Polym Sci (1994) 97:166–170
© Steinkopff-Verlag 1994

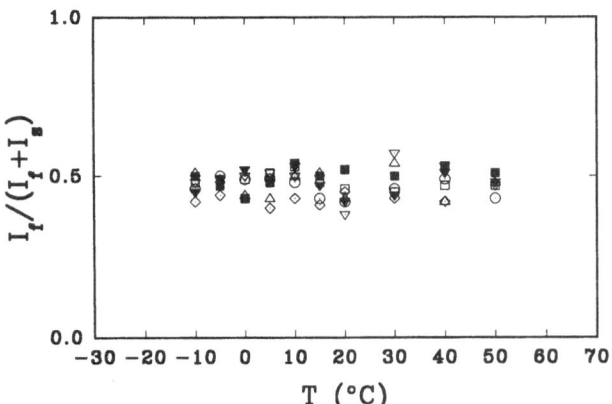

Fig. 4 Ratio of the area of the *fast* Lorentzian mode with the total area for the several studied water-BE concentrations versus the temperature

the same of the bulk water at all concentrations studied, τ_s is the same only for $X = 0.015$. For the other concentrations this latter time is larger (30%) than the corresponding one in water, but shows the same Arrhenius temperature dependence. In such a case, as shown by the surface tension data, the concentration X is above the CMC and, as verified by SANS [13], a well defined micellar structures, with a gyration radius corresponding to the length of the alcohol molecule, are present in the system. Such a behavior for τ_s can be explained if we take into account the results of recent scattering studies [20, 21] on the aqueous suspension of a long chain amphiphile ($C_{10}E_5$). The data analysis clearly suggests that the water molecules interact with the hydrophilic groups of the amphiphile molecules. More precisely, water is partially bound to the oxyethylene groups of the amphiphile, forming a layer that surrounds the micellar structure. The structure of such bound water presents a local, low density, four coordinated environment typical of the supercooled water. A behavior verified in many experiments for transport properties of water in confined geometries (water in confined geometries at room temperature has a behavior similar to the supercooled bulk water). This latter fact is reflected in the OH stretching vibrational modes with a behavior that corresponds to water at a lower temperature than the actual one in which the measurements is performed [21]. This phenomenon is analogous to those observed in the actual results for τ_s, and considering that the T behavior (Arrhenius) of this relaxation time is the same of the corresponding one in bulk water, we can consider that the data of the present analysis confirms, through the study of the water (hydrogen bond) dynamics, that micellar structures are present in water-BE suspensions. Additional confirmations for this are given by the results on both the relaxation times at the concentration $X = 0.015$ (below the CMC). In fact, the

system is unable to build-up alcohol structures at this concentration value (as verified also from SANS data [12]), the observed dynamics corresponds (within the uncertainty) to the hydrogen bonds dynamics in bulk water.

The two characteristic times noticeably increase with T, for $(T > 20\,°C)$, showing water-dynamics in the BE solutions to be very different in respect to the pure bulk water. This behavior in both τ_s and τ_f is related to the demixing phenomenon that takes place in the system which behaves as a critical one. Scattering data [10, 11, 13] give information, in this temperature range, of an increase in the long range correlation length ξ of the fluctuations. This quantity, as shown by light scattering data [22], obeys the laws for critical phenomena showing the well known characteristic divergence, approaching the critical temperature T_c as $\xi = \xi_0 \varepsilon^{-\nu}$, where $\varepsilon = |T - T_C|/T_C$ is the reduced temperature and ν is the critical exponent. In addition, whereas in the low temperature region the SANS data are well fitted with a modified Guinier form, for the present temperatures a good fit is obtained only with the use of the simple Ornstein–Zernike relation [13]. Both these scattering results give, therefore, the information that the onset of the critical phenomenon can be detectable at temperatures near 20 °C. The increase with temperature in the correlation length ξ has been ascribed [13] to the presence in the system of concentration fluctuations together with the spherical micelles; these fluctuations increase with temperature and are dominant when approaching the demixing curve. This implies that the stability of the micelles decreases with increasing temperature, and can be partly due to a possible exchange of alcohol molecules between micelles. In such a case, as shown by x-ray [20] and Raman [21] data in water solutions of a long chain amphiphile ($C_{10}E_5$), the isolated alcohol molecules are hydrated in the oxyethylene group with an average number of bound water molecules for group n_w larger than that for amphiphile molecules aggregated in the micellar structure. In particular, $n_w > 2$ for free alcohol molecules and $n_w \simeq 1$ for molecules within the micelles.

Another possible explanation for this high temperature dynamics can be due to a clustering of the micelles in the critical region with ordering effect in water outside the micellar structure. The increase in ξ approaching T_c can be connected to a percolation-like phenomenon, similar to that observed in microemulsion systems [23] with a LCST. By using this model, one explains the critical behavior of the shear viscosity and the relaxation rate in the density–density correlation function of the present mixture. Such data can be analyzed in terms of the mode-coupling theory only if background effects (due to a persisting presence of micellar aggregates) on the transport coefficient are taken into account [22]. Both these two possibilities, for the explanation of the physical origin in the increases of the two observed rotational times, are accounted for by the results represented in Fig. 4 for the ratio of scatterers involved in both the relaxation processes. We conclude with the suggestion that additional measurements of surface tension and of small angle x-ray scattering could lead to a definitive explanation of the high temperature behavior of the structural and dynamical properties of the water-BE system.

References

1. Degiorgio V (1985) In: Degiorgio V, Corti M (eds) Physics of Amphiphiles, Micelles, Vesicles and Microemulsions. North-Holland, Amsterdam, p 303; and refs. cited therein
2. D'Arrigo G, Mallamace F, Micali N, Paparelli A, Teixiera J, Vasi C (1991) Progress in Colloid Polymer Science 84:177; Mallamace F, Micali N, D'Arrigo G (1991) Phys Rev A 44:6652
3. Roux G, Roberts D, Perron G, Desnoyers JE (1980) J Solution Chem 9:29
4. Kilpatric PK, Davis HT, Scriven LE, Miller WG (1987) J Coll and Interface Sci 118:270
5. Musbally GM, Perron G, Densoyers JE (1974) J Coll and Interface Sci 48:494
6. Puvvada S, Blankschtein D (1990) J Chem Phys 92:3710
7. Shindo Y, Nabu M, Harada Y, Ishida Y (1981) Acoustica 48:186
8. Kato S, Jobe D, Rao NP, Ho CH, Verrall RE (1986) J Phys Chem 90:4167
9. Arrigo G, Paparelli A (1988) J Chem Phys 88:405
10. D'Arrigo G, Teixiera J (1990) J Chem Soc Faraday Trans 86:1503; D'Arrigo G, Teixiera J, Mallamace F, Giordano R (1992) J Chem Phys 95:2732
11. Mallamace F, Micali N, Vasi C, D'Arrigo G (1992) II Nuovo Cimento D, 14:333
12. G. Onori, A. Santucci, private communication.
13. D'Arrigo G, Giordano R, Teixiera J (1992) Physica Scripta T45:248
14. Langevin D (1993) in "Micelles and Microemulsions" to appear on Advanced in Phsical Chemistry
15. Berne B, Pecora R (1976) in "Dynamic Light Scattering". Wiley, New-York
16. Aliotta F, Vasi C, Maisano G, Majolino D, Mallamace F, Migliardo P (1986) J Chem Phys 84:4731
17. Mazzacurati V, Nucara A, Ricci MA, Ruocco G, Signorelli G (1990) J Chem Phys 93:7767
18. Montrose CJ, Bucaro JA, Marchall-Coakley J, Litovitz TA (1974) J Chem Phys 60:5025; Conde O, Teixeira J (1983) Mol Phys 44:525; Sciortino F, Geiger A, Stanley HE (1992) J Chem Phys 96:3857
19. A. Geiger, private communications
20. Degiorgio V, Corti M, Piazza R, Cantu' L, Rennie AR (1991) Colloid Polym Sci 269:501; Barnes LS, Corti M, Degiorgio V, Zemb T (1993) (to be published)
21. Mallamace F, Micali N, Corti M, Degiorgio V (1993) Phys Rev E (in press)
22. Lombardo D, Mallamace F, Micali N, D'Arrigo G (1993) Phys Rev E (in press)
23. Chen SH, Mallamace F, Rouch J, Tartaglia P (1992) in Kawasaki K, Kawakatsu T, Tokuyama (eds) "Slow Dynamics in Condensed Matter". AIP Publications, New York. 256:301

Progr Colloid Polym Sci (1994) 97:171–173
© Steinkopff-Verlag 1994

SURFACTANTS

D. F. Anghel
C. Bobica
M. Moldovan
C. Albu
A. Voicu

The effect of cationic surfactant micelles upon the hydrolysis of p-Nitrophenyl esters

Received: 16 September 1993
Accepted: 31 October 1993

Camelia Petruta Bobica (✉)
D. F. Anghel · M. Moldovan · C. Albu
A. Voicu
Institute of Physical Chemistry
Department of Colloids
Spl. Indenpendentei 202
79611 Bucharest, Romania

Abstract This work presents the results obtained by means of UV-VIS spectrophotometry in the kinetics study of p-nitrophenyl acetate, propionate, and butyrate hydrolysis in the presence of hexadecyl-pyridinium chloride. The rate enhancements were treated in terms of pseudo-phase ion-exchange model, which explains the results. From the fitting of data, the solubilization, ion-exchange and micellar second-order rate constants were determined. Micellar binding of OH^- is promoted when the substrate is less hydrophobic. The solubilization constants became higher with increasing substrate's hydrocarbon chain length. Because the micellar second-order rate constants were smaller than those in water for all the substrates, one may conclude that the rate enhancements were due to the higher relative concentration of the reactants into the micelles.

Key words Micellar catalysis – cationic surfactants – micelles – esters hydrolysis

Introduction

Micellar catalytic effects in bimolecular reactions are generally explained in terms of a favorable partition of the substrate between the aqueous and micellar phases. The analysis of the surfactant concentration-rate profiles is the first step in comprehending the mechanism of micellar catalysis.

In addition to the previous papers which had studied the basic hydrolysis of p-nitrophenyl esters in the presence of cationic micellar solutions [1, 2], the present work was concerned with the effect of substrate hydrophobicity upon the reaction rate.

Theoretical approaches

The experimental results were adapted to the kinetic model proposed by Menger [3] and developed by Bunton [4] and Romsted [5]. The model provides an equation for the pseudo-first-order rate constant, k_ψ

$$k_\psi = \frac{k_W[OH_T] + (k_M K_S - k_W)m_{OH}[Dn]}{1 + K_S[Dn]} \qquad (1)$$

where the subscripts M and W denote the micellar and the aqueous pseudo-phases, respectively, the subscript T denotes the total concentration, $[Dn]$ is the concentration of the micellized surfactant:

$$[Dn] = [D] - cmc, \qquad (2)$$

where $[D]$ is the total surfactant concentration, cmc is the critical micelle concentration, k_W is the second-order rate constant, m_{OH} is the micellar concentration of OH^- expressed as a mole ratio:

$$m_{OH} = [OH_M]/[Dn], \qquad (3)$$

K_S is the binding constant of the substrate to micelle written in terms of micellized surfactant:

$$K_S = [S_M]/([S_W][Dn]). \qquad (4)$$

Table 1 The values of the constants used in the fitting of data to Eq (1): aqueous second-order rate constant (k_W), solubilization constant (K_S), ion-exchange constant (K_{Cl}^{OH}), micellar rate constant (k_M), micellar second-order rate constant (k_m^2), and the values of the ratio k_W/k_m^2

Substrate	k_W (s^{-1}mol^{-1}L)	K_{Cl}^{OH}	K_S (mol^{-1}L)	k_M (s^{-1})	k_m^2 (s^{-1}mol^{-1}L)	k_W/k_m^2
PNPA	5.30	2	28	1.50	0.21	25.24
PNPP	7.73	60	58	8.23	1.15	6.71
PNPB	4.28	60	80	8.34	1.17	3.67

In some cases the distribution of both reactants can be measured directly [4, 6, 7]. The problem is more difficult for hydrophilic ions, so that many workers have used an ion-exchange model [5], which in our case is expressed by:

$$OH_M^- + Cl_W^- \rightleftharpoons OH_W^- + Cl_M^-, \tag{5}$$

where Cl^- is the surfactant counterion.

Materials and methods

p-Nitrophenyl acetate (PNPA), propionate (PNPP) and butyrate (PNPB) have been synthesized from p-nitrophenol and the corresponding acids. Hexadecylpyridinium chloride (HPyCl) from Fluka Chemie AG was used without further purification.

Kinetic measurements were carried out spectrophotometrically using a SPECORD M 40 spectrophotometer equipped with thermostatted cell. All the measurements were done at 298 K. The reactions were followed at 344 nm, the isosbestic point of p-nitrophenol between the protonated and deprotonated form.

Results and discussion

First-order rate constants were obtained from the plots of $\log(A_\infty - A_t)$ against time. The aqueous second-order rate constants k_W were obtained as the slopes of the NaOH concentration vs. pseudo-first-order rate constants profiles (see Fig. 1). Their values are given in Table 1.

To fit our experimental results to Eq. (1) the values of cmc and β can be taken as constants according to the literature [8] with the values: $cmc = 0.9 \times 10^{-3}$ mol L^{-1} and $\beta = 0.8$.

From the fitting of data (Fig. 2) to the theoretical Eq. (1), one may determine the values for the solubilization constants K_S, the ion-exchange constants K_{Cl}^{OH}, and the micellar rate constants k_M. The obtained values are presented in Table 1.

Since the rate constants k_M have the dimensions of reciprocal time they cannot be compared with the second-order rate constants k_W (mol^{-1}s^{-1}L) in water. The comparison can be made considering the volume of Stern

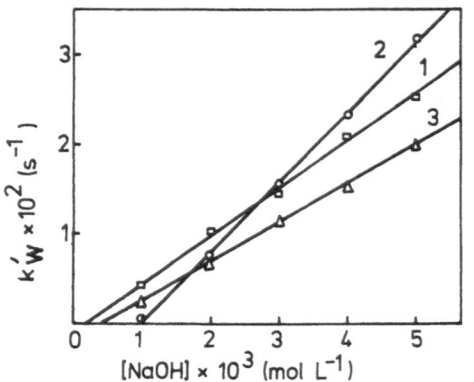

Fig. 1. Pseudo-first-order rate constants versus OH$^-$ concentration for the hydrolysis of PNPA (1), PNPP (2), and PNPB (3) in water

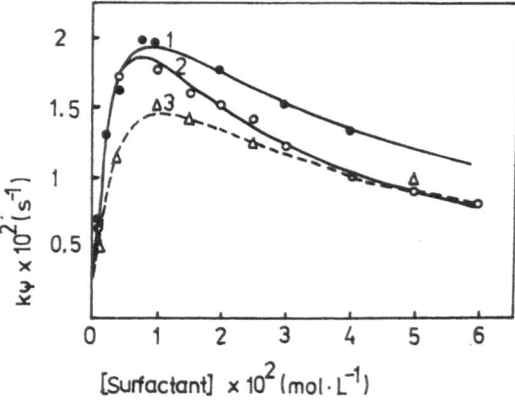

Fig. 2. Variation of the pseudo-first-order rate constant, k_ψ, with surfactant concentration for PNPA (1), PNPP (2), and PNPB (3). The lines are obtained by means of Eq (1)

layer as 0.14 L·mol^{-1} [2]. The second-order rate constants in micellar pseudo-phase are given in Table 1.

They were calculated using the equation:

$$k_m^2 = 0.14 \, k_M \tag{6}$$

For comparison, the values of k_W/k_m^2 ratio were also included in Table 1.

Progr Colloid Polym Sci (1994) 97:171–173
© Steinkopff-Verlag 1994

Conclusions

All the hydrolyses are catalyzed by aqueous solutions of HPyCl.

The solubilization constants K_S and the micellar rate constants k_ψ increased with increasing ester hydrophobicity. The values of K_{Cl}^{OH} show that the micellar binding of OH^- is promoted when the substrate is less hydrophobic.

The ratio k_W/k_m^2 is always above unity and decreases with increasing hydrocarbon chain length. Taking into account this fact and the observation that the hydrolyses in micellar solutions at optimum surfactant concentration are 5–25 times faster than those in water, one may conclude that the increase of the observed rate is due to the higher relative concentration of the reactants in micelles.

References

1. Hong YS, Kim JB, Park HH, Lee DR (1989) J Kor Chem Soc 33: 97–105
2. Rodenas E, Vera S (1985) J Chem Educ 62:1120–1121
3. Menger FM, Portnoy CE (1967) J Am Chem Soc 89: 4968–4972
4. Bunton CA (1979) Catal Rev Sci Eng 20: 1–56
5. Romsted LS (1977) In: Mittal KS (ed) Micellization, Solubilization and Micro-emulsions. Plenum Press, New York, pp 509–532
6. Al-Lohedan HA, Bunton CA, Romsted LS (1982) J Org Chem 47:3528–3532
7. Bunton CA, Cerichelli G, Ihara Y, Sepulveda L (1979) J Am Chem Soc 101:2429–2435
8. Fendler JH, Fendler EH (1975) Catalysis in Micellar and Macromolecular Systems. Academic Press, New York, pp 20–21

Progr Colloid Polym Sci (1994) 97:174–178
© Steinkopff-Verlag 1994

SURFACTANTS

Surfactant effects in crystallization: nucleation and crystal habit of γ-aminobutyric acid

C.H. Lin
N. Gabas
J.P. Canselier
J. Tanori
I. Pezron
D. Clausse
G. Pèpe

Received: 13 December 1993
Accepted: 10 February 1994

J.P. Canselier (✉)
ENSIGC
18 chemin de la Loge
31078 Toulouse Cedex, France

C.H. Lin · N. Gabas
Laboratoire de Génie Chimique (URA
CNRS 192) ENSIGC
31078 Toulouse, France

J. Tanori · I. Pezron · D. Clausse
Départment de Génie Chimique
Université de Technologie de Compiègne
60206 Compiègne, France

G. Pèpe
Centre de Recherche sur les Mécanismes de
la Croissance Cristalline
Campus de Luminy
13288 Marseille, France

Abstract The influence of ionic and nonionic surfactants on the solubility, nucleation, and growth habit of γ-aminobutyric acid (GABA) in water has been studied. Between 25° and 55 °C, no effect of the surface-active agents on GABA solubility was detected. On the other hand, it has been found by two experimental techniques (laboratory-scale batch crystallization and DSC), that these additives tend to delay homogeneous nucleation. Attachment energies of the main crystallographic faces of GABA (F faces in the PBC theory) were calculated via molecular/crystal modeling by means of the GenMol software, in order to predict the theoretical habit of GABA crystal in vacuo. Predictions on habit modification are found in good agreement with the shape of crystals grown from aqueous solution with or without additive.

Key words Surfactants
– crystallization – nucleation
– crystal habit – molecular modeling
– γ-aminobutyric acid

Introduction

Poorly defined crystal morphology can have serious detrimental effects on an industrial process. Since impurities or additives often play an important role in crystallization phenomena, voluntary addition of an adequate foreign substance is commonly used nowadays to improve crystal shape through modifying nucleation and growth kinetics: additives can reduce the supply of material to crystal faces, lower specific surface energies and block surface sites [1, 2]. Because molecular characteristics can differ from one crystal face to another, additives adsorb preferentially onto certain faces, causing a nonuniform reduction of growth rates, hence a change of the crystal habit. Nevertheless, a number of such applications and even research work in this domain remain largely empirical. In order to design more effective or special-purpose additives, a better insight in the underlying interactions on a molecular level is needed via molecular modeling software. Based on atom–atom potential energy calculations, quantitative models emerged to predict the morphology of crystals [3–5].

Surface-active agents, having an amphiphilic character (hydrophobic tail and polar head), are a special class of additives whose strong, specific effects are frequently turned to account in industrial crystallization [6]. In fact, even a small amount (about 100 ppm) of a surfactant is often able to affect nucleation and growth kinetics and to induce crystal habit modification. Surfactants adsorb onto surfaces by two different ways: if the surface is charged, the surfactant can adsorb "head first", through Coulombic

Progr Colloid Polym Sci (1994) 97:174–178
© Steinkopff-Verlag 1994

forces. If the surface is hydrophobic, the surfactant will bond physically to the surface. Although electrostatic interactions are thought to predominate with ionic or polar molecular crystals and ionic surfactants, it has been shown that non-surfactant ionic additives, like organic salts, were not so efficient [7].

ω-Aminoacids form a family of strongly polar organic compounds crystallizing readily from water and likely to present crystal faces of various nature: hydrophilic (positively or negatively charged) or hydrophobic. γ-Aminobutyric acid (GABA), one of the first terms of this series, is of biological and pharmaceutical interest, as a neurotransmitter and part of the formulation of psychostimulating drugs. It crystallizes under the zwitterionic form ($N^+H_3(CH_2)_3COO^-$) from aqueous solution at the isoelectric point (pH = 7.33), giving monoclinic prisms elongated along the c axis (space group $P2_1/a$, 4 molecules in the unit cell, crystallographic parameters: $a = 8.214$ Å; $b = 10.000$ Å; $c = 7.208$ Å; $\beta = 110.59°$) [8]. The surfactants chosen are typical C_{12} anionic and cationic species: sodium dodecylsulfate (SDS) and dodecyltrimethylammonium bromide (DTAB) and a less common nonionic one, hexane-1,2-diol (HD) some of whose properties have been reported in [9]. They were all used below their CMC.

The objective of the present work is to determine the influence of SDS, DTAB, and HD on the solubility, nucleation, and growth of GABA crystals from aqueous solution.

Theoretical considerations

After the simplest Bravais–Friedel–Donnay–Harker (BFDH) theory, the apparent faces of a crystal possess the largest interplanar distances d_{hkl}, therefore the lowest hkl values, after corrections for the symmetry of the space group [4]. The Periodic Bond Chain (PBC) theory [10] allows to predict the most developed, or so-called F (flat) faces, with better certainty. But crystal shape, obviously relying on crystal structure, is assumed to be mainly determined by the kinetic processes of growth: a relation exists between crystal habit and certain energy quantities. Hartman and Bennema [11] have demonstrated that, at least at low supersaturation below the roughening transition, the linear growth velocity perpendicular to the (hkl) face (v_{hkl}) is an increasing function of the attachment energy ($E_{att(hkl)}$) of this face. This means that the weaker $E_{att(hkl)}$, the more developed the (hkl) face is. $E_{att(hkl)}$, defined as the energy released *per* mole when a crystal slice is deposited on the (hkl) face, is calculated as the difference between the total lattice energy of the crystal (E_{cr}) and the energy released on the formation of a growth slice of thickness d_{hkl}, parallel to

the (hkl) face ($E_{slice(hkl)}$) [4]:

$$E_{att(hkl)} = E_{cr} - E_{slice(hkl)} . \tag{1}$$

Let us consider a central molecule with n atoms and N surrounding molecules in the crystal (each containing n atoms). The total lattice energy is given by:

$$E_{cr} = \sum_{k=1}^{N} \sum_{i=1}^{n} \sum_{j=1}^{n} E_{kij} = E_{vdW} + E_C + E_H , \tag{2}$$

where the E_{kij} are the atom-atom interaction energies and E_{vdW}, E_C, and E_H are the van der Waals, Coulombic, and supplementary hydrogen bond energies, respectively. The molecular/crystal modeling software GenMol takes these energy terms into account, using adapted potential expressions to calculate the non-bonded interaction energies leading to $E_{att(hkl)}$ in vacuo [2]. Attachment energies will be modified by solvent (resp. additive) interaction or "adsorption".

Experimental

Solubility of GABA (Fluka Biochemika) in pure water and in aqueous solutions of surfactants (Fluka or Aldrich) was measured according to the classical liquid-solid equilibration method (maintaining solutions with excess solid solute at a given temperature), followed by refractometry analysis (Zeiss, five decimal positions). Conductivity measurements (Knick 702 conductometer) were used to determine critical micelle concentrations (CMC) of the ionic surfactants. A Lauda TD1 tensiometer was employed for surface tension measurements (Wilhelmy plate method).

Laboratory-scale batch crystallization experiments to estimate metastable zone widths were performed in an apparatus consisting of a 600 mL baffled jacketed glass crystallizer equipped with a three-blade marine-type propeller, a Pt 100 probe feeding back to a cooling unit programmed by a 386 PC, and two infrared lamps intended to prevent the advent of moisture on the inner walls of the crystallizer. Each run was started by heating the stirred solution of GABA with or without additive up to 10 °C above the saturation temperature during 30 min., in order to dissolve the crystals completely. Then, the solution was cooled down at a constant rate (0.4 °C/min.) until the first crystals appeared (at θ_n). The difference between the saturation temperature (about 45 °C, precisely known from the refractive index value) and θ_n was taken as the metastable zone width or critical subcooling ($\Delta\theta_c$) (Fig. 1). This procedure was repeated 31 times in the absence of additive. Four runs were performed for each additive concentration.

Differential scanning calorimetry experiments were
carried out in a Setaram DSC 1 1 1 flux calorimeter with
sample masses of 50 to 100 mg; cooling and heating rates
were varied from 0.4 to 5 °C/min.

Larger crystals were obtained in two steps: nucleation,
then seeding of a supersaturated solution. Crystal habit
was determined by means of an optical goniometer.

Results and discussion

Phase equilibria, solution properties

The aqueous solubility of GABA over the temperature
range 25–55 °C can be plotted as a straight line (Fig. 1). Up
to a concentration of 500 ppm, the three surfactants
chosen show no detectable effect on this property (the
accuracy of the measurement does not exceed 0.001 kg
GABA/kg H_2O). Simon et al. [13] and Zumstein et al.
[14] also observed a negligible effect of the surfactant (at
a rather low concentration) on substrate solubility. On the
reverse, it is well known that surfactant properties (surface
tension of solutions, CMC, ...) can be strongly affected by
the presence of solutes, especially electrolytes [15]. A low
initial conductivity and the constancy of the CMC (on
a wt./solvent wt. basis) of ionic surfactants in saturated
GABA solutions (SDS: 2360 ppm at 30 °C, same as in pure
H_2O at 25 °C; DTAB: 5550 ppm at 30 °C instead of
5300 ppm in pure H_2O) confirmed the zwitterionic character of this solute. Besides, the experiment involving DTAB
pointed out a seldom noticed phenomenon [16], that is
a higher value of the specific conductivity *vs.* concentration
slope ($\Delta\kappa/\Delta c$) above the CMC (Fig. 2). Specific conductivities are always much lower in the presence of GABA than
in pure water, which is probably due to viscosity and
shielding effects ($\eta = 22.8$ cp for 1.3 g GABA/g H_2O). But
it appears that, in the case of SDS, $\Delta\kappa/\Delta c$ values are
divided by about 12 both below and above the CMC,

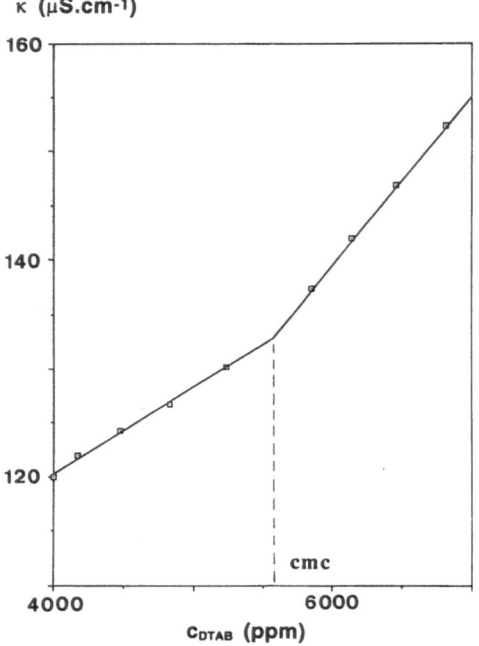

κ (μS.cm⁻¹)

Fig. 2 Conductometric determination of the CMC of DTAB in the
presence of GABA ($\theta = 30$ °C; $W = 1.3$ g GABA/g H_2O)

whereas with DTAB the reduction factor is 35 below the
CMC and only 4 above. These results suggest a much
stronger interaction between the GABA carboxylate function and the quaternary ammonium group of DTAB than
between the GABA positive end and the sulfate group.
Conversely, the DTAB micelle seems more protected from
the GABA influence.

The surface tension of a saturated aqueous solution of
GABA at room temperature was found to be slightly
higher than that of pure water, but surfactants showed
about the same efficiency in lowering surface tension in the
presence of GABA.

Metastable zone width measurement

Figure 3 shows the distribution of the subcoolings $\Delta\theta_c$
for pure GABA solutions in the laboratory-scale crystallizer. This distribution seems to be bimodal: two peaks
appear at about 12° and 16 °C. This behavior can be
compared with Beckman et al.'s results [17]. These
authors assume that the first peak can be assigned to
heterogeneous nucleation while the second one may be
related to homogeneous nucleation. Though it may not be
significant to calculate an average subcooling value with
so few data, we use the characteristic ratio
$r(r = \Delta\theta_{add}/\Delta\theta_{pure})$ where $\Delta\theta_{pure}$ and $\Delta\theta_{add}$ are the average
subcoolings in the pure solution and in the presence of

Fig. 1 Solubility of GABA in water

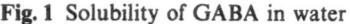

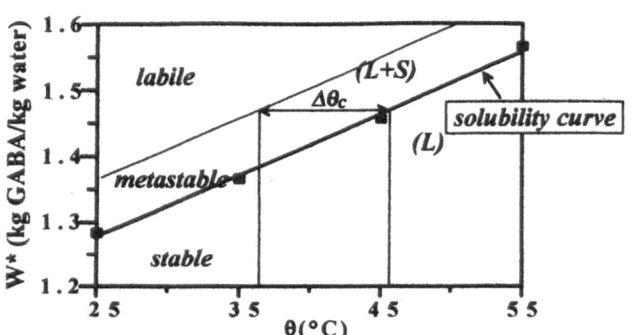

Progr Colloid Polym Sci (1994) 97:174–178
© Steinkopff-Verlag 1994

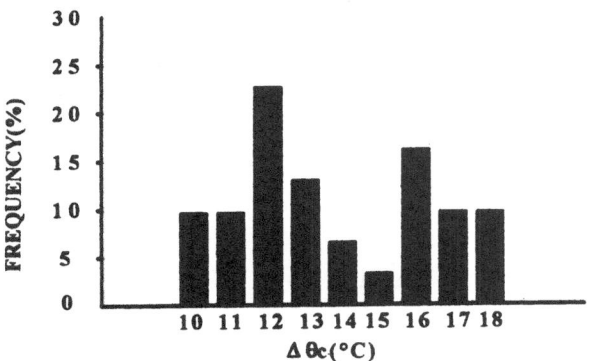

Fig. 3 Critical subcoolings for pure GABA solutions

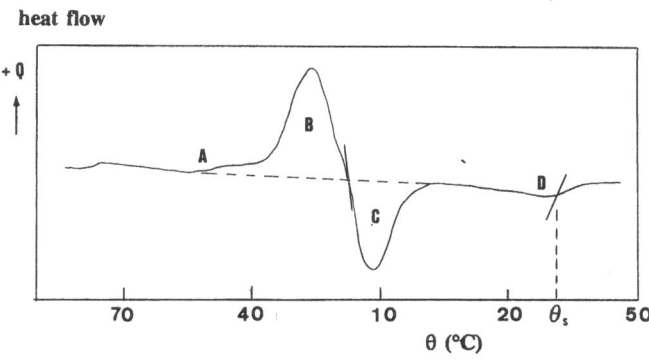

Fig. 4 DSC thermogram of heating a glass-like GABA solution. **A**: glass transition; **B**: crystallization; **C**: eutectic melting; **D**: melting of the remaining solid; θ_S: saturation temperature

Table 1 Critical subcoolings, supersaturations and r ratios for pure and surfactant-containing GABA solutions

Solution	$\Delta\theta_c$ (°C)	ΔW (kg/kg H$_2$O)	$r = \Delta\theta_{add}/\Delta\theta_{pure}$
pure	14.4	0.14	1.00
SDS 100 ppm	10.8	0.10	0.75
SDS 200 ppm	10.6	0.10	0.74
SDS 500 ppm	7.9	0.07	0.55
DTAB 100 ppm	13.1	0.12	0.91
DTAB 200 ppm	8.0	0.08	0.56
DTAB 500 ppm	9.8	0.09	0.68
HD 100 ppm	13.6	0.13	0.94
HD 1000 ppm	12.8	0.12	0.89

Table 2 Calculated attachment energies of individual GABA crystal faces

Face	d_{hkl} (Å)	Apparent groups	E_{att} (kcal/mol)
(1 2 0)	4.19	–a)	−15.2
(0 0 1)	6.75	COO$^-$	−16.4
(1 1 0)	6.10	NH$_3^+$, COO$^-$	−17.2
(0 2 0)b)	5.00	–a)	−18.2
(2 0 $\bar{1}$)	4.00	COO$^-$	−29.6

a) polar groups buried under hydrophobic chains
b) instead of the geometrically equivalent (0 1 0) face, due to space group symmetry

additive, respectively. Table 1 gives the critical subcoolings for the three surfactants at various concentrations, as well as the corresponding supersaturations ΔW ($\Delta W = W - W^*$) (Fig. 1) and r ratios. In these conditions and within the experimental concentration range (100–1000 ppm), the three surfactants do not show a strong influence, but seem to narrow the metastable zone. A lower supersaturation is then necessary to cause nucleation. At first sight, this unexpected effect could be due to a predominant influence of the reduction of the surface free energy on the nucleation rate [2]. However, although rather few data are available in the presence of surfactants, careful observation shows that nucleation never takes place in the bulk of the solution but always on non-glass parts. Therefore, it is quite probable that $\Delta\theta_{add}$ values are related to the first peak of Fig. 3, so that additives can be considered as delaying homogeneous nucleation without affecting heterogeneous nucleation.

In DSC experiments, pure GABA has been shown to be very difficult to crystallize ($\Delta\theta_c = 195$ °C). Most of the time, saturated aqueous solutions of GABA showed no crystallization during the cooling period (0.4 °C/mn) down to − 90 °C, but only during the heating period (Fig. 4).

The presence of HD did not change this behavior, but addition of one of the ionic surfactants often suppressed crystallization during the heating period, so that, in these conditions, SDS and DTAB again appear as nucleation inhibitors. In fact, while cooling, nuclei may form at so low a temperature that they cannot grow, due to the high viscosity of the medium; they can do so only while heating, above the glass transition temperature [18].

Habit of GABA crystals

The crystal faces likely to appear were identified and their attachment energies calculated by means of the current revision of the GenMol software (Table 2). The (120) face has the lowest attachment energy directly followed by the (0 0 1) and (1 1 0) faces. If it is assumed that v_{hkl} is proportional to $E_{att(hkl)}$, these three faces are likely to be well-developed. The resulting theoretical habit (Fig. 5) may be modified by the presence of the solvent (not taken into account previously) or an additive. As regards solvent effect, calculated attachment energies $E_{att(hkl)}$ in vacuo are

178

C.H. Lin et al.
Surfactants in crystallization of γ-aminobutyric acid

Fig. 5 Habit of GABA crystals (SHAPE software) *left*: from d_{hkl} values (BFDH theory); *middle*: from attachment energies; *right*: experimental

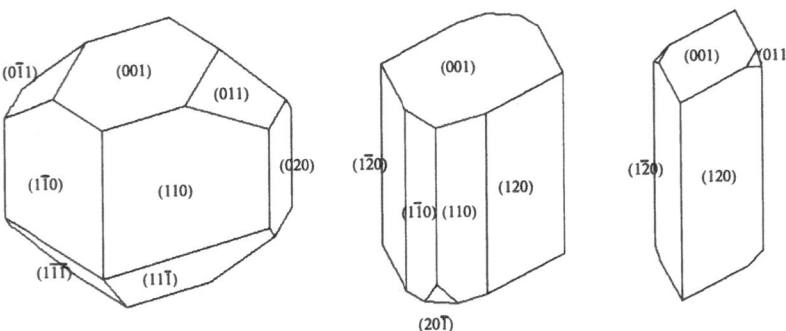

in fair agreement with the experimentally determined shape of GABA crystals, grown from pure aqueous solution (Fig. 5). According to a first set of experiments conducted with surfactants, no new face appears, and only the relative surface area of the two main faces, expressed by the L/W aspect ratio, is modified by the additive. The cationic species will adsorb preferentially onto the (001) face, which exposes COO^- groups: thus, it makes sense that 500 ppm DTAB reduce the L/W value (1.7 instead of 2.9 in pure water).

Conclusion

Although it is difficult to isolate nucleation as a single event, batch crystallization of GABA in the presence of model surfactants and DSC thermograms of the same systems confirm that such additives delay homogeneous nucleation, as observed frequently. A preliminary study of crystal habit through molecular modeling yields encouraging results, since theoretical and experimental shapes of GABA crystals grown from pure aqueous solutions are rather similar. It therefore seems possible to predict crystal morphology in the presence of additives from purely structural and energetic considerations, which is an important step in crystal engineering.

Acknowledgements Dr. P. Costessèque for crystal habit determination; Dr. C. Laguérie for a fruitful discussion; V. Morin and O. Devoivre for taking part in the experimental work.

References

1. Mullin JW, Crystallization (1993) 3rd ed Butterworth London pp 248–257
2. Boistelle R (1982) In: Mutaftschiev (ed) Interfacial aspects of phase transformations, D. Reidel Dordrecht pp 621–638
3. Addadi L, Berkovitch-Yellin Z, Weissbuch I, van Mil J, Shimon LJW, Lahav M, Leiserowitz L (1985) Angew Chem Int Ed Engl 24:466–485
4. Docherty R, Clydesdale G, Roberts KJ, Bennema P (1991) J Phys D: Appl Phys 24: 89–99
5. Saska M, Myerson AS (1983) J Cryst Growth 61:546–555
6. Canselier JP (1993) J Disp Sci Technol 14:625–644
7. Hiquily N, Canselier JP, unpublished results
8. Weber HP, Craven BM, McMullan RK (1983) Acta Cryst B39:360–366; Craven BM, Weber HP id 743–748
9. Hajji SM, Errahmani MB, Coudert R, Durand RR, Cao A, Taillandier E (1989) J Phys Chem 93:4819–4824
10. Hartman P (1973) In: Hartman (ed) Crystal growth: an Introduction, North Holland, Amsterdam
11. Hartman P, Bennema P (1980) J Cryst Growth 49:145–156
12. Pèpe G, Siri D (1990) Studies in Physical and Theoretical Chemistry 71:93–101
13. Simon B, Grassi A, Boistelle R (1974) J Cryst Growth 23:90–96
14. Zumstein RC, Rousseau RW, Turchi C (1989) Process Technol Proc 6:507–510
15. Corrin ML, Harkins WD (1947) J Am Chem Soc 69:683–688
16. Escoula B, Hajjaji N, Rico I, Lattes A (1984) J Chem Soc Chem Comm 1233–1234
17. Beckman W, Behrens M, Lacmann R, Rolfs J, Tanneberger U (1990) J Cryst Growth 99:1061–1064
18. Clausse D, Babin L, Sifrini I, Broto F, Dumas JP (1980) In: Straub J, Scheffer K (eds) Water and Steam, Pergamon Press, Oxford, pp. 664–671

Progr Colloid Polym Sci (1994) 97:179–182
© Steinkopff-Verlag 1994

EMULSIONS AND RHEOLOGY

D. M. Heyes
P. J. Mitchell
P. B. Visscher

Viscoelasticity and near-Newtonian behaviour of concentrated dispersions by Brownian dynamics simulations

Received: 14 September 1993
Accepted: 1 December 1993

Dr. D. M. Heyes (✉)
P. J. Mitchell
Department of Chemistry
University of Surrey
Guildford GU2 5XH
United Kingdom

P. B. Visscher
Department of Physics and Astronomy
University of Alabama
Tuscaloosa, AL 35487-0324, USA

Abstract We have developed the Brownian Dynamics simulation technique to calculate the viscoleastic behaviour of model colloidal dispersions. The *linear* or Newtonian behaviour of the liquid has been obtained using the Green–Kubo formula which incorporates the stress relaxation time autocorrelation function calculated from an *unsheared* model colloidal liquid. The viscoelastic behaviour, characterised in terms of the complex dynamic modulus (G', G'') and complex dynamic viscosity (η_r') of the liquid was obtained by Fourier transformation of the stress autocorrelation function. We also used the direct application of an oscillating shear strain at constant strain amplitude to obtain the dynamic moduli. Two variants on the method were used, one progressively (descending from high to low frequency) applying a series of widely spaced discrete oscillation frequencies. Another more efficient approach was also used, employing a continuously varying sweep through frequency space with a broad Gaussian smoothing window function. Results using the r^{-36} pair potential model for stable colloidal dispersions at high volume fractions are in agreement with experimental trends.

Key words Rheology – particle simulation

Introduction

The Brownian Dynamics (*BD*) computer simulation technique (a discrete particle model) has proved useful in predicting colloidal liquid rheology and giving insights into its microscopic origins. Recent applications of the *BD* technique to the rheology of colloidal suspensions by this group include stable suspensions [1], and flocculated suspensions (electro-rheological fluids and depletion flocs) [2, 3]. All of these simulations were carried out applying a constant homogeneous shear rate to the model colloidal particles to explore mainly non-Newtonian behaviour. In this report we extend the technique to consider viscoelastic behaviour. One method to achieve this is the Green–Kubo (*GK*) method [4] which uses the integral of the shear stress time autocorrelation function to obtain the Newtonian Viscosity. This technique has the advantage that, as no shear is applied to the sample, linear response or here, Newtonian behaviour is guaranteed. The other method we investigate is that of directly applied oscillatory shear within a no-equilibrium *BD* simulation. In this case, an oscillating homogeneous shear strain is applied to the sample, in a analogous fashion to the operation of oscillatory shear rheometers. In the limit of zero strain amplitude the Newtonian response is obtained. The technique can be used to investigate non-linear response.

The model

The model stabilised colloid particles interact through a hard "soft-sphere" interaction,

$$\phi(r) = \varepsilon(\sigma/r)^{36} , \tag{1}$$

where σ is the equivalent hard-core diameter of the colloid molecule and r is the centre-to-centre separation between the two model particles.

The volume fraction of the N colloidal particles in volume V is, $V_f = \pi N/6V$. The details of the BD model for updating the particle coordinates have been described elsewhere [1]. The particles experience forces from the solvent (drag and Brownian) and from other particles. The Langevin equations of motion for interacting particles in the free-draining and large particle limit are used. The particle positions are evolved through time and space using a forward difference integration algorithm.

At each time step in the simulation we make use of the coordinates of the particles at that time to compute the instantaneous value of the stress tensor, $\underline{\sigma}$, whose components are given by,

$$\sigma_{\alpha\beta} = \frac{1}{V} \sum_{i=1}^{N-1} \sum_{j=1+1}^{N} (r_{\alpha ij} r_{\beta ij}/r_{ij}) \phi'_{ij} . \tag{2}$$

In the Green–Kubo method, the shear-stress time autocorrelation function is required $C_s(t)$, [4]

$$C_s(t) = \langle \sigma_{xy}(0)\sigma_{xy}(t) \rangle , \tag{3}$$

where $\langle \cdots \rangle$ indicates an average over time origins. The infinite frequency shear modulus using the Green–Kubo formula is,

$$G_\infty = \frac{V}{k_B T} C_s(t = 0) , \tag{4}$$

For the current Brownian dynamics algorithm (which omits many-body hydrodynamics) the Green–Kubo formula gives the difference between the Newtonian viscosity η_0 (the zero-shear-rate limit) and the so-called infinite shear-rate viscosity, η_∞ (which is the limiting viscosity at high shear rate) of the model colloidal liquid,

$$\eta_0 - \eta_\infty = \frac{V}{k_B T} \int_0^\infty C_s(t) \, dt . \tag{5}$$

It is convenient to present the colloid liquid's viscosity in terms of the relative viscosity, $\eta_r = \eta/\eta_s$, so we have, $\eta_{r0} = \eta_0/\eta_s$ and $\eta_{r\infty} = \eta_\infty/\eta_s$.

The time correlation function can also be used to calculate the linear viscoleasticity of the colloidal liquid by Fourier transformation. The complex shear modulus is [6]

$$G^*(\omega) = G'(\omega) + i G''(\omega) , \tag{6}$$

where $G'(\omega)$ is the storage modulus and $G''(\omega)$ is the loss modulus. In terms of the stress time-correlation function we have,

$$G^*(\omega) = i \frac{V}{k_B T} \int_0^\infty C_s(t) \exp(-i\omega t)\omega \, dt . \tag{7}$$

We now consider the method for obtaining dynamic moduli more directly using non-equilibrium Brownian dynamics, in which we applied an oscillatory shear flow to the sample. We apply a series of well-separated oscillatory frequencies at fixed amplitude to the model liquid, usually descending from high to low frequency.

The strain on the sample, $\gamma(t)$ is given by,

$$\gamma(t) = \gamma_0 \cos \omega t , \tag{8}$$

where γ_0 is the strain-amplitude. The analytic expressions for the dynamic moduli are, for the storage modulus, G'',

$$G' = \frac{\omega}{n\pi\gamma_0} \int_0^{2n\pi/\omega} \sigma_{xy}(t') dt' \cos(\omega t') , \tag{9}$$

and for the loss modulus G'',

$$G'' = \frac{\omega}{n\pi\gamma_0} \int_0^{2n\pi/\omega} \sigma_{xy}(t') dt' \sin(\omega t') , \tag{10}$$

The contents of the cell are homogeneously strained in an oscillatory fashion over a whole number of cycles. Another approach called the "Chirp" method was also used, which enabled a continuously varying frequency to be applied to the sample [5]. This method was found to have number of advantages over the discrete frequency method; in particular, it is possible to scan rapidly through a wide frequency range eliminating start-up artefacts.

Result and discussions

In Fig. 1, we show the $C_s(t)$ for different volume fractions. The time correlation functions decay rapidly at short time, but their rate of descent decreases with time. In relaxation spectrum terminology, there is a wide spread of relaxation times associated with the temporal evolution of the stress fluctuations. As V_f increases the correlation function develops a "long-time tail". As the viscosity difference $\eta_0 - \eta_\infty$ is proportional to the area under $C_s(t)$, then this indicates an increase in the Newtonian viscosity. In Fig. 2, we show the frequency dependent viscosity $\eta' = G''(\omega)/\omega - \eta_\infty$, compared with that obtained by the direct oscillation route (see below). Within the statistical uncertainty of the non-equilibrium technique, the agreement is excellent at high frequency, with the GK approach having better statistics in the low frequency regime as $\omega \to 0$.

Progr Colloid Polym Sci (1994) 97:179–182
© Steinkopff-Verlag 1994

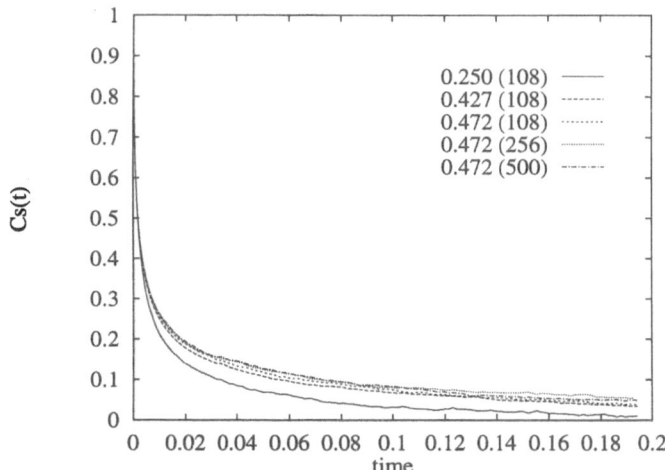

Fig. 1 The time correlation functions for a series of volume fractions given on the figure. The number of particles used in the simulation is given in bracket. Time is in units of τ_r

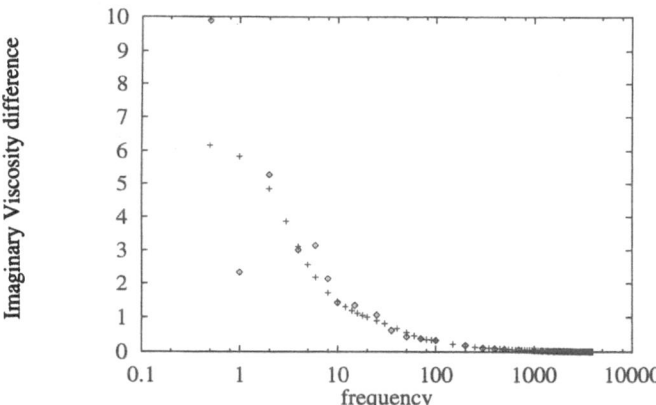

Fig. 2 The $\eta'(\omega\tau_r)$ derived from the $C_s(t)$ (crosses) and direct oscillation method (diamonds) with $\gamma_0 = 0.02$, $N = 108$ and $V_f = 0.472$

Over the complete volume fraction range the following analytic expression fit the experimental relative viscosity data of near hard-sphere dispersions within experimental uncertainty, [7]

$$\eta_{r0} = (1 - V_f/0.63)^{-2} , \tag{11}$$

and

$$\eta_{r\infty} = (1 - V_f/0.71)^{-2} . \tag{12}$$

These Krieger–Dougherty expressions produce values for $\eta_{r0} - \eta_{r\infty}$, which are very close (within the approximate statistical error) to the simulation results, which is quite remarkable as the model has no many-body hydrodynamics in the equations of motion in the present model. For

example, at the $V_f = 0.472$ state point of Fig. 2, $\eta_{r0} - \eta_{r\infty} = 7.0$ using the above formulae, agreeing well with the value of 6 ± 1 obtained by extrapolation of $\omega \to 0$ of the simulation data.

Oscillatory shear non-equilibrium simulations were carried out using the stepped frequency and "chirp" methods. We use a dimensionless frequency $\omega\tau_r$, where $\tau_r = 3\pi\sigma^3\eta_s/4k_BT$, is the time it takes a colloidal particle at infinite dilution to diffuse a distance $\sigma/2$. In Fig. 3, we show an example of the stress vs strain profile for an $\omega\tau_r = 1000$ with simulation at a strain amplitude of 0.2.

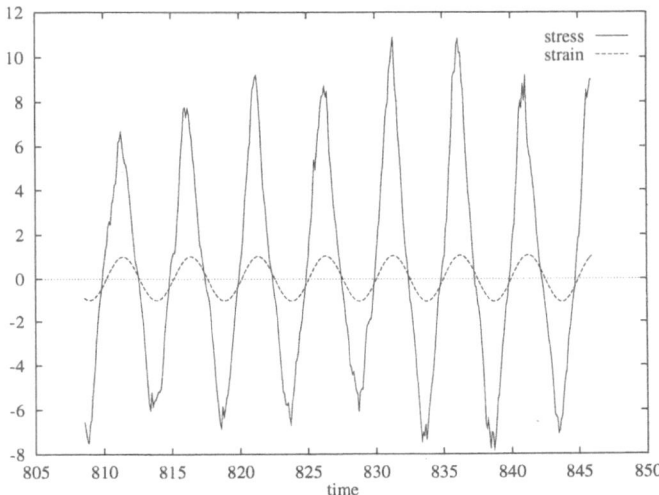

Fig. 3 The stress vs. strain profile for $\gamma_0 = 0.2$, $N = 108$ and $V_f = 0.472$ taken over an arbitrary number of cycles at $\omega\tau_r = 1000$. Time is in particle units, $\sigma(m/\varepsilon)^{1/2}$. The solid curve is the stress and the dashed line is the sinusoidal strain. The strain is arbitrarily scaled to assist comparison with the stress response, $\sigma_{xy}(t)$

Fig. 4 The strain amplitude dependence of the storage modulus. The state point is $V_f = 0.472$, $N = 108$ with $\omega\tau_r = 5$ (diamonds), $\omega\tau_r = 50$ (crosses) and $\omega\tau_r = 500$ (squares)

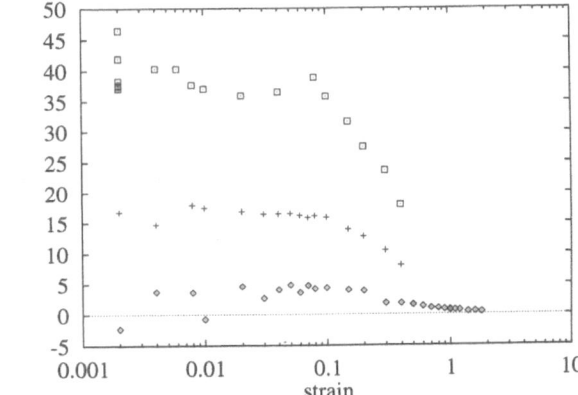

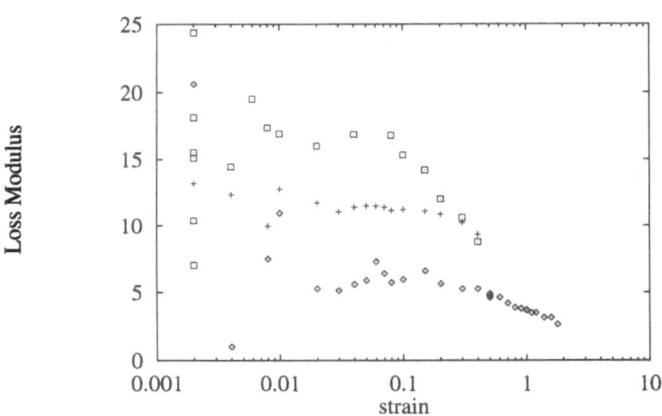

Fig. 5 As for Fig. 4, except the loss moduli, G'' are shown.

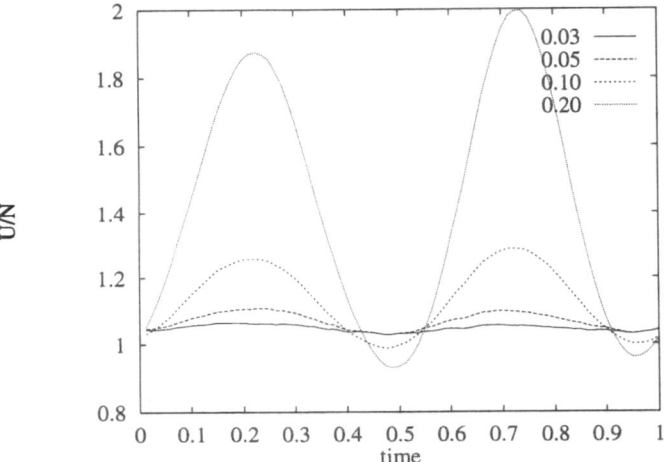

Fig. 6 The time variation of the interaction energy per particle u for a $V_f = 0.527$ state at $\omega\tau_r = 1000$ for a range of strain amplitudes, γ_0. The cycle time is normalised to go from 0 to 1

The response at this amplitude is non-linear, which is indicated by the stress profile developing a spikey appearance, rather than a more sinusoidal response obtained at lower amplitudes. In Fig. 4, we show the storage moduli, G' for $V_f = 0.472$ states at a series of strain amplitudes for three frequencies. In Fig. 5, we show the corresponding G'' as a function of the strain amplitude. We note that with increasing frequency the storage modulus increases to $G_\infty = 64$ for this state. A distinct non-linearity (decrease) in response becomes evident for strain amplitudes in excess of ca. 0.03, indicative of a shear thinning of the system.

In Fig. 6 the variation of the interaction energy per particle u, averaged over many cycles for a $V_f = 0.527$ state point with $\omega\tau_r = 1000$ is shown. We have,

$$\langle u \rangle = \frac{1}{2N} \sum_{i=1}^{N} \sum_{i \neq j} \langle \phi_{ij}(r_{ij}) \rangle . \tag{13}$$

This and other thermodynamic properties oscillate at twice the imposed shear oscillation frequency. This is because the positive and negative strain halves of the cycle produce a thermodynamically equivalent distortion to the radial distribution function and therefore derived thermodynamic properties of the system. At high amplitudes there is a dramatic variation in thermodynamic values over the cycle, up to twice the equilibrium value at the point of maximum strain.

To conclude, we have shown that the viscolastic behaviour of colloidal dispersions can be modelled routinely now using Brownian Dynamics simulations. The linear ("Newtonian") response can be obtained unambiguously without applying any shear using the Green–Kubo technique. The non-linear response can also be followed by application of an oscillatory shear strain history to the model sample.

Acknowledgements P.B.V. thanks the SERC for a visiting fellowship (grant number GR/H31554). P. J. M. thanks the SERC and ECC International for a research fellowship (grant number GR/H80644). Computations were carried out on the CONVEX C3 at the University of London Computer Centre.

References

1. Heyes DM, Melrose JR (1993) J Nonnewt Fl Mech 46:1–18
2. Melrose JR, Heyes DM (1993) J Chem Phys 98:5873–5886
3. Melrose JR, Heyes DM (1993) J Coll & Interface Sci 157:227–234
4. Levesque D, Verlet L, Kurkijarvi J (1973) Phys Rev A 7:1690–1700
5. Visscher PB, Mitchell PJ and Heyes DM (1993) J Rheol 38:465–483
6. Ferguson J, Kemblowski, Z Applied Fluid Rheology, (Elsevier, London, 1991)
7. Russel WB, Saville DA, Schowalter WR "Colloidal Dispersions", Cambridge Univ Press, 1989, p. 466

Progr Colloid Polym Sci (1994) 97:183–187
© Steinkopff-Verlag 1994

J. Krägel
S. Siegel
R. Miller

Surface shear rheological studies of protein adsorption layers

Received: 16 September 1993
Accepted: 28 February 1994

Dr. J. Krägel (✉) · S. Siegel
KAI e.v.
WIP
Rudower Chaussee 5
12489 Berlin-Adlershof, FRG

R. Miller
MPI für Kolloid- und
Grenzflächenforschung
Rudower Chaussee 5
12489 Berlin, FRG

Abstract The surface shear properties of protein adsorption layers at the air/water interface have been studied at small periodic deformations by means of a modified torsion pendulum instrument. Surface shear rheological investigations of different proteins and protein/surfactant mixtures are discussed. As proteins, human albumins and gelatin are used without and in the presence of the anionic surfactant sodium dodecyl sulphate. The results are discussed in terms of adsorption layer structure and the interaction between protein molecules and of proteins with surfactants.

Key words Protein adsorption layers – human albumin – gelatin – sodium dodecyl sulphate

Introduction

There are various parameters controlling the interfacial mechanical properties, such as interfacial tension, and the four rheological parameters: interfacial dilational viscosity and elasticity, and shear viscosity and elasticity. Changes in these properties are induced by adsorption of surface active substances, such as surfactants, polymers or their mixtures [1]. The interactions of proteins with surfactants in the bulk and the properties are of particular interest in many applications. For instance, the dynamic interfacial mechanical properties play an important role in foam and emulsion stability and break-down [2–8]. Therefore, shear rheological studies of adsorption layers along with other interfacial investigations are useful for a discussion of the structure of proteins and mixed protein/surfactant systems at liquid/gas and liquid/liquid interfaces.

Rheological investigations of adsorption layers at fluid interfaces require equipment which does not disturb their structure during the measuring procedure. Therefore, in surface shear rheology the torsion pendulum technique is preferred. This method allows experiments with very small mechanical deformations of the adsorbed layer.

In the present paper, surface rheological studies are carried out with a modified torsion pendulum set-up developed recently [9]. The principle of the rheometer is based on a ring with a sharp edge hanging at a torsion wire. When applying an impulsive torque by an instantaneous movement of the torsion head the pendulum performs damped oscillations with a damping factor α and angular frequency β. This kind of experiment provides information on the surface shear coefficient of viscosity and the surface shear modulus of rigidity from a single experiment. It is the purpose of this paper to present a modified torsion pendulum apparatus and to show the sensitivity and accuracy of the instrument by model experiments.

Experimental

The damping of a torsion pendulum is one of the oldest methods of measuring surface rheological properties [10, 11]. The method offers unusual simplicity in construction and operation [12]. Earlier experimental designs which are described in the literature need higher deflection angles to set the pendulum in motion and to record the oscillation [13, 14]. At such high deflection angles the structure of the adsorption layer can be perturbed and in some cases destroyed. Rheological investigations of adsorption layers at fluid interfaces require equipment which does not disturb their structure during the measuring procedure. With modern electronic components and sensitive sensors an instrument was designed which allows experiments to be carried out with very small mechanical deformations of the adsorption layer.

The scheme of this new equipment is shown in Fig. 1. The main parts of the surface shear rheometer are the drive for the deflection (stepper motor, transmission, motor controller), the torsion wire with a circular measuring body, laser light source, and a circular measuring vessel. The measuring body has a sharp edge which touches the interface of the solution. A thin tungsten wire transfers the deformation, produced by the steeper motor, via the edge onto the interface. The movement of the edge is registered by a position-sensitive photo diode. A detailed description of the measuring procedure is given elsewhere [15]. Due to the sensitivity of the photo diode and the analog/digital converter the circular movement of the edge can be measured with an accuracy of ± 0.01 degrees at a deflection angle of 2 degrees. By deflecting the torsion head, a torque is applied to the interface. The transferred torque leads to a shear of the interface in the slit between the edge and the wall of the measuring vessel. The shear stress is given by

$$S = \frac{M}{4\pi}\left(\frac{1}{r_1^2} + \frac{1}{r_2^2}\right) \tag{1}$$

with, M – transferred torque, r_1 – outer radius of the edge, and r_2 – inner radius of the measuring vessel.

The mathematical relations for the oscillating torsion pendulum for the study of surface films were derived by Tschoegl [16]. The concept of linear viscoelastic theory has been used for two-dimensional systems. Under these conditions the viscoelastic behaviour of the film may be described adequately by a Voigt model (spring and dash pot in parallel). Figure 2 shows the torque circuit diagram for the torsion pendulum. Eq. (2) describes the motion of the torsion pendulum

$$I_r \cdot y + (F_r + \eta_s/H_s) \cdot y + (E_r + G_s/H_s) \cdot y = \Theta(t), \tag{2}$$

where $\Theta(t)$ is the oscillation impulse, I_r is the moment of inertia of the measuring system, E_r is the elasticity of the torsion wire, F_r is the friction of the clean solvent interface, H_s is an apparatus constant which depends on the slit geometry, η_s is the surface shear coefficient of viscosity,

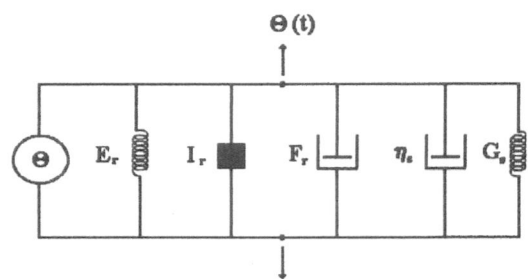

Fig. 2 The torque circuit diagram for the torsion pendulum

Fig. 1 A schematic diagram of the torsion pendulum apparatus

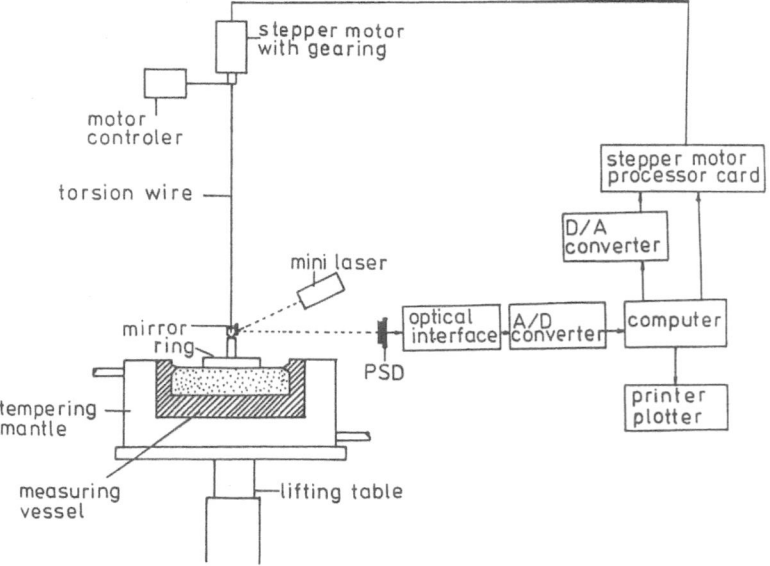

and G_s is the surface shear modulus of rigidity. The general solution of the equation of motion has the following form:

$$y(t) = y_0 \cdot \exp(-\alpha \cdot t) \cdot \sin(\beta \cdot t - \varphi) + c, \qquad (3)$$

with y_0 – amplitude, α – damping coefficient, and β – period of oscillation, φ – phase shift, and c – offset. The parameters y_0, α, β, φ and c are determined by least square fitting. The parameters α and β are necessary for the calculation of the rheological coefficients, while the other parameters depend on the geometry (sensor calibration, starting point and others) and the measuring conditions (deflection angle). The parameters I_r, E_r, F_r and H_s must be determined by separate experiments. The surface shear coefficient of viscosity η_s and the surface shear modulus of rigidity G_s will be calculated via Eq. (4) and (5).

$$\eta_s = H_s(2 \cdot I_r \cdot \alpha - F_r) \qquad (4)$$

$$G_s = H_s \cdot I_r(\alpha^2 + \beta^2) - H_s \cdot E_r \qquad (5)$$

Figure 3 shows a typical measuring curve for a simple oscillation experiment with a gelatin solution. The experimental points describing the momentary position of the light spot on the sensor are in excellent agreement with the fitted curve which is calculated by the software. Both rheological parameters are obtained in one experiment. The measuring procedure and the data interpretation are fully automated. Through the appropriate choice of the torsion wire, the measuring body, the measuring vessel, and the deflection angle, it is possible to determine the rheological parameters over a wide range: the shear viscos-

ity form 0.1 μNs/m to 1 mNs/m, the shear elasticity from 0.1 μPa m to 1 mPa m. The software controls the stepper motor, records the motion of the measuring body and calculates the rheological parameters.

Results and discussion

The modified torsion pendulum instrument is used for the determination of the surface shear properties of macromolecular adsorption layers at liquid/fluid interfaces. In particular, adsorbed proteins are studied in the present paper. Measurements are carried out with two different proteins: gelatin (blend from Calbe) and human serum albumin (HA from Serva). All aqueous solutions are prepared with doubly distilled water. The experiments are performed at 23 °C.

The dependence of surface shear viscosity and elasticity of aqueous gelatin solutions on adsorption time are shown in Figs. 4a and 4b. After 60 min adsorption time the shear viscosity reaches a value of 60 μNs/m, the limit of measurements with the present set-up using a 30 μm tungsten wire. To follow the time dependence further on a thicker wire must be used.

The addition of sodium dodecyl sulphate (SDS, synthesized and purified by Dr. G. Czichocki from the Max-Planck-Institut für Kolloid- und Grenzflächenforschung, Berlin) leads to an initial higher shear viscosity which then

Fig. 3 A typical measuring curve of a pendulum experiment, performed with a 0.5 wt-% gelatin solution after an adsorption time of 60 min

Fig. 4A Surface shear viscosity of gelatin-SDS-mixtures measured with a 30 μm wire * 0.5 wt-% gelatin without SDS + 0.5 wt-% gelatin with $9 \cdot 10^{-4}$ mol/l SDS ■ 0.5 wt-% gelatin with $4 \cdot 10^{-3}$ mol/l SDS

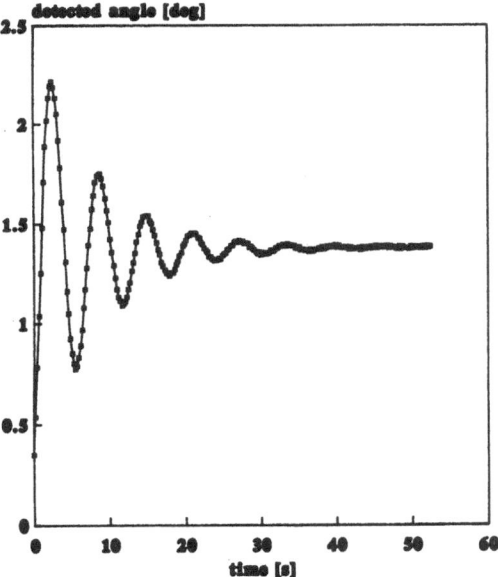

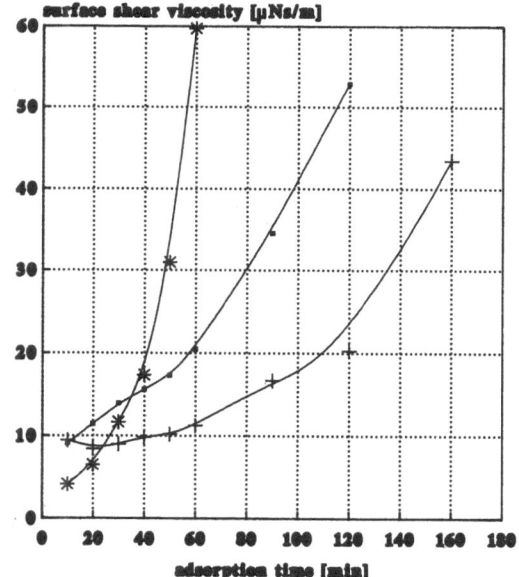

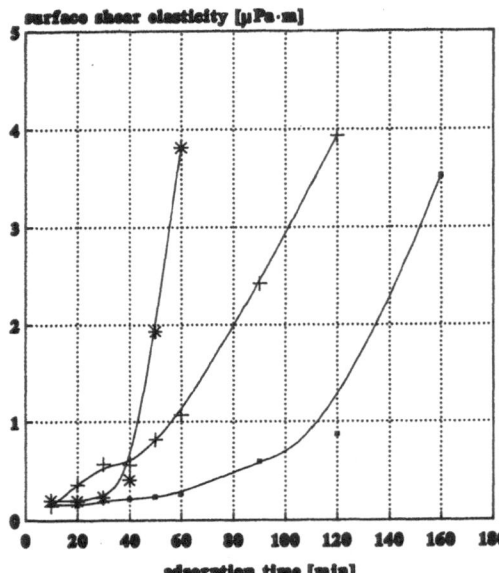

Fig. 4B Surface shear elasticity of gelatin SDS-mixtures measured with a 30 μm wire * 0.5 wt-% gelatin without SDS + 0.5 wt-% gelatin with 9·10^{-4} mol/l SDS ■0.5 wt-% gelatin with 4·10^{-3} mol/l SDS

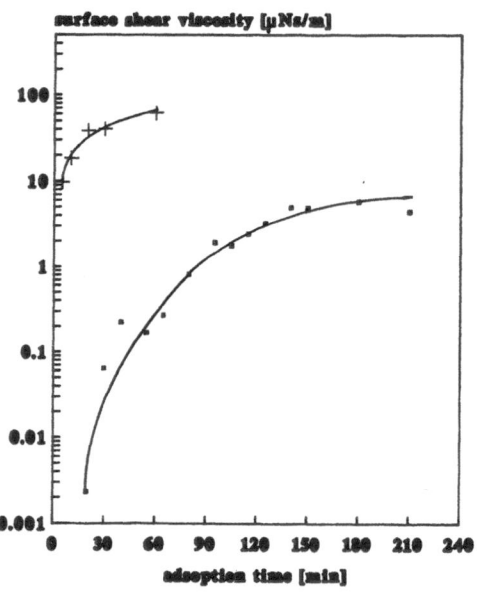

Fig. 5A Surface shear viscosity of human serum albumin in an aqueous buffer solution ■0.001 mg/ml HA measured with a 30 μm wire + 0.1 mg/ml HA measured with a 100 μm wire

increases more slowly than that of the gelatin solution without surfactant. The higher the surfactant concentration, the smaller is the slope of the shear viscosity changes (Fig. 4a). The same picture results for the shear elasticity (Fig. 4b). The lower values of both rheological parameters at longer adsorption times can be explained by a partial displacement of adsorbed protein molecules by surfactants with increasing surfactant concentration. The slightly higher elasticity and viscosity values at short adsorption times are real and cannot be explained so far.

Experimental results for human albumin in an aqueous buffer solution (pH 6.5, 0.99 g/l $Na_2HPO_4 \cdot 2H_2O$ and 1.76 g/l KH_2PO_4; cf. [17]) are displayed in Figs. 5a and 5b. The surface shear viscosity of the 0.001 mg/ml HA solution increases with adsorption time and levels off at about 150 min while the surface shear elasticity still increases. The same picture is obtained for the higher concentration of 0.1 mg/ml HA. The absolute values are about three orders of magnitude higher than those of the 0.001 mg/ml HA solution. Therefore, a thicker tungsten wire with a diameter of 100 μm was used.

The results confirm the good reproducibility and accuracy of the surface shear rheometer described here. The advantage of the present instrument compared to others using a constant shear stress or shear rate is the possibility of simultaneously studying the time dependence of both shear elasticity and viscosity. Depending on the wire and

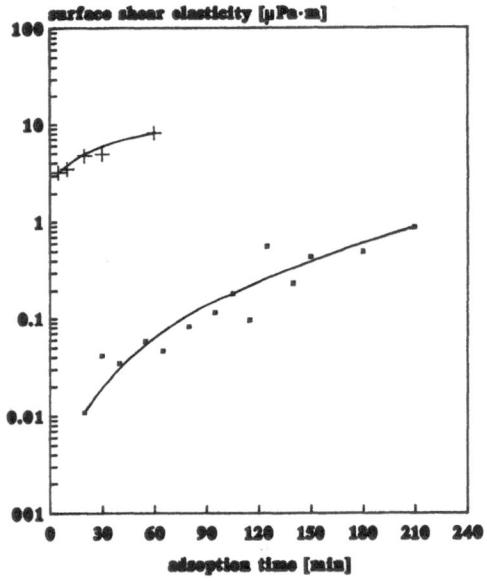

Fig. 5B Surface shear elasticity of human serum albumin in an aqueous solution ■0.001 mg/ml HA measured with a 30 μm wire + 0.1 mg/ml HA measured with a 100 μm wire

the body used, subsequent measurements in time intervals of 10 min down to about 2 min are possible.

The instrument can be applied to liquid/liquid interfaces as well. In this case, a disk-like body is used instead of the ring with a sharp edge. In a forthcoming paper, we will report about these experiments.

Progr Colloid Polym Sci (1994) 97:183–187
© Steinkopff-Verlag 1994

References

1. Edwards DA, Brenner H, Wasan DT (1991) Interfacial Transport Processes and Rheology, Butterworth-Heineman Publishers, Stoneham
2. Dickinson E, Murray BS, Stainsby G (1985) J Colloid Interface Sci 106:259–262
3. Krugljakov PM, Exerowa D (1990) Foam and Foam Films (Russ), Khimija, Moscow
4. Nikolov AD, Wasan DT, Denkov ND, Kralchewsky P, Ivanov IB (1990) Progr Colloid Polym Sci 82:87
5. Wasan DT, Sampath K, Aderangi N (1980) AlChE Symposium Series No. 192, 76:93
6. Fletcher PDI, Horsup DI (1992) J Chem Soc Faraday Trans 88:855
7. Morita M, Matsumoto M, Usui S, Abe T, Denkov N, Velev O, Ivanov IB (1992) Colloids Surfaces 67:81
8. Klahn JK, Agterof WGM, van Voorst Vader F, Groot RD, Groeneweg F (1992) Colloids Surfaces 65:151
9. Krägel J, Siegel S, Miller R, Born M, Emke B, Schano KH (1993) Prog Colloid Polym Sci (in press)
10. Mouquin H, Rideal EK (1927) Proc Roy Soc A114:690
11. Myers RJ, Harkins WD (1937) J Chem Phys 5:601
12. Joly M (1972) In: Matijevic E (ed) Surface and colloid science. Wiley Interscience, New York, Vol 5, pp 1–193
13. Wüstneck R, Fruhner H (1981) Colloid Polym Sci 259:1228
14. Dickinson E, Robson EW, Stainsby G (1983) J Chem Soc Faraday Trans 79:2937
15. Krägel J, Siegel S, Born M, Miller R, Schano K-H, Rev Sci Instrum (submitted)
16. Tschoegl NW (1961) Kolloid Z. 181:19–29
17. Benjamin J, van Voorst Vader F (1992) Colloids and Surfaces 65:161–174

Progr Colloid Polym Sci (1994) 97:188–193
© Steinkopff-Verlag 1994

R. Miller
P. Joos
V. B. Fainerman

Dynamic studies of soluble adsorption layers

Received: 16 September 1993
Accepted: 29 March 1994

Dr. R. Miller (✉)
MPI für Kolloid-und Grenzflächenforschung
Rudower Chaussee 5
12489 Berlin-Adlershof, FRG

P. Joos
Universitaire Instelling Antwerpen
Dep. Scheikunde
Universiteitsplein 1
2600 Antwerpen, Belgien

V. B. Fainerman
Institute of Technical Ecology
Blvd. Shevcheko 25
340017 Donetsk
Soviet Union

Abstract The dynamic surface tension of surfactant and polymer solutions can be measured by different experimental techniques, each of them having a specific time window, from the range of milliseconds up to seconds, minutes, and hours. In the present paper, the application is described of a new design of maximum bubble pressure instrument for measuring the dynamic surface tension of surfactant solutions in the millisecond time interval. The results obtained with the MPT1 for different surfactant systems are compared with data from other methods. The presented experimental results show the high accuracy and good reproducibility of the maximum bubble pressure measurements. The dynamic surface tension data show excellent agreement with those from drop volume, oscillating jet and inclined plate methods. In all methods the effective surface age was calculated via the corresponding theories.

Key words Maximum bubble pressure – dynamic surface tension – surfactants – drop volume method

Introduction

The dynamic surface tension of surfactant solutions is an important physical parameter which can be used to characterize the adsorption process at the solution/air surface and in the bulk of a liquid. After Rehbinder [1] applied the maximum bubble pressure method to measure the dynamic surface tension of surfactant solutions, further developments of this method were undertaken by different authors [2–24]. Several important steps were made during the development of the method to its present state. Kloubek derived a simple experimental procedure for the determination of the dead time [10] and gave an estimate of the effective bubble surface lifetime [9]. The use of electric pressure sensors for measuring the pressure and the bubble formation frequency [7, 11–15] simplified the measurement procedure substantially.

The maximum bubble pressure method in its present form allows investigations up to the region of high bubble formation frequencies by using a system volume, which is large in comparison with the bubble volume, and an electric and acoustic sensor to determine the bubble frequency. To separate the surface lifetime from the total time interval between subsequent bubbles a critical point in the pressure/gas flow rate dependence is defined. This point corresponds to a charge in the flow regime from individual bubble formation to a gas jet regime. The problem of recalculation of bubble surface lifetime to the so-called effective age of the surface (effective adsorption time) was discussed in [19–24].

In the present paper a bubble pressure instrument, the MPT1 from LAUDA [24, 28], is applied for measurements of dynamic surface tensions. Beside results obtained with this instrument, a comparison with data of different surfactant systems from other methods is presented: oscillating jet [26–28], inclined plate [25, 28], and drop volume [29–34].

Material and methods

The dynamic surface tension measurements are performed with four experimental set-ups which are described in detail elsewhere: bubble pressure method in [19], drop volume method TVT1 from LAUDA in [34], inclined plate method (IP) in [25], and oscillating jet method (OJ) in [26]. The methods work in different time intervals: TVT1: 0.5 s–1000 s; MPT1:1 ms–10 s: IP: 50 ms–1 s; OJ: 3 ms–25 ms.

The measurements were performed with two surfactants. Triton X-100 (octylphenol polyglycol ether, $C_{14}H_{21}O(C_2H_4O)_{10}H$) was purchased from Serva and used without further purification. The ethoxylated para-tertiary butyl phenol with 10 EO-groups (pt-BPh-E010) was synthesized and purified in the Max-Planck-Institute of Colloid and Surface Science in Berlin by Dr. G. Czichocki. All solutions were prepared with doubly distilled water.

Determination of surface tension and effective surface age

The four experimental methods used in the present study are based on different physical principles. Therefore, the appropriate theories have to be used to calculated the surface tensions and the effective adsorption time. Only then can the data be compared.

The surface tension value in the maximum bubble pressure method is calculated via the Laplace equation. As the capillary radius in the MPT1 is small, the bubble shape is spherical and correction factors are not necessary. Thus, the following equation results:

$$p = \frac{2\sigma}{r} + \rho g h + \Delta p \,, \tag{1}$$

where ρ is the density of the liquid, g is the acceleration of gravity, h is the immersion depth of the capillary of radius r, and Δp is a correction value caused by hydrodynamic effects. $\Delta p < 0$ leads to a correction of $\Delta \sigma = \sigma_{app} - \sigma_{corr} > 0$ (indices "app" and "corr" stand for apparent and corrected surface tensions, respectively) which can be estimated according to the following relation:

$$\Delta \sigma = \frac{3 \mu r}{2 t} \,. \tag{2}$$

Recent experimental studies [24] confirmed qualitatively the validity of Eq. (2): the value increases with increasing liquid viscosity μ, increasing capillary radius r and decreasing surface lifetime t.

The calculation of the effective surface age is possible only if the dead time of the bubble and the relative surface area deformation can be determined exactly. The value τ_b, the time interval necessary for the formation of a bubble with radius R, is related to the dead time via the Poiseuille law [16, 18, 19]:

$$\tau_d = \frac{\tau_b L}{K p} \left(1 + \frac{3r}{2R} \right), \tag{3}$$

where $K = \pi r^4/8 l \eta$ is the Poiseuille law constant, η is the gas viscosity, R is the radius of the detaching bubble, L is the gas flow rate, $p =$ the pressure, and l is the capillary length. The calculation of τ_d can be simplified when taking into account the existence of two gas flow regimes for the gas flow leaving the capillary [10, 16, 19]: bubble flow regime when $t > 0$ and jet regime, when $t = 0$ and hence $\tau_b = \tau_b$. Under the condition of constant bubble radius R the following simplified equation results [19]:

$$\tau_d = \tau_b \frac{L p_c}{L_c p} \,, \tag{4}$$

where L_c and p_c are related to the critical point, and L and p are the actual values of the pressure and gas flow rate below the critical point.

The surface lifetime can be calculated via the formula [19]:

$$t = \tau_b - \tau_d = \tau_b \left(1 - \frac{L p_c}{L_c p} \right), \tag{5}$$

The critical point in the $p(L)$-dependence can easily be located. The effective surface age is calculated by [22]:

$$\tau_a = \frac{t}{2\xi + 1} \,, \tag{6}$$

where $\xi = \dfrac{\sin \phi_0}{1 + \sin \phi_0}$ is the relative deformation rate of the bubble surface area in the first stage of its growth with $\phi_0 = \text{across} \left(\dfrac{\sigma}{\sigma_0} \right)$. σ is the dynamic surface tension of the liquid at time t, and σ_0 is the surface tension of the pure solvent. For surface tensions $\sigma_0 - \sigma > 10$ mN/m the relative surface deformation rate is approximately equal to $\xi = 0.5$, i.e., $\tau_a = t/2$.

The other three methods used in the present study also yield data in the form of surface tension dependent on a specific time function. In the drop volume method, the result is surface tension as a function of drop formation time t, which is larger than the effective age of the drop surface because of the continuous growth of the drop. As a first approximation, the effective age τ_a is obtained from t via the relation

$$\tau_a = \frac{3t}{7} \approx 0.43\, t \ . \tag{7}$$

At small drop times the data from the drop volume method are affected by the so-called hydrodynamic effect described by different authors [34–37]. This effect, caused by the process of drop detachment, simulates higher surface tensions and can be corrected by one of the proposed relations given in [31, 35, 37]. In the present study, the experimental drop volume data $V(t)$ are corrected via the relation

$$V_c = V(t)\left(1 - \frac{\alpha + \beta r_{cap}}{t}\right), \tag{8}$$

yielding the corrected drop volume V_c, where r_{cap} is the capillary tip radius and α and β are coefficient given in [37], having values of $\alpha = 0.008$ s and $\beta = 0.41$ s· cm^{-1}, respectively.

The oscillating jet and inclined plate methods need relations to calculate the effective surface age from the geometric length of the jet or the flowing film, measured from their inlet, and the liquid flow. The theoretical derivations are given elsewhere [26, 27, and 25, respectively].

Results and discussion

The aim of the present paper is to compare the maximum bubble pressure apparatus MPT1 with other methods having an overlapping time window. Therefore, experiments with the MPT1 and the drop volume method, the oscillating jet and inclined plate set-ups are performed with the same surfactant solutions. The experimental de-

tails of these methods are given elsewhere [19, 25, 26, 34]. The four methods have different overlapping time intervals. The time windows of the drop volume ad bubble pressure method show only a small overlap while the time windows of the inclined plate and oscillating jet methods are completely within that of the bubble pressure instrument.

Experiments with the MPT1 and TVT1 are performed with aqueous solutions of an ethoxylated paratertiary butyl phenol with 10 EO-groups (pt-BPH-E010). The dynamic surface tension of a 0.025 mol/l solution of pt-BPh-E010 is shown in Fig. 1. The figure contains the original data as well a recalculated results in form of surface tension as a function of the effective surface age. The original data $\sigma(t)$ of the bubble pressure method are transferred into $\sigma(\tau_a)$ by a shift in the $\sigma/\log t$-plot, according to Eq (6). The drop volume data were corrected first with respect to the hydrodynamic effect at drop formation times $t < 30$ s using Eq (8) and then the effective surface age was calculated using Eq. (7). It is clear that the apparent surface tension is significantly increased by the hydrodynamic effect in the drop time interval up to about 10 s and amounts up to 1 mN/m. Only the corrected dynamic surface tenions σ as functions of the recalculated effective surface age τ_a are displayed in the following figures.

The dynamic surface tensions of aqueous solutions of pt-BPh-E010 at five concentrations are shown in Fig. 2. To demonstrate how the results from the MPT1 and TVT1 complement each other, a $\sigma/(\log\tau_a$-plot has been used. The curves show the typical course of $\sigma(\log\tau_a)$-behavior for a diffusion-controlled adsorption process. At medium concentrations a slight shoulder is observed, which is not present for a pure surfactant system. A simulation of the diffusion-controlled adsorption of a surfactant is shown in Fig. 3 using the same $\sigma(\log t)$-plot. This shoulder can be attributed to a second surface active

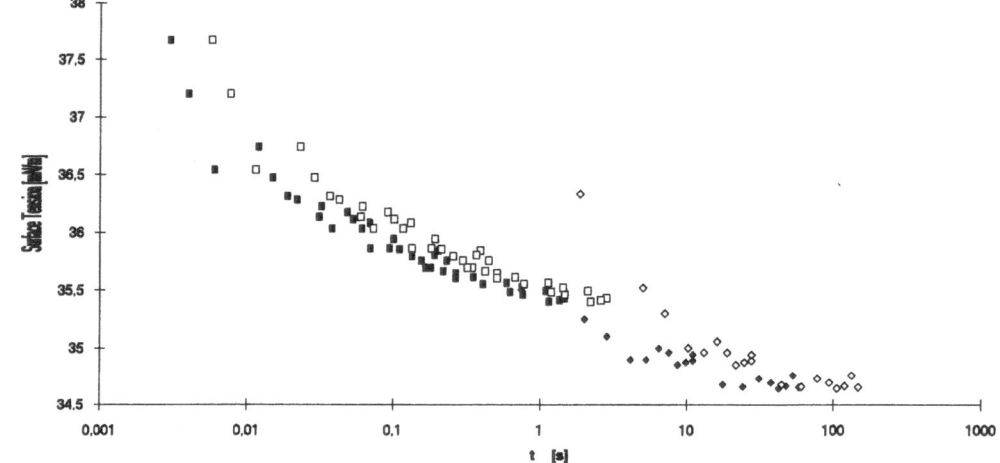

Fig. 1 Dynamic surface tension of a 0.025 mol/l pt-BPh-E010 solution measured using the maximum bubble pressure (■ □) and drop volume (◆ ◇) methods; original data (□ ◇), corrected data (■ ◆)

Progr Colloid Polym Sci (1994) 97:188–193
© Steinkopff-Verlag 1994

minor component in the sample or to changes in the adsorption mechanism.

The quantitative analysis of the adsorption mechanism shows a diffusion-controlled adsorption [38] over the whole concentration range with a slight change of the diffusion coefficient D with adsorption time and surfactant concentration. This can again be explained by the effect of a potential surface active impurity in the sample. Another explanation would certainly be the overlapping of two adsorption mechanisms. A detailed data analysis is made in [39] with butyl phenols of different chemical structure.

When analyzing the data in the range of short adsorption times a $\sigma(\sqrt{t})$-plot is useful [28, 40]. The results from

Fig. 2 are shown in this form in Fig. 4. Only at low concentrations can a reasonable diffusion coefficient be calculated. For higher concentration the final slope of $\sigma(\sqrt{t})$, needed for the approximate determination of D, is located outside the experimental range.

The MPT1 completely overlaps the time intervals of the oscillating jet and inclined plate methods. A comparison of the bubble pressure with the inclined plate method was performed with aqueous solutions of Triton X-100 (Fig. 5). The time interval of the inclined plate overlaps the one of the bubble pressure method at surface age between 50 and 1000 ms. The agreement is excellent.

A comparison of the bubble pressure with the oscillation jet method was also performed with aqueous solutions of Triton X-100 (Fig. 6). In contrast to the inclined plate, the oscillating jet only works in the time interval of few milliseconds. Also in this time interval the agreement with the maximum bubble pressure method is excellent and shows deviations in the range of the accuracy of the two methods only.

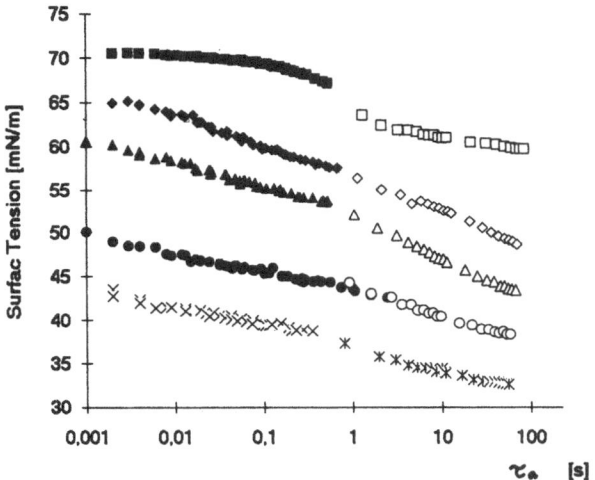

Fig. 2 Dynamic surface tension of five pt-BPh-E010 solution measured using the maximum bubble pressure (■ ◆ ● ▲ ×) and drop volume (□ ◇ ○ △ ∗) methods; $c_0 = 0.0001$ (■ □); 0.0005 (◆ ◇); 0.001 (△ ▲); 0.0025 (● ○); 0.005 (× ∗) mol/l

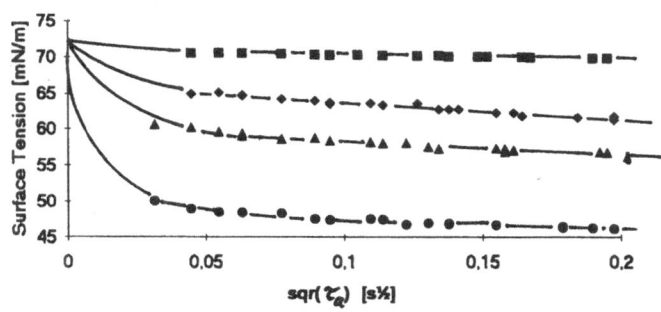

Fig. 4 Dynamics surface tension of four pt-BPh-E010 solution measured using the maximum bubble pressure method: $c_0 = 0.0001$ (■); 0.0050 (◆); 0.001 (△); 0.0025 (●) mol/l

Fig. 3 Dynamic surface tension calculated from the diffusion control adsorption model with a Langmuir isotherm

$$\Gamma = \Gamma_\infty \frac{c_0}{a_L + c_0} \text{ with}$$
$\Gamma_\infty = 4.10^{-10} \text{ mol/cm}^2$,
$a_L = 5.10 \text{ mol/cm}^3$;
$c_0 = 2.10^{-8}$ (■ ◆), 3.10^{-8}
(□ ◇) mol/cm³; $D = 1.10^{-5}$
(◆ ◇), 2.10^{-5} (■ □) cm²/s

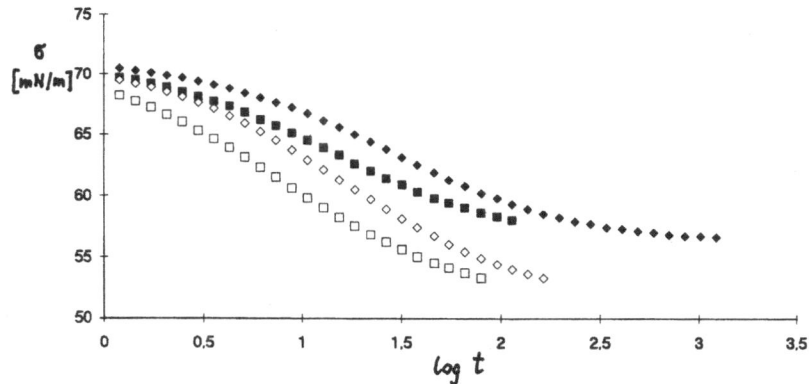

Fig. 5 Dynamic surface tension of two TRITON X-100 solutions measured using the maximum bubble pressure ($\square$ $\diamond$) and inclined plate ($\blacksquare$ $\blacklozenge$) methods; $c_0 = 0.2$ ($\blacksquare$ $\square$); 0.5 ($\blacklozenge$ $\diamond$) g/l

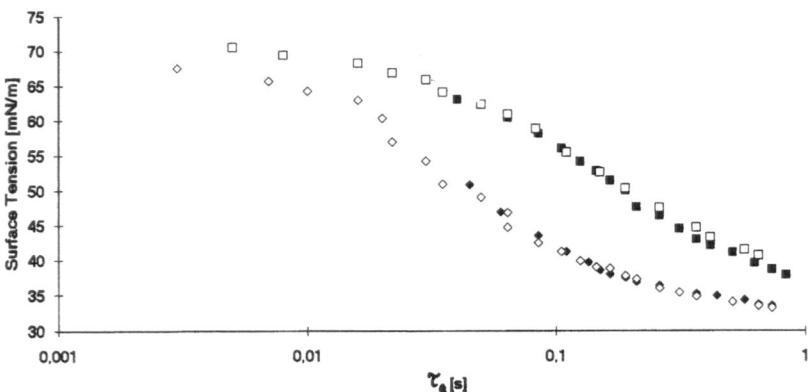

Fig. 6 Dynamic surface tension of four TRITON X-100 solutions measured using the maximum bubble pressure ($\square$ $\diamond$ $\bigcirc$ $\triangle$) and oscillating jet ($\blacksquare$ $\blacklozenge$ $\bullet$ $\blacktriangle$) methods; $c_0 = 0.2$ ($\triangle$ $\triangle$), 0.5 ($\bullet$ $\bigcirc$), 2.0 ($\blacksquare$ $\square$), 5.0 ($\blacklozenge$ $\diamond$) g/l

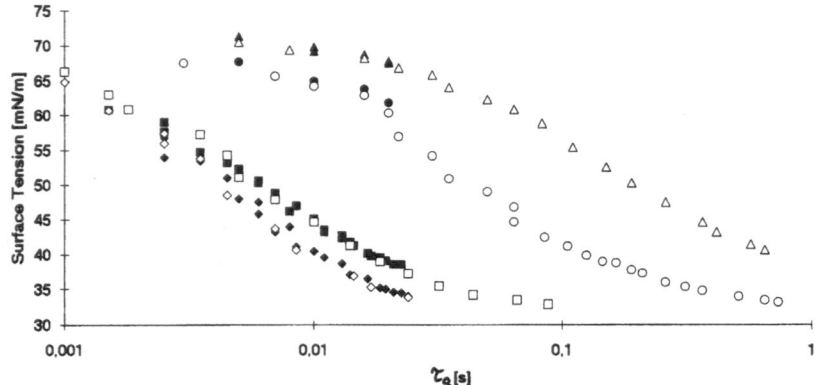

Summary

The aim of the study was a comparison of four experimental methods, having different time windows, for measuring the dynamic surface tension of surfactant solutions. The results of all four methods show very good agreement. To study the dynamics of adsorption in a time interval from milliseconds up to some minutes commercial instruments using the drop volume and maximum bubble pressure methods are available. They yield data which show an excellent agreement, within the accuracy of the individual instruments. Comparison with the other two methods, oscillating jet and inclined plate, also gave very good agreement within the range of accuracy, even in the milliseconds time range.

Acknowledgements This work was supported by the Deutsche Forschungsgemeinschaft (grant 478/199/92 and 436 UKR) and the "Human Capital and Mobility Programme" of the European Community, grant no. ERB4050PL930241. The support of the "Fonds der Chemischen Industrie" is also gratefully acknowledged

References

1. Rehbinder PA (1924) Z Phys Chem 111:447; (1927) Biochem Z 187:19
2. Adam NK, Shute HL (1935) Trans Faraday SOC 31:204; (1938) Trans Faraday Soc 34:758
3. Kuffner RJ (1961) J Colloid Sci 16:797
4. Kragh AM (1964) Trans Faraday Soc 60:225
5. Austin M, Bright BB, Simpson EA (1967) J Colloid Interface Sci 23: 108
6. Kloubek J (1968) Tenside 5:317
7. Bendure RL (1971) J Colloid Interface Sci 35:238
8. Finch JA, Smith GW (1973) J Colloid Interface Sci 45:81
9. Kloubek J (1972) J Colloid Interface Sci 41:1
10. Kloubek J (1972) J Colloid Interface Sci 41:7
11. Razouk R, Walmsley D (1974) J Colloid Interface Sci 47:515
12. Miller TE, Meyer WC (1984) American Laboratory February:91
13. Woolfrey SG, Banzon GM, Groves MJ (1986) J Colloid Interface Sci 112:583
14. Hua XY, Rosen MJ (1988) J Colloid Interface Sci 124:652
15. Mysels KJ (1989) Langmuir 5:442
16. Fainerman VB (1979) Koll Zh 41:111

17. Fainerman VB, Lylyk SV (1982) Koll Zh 44: 598
18. Fainerman VB (1990) Koll Zh 52:921
19. Fainerman VB (1992) Colloids Surfaces 62:333
20. Joos P, Rillaerts E (1981) J Colloid Interface Sci 79:96
21. Joos P, Fang JP, Serrien G (1992) J Colloid Interface Sci 151:144
22. Fainerman VB, Makievski AV, Joos P (1993) J Phys Chem (Russia) 67:452
23. Garrett PR, Ward DR (1989) J Colloid Interface Sci 132:575
24. Fainerman VB, Makievski AV, Miller R (1993) Colloids Surfaces 75:229
25. Van den Bogaert P, Joos P (1979) J Phys Chem 83:2244
26. Defay R, Petre G (1971) Surface and Colloid Science, Ed. E. Matijevic, Vol. 3, Wiley, New York, p. 27
27. Hansen RS (1964) J Phys Chem 68:2012
28. Fainerman VB, Miller R, Joos P (1994) Colloid Polymer Sci 272:731
29. Addison CC (1946) J Chem Soc:570
30. Davies JT, Rideal EK (1969) in "Interfacial Phenomena", Academic Press, New York
31. Kloubek J, Friml K, Krejci F (1976) Czech Chem Commun 41:1845
32. Tornberg E (1978) J Sci Fd Agric 29:762
33. Miller R, Schano K-H (1990) Tenside Detergents 27:238
34. Miller R, Hofmann A, Hartmann R, Schano K-H, Halbig A (1992) Advanced Materials 4:370
35. Jho C, Burke R (1983) J Colloid Interface Sci 95:61
36. Van Hunsel J, Bleys G, Joos P (1986) J Colloid Interface Sci., 114:432
37. Miller R, Schano K-H, Hofmann A (1994) Colloids Surfaces in press
38. Miller R, Kretzschmar G (1991) Adv Colloid Interface Sci 37:97
39. Miller R, Czichocki G (in preparation)
40. Van Hunsel J, Joos P (1987) Colloids Surfaces 24:139

Progr Colloid Polym Sci (1994) 97:194–198
© Steinkopff-Verlag 1994

EMULSIONS AND RHEOLOGY

G. Lundsten
S. Backlund
G. Kiwilsza

Solubility limits of water in systems of aromatic oils and non-ionic surfactants

Received: 29 September 1993
Accepted: 3 December 1993

G. Lundsten (✉) · S. Backlund
G. Kiwilsza
Department of Physical Chemistry
Åbo Akademi University
Porthansgatan 3-5
FIN-20500 Åbo
Finland

Abstract The aim of this study was to formulate an aromatic oil-based concentrate stabilized by a non-ionic surfactant with a high ability to solubilize water, i.e., to produce a single-phase water-in-oil (W/O) microemulsion. The oils used were benzene, methylbenzene, 1,4-dimethylbenzene and 1,3,5-trimethylbenzene. The surfactants used were commercial nonyl phenyl polyoxyethylenes, Berol 02 with six ethoxy groups and Berol 268 with 11 ethoxy groups, and a surfactant mixture, Berol 223 with a mass fraction dipropylene glycol monomethyl ether equal to 0.2. The extension of the W/O-microemulsion phase in the systems was determined visually at 298.2 K. At low surfactant contents the solubility of water in surfactant-aromatic oil solutions is small but increases with increasing surfactant content and reaches a maximum at a given oil to surfactant ratio. At high surfactant contents the water solubility decreases again and a lamellar liquid crystalline phase is formed. As the non-ionic surfactants are mixed with sodium 1,4-bis(2-ethylhexyl) sulfosuccinate (AOT) the water solubility is obviously enhanced at low surfactant contents.

Key words W/O-microemulsion – aromatic oil – alkyl phenyl polyethylene oxide – mixed surfactant

Introduction

The role of water is of great importance for the understanding of different structures in microemulsions. Water, itself, cannot stabilize or form steric interfaces between the different domains in these microheterogeneous systems. However, at small contents, water can have a great effect on the properties of non-aqueous solutions and dispersions. In this work, we report on the solubility and role of water in water-in-oil (W/O) microemulsions composed of non-ionic surfactants of alkyl phenyl polyoxyethylene type, water and aromatic oils.

The solubility of water and the aggregation of species in mixtures of non-ionic surfactants of alkyl polyoxyethylene or alkyl phenyl polyoxyethylene type and aromatic oils have been under investigation for the last four decades [1–8]. Already in 1948, Marsden and McBain [1] determined the solubility of water in a mixture of octyl phenyl nonylethoxylate (Triton X-100) and benzene. In general, at high oil contents the solubility of water is low. As the surfactant content increases the water solubility also increases very rapidly and for surfactants with less than eight carbon atoms in the alkyl chain the solubility limit of water extends toward the water corner in a triangle diagram. The surfactant-aromatic oil-water solution is in equilibrium with an almost pure water solution. When the alkyl chain of the surfactant increases the water solubility reaches a maximum and decreases again as the surfactant content increases. However, in this case the solution phase is now in equilibrium with a lamellar liquid crystalline phase.

The length of the oxyethylene chain has a profound influence on the aromatic oil to surfactant ratio for maximum water solubility. The longer oxyethylene part permits more aromatic hydrocarbon to be present without destabilization of the lamellar structure [6] and thus surfactants with longer oxyethylene chains show higher aromatic oil to surfactant ratios for maximum water solubility.

The formation of reversed micelles of non-ionic surfactants shows a strong oil solvent dependence [9]. For instance, Triton X-100 aggregates in cyclohexane and forms micelles [10], but not in benzene. The highly polarizable benzene molecules have through charge transfer[6], such a strong affinity for the polyoxyethylene chain that the surfactant monomers cannot aggregate in benzene. However, small amounts of water may, in some systems promote micellization [11]. The added water gives a gradual retraction of the aromatic hydrocarbon from the polar chain caused by the advancing water[5], thus promoting aggregation. The solvents can be characterized by means of their "'hydrophobicity" [12, 13], $\log P$, which is defined as the logarithm of the partition coefficient, P between octanol and water. The numerical values of $\log P$ for benzene, methylbenzene, 1,4-dimethylbenzene and 1,3,5-trimethylbenzene are 2.13, 2.69, 3.15 and 3.84, respectively.

The questions that now arise are whether the "hydrophobicity" of the aromatic oils has an influence on the water solubility and whether the addition of an ionic surfactant is capable of enhancing the water solubility at high aromatic oil to surfactant ratios. To answer these questions, the solubility limits of water in mixtures of benzene, methylbenzene, 1,4-dimethylbenzene and 1,3,5-trimethylbenzene, respectively, and surfactants of nonyl phenyl polyoxyethylene type were determined at 298.2 K. Furthermore, sodium 1,4-bis(2-ethylhexyl)sulfosuccinate was mixed with the non-ionic surfactants and the solubility limits of water were determined.

Experimental

Chemicals

Benzene (> 99.7% purity) and 1,3,5-trimethylbenzene (mesitylene, > 98% purity) were supplied by Merck, methylbenzene (toluene, > 99.5% purity), 1.4-dimethylbenzene (p-xylene, > 99% purity), dipropylene glycol monomethyl ether (> 97% purity) and sodium 1,4-bis(2-ethylhexyl)sulfosuccinate (AOT, > 98% purity) were supplied by Fluka. The non-ionic surfactants were of the alkyl phenyl polyoxyethylene type, $(C_nPh(EO)_mOH)$. Berol 02 $(C_9Ph(EO)_6OH)$, Berol 268 $(C_9Ph(EO)_{11}OH)$ and Berol

223 (surfactant mixture containing dipropylene glycol monomethyl ether at a mass fraction equal to 0.2) were supplied by Berol Nobel. The mass fraction of water determined by Karl Fisher titration was 0.0002 in Berol 02, 0.0007 in Berol 268, and 0.004 in Berol 223. All chemicals were used as supplied. The water was twice distilled.

Phase regions

About 20 stock solutions on the binary axis aromatic oil-surfactant were prepared. To these stock solutions water was added and the solubility regions were determined by visual inspection of samples weighed into glass vials with screw caps. Before the inspection the samples were thermostatted at 298.2 K in a water bath for at least 24 h. The presence of a liquid crystalline phase was detected from its appearance between crossed polarizers.

Results and discussion

The solubilities of water at 298.2 K, i.e., the extensions of the isotropic liquid phase, L_2, in mixtures of Berol 02, Berol 268, and the different aromatic oils are shown in Figs. 1 and 2, respectively. Both surfactants are completely soluble in the aromatic oils. They are almost insoluble in water but show a swelling into a liquid crystalline phase in equilibrium with almost pure water. The surfactants dissolve water, but the number of water molecules per EO-group is as low as 0.9 at the solubility border. One can assume that both surfactants like Triton X-100 do not aggregate in pure aromatic oils[11]. In the oil-rich corner of the phase diagrams (Figs. 1 and 2) the solubility of water is very low. The solubility increases with increasing surfactant content and reaches a solubility maximum at a given oil to surfactant ratio. With further increased surfactant content the water solubility decreases again and a lamellar liquid crystalline phase is formed.

The calculated number of aromatic oil molecules to the number of surfactant molecules, N_0/N_s, giving maximum water solubility are presented in Table 1. Independent of the oil, this ratio is almost constant for a given surfactant but increases as the number of oxyethylene groups increases. The latter observation is in accordance with the results for dodecyl polyoxyethylene-benzene systems investigated by Christenson et al.[6]. The larger oxyethylene moiety permits more aromatic hydrocarbon to be accumulated along the polar EO-chain without destabilization of the lamellar structure [5, 6]. The number of dissolved water molecules, calculated per EO-group, at maximum water solubility, N_W/N_{EO}, varies between two and three, as

Fig. 1 The extension of the
isotropic liquid phase, L_2, in
systems of Berol 02
($C_9Ph(EO)_6OH$), aromatic oils
and water at 298.2 K. L.C.
denotes liquid crystalline phase

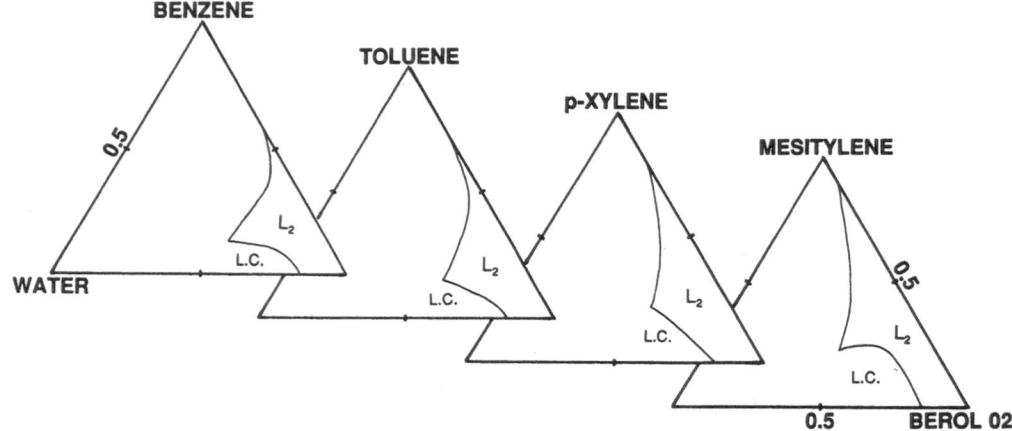

Fig. 2 The extension of the
isotropic liquid phase, L_2, in
systems of Berol 268
($C_9Ph(EO)_{11}OH$), aromatic oils
and water at 298.2 K. L.C.
denotes liquid crystalline phase

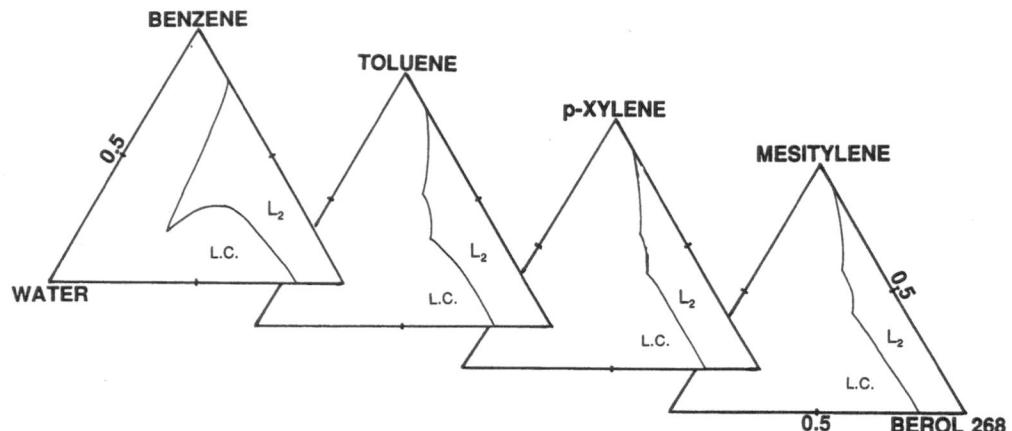

Table 1. The number of oil to surfactant molecules, N_O/N_S, and the number of water molecules to EO-groups, N_W/N_{EO}, at maximum water solubility in solutions of Berol 02 ($C_9Ph(EO)_6OH$) and Berol 268 ($C_9Ph(EO)_{11}OH$), respectively, and aromatic oils at 298.2 K.

	Benzene N_O/N_S	N_W/N_{EO}	Toluene N_O/N_S	N_W/N_{EO}	p-Xylene N_O/N_S	N_W/N_{EO}	Mesitylene N_O/N_S	N_W/N_{EO}
Berol 02	1.5	2.8	1.5	2.5	2	2.4	2	3.3
Berol 268	6	6.2	6	1.9	6	1.6	6	1.6

can be seen in Table I. This is in accordance with the hydration number of the EO-group in water [14]. These results are about one or two water molecules (calculated per EO-group) lower than the results by Christenson et al.[6] (Table 2). They conclude that, at small water contents, the water molecules are distributed along the EO-chains hydrogen bonded to the ether oxygen atoms. At water saturation of the chains, additional water induces surfactant aggregation. The formed aggregates are capable of solubilizing water in the core. This must also be the case

for the system Berol 268-benzene-water, where N_W/N_{EO} exceeds six. In aqueous solutions the interaction between the EO-group and water is very sensitive to temperature. Water changes from a good to a poor solvent for the EO-group of the surfactant, within an accessible temperature range[15, 16], which means that aggregates will be formed more easily at higher temperatures. This can explain the lower water solubility at 298.2 K in our systems compared to the system at 303.2 K studied by Christenson et al. [6]. A further explanation is that the phenyl-group in

Table 2. The number of oil to surfactant molecules, N_O/N_S, and the number of water molecules to EO-groups, N_W/N_{EO}, at maximum water solubility in solutions of dodecyl polyoxyethylenes with different numbers of EO-groups and benzene at 303.2 K. The results are determined from Fig. 1 in ref. [6].

	$C_{12}(EO)_3OH$	$C_{12}(EO)_4OH$	$C_{12}(EO)_5OH$	$C_{12}(EO)_6OH$	$C_{12}(EO)_7OH$	$C_{12}(EO)_8OH$
N_O/N_S	2/3	4/3	2	3	4	6
N_W/N_{EO}	3.5	3.9	7.6	4.2	4.0	4.1

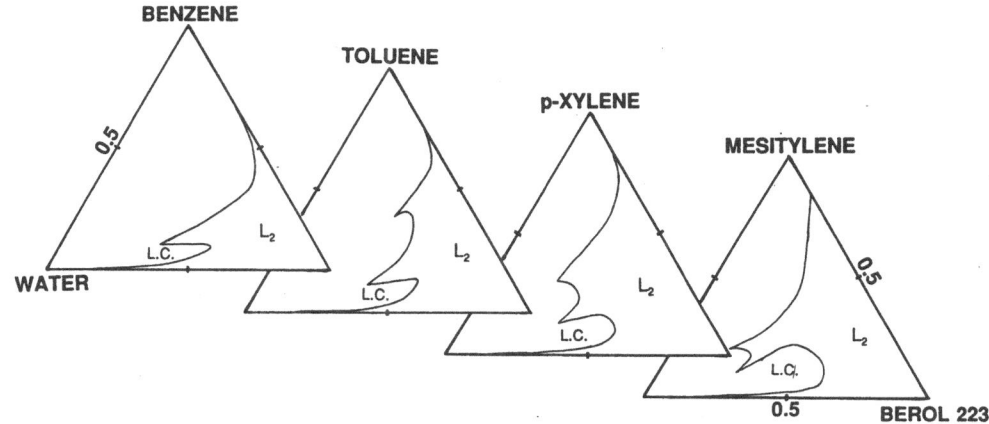

Fig. 3 The extension of the isotropic liquid phase, L_2, in systems of Berol 223 (a mixture of non-ionic surfactants), aromatic oils and water at 298.2 K. L.C. denotes liquid crystalline phase

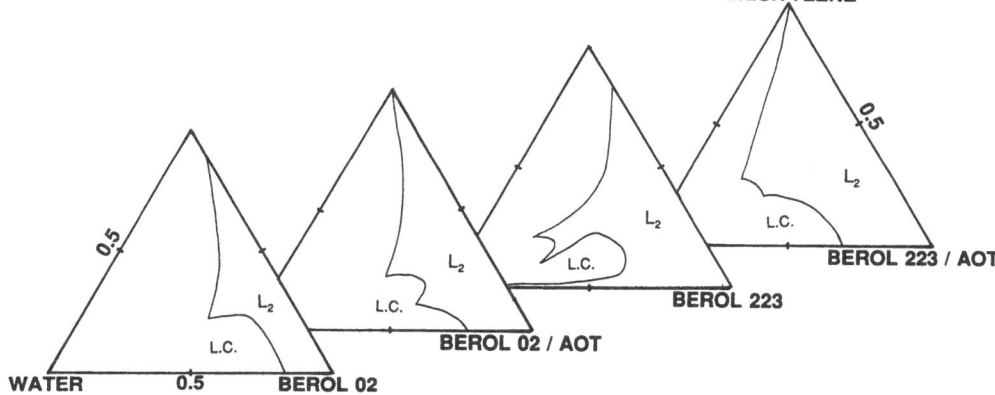

Fig. 4 The extension of the isotropic liquid phase, L_2, in systems of surfactants, mesitylene and water at 298.2 K. L.C. denotes liquid crystalline phase

the alkyl chain, can have a negative effect on the formation of reversed micelles through its interaction with the aromatic oils in our systems.

The solubility of water in Berol 223 and the different oils follows the same pattern as for Berol 02, Fig. 3. At low surfactant contents the solubility of water is very low, but increases with increasing surfactant content. On the other hand, there are two local solubility maxima, which probably arise from two different components in the surfactant. As the surfactant content is further increased, the solubility of water decreases again as a lamellar liquid crystalline phase is formed. Furthermore, the solubility of water is lowest in benzene and highest in mesitylene. From these

results and earlier investigations [1, 3, 6, 8], one can assume that Berol 223 contains components of alkyl phenyl polyoxyethylene type with at least eight carbon atoms in the alkyl chain and a couple of different ethoxylate chains.

To investigate the influence of the dipropylene glycol monomethyl ether on the water solubility capacity the ether was added to Berol 02 at a massfraction of 0.2 and the water solubilization limit was determined in the mixture Berol 02-ether-mesitylene; the solubility capacity of water was only slightly enhanced.

The extension of the L_2-phase as function of temperature has been determined for Berol 223, water and mesitylene at a mesitylene/Berol 223 ratio equal to 1 by weight.

198
Lundsten et al.
Solubility limits of water

The L_2-phase is largest at room temperature and shrinks drastically with increasing and decreasing temperature.

To investigate the influence of an ionic surfactant, AOT was mixed with Berol 02 and Berol 223 in equal masses (for Berol 02 and AOT that means almost equal amounts on a molecular scale). The solubility of water in the surfactant mixture-mesitylene solutions was determined at 298.2 K, Fig. 4. For Berol 02/AOT the water solubility is slightly enhanced and resembles the water solubility in the AOT-mesitylene system. But for Berol 223/AOT the water solubility capacity is obviously enhanced also at low surfactant contents, indicating that aggregates solubilizing water are formed.

In conclusion, it can be said that the "hydrophobicity" of the oils has a slight effect on water solubility in this case. On the other hand, the choice of surfactant is important. A mixture of alkyl phenyl ethoxylates with different numbers of EO-groups seems to give the highest water solubility in aromatic oil-non-ionic surfactant systems. When the non-ionic surfactants are mixed with the ionic surfactant AOT, the water solubility is obviously enhanced also at low surfactant contents.

Acknowledgements G. L. thanks Suomen Akatemia – Finlands Akademi and Svenska Kulturfonden in Finland for financial support.

References

1. Marsden SS, McBain JW (1948) J Phys Chem 52:110–130
2. Mulley BA, Metcalf AD (1964) J Colloid Sci 19:501–515
3. Kumar C, Balasubramanian D (1979) J Colloid Interface Sci 69:271–279
4. Kumar C, Balasubramanian D (1980) J Colloid Interface Sci 74:64–70
5. Christenson H, Friberg SE (1980) J Colloid Interface Sci 75:276–285
6. Christenson H, Friberg SE, Larsen DW (1980) J Phys Chem 84:3633–3638
7. Zhu D-M, Wu X, Schelly ZA (1992) Langmuir 8:1538–1540
8. Zhu D-M, Feng K-I, Schelly ZA (1992) J Phys Chem 96:2382–2385
9. Friberg SE (1987) In Interfacial Phenomena in Apolar Media Eicke H-F, Parfitt GD Eds, Marcel Dekker, New York
10. Zhu D-M, Schelly ZA (1992) Langmuir 8:48–50
11. Shinoda KH, Saito H (1971) J Colloid Interface Sci 35:359–361
12. Valsaraj KT, Thibodeaux LJ (1990) Separation Sci Techn 25:369–395
13. El Tayar N, Testa B, Carrupt P-A (1992) J Phys Chem 96:1455–1459
14. Xenacis A, Tordre C (1987) J Colloid Interface Sci 117:442–447
15. Olsson U, Wurz U, Strey R (1993) J Phys Chem 97:4535–4539
16. Bedö ZS, Berecz E, Lakatos I (1987) Colloid Polym Sci 265:715–722

Progr Colloid Polym Sci (1994) 97:199–203
© Steinkopff-Verlag 1994

P. Taylor
R.H. Ottewill

Ostwald ripening in O/W miniemulsions formed by the dilution of O/W microemulsions

Received: 15 September 1993
Accepted: 30 November 1993

P. Taylor (✉)
Zeneca Agrochemicals
Jealott's Hill Research Station
Bracknell
Berkshire RG12 6EY
United Kingdom

R.H. Ottewill
School of Chemistry
University of Bristol
Cantock's Close
Bristol BS8 1TS
United Kingdom

Abstract The growth, via Ostwald ripening, of O/W miniemulsions (formed from pentan-1-ol, sodium dodecyl sulphate, and water) has been investigated using a turbidimetric technique. It was found that the ripening rate was linearly dependent upon the interfacial tension within the emulsion below the CMC of the surfactant, as predicted by the LSW theory for Ostwald ripening.

Keywords Ostwald ripening – microemulsions – emulsions

Introduction

Emulsions are thermodynamically unstable systems and as such require large amounts of energy to form them [1]. The energy required increases with decreasing radius as a result of the increasing Laplace pressure that must be overcome to form the highly curved interface. Emulsions comprised of droplets of very small radius (miniemulsions), typically between 50 and 150 nm radius, have many potential uses in both industrial (e.g., pesticide delivery) and pharmaceutical (drug delivery, or artificial blood [2]) applications. However their formation is limited by the high energy needed for their production.

In this work O/W microemulsions [3], with radii typically < 5 nm, were used as a precursor in the formation of miniemulsions and also to investigate the factors determining the stability of small particle size emulsions. Microemulsions are thermodynamically stable since the entropy of formation outweighs the interfacial free energy changes associated with formation, thus giving a negative free energy of formation. Dilution into water results in a rapid rise in interfacial tension from around 10^{-5} mN m^{-1} up to 10^{-3} mN m^{-1} causing the free energy to become positive and the system to become unstable and prone to ageing. It was found that the ageing process was Ostwald ripening. In this process, the smaller droplets, as a result of their smaller radius and consequently higher chemical potential, dissolve and this material diffuses to and deposits onto the larger drops. This results in an overall increase in the average size of the particles in the emulsions. The rate of this process is dependent on the solubility of the dispersed phase in the bulk phase and on the interfacial tension between the two phases, as predicted by the Lifshitz, Slezov and Wagner theory of Ostwald ripening [4,5].

Experimental

A basic microemulsion composition of 15% n-dodecane, 15% pentan-1-ol, 10.5% SDS and 59.5% water was chosen to be the basis of the experiments. This was found to be an O/W microemulsion.

The emulsion preparation method was simply to add a known mass of the microemulsion to water with vigourous stirring to give a dilution of either 50 or 210 fold. Once prepared, the emulsions were stored at 25 °C for 5–20 days, dependent upon the stability of the emulsions. The emulsions were sized using the specific turbidity technique described by Heller et al. [6]. The turbidity (τ) of the emulsions was measured as a function of dilution using a Unicam SP600 spectrophotometer (with a modified transmitted light acceptance angle) at six wavelengths (400–600 nm at 40 nm intervals), and the specific turbidity (τ/ϕ) was plotted as a function of droplet concentration. Extrapolation to zero concentration gave the true specific turbidity in the absence of multiple or secondary scattering effects. The resulting turbidity spectra were fitted using Mie theory (using the appropriate refractive indices) and the LSW particle size distribution [4]. This gave the number average radius, $\bar{r}_n$. The volume fraction of the emulsion was taken to be that of the dodecane alone, the pentanol essentially being situated in the bulk phase. For systems above the CMC of the SDS/pentanol mixture, allowance was made for the reduction in volume fraction of the emulsion droplets due to solubilization of dodecane into the micelles. This was important due to the low dodecane volume fractions used (ca. 0.0009–0.004). In some cases the wavelength exponents [7] of the turbidity spectra were used to size the emulsions, the exponents being calculated from the slope of log-log plots of specific turbidity vs. wavelength for both experimental and theoretical spectra.

Interfacial tension measurements were made on the dodecane/aqueous SDS (in the presence of pentanol, 0.071% or 0.3%) using a spinning drop tensiometer (Bailey Engineering, Windsor). The extent of solubilization of dodecane in aqueous solutions of SDS in 0.071% pentanol was determined by gas liquid chromatography (Pye 104 Gas Chromatograph).

Results

A plot of number average radius ($\bar{r}_n$) vs time is shown in Fig. 1 for the "basic" emulsion (dilution 1/50). The radius was found to increase with time from 60 nm up to 150 nm over a 5-day period. This showed that dilution of microemulsions could produce emulsions in the required size range. The ageing vs. rate was determined as the slope (ω) of a plot of $(\bar{r}_n)^3$ time, which was found to be linear. The ageing rate obtained of 5.9×10^{-27} m^3 s^{-1} agreed well with 1×10^{-26} m^3 s^{-1} reported by Kabal'nov et al. for emulsions of dodecane stabilized by 0.1 moldm^{-3} SDS [8].

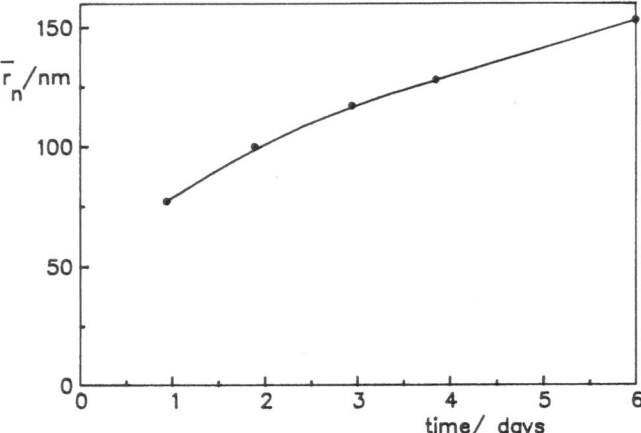

Fig. 1 Variation of average radius with time for the basic emulsion composition

The effect of droplet volume fraction was determined by diluting an emulsion (original dilution 1/50) with an SDS (0.21%)/pentanol (0.15%) aqueous solution to maintain the composition of the bulk phase. The ageing rate was found to be essentially independent of volume fraction in the range of $\phi = 0.008$–0.004, varying from 7.9×10^{-27} m^3 s^{-1} to 6.9×10^{-27} m^3 s^{-1}. Coalescence would show only a minor effect due to the low particle concentration, while Ostwald ripening is independent of $\phi \leq 0.1$. Above that point, droplet interactions affect the concentration gradients surrounding the droplets causing enhanced rates of ripening [9].

The effect of SDS concentration was determined by addition of additional SDS to an emulsion (dilution of 1/211) giving a concentration range of 0.005–0.74% (wt%). Below the CMC (0.19% SDS) of the SDS/pentanol (0.071%) mixture the ageing rate fell linearly with log[SDS], Fig. 2. In this region, the interfacial tension (γ) also fell with log[SDS], suggesting that the rate was strongly correlated with γ. The LSW theory gives the ripening rate (ω) as [4,10]:

$$\omega = \frac{d\bar{r}_n^3}{dt} = \frac{8}{9} \frac{D C_\infty \gamma V_{lm}}{\rho R T}, \tag{1}$$

where D and c_∞ are the diffusion coefficient and the solubility of the droplet phase material in the bulk phase respectively and V_{lm} its molar volume. This equation clearly predicts the observed linearity between ω and γ, showing that the ageing process was Ostwald ripening, Kabal'nov et al. have also demonstrated this for SDS stabilized emulsions, albeit of a greater droplet radius [8]. When suitable values for the solubility of dodecane in water are taken, reasonable agreement is found between the experimental data and the theoretical values. Plots of

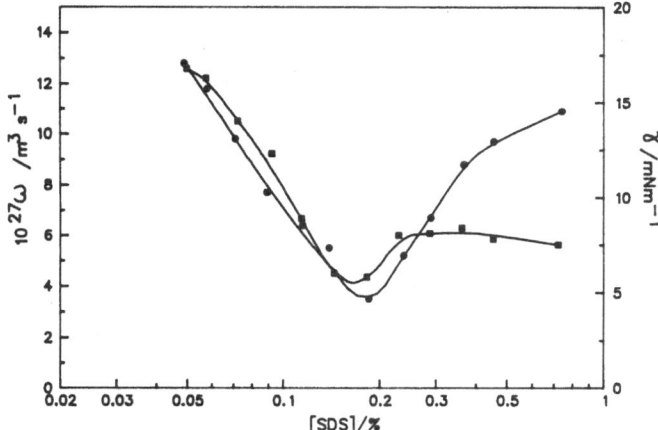

Fig. 2 Plots of Ostwald ripening rate, ●, and interfacial tension, ■, as a function of SDS concentration (0.071% pentanol)

the experimental rate (ω_E) divided by the theoretical rate (ω_T) against log[SDS] are shown in Fig. 3.

The agreement is closer when Franks' value for the aqueous solubility of dodecane is used with ω_E/ω_T close to 1, however, work by other groups suggest that the value of

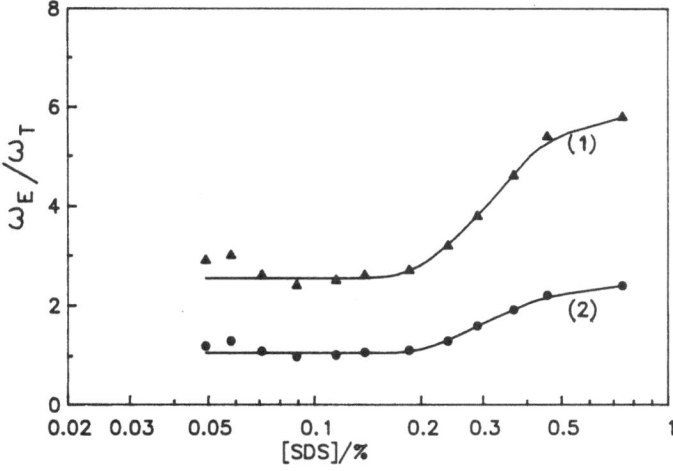

Fig. 3 Comparison of experimental rates with those predicted by the LSW equation. Curve (1) $C_\infty = 3.9 \times 10^{-6}$ kgm^{-3} (Coates, [13]); Curve (2) $C_\infty = 8.4 \times 10^{-6}$ kgm^{-3} (Franks, [14]); $D = 6.1 \times 10^{-10}$ m^2 s^{-1} [15,16], $\rho = 750$ kgm^{-3}, and $V_{1m} = 2.28 \times 10^{-4}$ m^3 mol^{-1}

3.9×10^{-6} kgm^{-3} is the more accurate. In this case the rates below the CMC agree to within a factor of 3, which probably represents good agreement.

Above the CMC, the rate was found to increase once more, contrary to the prediction of Eq. (1), Fig. 3, since in this region the interfacial tension is essentially independent of [SDS]. The discrepancy is due to the presence of micelles in the system. Ostwald ripening is governed by the solubility of the disperse phase in the bulk phase. Micelles increase the dodecane solubility through solubilization; however, substitution of the relevant solubilization concentrations into Eq. (1) gives values much in excess of those measured, Table 1. The values obtained experimentally agree much more closely with those calculated for the molecularly dissolved dodecane. The solubilized dodecane is effectively hidden from the droplets within the micelles. Micelles are known to be highly dynamic species, thus, if a micelle breaks up in solution, its solubilizate is temporarily stranded in the bulk phase prior to being incorporated into a newly formed micelle. Consequently, the solubilizate has only a limited period during which it can affect the rate of ripening, hence the smaller dependence on micelle concentration than might be expected and the closer relation to the molecularly dissolved case. This is in agreement with that found by Kabal'nov et al. [8].

The effects of progressively replacing the dodecane with a longer chain alkane produced very large changes in the rate of ripening of the emulsions (dilution 1/50). Tetradecane reduced the rate by over a factor of 10 in the mole fraction range (with respect to dodecane) of $x_2 = 0 - 1$, Fig. 4.

The mechanism for the two-component case is that initially the dodecane diffuses from the smaller droplets to the larger droplets, until its chemical potential in all the drops is equal as a result of the change in both size of the droplets and the accompanying change in its concentration within different droplets. At this point, there is no net driving force for any further transfer of dodecane and so only the longer chain alkane may diffuse and the rate is now governed by this transfer which is much slower as a result of its lower solubility. Schuckin et al. [10] derived an approximate equation for the effect of a second component on the rate of ripening. This equation, though not strictly applicable to the emulsions formed here, predicted the correct form for rate vs mole fraction, Fig. 4.

Table 1 Comparison of the experimental ripening rates with those calculated on the basis of micellar transport, [dod]$_{sol}$ is the equilibrium concentration of solubilised dodecane. $D = 1.02 \times 10^{-10}$ m^2 s^{-1} (SDS micelles [17])

[SDS]/%	[dod]$_{sol}$/kg m^{-3}	ω_E/m^3 s^{-1}	ω_T/m^3 s^{-1}	ω_T/ω_E
0.238	0.015	5.2×10^{-27}	1.3×10^{-24}	250
0.289	0.030	6.7×10^{-27}	2.6×10^{-24}	388
0.367	0.055	8.8×10^{-27}	4.8×10^{-24}	544
0.456	0.082	9.6×10^{-27}	7.3×10^{-24}	760
0.742	0.17	10.9×10^{-27}	15×10^{-24}	1376

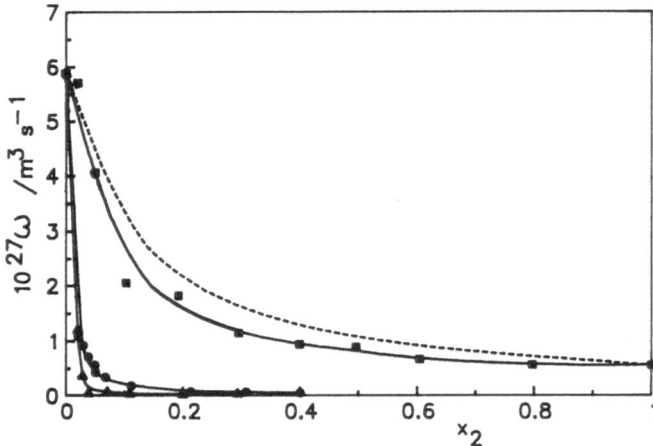

Fig. 4 Effects of added tetradecane, ○, hexadecane, ■, and octadecane, ▲, on the rate of ripening in dodecane emulsions. Dotted line represents the prediction for the effect of tetradecane made by Schuckin's equation [8]

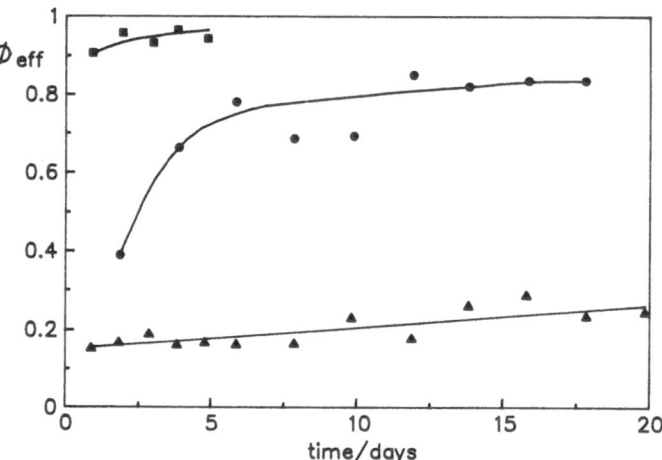

Fig. 5 Variation of the fraction of disperse phase (ϕ_{eff}) contained within the upper peak with time for three additives; tetradecane ■; hexadecane ●; and octadecane ▲

As the chain length was increased (hexa- and octadecane) the rate decreased even further, showing that these alkanes possessed much lower aqueous solubilities than that of dodecane. Much of the literature on alkane solubility shows the solubility of alkanes to stop decreasing with increasing chain length once dodecane is reached [15]. These results show that this is not the case, and that the apparently high solubility of n-alkanes in water is probably due to aggregate formation.

Comparisons between the expected specific turbidity calculated from the average radii determined from the wavelength exponent, n, where ($\tau/\phi \propto \lambda^{-n}$), and the measured specific turbidity showed marked discrepancies in some cases. This was attributed to the formation of bimodal systems. In such a system, the peak at smaller radii would be the remnants of the original microemulsion droplets, and as such would be effectively transparent as a result of their very small size, typically less than 5 nm. Thus, comparison of the measured and expected specific turbidities allowed an estimate of the volume of material in the upper peak, ϕ_{eff}, compared to that in the lower peak. Plots of ϕ_{eff} versus time are shown in Fig. 5. In general, it was found that at a given additive mole fraction the value for ϕ_{eff} decreased with increasing additive chain-length. Moreover, the time dependence of ϕ_{eff} decreased with increasing chain length, as did the variation of droplet radius with time. In the presence of octadecane, the droplet size (upper peak) varied little over a period of up to 18 days (Fig. 6), even at relatively low concentration. Typically average radii were of the order of 40–70 nm, showing that it was possible to stabilize miniemulsion droplets, forming a pseudoequilibrium system, though it must be remem-

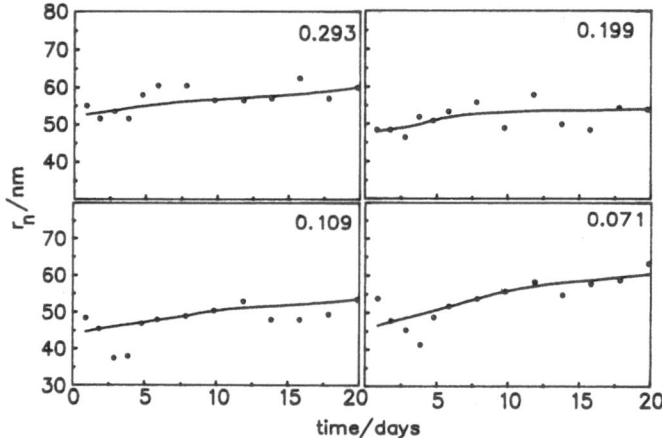

Fig. 6 Variation of upper peak radius with time for various mole fractions of octadecane (0.293, 0.199, 0.109, 0.071). All graphs have the same axes ranges

bered that this is only achieved through the formation of a bimodal distribution of sizes.

Kabal'nov et al. deduced that the formation of a bimodal emulsion in this situation could be explained in terms of the excess chemical potential of the dodecane being a competition between a Laplace pressure term and a term involving the concentration of dodecane in each individual droplet [10]. Their analysis suggested that in the emulsions formed here, an initial additive mole fraction of 0.08 should produce a monomodal emulsion. This was not the case, as shown, for instance, by the system containing a mole fraction of 0.2 (Fig. 5) octadecane, while systems containing tetradecane at any concentration showed effec-

Progr Colloid Polym Sci (1994) 97:199–203
© Steinkopff-Verlag 1994

tive volume fractions close to one. Clearly, there are other factors involved in the process.

Firstly, the octadecane/dodecane mixture in the smaller droplets might solidify, octadecane being a solid at room temperature. Diffusion, and hence Ostwald ripening, from such a phase would be severely reduced compared to a liquid droplet phase. Any subsequent ageing would be mostly within the upper peak droplets, which would contain less octadecane and still be in a liquid state.

Hexadecane has a lower melting point than octadecane, and would reach a solid phase at a higher concentration in dodecane, whilst tetradecane is a liquid at room temperature and so this limitation would not apply. Secondly, as the radius of the lower peak is reduced then it approaches that of a swollen micelle and this would also affect the form of the excess chemical potential vs radius curve. These effects will be discussed more fully in later publications.

References

1. Walstra P (1983) In: Becher P (ed) Encyclopedia of emulsion technology. Volume 1, Marcel Dekker, New York, pp 58–127
2. Kabal'nov AS, Shchukin ED (1992) Adv Coll Int Sci 38:69
3. Langevin D (1988) Acc Chem Res 21:255
4. Lifshitz IM, Slezov VV (1961) J Phys Chem Solids 19:35
5. Wagner C (1961) Ber Bunsenges Phys Chem 16:581
6. Heller W, Pangonis WJ (1957) J Chem Phys 26:498
7. Heller W, Bhatnagar HC, Nakagaki M (1962) J Chem Phys 36:1163
8. Kabal'nov AS, Makarov KN, Pertzov AV, Shchukin ED (1990) J Coll Int Sci 138:98
9. Voorhees P (1985) J Stat Phys 38:231
10. Kabal'nov AS, Pertzov AV, Shchukin ED (1987) Colloids Surfaces 24:19
11. Coates M, Connel DW, Barron DM (1985) Environ Sci Technol 19:628
12. Franks F (1966) Nature 210:87
13. Hayduk W, Laudie H (1974) AIChEJ 20:611
14. Stigter D, Williams RJ, Mysels KJ (1955) J Phys Chem 59:330
15. Abraham MH (1984) J Chem Soc Far Trans I 80:153

Progr Colloid Polym Sci (1994) 97:204–209
© Steinkopff-Verlag 1994

EMULSIONS AND RHEOLOGY

L.-J. Chen
M.-C. Hsu
S.-T. Lin

Salt effects on interfacial behavior at liquid–liquid interfaces in the water + N-tetradecane + C₆E₂ system

Received: 16 September 1993
Accepted: 2 March 1994

L.-J. Chen (✉) · M.-C. Hsu · S.-T. Lin
Department of Chemical Engineering
National Taiwan University
Taipei, Taiwan 106
Republic of China

Abstract In a system with three coexisting phases (α, β and γ) at equilibrium and having densities in the order $\rho_\alpha < \rho_\beta < \rho_\gamma$, the interfacial tensions ($\sigma_{\alpha\beta}$, $\sigma_{\beta\gamma}$ and $\sigma_{\alpha\gamma}$) either obey Antonow's rule, which gives wetting behavior, or conform to Neumann's inequality, which gives non-wetting behavior. The validity of these implications has been experimentally investigated and the influence of salt concentration and salt type determined.

Key words Wetting transition – interfacial tension – lyotropic salt – hydrotropic salt

Introduction

Consider a ternary mixture of water, an oil, and a nonionic surfactant, C_iE_j, i.e., $C_iH_{2i+1}(OCH_2CH_2)_jOH$. Within a certain temperature range, such a mixture may separate into three coexisting liquid phases, namely, an oil-rich α phase, a surfactant-rich β phase, and a water-rich γ phase. The densities of these three phases are in the order $\rho_\alpha < \rho_\beta < \rho_\gamma$.

It is found that in this ternary mixture when the surfactant chain length is short ($i \leq 4$), the middle β phase wets the α–γ interface, Fig. 1(d), over the entire three-liquid-phase region [1]. On the other hand, for the relatively long chain surfactant ($i \geq 8$), the middle β phase does not wet the α–γ interface and it forms a lens suspended at the α–γ interface, Fig. 1(c), over the entire three-liquid-phase region [2]. While the systems with an intermediate chain length surfactant, say $i = 5$ or 6, exhibit a wetting transition [3] from a nonwetting to a wetting β phase, or vice versa, simply by varying temperature [4]. Such an interfacial phase transition has also been experimentally observed, not only at liquid–liquid interfaces [5], but also at vapor–liquid interfaces [6] and at solid–liquid interfaces [7].

The critical wetting theory [8, 9] predicts that in such a three-component surfactant system one should find a wetting transition from a nonwetting to a wetting behavior as the system approaches either one of its critical endpoints, i.e., either an upper critical consolute temperature or a lower critical consolute temperature. Aratono and Kahlweit [5] have found that the middle β phase of the mixture water + n-octane + C_5E_2 does exhibit a wetting transition at the α–γ interface as approaching either one of the critical endpoint temperatures, consistent with the prediction of the critical wetting theory.

A much more intriguing and less obvious interfacial phenomenon is the fact that the system, water + n-tetradecane + C_6E_2 exhibits two wetting transitions occurring at two different liquid–liquid interfaces in the region of three coexisting liquid phases [10]. The middle β phase exhibits a wetting transition at the α–γ interface as temperature increases towards its upper critical consolute temperature. With a decreasing temperature towards its lower critical consolute temperature, the lower γ phase exhibits a wetting transition from suspending beads to an intruding layer at the α–β interface before the critical endpoint is reached. The evolution of interfacial behavior of the system water + n-tetradecane + C_6E_2 with increasing temperature is schematically illustrated in Fig. 1.

Progr Colloid Polym Sci (1994) 97:204–209
© Steinkopff-Verlag 1994

Fig. 1 Evolution of qualitative interfacial behaviors of the system water + n-tetradecane + C_6E_2 as increasing temperature, or as adding a lyotropic and a hydrotropic salt. The expected condition for only a small amount of the β phase at the α–γ interface is shown in the upper row, the condition for a larger amount of the β phase shown in the lower row

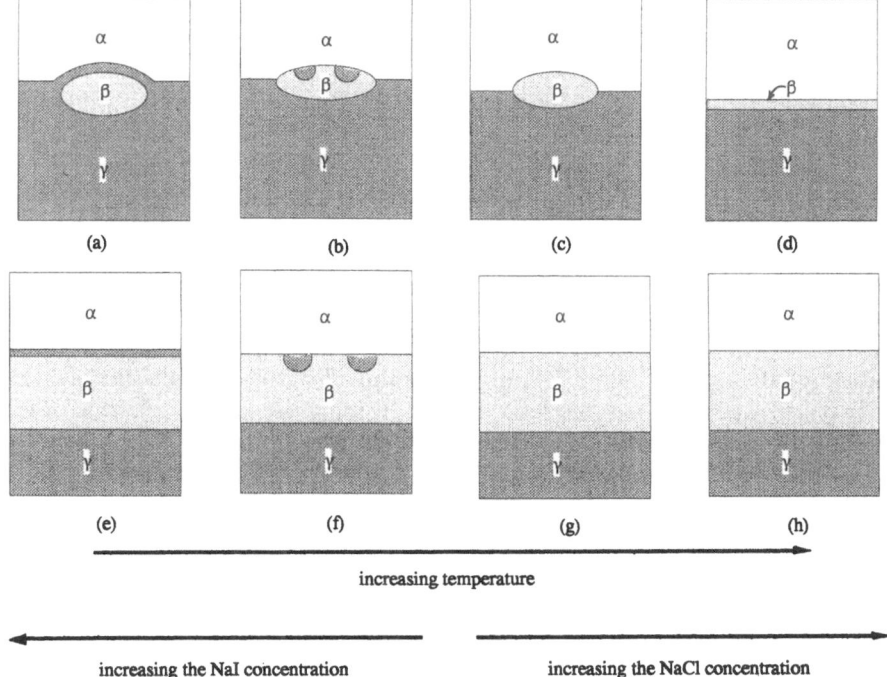

Now, consider the fourth component, a salt, in the system water + n-alkane + C_iE_j. It is well understood [11] that both temperature and salt concentration have the same effect on the phase behavior of such systems. More precisely, adding a lyotropic (or hydrotropic) salt to these three-component surfactant systems at a constant temperature is equivalent to increasing (or decreasing) temperature in these systems without any salt.

It is also believed that the properties of interface are directly related to those of the coexisting bulk phases, especially close to its critical endpoint. Since temperature and salt concentration have the same effect on phase behaviors of these systems, it is natural for us to conjecture that both temperature and salt concentration also have the same effect on their interfacial behaviors, i.e., instead of increasing temperature, the middle β phase of the water + n-alkane + C_iE_j system with a intermediate chain length surfactant should also exhibit an wetting transition at the α–γ interface at a constant temperature by increasing a lyotropic salt concentration. This conjecture has been verified by raising the NaCl, a lyotropic salt, concentration in the system water + n-octadecane + C_6E_2 at 35.0 °C to observe a wetting transition from a nonwetting to a wetting middle β phase [12].

A more rigorous verification of our conjecture is reported in this manuscript. Instead of adding a lyotropic salt only, a hydrotropic salt is also systematically added into the water + n-tetradecane + C_6E_2 system to observe

a γ phase wetting transition at the α–β interface that corresponds to the temperature effect on the occurrence of a γ phase wetting transition as decreasing temperature in the system with no salt.

In this contribution, we present experimental results on wetting transitions in a four-component system, water + n-tetradecane + C_6E_2 + salt (NaCl and NaI) system at 30 °C and at a fixed weight ratio of water:n-tetradecane:C_6E_2 (= 2:2:1), in which the salt concentration is varied as the system parameter. In the next section, we briefly describe the relationship between interfacial tensions and interfacial (wetting and nonwetting) behavior in three-phase coexisting systems, as well as the wetting transition which may occur in such systems. The experimental results confirm our conjecture that both temperature and salt concentration have the same effect on their interfacial behavior, and further discussions are given in Section III.

Wetting behavior and interfacial tensions

For a three-phase α, β, and γ coexisting system, there exist three interfacial tensions: $\sigma_{\alpha\beta}$, $\sigma_{\alpha\gamma}$, and $\sigma_{\beta\gamma}$, where σ_{ij} stands for the interfacial tension of the i–j interface. Whether the middle β phase (or the lower γ phase) wets or does not wet the interface separating the other two phases is determined by these three interfacial tensions.

L.-J. Chen et al.
Salt effects on interfacial behavior at liquid–liquid interfaces

When the middle β phase wets the $\alpha-\gamma$ interface, i.e., the β phase completely spreads across the $\alpha-\gamma$ interface as illustrated in Fig. 1(d), the interfacial tensions are related by Antonow's rule: $\sigma_{\alpha\gamma} = \sigma_{\alpha\beta} + \sigma_{\beta\gamma}$. When the middle β phase does not wet the $\alpha-\gamma$ interface, i.e., the β phase only partially wets the $\alpha-\gamma$ interface, Fig. 1(c), the interfacial tensions satisfy Neumann's inequality: $\sigma_{\alpha\gamma} < \sigma_{\alpha\beta} + \sigma_{\beta\gamma}$.

As a consequence, a β phase wetting transition from a nonwetting (partially wetting) regime to a wetting regime can be interpreted to be a transition of the relation of interfacial tensions from $\sigma_{\alpha\gamma} < \sigma_{\alpha\beta} + \sigma_{\beta\gamma}$ (Neumann's inequality) to $\sigma_{\alpha\gamma} = \sigma_{\alpha\beta} + \sigma_{\beta\gamma}$ (Antonow's rule). Similarly, a γ phase wetting transition can also be recognized as a transition of the relation of interfacial tensions from $\sigma_{\alpha\beta} < \sigma_{\alpha\gamma} + \sigma_{\beta\gamma}$ (Neumann's inequality) to $\sigma_{\alpha\beta} = \sigma_{\alpha\gamma} + \sigma_{\beta\gamma}$ (Antonow's rule), or vice verse [13].

Note that when $\sigma_{\alpha\beta} = \sigma_{\alpha\gamma} + \sigma_{\beta\gamma}$, the $\alpha-\beta$ interface is thermodynamically unstable. Under this condition, the surface forces overwhelm the earth's gravitational forces and a very small amount of γ phase of greatest density forms a thin intruding layer separating two other phases α and β to minimize the total system energy, as shown in Fig. 1(e). We exaggerate the thickness of the intruding γ layer in Figs. 1(a) and (e). In reality, the thickness of this intruding layer is impossible to observe by the naked eye. For example, an intruding layer's thickness at gas–liquid interface of two binary systems: methanol + cyclohexane and methylcyclohexane + perfluoromethyl cyclohexane is experimentally found to be a few hundred Angstroms only by the ellipsometry technique [14]. It should also be pointed out that when the system has only a small amount of β phase, the middle β phase would form, instead of a thick layer, a lenticular droplet and a thin film of γ phase separates the α and β phases, as shown in Figs. 1(a) and (b).

In this study, we perform experiments by direct observation via an enhanced video microscopy system and interfacial tension measurements to verify the existence of both β phase and γ phase wetting transitions in the water + n-tetradecane + C_6E_2 system by tuning the salt concentration. The experimental procedure can be found in our previous papers [10, 12]. To support and cross-check the results from our direct visual observations, the experimental results of interfacial tensions can be used to verify whether the tensions undergo a transition from Neumann's inequality to Antonow's rule when a wetting transition occurs.

Results and discussion

According to the Gibbs' phase rule, there are three degrees of freedom for a three-coexisting-liquid-phase quaternary

system. In this study, all the experiments are performed in three-liquid-phase coexisting region of the quaternary system H_2O + n-tetradecane + C_6E_2 + salt under the atmospheric pressure and at 30.0 °C. Consequently, there is only one degree of freedom left, and properties such as densities and interfacial tensions uniquely depend on mean composition. The weight ratio of water : n-tetradecane : C_6E_2 is fixed at 2:2:1, and the mean composition is changed by varying the amount of salt. Therefore, we simply adjust the amount of salt to search for the wetting transitions at liquid–liquid interfaces in a three-coexisting-liquid-phase region of the system water + n-tetradecane + C_6E_2 + salt.

For the water + n-tetradecane + C_6E_2 + NaCl system at 30.0 °C, our experimental phase diagram shows that the three liquid-phase coexistence region ranges from 0% to 8.6% NaCl weight percent in brine, which is the region where we search for a wetting transition. At low NaCl concentrations, the middle β phase always exhibits a nonwetting behavior. While the NaCl weight percentage is close to, however, below 8.6%, the β phase exhibits a wetting behavior. It is found from visual observations via video microscopy system that the water + n-tetradecane + C_6E_2 system does exhibit a wetting transition lying at 7.37% weight percent of NaCl. The concentration at which the wetting transition occurs is known as the wetting transition concentration C_w. This result is consistent with that resulting from interfacial tension measurements.

The variation of interfacial tensions as a function of NaCl concentration is illustrated in Fig. 2. The wetting transition occurs when the three interfacial tensions ($\sigma_{\alpha\beta}$, $\sigma_{\alpha\gamma}$, and $\sigma_{\beta\gamma}$) change from satisfying Antonow's rule to Neumann's inequality, or vice versa. It is clear in Fig. 2

Fig. 2 Variation of the interfacial tensions as a function of NaCl concentration for the system water + n-tetradecane + C_6E_2 at 30 °C

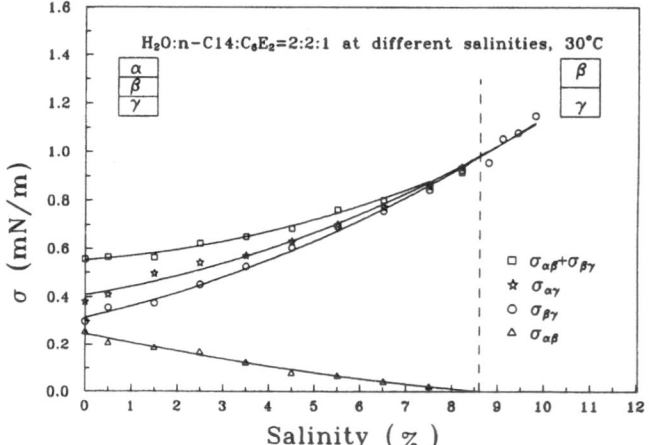

that at a particular concentration (wetting transition concentration C_w), the sum of $\sigma_{\alpha\beta}$ and $\sigma_{\beta\gamma}$ becomes equal to $\sigma_{\alpha\gamma}$, i.e., a wetting transition occurs at the concentration where the curves of $\sigma_{\alpha\beta} + \sigma_{\beta\gamma}$ and $\sigma_{\alpha\gamma}$ coincide in Fig. 2. The wetting transition concentration C_w is found to be 7.67%, which is slightly larger than the value obtained from eye observations due to experimental uncertainty.

Our experimental results of the wetting/nonwetting behaviors confirm that increasing the NaCl concentration in the water + n-tetradecane + C_6E_2 system does have the same effect on interfacial behaviors as raising temperature. However, the salt effect on such systems is not unique, and also depends on the nature of salts.

According to the effect of anions, hydrotropic salts, such as NaI, increase the mutual solubility between water and nonionic surfactants, whereas lyotropic salts, such as NaCl, decrease it. Consequently, adding a lyotropic salt NaCl to a water + oil + C_iE_j system makes the surfactant move systematically from the water-rich to the oil-rich phase, which is similar to the effect of increasing temperature to enhance the hydrophobicity of surfactant. While adding a hydrotropic salt NaI to such systems makes the surfactant move from the oil-rich to the water-rich phase, which is equivalent to the effect of decreasing temperature to enhance the hydrophilicity of surfactant [11].

Critical wetting theory [9, 10] directly predicts that in a three-phase system with an incomplete wetting phase, a wetting transition can be induced by tuning a system parameter to drive the system close to its critical point. Here, we increase the salinity (NaCl concentration) to bring the ternary mixture with a nonwetting middle β phase close to its critical endpoint, more precisely, the upper critical solution point, and a β phase wetting transition does occur. Besides the upper critical endpoint,

we also expect the occurrence of another wetting transition in the water + n-tetradecane + C_6E_2 system when approaching its lower critical endpoint by properly adjusting a system parameter, either decreasing temperature or adding a hydrotropic salt.

It was recently found that the system water + n-tetradecane + C_6E_2 does exhibit a γ phase wetting transition, instead of a β phase wetting transition, by decreasing temperature, approaching its lower critical consolute temperature. On the other hand, when we add the hydrotropic salt NaI to the system to bring it close to its lower critical endpoint, we expect a γ phase wetting transition to occur before the critical endpoint is reached.

At a constant weight ratio of water:n-tetradecane:C_6E_2 (= 2:2:1) and constant temperature 30 °C, the three liquid-phase coexistence region ranges from 0% to 12.72% NaI weight percent in water, which is the region where we search for a wetting transition. The middle β phase is found to exhibit nonwetting behavior for the NaI weight percentage over all the three liquid-phase coexistence region. From the direct visual observation via video microscopy system, the γ phase wetting transition is found to be at 9.0% NaI weight percent in water, in accord with the result of interfacial tension measurements.

Figure 3 shows the variation of the interfacial tensions as a function of NaI concentration. It is obvious that the interfacial tensions have a transition between $\sigma_{\alpha\beta} = \sigma_{\alpha\gamma} + \sigma_{\beta\gamma}$ and $\sigma_{\alpha\beta} < \sigma_{\alpha\gamma} + \sigma_{\beta\gamma}$ at the NaI concentration 9.2% weight percent in water, which is slightly larger than the value resulting from direct visual observation due to experimental uncertainty. As a consequence, here we also confirm that the effect of increasing the NaI concentration in the water + n-tetradecane + C_6E_2 system on the wetting/nonwetting behaviors, as well as the γ phase wetting transition, is equivalent to the effect of decreasing temperature.

It should be pointed out that the temperature effect on the interfacial behavior of water + n-tetradecane + C_6E_2 systems is not exactly the same as adding a salt to this system. In the ternary water + n-tetradecane + C_6E_2 system, there exists a unique wetting transition temperature. While in the quaternary water + n-tetradecane + C_6E_2 + salt system at a constant temperature, the wetting transition concentration is, instead of an unique value, a function of the weight ratio of water:n-tetradecane:C_6E_2. Consequently, these wetting transition concentrations at different weight ratios of water:n-tetradecane:C_6E_2 will form a wetting transition concentration surface inside a tetrahedron phase diagram of the quaternary water + n-tetradecane + C_6E_2 + salt system at a constant temperature and pressure.

In summary, Fig. 4 shows the photographs taken from the video microscopy system to illustrate the evolution of

Fig. 3 Variation of the interfacial tensions as a function of NaI concentration for the system water + n-tetradecane + C_6E_2 at 30 °C

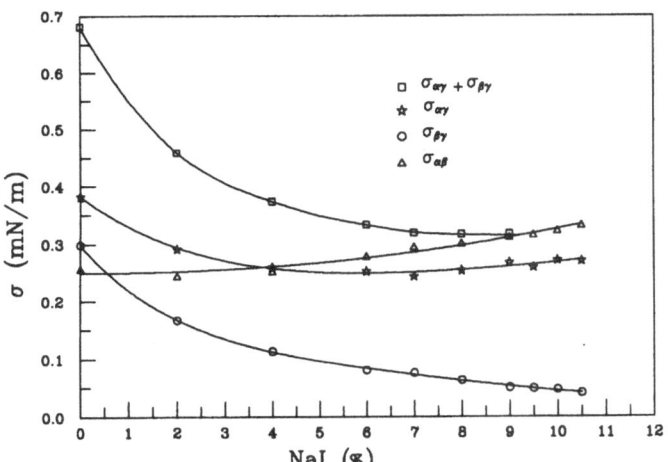

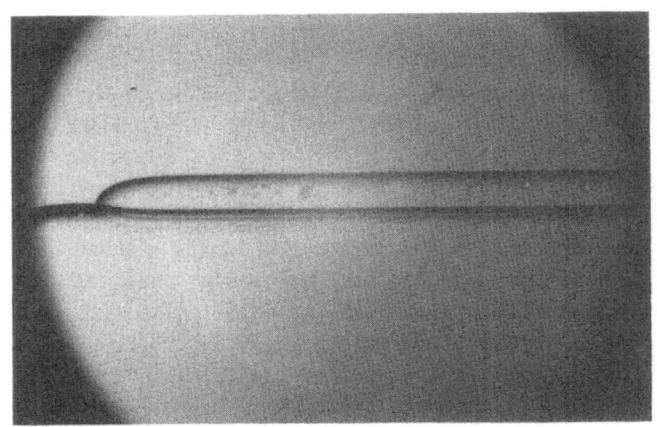

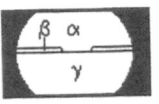

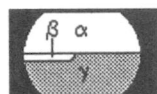

(c) NaCl 6.53%

(a) NaI 9.00%

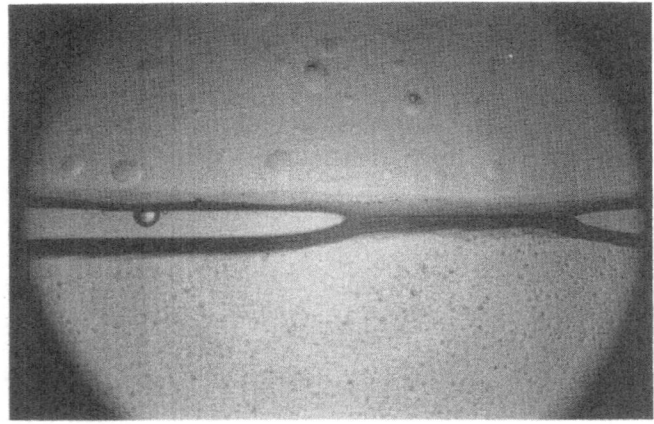

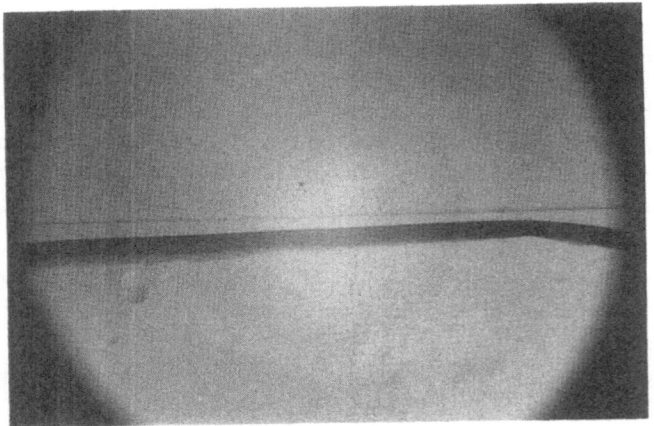

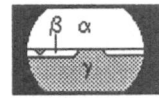

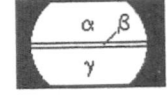

(b) NaI 3.49%

(d) NaCl 7.48%

Fig. 4 Photographs of the interfacial behaviors of the system water + n-tetradecane + C_6E_2 at four different salt concentrations: a) NaI 9.00%; b) NaI 3.49%; c) NaCl 6.53%; d) NaCl 7.48%

Progr Colloid Polym Sci (1994) 97:204–209
© Steinkopff-Verlag 1994

interfacial behavior of this particular system at different concentrations of NaCl or NaI, in accord with the schematic illustration shown in Figs. 1(a–d). It is found that the water + n-tetradecane + C_6E_2 system does exhibit two wetting transitions: i) a β phase wetting transition at the α–γ interface as increasing the lyotropic salt NaCl concentration; ii) a γ phase wetting transition at the α–β interface as increasing the hydrotropic salt NaI concentration.

Acknowledgments This work was supported by the National Science Council of Taiwan, Republic of China under the grant number NCS82-0402-E002-215.

References

1. Seeto Y, Puig JE, Scriven LE, Davis HT (1983) J Colloid Interface Sci 96:360–372
2. Kahlweit M, Strey R, Firman P, Haase D, Jen J, Schomacker R (1988) 4:499–511
3. For a review, see: Sullivan DE, Telo da Gama MM (1986) In: Croxton CA (ed) Fluid Interfacial Phenomena, John Wiley & Sons, pp 45–134
4. Chen LJ, Jeng JF, Robert M, Shukla KP (1990) Phys Rev A 42:4716–4723
5. Aratono M, Kahlweit M (1991) J Chem Phys 95:8578–8583
6. Moldover MR, Cahn JW (1980) Science 207:1072–1075
7. Pohl DW, Goldburg WI (1982) Phys Rev Lett 48:1111–1114
8. Cahn JW (1977) J Chem Phys 66:3667–3672
9. Ebner C, Saam WF (1977) Phys Rev Lett 38:1486–1489
10. Chen LJ, Yan WJ (1993) J Chem Phys 98:4830–4837
11. Kahlweit M, Lessner E, Strey R (1984) J Phys Chem 88:1937–1944
12. Chen LJ, Hsu MC (1992) J Chem Phys 97:690–694
13. Rowlinson JS, Widom B (1982) Molecular Theory of Capillarity, Clarendon, Oxford
14. Kwon OD, Beaglehold D, Webb WW, Widom B, Schmidt JW, Cahn JW, Moldover MR, Stephenson B (1982) Phys Rev Lett 48:185–188

Progr Colloid Polym Sci (1994) 97:210–212
© Steinkopff-Verlag 1994

D.J. Morantz

Entropic aspects of the viscosity of a polymer-resin monolayer

Received: 2 October 1993
Received after revision: 15 February 1994
Accepted: 21 February 1994

D.J. Morantz
Pira International
Randalls Road
Leatherhead
Surrey KT22 7RU
United Kingdom

Abstract The surface pressure–area isotherms for a polymer resin spread on an aqueous subphase were investigated. A time-dependent hysteresis effect was observed on re-expansion of the film. An explanation for this effect is proposed based on the concept of two "welded" monolayers.

Key words Monolayer – hysteresis – bilayer model – entropy – viscosity – resin – ink

Introduction

A recent study [1], using a Laude Filmwaage FW2, Langmuir Trough, has reported on the monolayer behaviour of an ink varnish. The varnish contains a synthetically modified polymer rosin, which is derived from natural tree resins. The resin is responsible for the key hysteresis effects observed; these effects included a time-dependent transient pressure increase on re-expansion of the monolayer. Application of low surface pressures, < 20 mN/m, to a monolayer spread from the resin resulted in a molecular cross-sectional area compression of 30%. The monolayer collapse pressure, 75 mN/m, was found to exceed the surface tension of the water subphase (73 mN/m) by an amount, 2 mN/m, which is probably due to local (non-equilibrium) stresses. Such over-pressures have been attributed [2] to stress inhomogeneities, for the case of low molecular weight systems. These, latter, inhomogeneities comprise (crystalline) nuclei from which collapse regions may grow. In the case of the varnish or the resin, such inhomogeneities may be attributed to incipient bi-layer/multilayer "nuclei" as opposed to crystalline nuclei. The partially collapsed film (Fig. 1) led to irreversible bulk formation, estimated [1] at 40% of the original monolayer.

In the example shown (Fig. 1), just below collapse, the ink varnish compressibility decreased at an accelerating rate. On expansion, after a preset time delay > 0.1 min; the observed pressure drop was temporarily reversed before falling to zero surface pressure.

The example shown is for a film which was at the point of collapse, here the hysteresis cycle was not reproducible. Larger pressure reversals were seen for reproducible hysteresis cycles where the maximum pressure applied was below 60mN/m, i.e., where an irreversible collapse had not been initiated (see reference [1]). The initial monolayer expanded state is assumed to be anchored in the subphase by its polar head groups, these being linked by high molecular weight alkane chains which are horizontally deployed; whereas in the compressed state the head groups are forced to approach each other, in turn forcing the alkane moieties into more vertical alignment.

When isopropanol was added to the subphase, the surface area of a resin monolayer decreased by more than 25% before the application of surface pressure. This substantial decrease was assumed [1] to arise by redeployment of the resin head groups by some nonplanar displacement from the subphase.

Thus, compression at higher pressures was postulated to involve changes in the vertical displacement of head groups and, at collapse, disruption would pervade the

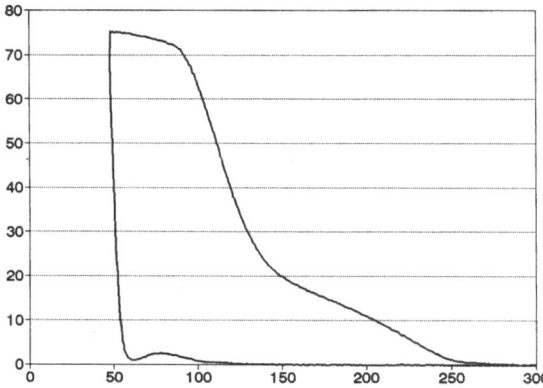

Fig. 1 Langmuir Trough plot for an ink varnish. Surface pressure (mN/m) vs surface area (cm²); the hysteresis cycle indicates the collapse above 70 mN/m

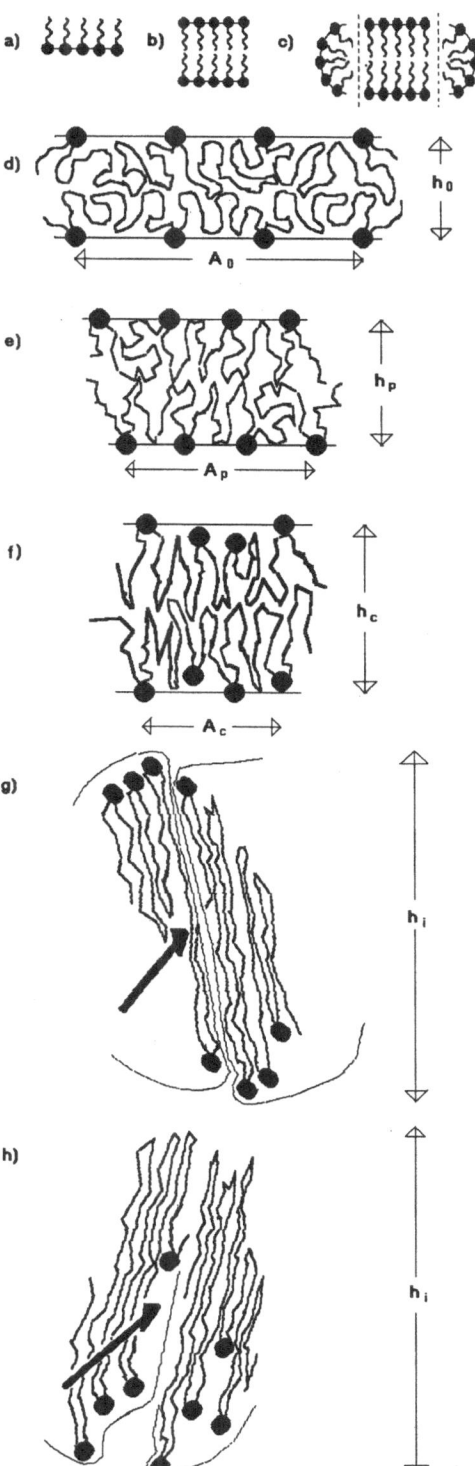

whole of the z direction of the monolayer. The alkane moieties interfacing with air no longer have a surface energy constraint, as this energy becomes reduced through the value of zero.

Time-dependent effects were considered for each of the main features in the hysteresis phenomena [1]. A quantitative analysis of such effects is planned. Meanwhile, a qualitative framework for that discussion can be outlined.

Unstable and stable bilayer models

A stable bilayer would result by joining two monolayers at their headgroups, as in a Langmuir Blodgett preparation.

However, it is of interest to consider an unstable conceptual bilayer model where the monolayers are welded, in the absence of air, at what had been the alkane/air monolayer interface. In Fig. 2, a monolayer segment, a, is so joined to produce a bilayer, b. It might seem unrealistic to construct such an entity from a real monolayer, but the concept is useful and such a bilayer does exist as part of a lamellar micelle {Fig. 2c}.

A resin "bilayer" would possess longer and less orientated alkane moieties and could be represented as in Fig. 2d.

Assume, now, that horizontal pressure, < 20 mN/m, may be applied to a bilayer resin segment of cross-sectional area proportional to A_0, where its thickness is h_0. The bilayer would compress to A_p and the thickness would increase to h_p. The alkane moieties would become more vertically orientated, but with some randomness persisting at the weld plane, as indicated in Fig. 2e.

As the lateral compression increases, the head groups will increasingly dislodge from a planar orientation and

Fig.2. Monolayer/bilayer models. Sketches a) and g) represent monolayer and b) → h) represent bilayer models; the thin continuous line represents a water interface; heavy arrows in g) and h) indicate regions of rupture. The arrowed dimensions A and h represent interface area and thickness respectively; these are taken from Fig. 1, where sketches d) → f) correspond to sample pressures of 0, 20 and 60 mN/m respectively; g) and h) each correspond to 70 mN/m

this will be associated with increasing crowding and disruption at the weld. This process, illustrated in Fig. 2f, will progress through the (equivalent monolayer) pressure range of, say, < 20 mN/m $< p < 60$ mN/m, where p has a time-dependent relationship with the lateral compression.

At the incipient collapse pressure corresponding to $p = 75$ mN/m, rupture "nuclei" may form. This is illustrated in Fig. 2g, indicating the possibility that head groups can be displaced sufficiently to bridge the bilayer thickness $h_i > h_c$. At such a "nucleus" a bulk multilayer phase will form.

As the pressure is increased laterally, until incipient film collapse, the lateral cohesion may be said to vanish across planes which intersect the weld-plane; thus, the head group moieties can pass freely and rapidly through the bilayer.

Each of these configurations, Fig. 2d, e and f is directly related to a corresponding monolayer model configuration as described in [1].

A collapse "nucleus" as depicted in Fig. 2g would correspond to Fig. 2h, where disrupted head groups may be forced into the air phase leading to a variety of bulk configurations. The case where the monolayer buckles symmetrically about a vertical plane would result in bilayers and multilayers of the Langmuir–Blodgett type; the latter structures, obtained on compression of a polymer bilayer, are implied by Malcolm's work [3].

There would seem to be no conceptual problems in describing macroscopic properties of such a bilayer in terms of molecular constituents. In that event bilayer viscosity properties may be discussed in the same terms as bulk viscosity; in contrast with the apparent difficulties in describing a monolayer viscosity.

A formal means for reconsidering the properties of a monolayer, in terms of bulk properties, may therefore be available via the welded monolayer/bilayer concept. This could be applied, for example in discussing the reported [4] polymer monolayer surface pressure gradients.

A bilayer model may be quantitatively developed in terms of time-dependent variations of free energy in each of the spatial coordinates. This would provide a model where there would be a computable, time-dependent, free energy gradient at all points in the model space; and for all stages approaching film collapse; these could then be transformed to a realistic monolayer model. The concept of a monolayer as half of a bilayer was discussed by Baret [5], who noted that "bilayers are made of two interacting monolayers"; and that it is their interaction which distinguishes their behaviour. Baret also pointed out, at that time, that "no universally accepted model allows us to describe both bilayer and monolayer transitions". At the point of film collapse, the model transformation appears to be more readily achievable.

The time dependence and reversibility of the gamut of possible monolayer configurations, up to collapse of the resin monolayer, have much in common with such properties of phospholipid monolayers. The results for the resins need further study and discussion in terms of quantitative data for the second order processes, and to test the suitability, for models of such examples of high molecular weight monolayers, of their integration with bilayer and bulk theoretical models.

It may be noted here that the collapse which commences at around 69 mN/m (Fig. 1) is a cumulative, irreversible transformation to a bulk multilayer-like material. There seems little resemblance at this point to a thermodynamically reversible phase change. Nevertheless, the bilayer model does provide a means for setting up a formal statement along the lines that:

entropy $s = f_1(x, y, z, t)$

chemical potential $u = f_2(x, y, z, t)$

viscosity $n = f_3(x, y, z, t)$

Acknowledgements I would like to thank Dr. Spencer E. Taylor of BP Research Centre, Sunbury-on-Thames, Middlesex, UK for providing facilities for this work; Dr. John H. Clint, now of the School of Chemistry, The University of Hull, UK for introducing me to the methods of the Langmuir Trough; and Mr. John Birkenshaw of Pira International, Leatherhead, Surrey, UK for supporting the research programme.

References

1. Morantz DJ (1994) Orientation and (reversible?) transitions incipient collapse of a polymer resin at an air/water interface, Colloids and Surfaces (in the press)
2. Nikomarov ES (1990) Langmuir 6:1994 410
3. Malcolm BR (1985) J Colloid Interface Sci 104:520
4. Peng JB, Barnes GT (1990) Langmuir, 6:578
5. Baret JF (1981) in Progress in Surface and Membrane Science Vol 14, Eds. Cadenhead DA, Danielli JF, Academic Press pp 292–351

Progr Colloid Polym Sci (1994) 97:213–217
© Steinkopff-Verlag 1994

D. Renoux
J. Selb
F. Candau

Aqueous solution properties of hydrophobically associating copolymers

Received: 16 September 1993
Accepted: 30 September 1993

D. Renoux · F. Candau (✉) · J. Selb
Institut Charles Sadron (CRM-EAHP)
6, rue Boussingault
67083 Strasbourg Cedex, France

Abstract Water-soluble polymers containing small amounts of hydrophobic groups have been synthesized in an aqueous medium by free radical copolymerization of a hydrophilic monomer (acrylamide) with a micelle-forming cationic polymerizable surfactant. Such hydrophobically modified water-soluble polymers exhibit particular aqueous solution properties due to attractive hydrophobic interactions and repulsive electrostatic interactions. The competition between these two effects, as well as the balance between intra- and intermolecular interactions, give rise to various rheological behaviors depending on polymer concentration, ionic strength and shear time. The kinetics and the reversibility of the association/dissociation phenomena have been monitored by studying time-effect on the rheological solution properties.

Key words Acrylamide-hexadecyldimethylvinylbenzylammonium chloride copolymers – polyacrylamide – cationic polymerizable surfactant – hydrophobically associating copolymers – rheology of associating copolymers

Introduction

Hydrophobically associating polymers consist of a water-soluble polymer containing a small amount of hydrophobic groups [1–5]. In aqueous solution (above the overlap concentration C^*), the hydrophobic units form intermolecular hydrophobic associations resulting in a strong increase in solution viscosity. Under high shear, the intermolecular hydrophobic links are disrupted, but reform when shear is stopped. This is one of the advantages of associating polymers compared to high molecular weight homopolymer solutions which, in some cases, do not recover their initial viscosity after being submitted to high shear rates, due to mechanical degradation (e.g., polyacrylamide). In addition, the reversible association/dissociation mechanism gives rise to interesting rheological properties such as rheopexy and thixotropy, i.e., time-dependent effects. Such a behavior is useful for all domains of application where a viscosity control is required, for example in latex paints, cosmetics, oil recovery, etc. [1–5].

The increase in viscosity can be enhanced by using a charged hydrophobic polymer [6]. There is an expansion of the coil due to repulsion between charges. Addition of salt screens charge-charge interactions. Therefore, by varying the ionic strength, it is possible to obtain a finer control of the rheological behavior of the solution. The balance between electrostatic repulsions and hydrophobic attractions then gives rise to particular aqueous solution properties.

There are different ways to obtain such associating polymers. The most commonly used methods are either micellar copolymerization [7, 8] or chemical modification of a water-soluble polymer precursor [9].

In this paper, we present some preliminary results on the rheological properties of a new class of associating polymers. These were synthesized by another route [10, 11] consisting of a free radical copolymerization of a water-soluble monomer (acrylamide) with a cationic micelleforming polymerizable surfactant (hexadecyl-dimethylvinylbenzylammonium chloride).

Chemical structure of hexadecyldimethylvinylbenzylammonium chloride

The aqueous solution properties of these samples were studied as a function of polymer concentration (for the range of molecular weight investigated, the overlap concentration C^* can be estimated to 0.1 wt/wt%) and of added salt concentration (sodium chloride). The effect of shear time at constant shear rate $\dot{\gamma}$ was also investigated. The rheological data thus obtained were correlated to fluorescence data in order to obtain more information about the nature of the interactions.

Experimental section

Materials

Acrylamide (Merck) was twice recrystallized from chloroform.

Dimethylhexadecyl amine (Genamin 16 R 302 D provided by Hoechst) was distilled under reduced pressure just prior to use.

Water-soluble initiator 4,4'-azobis (4-cyanovaleric acid) (ACVA) (Aldrich) and vinylbenzylchloride (VBC) (Kodak) were used as provided.

Water was deionized and distilled.

The amphiphilic monomer (hexadecyldimethylvinyl-benzylammonium chloride, called N16) was synthesized by quaternization of dimethylhexadecyl amine with VBC as described by Ikeda et al. [12, 13].

Synthesis and characterization of copolymers

The associating copolymers were obtained by free radical copolymerization in aqueous media of acrylamide and the cationic polymerizable surfactant in the micellar state (Fig. 1). The properties of this surfomer (**surfactant monomer**) were studied in detail by D. Cochin et al. [14–16] in our laboratory. The aggregation number of N16 at 50 °C is about 50.

Experimental conditions were as follows: the total concentration of monomers in water was 3 wt/wt%, the sur-

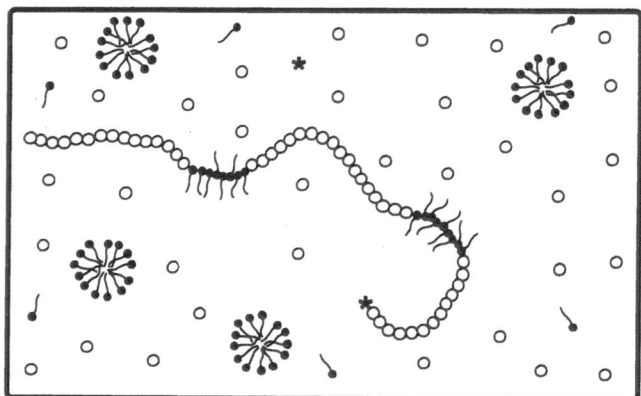

Fig. 1 Schematic representation of the copolymerization medium ○ Acrylamide ——● Polymerizable surfactant ★ Initiator

fomer mole percentage in the total monomer feed was varied from 0.5 to 3 mol.%. The reaction was initiated by ACVA (3 wt/wt% relative to the monomers) at 50 °C. After 7 h, the polymerization was stopped by cooling at room temperature, and the copolymer was rapidly precipitated into a large excess of methanol. The product was filtered and washed three times in fresh methanol before being dried under vacuum at 50 °C for several days.

The surfomer incorporation in the acrylamide polymer was determined by chloride ion elemental analysis. The copolymer molecular weights have been measured in formamide by static light scattering and were in the range 1.5×10^6 to 3×10^6 g/mol.

Rheological measurements

For polymer concentrations around 1% (wt/wt), the viscosity measurements of aqueous and brine solutions were performed at 25 °C using a Contraves Low Shear 30 (1-1 measuring system).

For higher concentrations, a Carri-Med CSL-100 controlled stress rheometer was used with a cone-plate geometry (diameter 6 cm, 1° angle acrylic cone).

Fluorescence study

Steady-state fluorescence measurements of the emission spectra of the solutions were carried out with a Hitachi F-4010 spectrophotometer at 25 °C. The I_3/I_1 ratio of the intensities of the third to the first peak of the emission spectrum of pyrene is sensitive to the local environment, i.e., hydrophobic or hydrophilic, sensed by the fluorescent probe [17].

Progr Colloid Polym Sci (1994) 97:213–217
© Steinkopff-Verlag 1994

Results and discussion

Polymer concentration effect

Figure 2 shows the effect of polymer concentration on the solution viscosities for a copolymer sample and a homopolyacrylamide prepared under similar conditions. As expected, the copolymer sample exhibits higher viscosity values than those of the homopolymer. The higher the concentration, the larger the difference between the respective viscosities. These improved thickening properties can be ascribed to two complementary effects: coil expansion due to charge-charge repulsions and intermolecular hydrophobic associations.

These data were supported by fluorescence experiments (Fig. 2): for homopolyacrylamide, the I_3/I_1 pyrene ratio at $C = 1\%$ (0.58) is close to that observed in pure water (0.53). The slight increase observed at higher concentration suggests that PAM possesses a certain hydrophobicity. The copolymer solution shows for C = 1% a much higher value of I_3/I_1 (0.74), and this value is seen to rise sharply with concentration. This behavior can be related to the formation of an increased number of hydrophobic microdomains in the solution.

Another interesting point to be noted is the difference in the shapes of the curves relative to the two samples. For homopolyacrylamide, one observes an exponential-like response to increasing concentration, a characteristic behavior of a neutral sample. On the other hand, the charged copolymer shows an inverse curvature shape, typical of

a polyelectrolyte [18]. The rapid increase is viscosity observed at low polymer concentration, is the result of both chain expansion due to the repulsion between charges and hydrophobic associations. At higher concentrations ($C > 3\%$), the auto-screening of the charges tends to reduce the coil expansion and therefore to level off the increase in viscosity.

Salt effect

The effect of addition of various amounts of sodium chloride on the solution viscosities has been investigated for two samples differing by their hydrophobe content (Fig. 3). For both samples, there is an initial loss in viscosity (for [NaCl] < 0.1 M) due to the effect of charge screening by the salt which causes a contraction of the hydrodynamic macromolecular volume. Beyond this threshold ([NaCl] > 0.1 M) the viscosity tends to increase, the higher the hydrophobe content, the larger the increase. Two additional phenomena can account for the latter result. On one hand, the screening of the repulsive electrostatic forces between the charged hydrophobic units facilitates the macromolecular interpenetration, thus promoting a greater number of intermolecular hydrophobic associations. On the other hand, addition of salt to the copolymer solution is known to lower the solubility of the hydrophobic moieties in water (salting-out effect), which enhances the formation of intermolecular aggregates, as previously shown in our laboratory on another type of associating polymers [18].

Fig. 2 Zero shear viscosity as a function of polymer concentration in pure water ◆ Polyacrylamide ● Copolymer (N16 content = 3 mol.%) Data point labels are I_3/I_1 pyrene fluorescence ratio

Fig. 3 Zero shear viscosity as a function of salt content for two copolymer samples (C = 1 wt%) ■ N16 content = 1 mol.% ● N16 content = 3 mol.%

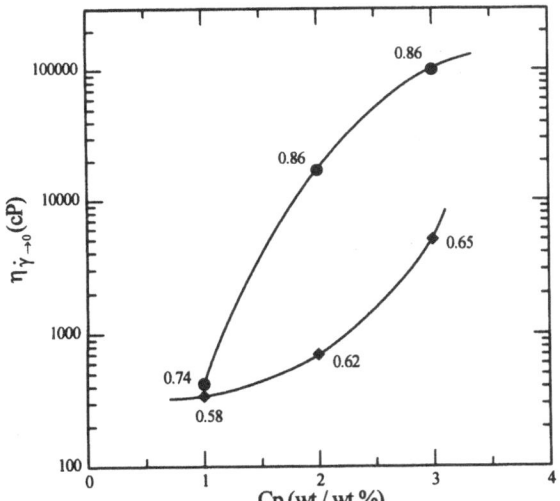

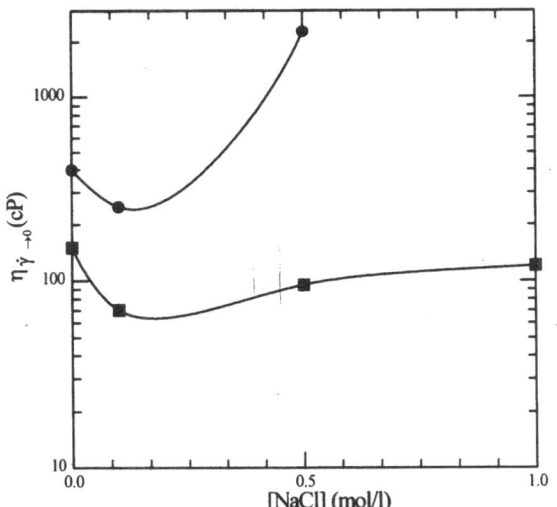

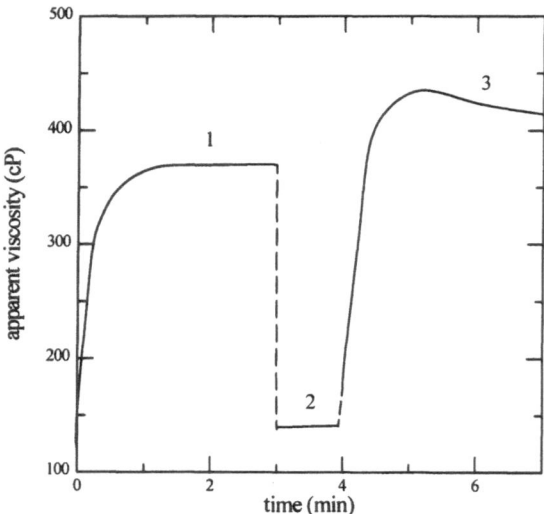

Fig. 4 Viscosity as a function of shear time under alternatively "low" shear rate ($\dot{\gamma} = 0.01\ \text{s}^{-1}$, curves 1 and 3) and "high" shear rate ($\dot{\gamma} = 15\ \text{s}^{-1}$, curve 2) for a copolymer sample (N16 content = 3 mol.%, C = 1 wt%)

duces the rupture of some intramolecular interactions, as their formation requires a more compact coil conformation. The newly liberated hydrophobic units have then the ability to associate intermolecularly with other hydrophobic groups.

Under higher shear (2nd step: $\dot{\gamma} = 15\ \text{s}^{-1}$), the viscosity value is much lower and the equilibrium is reached instantaneously: the hydrophobic associations are disrupted very rapidly.

When a low shear rate is applied again (3rd step: $\dot{\gamma} = 0.01\ \text{s}^{-1}$), a rheopectic effect is observed, followed by a thixotropic effect. The increasing viscosity is due to the reformation of intermolecular hydrophobic associations. The decrease in viscosity observed after 1 min can be explained by the fact that some intermolecular associations are broken in favor of intramolecular ones.

The conclusion which can be inferred from these data is that the destructuration of the system is a fast process while its restructuration is much slower. Further studies are required for a better understanding of this complex behavior.

According to these results, we can conclude that a low ionic strength ([NaCl] < 0.1 M), the rheological behavior is mainly controlled by the electrostatic repulsions, whereas at higher ionic strength ([NaCl] > 0.1 M) the hydrophobic effect becomes dominant.

Time effect

Rheopectic and thixotropic effects are shown in Fig. 4 where the rheological behavior has been studied as a function of time [18]. The viscosity of the copolymer solution was measured by applying alternatively low and high shear rates.

When a low shear rate is applied (1st step: $\dot{\gamma} = 0.01\ \text{s}^{-1}$), there is a rheopectic effect, i.e., an increasing viscosity with time under constant shear rate. A plateau viscosity value is reached after about 90 s. A possible explanation is that under low shear, the chain slightly extends, which pro-

Conclusion

The results reported here show that the copolymer formed from acrylamide and a cationic polymerizable surfactant are effective thickeners in aqueous solution. Their rheological behavior is controlled by the competition between electrostatic repulsions and hydrophobic attractions. Depending on the ionic strength, one or the other effect is predominant (electrostatic effect for [NaCl] < 0.1 M, hydrophobic effect for higher salt content). The samples exhibit some interesting rheological properties with reversible time-dependent viscosities (rheopexy, thixotropy) which are governed by the balance between intra- and intermolecular interactions.

Acknowledgements DR would like to thank PPG Industries France for their financial support.

References

1. Evani S, Rose GD (1987) Polym Mater Sci Eng 57:477
2. Glass JE (ed) (1989) Polymers in Aqueous Media: Performance Through Association Adv Chem Series 223, Am Chem Soc, Washington DC
3. McCormick CL, Bock J, Schulz DN (1989) Encyclopedia of Polymer Science and Engineering, 2nd ed, Mark HF, Bikales NM, Overberger CG, Menges G, Eds Wiley-Interscience New York, Vol 17, p 730
4. Glass JE (ed) (1986) Water-Soluble Polymers: Beauty with Performance Adv Chem Series 213 Am Chem Soc, Washington DC
5. Shalaby SW, McCormick CL, Buttler GB (eds) (1991) Water-Soluble Polymers. Synthesis, Solution Properties and Applications ACS Symposium Series 467, Am Chem Soc, Washington DC
6. McCormick CL, Middleton JC, Cummins DF (1992) Macromolecules 25 (4):1201

Progr Colloid Polym Sci (1994) 97:213–217
© Steinkopff-Verlag 1994

7. Valint PL Jr, Bock J, Schulz DN (1987) Polym Mater Sci Eng 57:482
8. Evani S (1984) US Patent 4 432 881
9. (a) Wang KT, Iliopoulos I, Audebert R (1988) Polym Bull 20:577; (b) Idem in ref 5 Chapter 14 p 218
10. Berret JF, Roux DC, Porte G, Lindner P (1994) Europhysics Lett 25:521–526 28:2110
11. Peiffer DG (1990) Polymer 31:2353
12. Ikeda T, Tazuke S (1984) Makromol Chem 185:869
13. Ikeda T, Tazuke S (1983) Makromol Chem Rapid Comm 4:459
14. Cochin D Thesis (1991) Université Louis Pasteur, Strasbourg, France
15. Cochin D, Zana R, Candau F (1991) Polymer International 30:491 Ibid (1993) Macromolecules 26:5765
16. Cochin D, Candau F, Zana R (1993) Macromolecules 26:5755
17. Kalyanasundaram K (1987) Photochemistry in microheterogeneous systems, Academic Press New York
18. Biggs S, Selb J, Candau F (1983) Polymer 34 (3):580

Progr Colloid Polym Sci (1994) 97:218–222
© Steinkopff-Verlag 1994

M.P. Pileni
F. Michel
F. Pitré

Synthesis of hydrophobic enzymes using reverse micelles. Enzymatic study of derivatives in AOT reverse micelles

Received: 30 September 1993
Accepted: 23 October 1993

M.P. Pileni (✉)
F. Micehl · F. Pitré
Université Pierre et Marie curie
Laboratoire S.R.S.I.
B.P. 52
4 place Jussieu
75231 Paris Cedex 05, France

C.E.N. Saclay
DRECAM/SCM
91191 Gif sur Yvette, France

Abstract Synthesis of hydrophobic ribonuclease and α-chymotrypsin is performed in AOT reverse micelles in order to covalently bind hydrophobic molecules on to the enzymes' surface. Enzyme characterization is described using high performance liquid chromatography and electrophoresis. Location of enzyme derivatives within reverse micelles is determined using fluorescence experiments. These indicate that modified enzymes are anchored on the inner surface of the host reverse micelle. Enzymatic assay of enzyme derivatives in reverse micelles is reported.

Key words Hydrophobic enzymes – reverse micelles – enzymatic study – AOT reverse micelles

Introduction

The system studied is composed of a dispersion of water droplets through an apolar medium which is isooctane [1]. The interface between oil and water is made with a double branched anionic surfactant, sodium di(2-ethylhexyl) sulfoscuccinate, commonly named AOT. The average radius of these spheroidal water droplets R_w is controlled by the number of water molecules per AOT molecule, called W and given by:

$$W = [H_2O]/[AOT] .$$

These microemulsions have the ability to host macromolecules, in particular, enzymes. Cytochrome c has been reported to strongly perturb the microemulsion: decrease in the micellar radius, stronger attractive interactions between droplets, decrease of the percolation threshold [2]. These phenomena have been attributed to the location of the protein at the interface of the droplet.

We have in this present investigation studied the effect of two small enzymes (ribonuclease A and α-chymotrypsin)

located in the droplet's water pool [3, 4]. We have made hydrophobic derivatives of enzymes to artificially anchor these at the droplet's interface. We observe that the hydrophobic character of the proteins have large effects on their activity in reverse micelles.

Experimental

Ribonuclease type II-A from bovine pancreas, AOT, myristoyl chloride ($C_{14}H_{27}ClO$), stearoyl chloride ($C_{18}H_{35}ClO$) and cholesteryl chloroformate were all obtained from Sigma, 9-fluorenylmethyl chloroformate from Aldrich and α-chymotrypsin from Fluka. Puriss p.a. isooctane was purchased from Merck.

A Waters 600E system equipped with sophisticated Waters 991 photodiode detector system is used for the high performance liquid chromatography (HPLC) analysis of enzyme modification.

Fluorescence experiments were performed with a SPEX fluorolog-2 device (Xe lamp, 150W).

Hydrophobic modification of enzyme

Synthesis

It has been previously shown that reverse micelles can be used to covalently modify enzymes[5]. This method has been chosen to increase the hydrophobic character of ribonuclease. Fatty acid chlorides (myristoyl and stearoyl) and chloroformates (cholesterol and 9-fluorenylmethyl) were used as reagents. They react with enzyme's amino groups by the following reaction:

$$ENZ\text{-}NH_2 + R\text{-}O\text{-}Cl \rightarrow ENZ\text{-}NH\text{-}O\text{-}R + HCl$$

Ribonuclease contains a total of 11 free amino groups (10 lysines plus the N-terminal amino acid), whereas α-chymotrypsin contains 11 (14 lysines plus 3 N-terminal amino acid).

Native enzyme (10 mg) is dissolved in 5 ml of AOT reverse micellar solution at fixed water content $W = 20$; 0.1 M borate buffer (pH = 9,5). Different aliquots of reagent, previously dissolved in isooctane ($5 \cdot 10^{-2}$ M), are then added to the micellar solution depending on the degree of modification desired. The ratio of reagent over enzyme concentration, α, has been varied from 3 to 20. The solution is continuously stirred and remains clear. After 15 min, the microemulsion is poured in a 10-fold volume of cold acetone ($-10\,°C$) in order to separate the enzyme from the AOT molecules and from free reagent. This solution is centrifugated at 5000 rpm, $t = -10\,°C$ for 20 mn. The precipitate is redispersed in 100 ml of acetone and again centrifugated. This washing step is repeated twice. After removing acetone, water is added to the precipitate. The solution is dialyzed (Spectrapore, mwco 6000–8000) against several exchanges of millipore water at $10\,°C$ to remove residual borate salt. The enzyme solution is then lyophilized.

The synthesis, which involves high ionic strength, organic solvent, dialysis, and lyophilization could have a negative effect on the enzyme. In order to investigate the influence of the extraction process and dialysis, we have carried out a control modification which was treated in the same way as the samples, but without addition of reagent. This enzyme was called control enzyme.

Ribonuclease characterization

When subjected to electrophoresis, the control RNAse sample was found to have the same mobility as the native enzyme. The ultra violet absorption spectrum was unchanged (maximum centered at $\lambda = 276$ nm due to the six

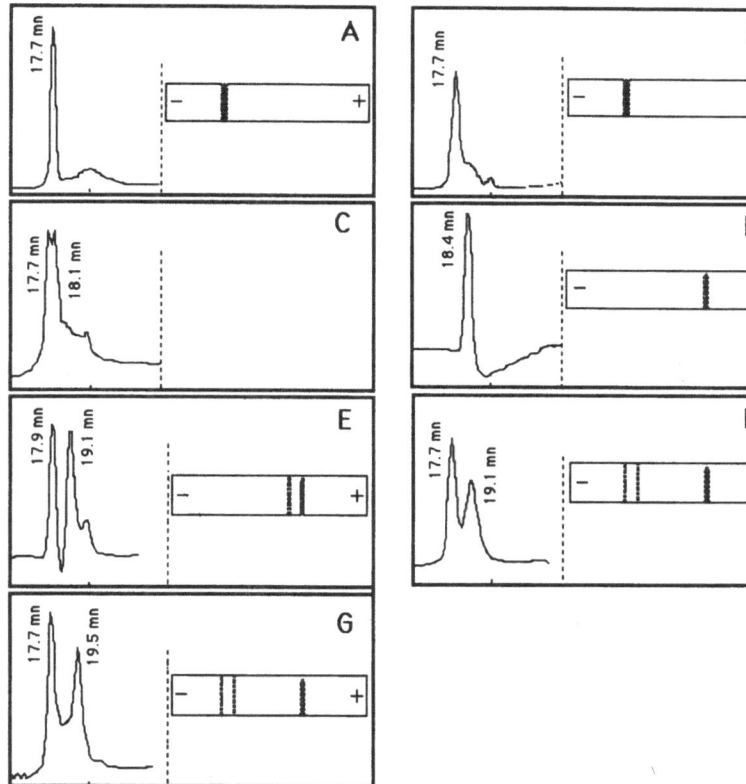

Fig. 1 Chromatograms (detection at $\lambda = 276$ nm) and electrophoretic patterns of: Reaction conditions: $\alpha = [\text{reagent}]/[\text{RNAse}] = 20$ A: native RNAse; B: control RNAse; C: FMOC-RNAse ($\alpha = 3$); D: FMOC-RNAse; E: Cholesterol-RNAse; F: Myristoyl-RNAse; G: Stearoyl-RNAse

Table 1 Specific activity constant $K_2 = K_{cat}/K_m$ ($M^{-1} \cdot s^{-1}$) for various enzymes in aqueous medium (Tris-HCl buffer, pH = 8.3) and reverse micellar medium (0.1 M AOT, W = 40 Tris-HCl buffer pH = 8.3), T = 30 °C. Hydrolysis of cytidine 2′:3′ monophosphate by ribonucleases and of N-glutaryl-L-phenylalanine-p-nitroaniline by α-chymotrypsins

		Native	Control	Cholesterol	Stearoyl	FMOC
Ribonuclease	Buffer	389	377	339	341	171
	Reverse Micelles	1502	1369	650	642	375
Chymotrypsin	Buffer	42	31	28	–	20
	Reverse Micelles	8	–	–	–	8

tyrosine residues). HPLC separation was identical to native enzyme (Fig. 1 A and B): one species is present and eluates at $t = 17{,}7$ mn. Furthermore, control RNAse in aqueous medium follows Michaelis–Menten kinetics and has the same specific constant K_2 as native ribonuclease (Table 1). This shows that synthesis treatment (reverse micellar solubilization, extraction by acetone and lyophilization) does not alter the enzyme's physical properties. Hydrophobocity, electrical net charge and biological activity of ribonuclease are maintained.

The reaction is strongly dependent on the ratio of reagent concentration over enzyme concentration, α, introduced during synthesis. This can be clearly seen for enzyme modification with chromophore 9-fluorenyl-methyl chloroformate (FMOC-Cl). Reverse phase high performance liquid chromatography (HPLC) separation of ribonuclease acylated by FMOC-Cl in a ratio α of 3 (Fig. 1C) shows that two species are present and eluate respectively at $t = 17.7$ mn and $t = 18.1$ mn. The first species is ascribed to native enzyme, whereas the second species to enzyme acylated by FMOC groups. Indeed, the absorption spectrum of the first derivative is characteristic of ribonuclease (maximum centered at 276 nm), while absorption spectrum of the second derivative is characteristic of FMOC-ribonuclease: two peaks centered at 265 nm and 298 nm with a shoulder at 276 nm. From reagent and enzyme extinction coefficients, the average number of FMOC groups bound to ribonuclease is found equal to 1. So, under these conditions (α = 3) the reaction is not total: modified samples of ribonuclease contain both native and acylated enzyme. Conversely, HPLC separation of ribonuclease acylated in FMOC-Cl in an α ratio of 20 shows that only one species is present and eluates at $t = 18.4$ mn (Fig. 1D). From the absorption spectrum this derivative is attributed to FMOC-ribonuclease with an average of two FMOC groups bound to the enzyme. Furthermore, enzyme modification is confirmed by electrophoresis: FMOC derivative has a mobility indicative of a decrease in positive charge, consistent with the substitution of positively charged groups (NH_3^+) by neutral ones (Fig. 1D).

To have maximum of acylated enzyme in the samples, all other modifications were carried out in an α ratio of 20. Chromatograms of the three other acylated ribonucleases are all consistent with their electrophoretic patterns (Fig. 1E, F&G). Cholesterol-ribonuclease is found free of any native enzyme whereas myristoyl and stearoyl-ribonuclease contain both native (40%) and modified enzyme (60%).

Table 1 reports the values of the specific constant K_2 of cytidine 2′:3′ monophosphate hydrolysis by FMOC, cholesterol, myristoyl, and stearoyl-ribonuclease in buffer. All of these derivatives are as active as native ribonuclease, except for FMOC-RNAse. In this latter case, diminution in activity is probably due to active site blocking by FMOC groups during enzyme modification. Indeed, ribonuclease's active site contains lysine residues[6]. Nevertheless, all ribonuclease derivatives retain activity after the modification process.

α-Chymotrypsin characterization

α-Chymotrypsin has been modified by 9-fluorenylmethyl and cholesterol chloroformate and has been obtained in the same way as ribonuclease derivatives. Chromatogram of FMOC-α-chymotrypsin (acylated in a ratio α in 20) shows that one peak is present. Its absorption spectrum is characterized by a principal maximum centered at 266 nm with shoulders and secondary maxima at 280, 290 and 298 nm. An average of 3.5 FMOC groups are bound to α-chymotrypsin. Native and modified α-chymotrypsin activities in buffer (Table 1) are compared using specific constant $K_2 = k_{cat}/K_m$ of N-glutary-L-phenylalanine hydrolysis[7]. K_m, Michaelis constant of native and modified α-chymotrypsins are rather similar. However, k_{cat}, catalytic kinetic rate constant of FMOC and cholesterol chymotrypsin is lower than k_{cat} constant of native enzyme. This could be due to binding of some reagents on isoleucine 16 terminal amino group [8]. Nevertheless, FMOC and cholesterol chymotrypsin retains about 60% of its initial activity after the modification process.

Progr Colloid Polym Sci (1994) 97:218–222
© Steinkopff-Verlag 1994

Location of enzyme derivative in reverse micelles

Structural study (small angle x-ray scattering and percolation) of reverse micelles containing these hydrophobic enzymes has been previously described [9, 10]. Table 2 reports the structural parameters using Baxter's [14] potential to model reverse micelles. They show no changes in the form or in the structure of the droplet, as compared to the system with native enzymes. Size and interactions between droplets are the same whether one works with the system containing native enzymes or the system containing hydrophobic enzymes.

Location of FMOC-α-chymotrypsin has been determined by a comparative analysis of the fluorescence spectra of FMOC-Cl reagent and a FMOC-α-chymotrypsin. In aqueous solution, FMOCl and FMOC-α-chymootrypsin have similar spectra (Figs 2A and B) characterized by two peaks centered at 305 and 315 nm. In pure isooctane, FMOC-Cl spectrum is characterized by one maximum centered at 305 nm and a shoulder at 315 nm. Spectra with same peak and shoulder are observed for pure reagent FMOC-Cl as well as for FMOC-α-chymotrypsin in reverse micellar media (Figs. 2A and B). Moreover, no change in the fluorescence spectra were detected by varying W from 10 to 40. Therefore, these fluorescence data indicate that the FMOC-α-chymootrypsin is located

Table 2 Polar radius R and sticky parameter τ^{-1} (characterizing attractive interactions between droplets) of AOT-water-isooctane microemulsions containing native or modified enzymes. W = 40 and [AOT] = 0.1 M. Values determined from small-angle x-ray scattering experiments

	R(Å)	τ^{-1}
Native chymotrypsin	67	2.5
FMOC chymotrypsin	69	2.7
Native Ribonuclease	68	2.4
FMOC Ribonuclease	70	2.5
Cholesterol Ribonuclease	70	2.4
Stearoyl Ribonuclease	66	2.3

at the inner oil-water interface, contrary to native enzyme which is located in the center of the water pool [3, 4].

Enzymatic activity of native and hydrophobic enzymes in reverse micelle

From a geometrical model [11] and quenching rate of hydrated electron by native enzymes [4], it has been demonstrated that native ribonuclease and α-chymotrypsin are located inside the droplet's water pool.

We have studied the enzymatic activity of ribonuclease's and α-chymotrypsin's hydrophobic derivatives in reverse micelles and compared it to native and control enzymes. All activity measurements were done at fixed water content $W = 40$ in order to fix the average size of the droplets ($R = 70$Å).

Hydrolysis of N-glutary-L-phénylalanine-p-nitroaniline by α-chymotrypsin derivatives has been performed in reverse micelles. Table 1 shows no changes in FMOC-chymotrypsin's enzymatic activity as compared to native enzyme. Activity assay of ribonuclease in reverse micelle has been described earlier by Luisi et al. [12]. All ribonuclease samples (native, control, Cholesterol-RNAse, Stearoyl-RNAse and FMOC-RNAse) follow Michaelis–Menten kinetics. Specific contents K_2 of cytidine 2':3' monophosphate hydrolysis by these various enzymes in reverse micelle are reported in Table 1. Enzymatic activity of ribonuclease is maintained in AOT reverse micelles and it is even higher than in buffer solution. This is consistent with results published by other groups [12]. Furthermore, control ribonuclease is as active as native ribonuclease. A significant decrease is observed for FMOC-RNAse, but it is also observed in buffer medium and has been explained by active site blocking during chemical modification of enzyme (cf. enzyme characterization paragraph).

Fig. 2 (A) Fluorescence spectra of FMOC-α-chymotrypsin in aqueous (dashed line) and in AOT reverse micelles (full line) solutions. (B) Fluorescence spectra of FMOC-Cl in isooctane (full line) and aqueous solutions (dashed line)

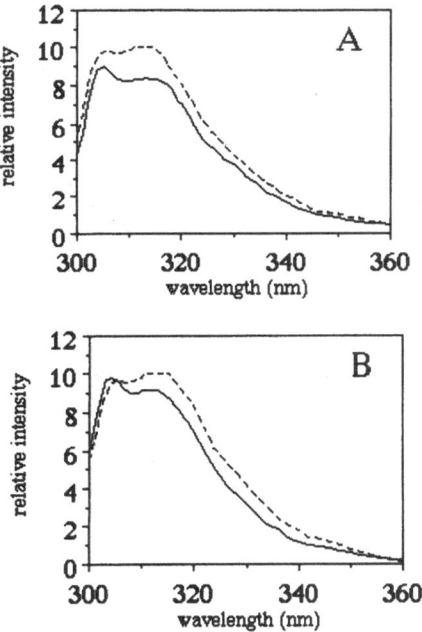

A 50% decrease in activity is obtained for Cholesterol-RNAse and Stearoyl-RNAse (as compared to native or control enzyme) while, on the contrary, in buffer medium these enzymes are as active as native ribonuclease. This diminution could be attributed to the average location of these enzymes inside the droplets. In fact, earlier studies have shown that in AOT reverse micelles quenching rate of hydrated electron by ribonuclease depends on the enzyme's location inside the water droplets [4]. It has been shown that enzyme is less reactive when located near the micellar interface than when located in the water pool. This has been explained by the inhibition of the enzyme's diffusion which interacts with the interface and is in good agreement with theory [13]. A possible explanation is that the anchoring of these hydrophobic enzymes at the interface can favor the active site's blocking. In fact, the active site is probably turned towards the micellar interface, preventing substrate to react with enzyme.

Conclusion

Water soluble enzymes, ribonuclease A, and α-chymotrypsin have been modified by hydrophobic reagents in AOT reverse micelles: the reaction medium offers the possibility to solubilize hydrophobic (reagent) and hydrophilic (RNAse) molecules. The method is rapid and employs mild conditions. It produces modified proteins with physical properties little different from those of the native proteins.

Fluorescence experiments on enzyme derivatives in reverse micellar system have shown that modified enzymes are located at the interface of the droplet.

References

1. Pileni MP (ed) (1989) Structure and reactivity in reverse micelles, Elsevier
2. Huruguen JP, Pileni MP (1991) Eur Biophys J 19:103; Huruguen JP, Authierr M (1991) Greffe JL, Pileni MP, Langmuir 7:243; Huruguen JP, Authier M, Greffe JL, Pileni MP (1991) J Phys: Cond Matt 3:865
3. Fletcher PDI, Robinson BH, Tabonyl J (1986) J Chem Soc Faraday Trans 1 82:2311
4. Petit C, Brochette P, Pileni MP (1986) J Phys Chem 90:6517
5. Kabanov A, Klyachko N, Martinek K, Levashov A (1988) Molek Biol USSR 22:473
6. Richards F, Wyckoff H (1973) Atlas of molecular structures in biology. Vol I, Clarendon, Oxford
7. Erlanger BF, Edel F, Cooper AG (1966) Arch Biochem Biophys 115:206
8. Oppenheimer HL, Labouesse B, Hess G (1966) J Biol Chem 241 11:2720
9. Pitré F, Regnault C, Pileni MP, Langmuir; Pitré F, Pileni MP, (1993) Prog Coll Polym Sci 93, PII
10. Michel F, Pileni MP, Langmuir; Michel F, Pileni MP (1993) Prog Coll Pol Sci 93, PII
11. Pileni MP, Zemb T, Petit C (1985) Chem Phys Lett 178:414
12. Wolf R, Luisi PL, Biochem Biophys Res Commun (1979) 89:209; Hagen A, Hatton T, Wang D (1990) Biotech & Bioeng 35:966
13. Gosele U, Klein U, Hauser M (1979) Chem Phys Lett 68:291
14. Baxter RJ (1968) J Chem Phys 49:2770

Progr Colloid Polym Sci (1994) 97:223–225
© Steinkopff-Verlag 1994

MICROEMULSIONS

F. Sicoli
D. Langevin

Shape fluctuations of microemulsions droplets: Role of surfactant film bending elasticity

Received: 5 October 1993
Accepted: 31 October 1993

D. Langevin (✉) · F. Sicoli
Laboratoire de Physique Statistique
  de l'ENS
24 rue Lhomond
75231 Paris Cedex 05, France

Abstract The surfactant film bending elasticity can be described by a spontaneous curvature C_0 and two elastic constants K and $\bar{K}$, associated with the mean curvature and the Gaussian curvature respectively. These parameters are very important in the determination of the structure of the dispersions stabilized by the surfactant (droplets or sponge-like structures). They also control the thermal fluctuations of the surfactant films. In the case of droplet microemulsions, the fluctuations create an equivalent droplet polydispersity. Recent neutron scattering determinations of the polydispersity for nonionic surfactant microemulsions are presented. The results are compared with previous data from ellipsometric experiments, whereby one determines the amplitude of the fluctuations of the flat surfactant films.

Key words Microemulsion – bending energy – Gaussian curvature elasticity – neutron scattering – droplet polydispersity

Microemulsion [1] are dispersions of oil and water stabilized by surfactant molecules. They are frequently made of droplets (oil in water (O/W) microemulsions, water in oil (W/O) microemulsions) surrounded by a surfactant monolayer and dispersed in a continuous phase (water or oil respectively). When the composition of the medium is known, the droplet radius can be predicted quite accurately by using the following relation,

$$R = \frac{3\phi}{c_s \Sigma}, \tag{1}$$

where ϕ is the dispersed volume fraction, c_s the number of surfactant molecules per unit volume, and Σ the area per surfactant molecule. This relation expresses the fact that virtually all the surfactant molecules sit at the oil-water interface and that each of them occupies a well defined area, independent of the composition. This is because, in order for the microemulsion to be thermodynamically stable, the surfactant monolayer must reduce the oil-water interfacial tension to about zero: its surface pressure must balance the tension of the bare interface, thus fixing the value of Σ. In the following, we will assume, as in recent microemulsion models, that the droplet surface tension γ_d is exactly zero [2].

The type of microstructure is closely related to the sign of the spontaneous curvature of the surfactant layer C_0. In many droplet microemulsions, the magnitude of C_0 also determines the maximum droplet size (maximum solubilization power) [3]. This can be simply established by introducing the surfactant film bending energy [4]:

$$F = \tfrac{1}{2} K (C_1 + C_2 - 2C_0)^2 + \bar{K} C_1 C_2$$

$$\text{(per unit area),} \tag{1}$$

where C_1 and C_2 are the two principal curvatures of the surfactant layer and K and $\bar{K}$ the mean and Gaussian bending elastic constants [3]. By convention $C_0 > 0$ for aqueous dispersions and $C_0 < 0$ for reverse systems. The second term is present because the system can change its topology: positive $\bar{K}$ favors saddle-splay structures as in bicontinuous cubic or sponge phases, while negative $\bar{K}$ favors lamellar or spherical structures. F is a surface energy usually negligible compared to the interfacial tension contribution. But in microemulsion systems, the interfacial tension being small or even zero, the bending energy becomes a very important term. For instance, the maximum droplet size can be shown to be:

$$\frac{R_m}{R_0} = \frac{2K + \bar{K}}{2K} + \frac{kT}{8\pi K} [\ln \phi - 1] , \quad (2)$$

where $R_0 = C_0^{-1}$ [5]. When by increasing the dispersed phase volume fraction ϕ, R exceeds R_m, the system phase separates into two phases: a microemulsion with droplet size R_m and an excess phase (emulsification failure) [3].

Experiments show that K is typically of order kT in microemulsion systems [6]. This means that in droplet microemulsions, the droplets are substantially distorted due to thermal motion. This has been theoretically analyzed by describing the droplet deformation with an expansion of spherical harmonics Y_{lm} [7]:

$$R(\theta, \phi) = R_m \left[1 + \sum_{lm} u_{lm} Y_{lm}(\theta, \phi) \right]. \quad (3)$$

It can then be shown that the main contribution comes from droplet size fluctuations ($l = 0$) and peanut-like deformations ($l = 2$). Both contribute to what is usually called droplet polydispersity. One finds at the emulsification failure limit ($\langle R \rangle = R_m$) [8]:

$$p^2 = \frac{\langle (R - R_m)^2 \rangle}{R_m^2} = \left\langle \left| \sum_{lm} u_{lm} Y_{lm}(\theta, \phi) \right|^2 \right\rangle$$

$$= \frac{\langle |u_0|^2 \rangle + 5 \langle |u_2|^2 \rangle}{4\pi} , \quad (4)$$

and

$$4\pi p^2 = \frac{kT}{2(2K + \bar{K}) + \dfrac{kT}{2\pi} [\ln \phi - 1]}$$

$$+ \frac{5kT}{4(4K - \bar{K}) + \dfrac{kT}{2\pi} [\ln \phi - 1]} \quad (5)$$

$\bar{K}$ plays a role here, because when changing the droplet radius at constant Σ, one changes the total number of droplets, and thus the topology. The polydispersity would of course also depend on the droplet surface tension γ_d.

Recent spin-echo neutron-scattering experiments suggest that $\gamma_d = 0$ for AOT microemulsions, i.e., the surfactant film is truly incompressible (constant Σ) [9].

We have measured the droplet polydispersity by static small-angle neutron-scattering experiments in Saclay (PAXE spectrometer). We have studied ternary oil-water-nonionic surfactant mixtures, on which previous determinations of the modulus K have been done with ellipsometry [10]. The surfactants are alkyl polyethylene glycol ether surfactants with alkyl chains of n carbon atoms and polar parts of m ethoxy groups: $C_{12}E_5$, $C_{10}E_4$ and C_8E_3. They have been mixed with deuterated water and respectively deuterated hexane, octane and decane (shell contrast). Their composition was such that at a given temperature, a dilute microemulsion phase ($\phi \sim 1\%$) is in equilibrium with an excess phase. O/W system were obtained at low temperature, W/O systems at high temperature. Details on the phase diagrams can be found in ref. [10] and [11]. Figure 1 shows typical spectra, together with the fit with a Gaussian distribution of shells. This type of fit allows an easy deconvolution from the resolution function (itself Gaussian). The "shell" contrast used here allows a more accurate determination of the polydispersity than the "sphere" contrast (protonated oils) [12].

Within the experimental accuracy, the polydispersity does not depend appreciably on the temperature (i.e., on

Fig. 1. Neutron scattered intensity I versus wave vector q (in Å^{-1}) for three different O/W microemulsions in the shell contrast; $T = 14°C$. The lines are fits: $C_{12}E_5$ $R_m = 100$ Å, $p = 0.24$; $C_{10}E_4$ $R_m = 96$ Å, $p = 0.29$; C_8E_3 $R_m = 73$ Å, $p = 0.40$.

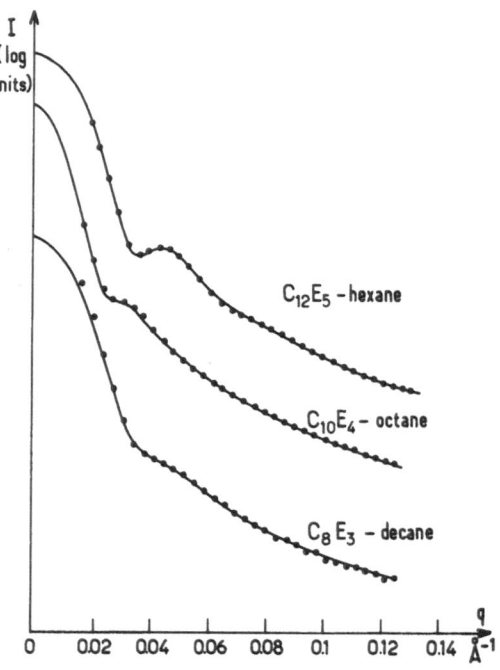

Prog Colloid Polym Sci (1994) 97:00–00
© Steinkopff-Verlag 1994

the type and radius of the droplets) and on the nature of oil: it slightly increases with increasing temperature and decreasing oil chain length. As expected, it is very sensitive to the surfactant chain length. The data analysis leads to $p \sim 0.2$ for $C_{12}E_5$, 0.3 for $C_{10}E_4$ and 0.4 for C_8E_3. The accuracy is about 20% for $C_{12}E_5$ and $C_{10}E_4$; it is poorer in the case of C_8E_3 for which only a limited number of experiments have been done up to now. K has been previously determined by studying the thermal fluctuations at the flat oil-water interface by ellipsometry: for $C_{10}E_4$-octane microemulsions, $K = 0.52\ kT$, independent of temperature, and for $C_{12}E_5$-hexane microemulsions K is larger: $K \sim kT$. [10] Equation (5) then leads to polydispersity values larger than the measured ones, whatever $\bar{K}$ is [13].

If instead of the directly measured values of K, we use the renormalized ones, 0.76 and $1.5\ kT$, respectively, we find values, respectively, of $-0.1\ kT$ and $-0.7\ kT$. These value are negative as expected, and as for K they increase in absolute value with increasing surfactant chain length. It is not yet clear why the renormalized value of K needs to be used here. It is also possible that K is different for planar and curved surfaces. The droplets surface tension might also be different from zero. Neutron spin-echo experiments will be performed to check these two last points. They will allow the determination of both γ_d and K for the curved surfaces. We expect then to be able to obtain more meaningful estimations of $\bar{K}$.

References

1. de Gennes, PG, Taupin C (1982) J Phys Chem 86:2294–2304
2. Safran SA, Roux D, Cates ME, Andlman D (1986) Phys Rev Lett 57:491–493
3. Safran SA, Turkevich LE (1983) Phys Rev Lett 50:1930–33
4. Helfrich W (1993) Z Naturforsch 28:693
5. Safran SA in "Modern Amphiphilic Systems" Ed. Ben-Shaul A, Gelbart W, Roux D, Springer 1993
6. Binks BP, Meunier J, Abillon O, Langevin D (1989) Langmuir 5:415–421
7. Milner ST, Safran SA (1987) Phys Rev A 36:4371
8. Sicoli F, Langevin D, Lee LT (1993) J Chem Phys 99:4759–65
9. Huang JS, Milner ST, Farago B, Richter D (1987) Phys Rev Lett 59:2600–2603 Farago B, Richter D, Huang JS, Safran SA, Milner ST (1990) Phys Rev Lett 65:3348–51
10. Lee LT, Langevin D, Meunier J, Wong K, Cabane B (1990) Prog Colloid Polym Sci 81:209–214
11. Kahlweit M, Strey R, Firman P, Hasse D, Jen J, Schomäcker R, 1988 Langmuir 4:499–511 Kahlweit M, Strey R, Firman P (1986) J Phys Chem 90:671–677
12. Sicoli F, Langevin D, Lee LT, Monkenbusch M (1993) Prog Colloid Polym Sci 93:105–107
13. Let us quote that in a previous paper (ref. [12]), we have used a simpler expression for the polydispersity (ref. [9]), where the contribution of the $l = 2$ modes is neglected; the $\bar{K}$ values obtained in this way need to be corrected.

Progr Colloid Polym Sci (1994) 97:226–228
© Steinkopff-Verlag 1994

MICROEMULSIONS

V. Papadimitriou
C. Petit
A. Xenakis
M. P. Pileni

Structural modifications of reverse micelles due to enzyme incorporation studied by SAXS

Received: 11 October 1993
Accepted: 15 December 1993

V. Papadimitriou · A. Xenakis
National Hellenic Research Foundation
Institute of Biological Research and
   Biotechnology
48, Vas. Constantinou Ave.
116 35 Athens, Greece

C. Petit · M.P. Pileni (✉)
CEA
Centre d'études nucléaires de Saclay
DRECAM-SCM
Yvette Gif survette, France

Université Pierre et Marie Curie
Loboratoire SRSI
URA CNRS, 1662
Paris, France

Abstract Small angle x-ray scattering (SAXS) measurements were carried out on w/o microemulsions containing two different enzymatic systems: trypsin or lipase. The results were compared to SAXS measurements in the absence of proteins. When relatively small reverse micelles were used, the presence of enzyme induced a decrease of the aqueous core radius. For larger droplets the presence of the biomolecules did not influence their size.

Key words Microemulsions – small-angle x-ray scattering – trypsin – lipase

Introduction

The aqueous core of reverse micelles can host various hydrophilic substrates, including biomolecules such as enzymes. The systems provide a new environment, a kind of microreactor where enzymatically catalyzed reactions can be studied. Over recent years it has been shown that in most cases enzymes retain their catalytic ability when entrapped in such microheterogeneous systems [1, 2].

The location of a protein within the different microdomains of the reverse micelles-aqueous core or surfactant interface–can be influenced by the size and degree of hydrophobicity of the considered protein [3]. The presence of enzyme molecules in these media possibly induces changes in the structure of the reverse micelles.

In the present work, the location of two enzymes, trypsin and lipase, in AOT reverse micelles was studied by small-angle x-ray scattering (SAXS) [4].

Experimental

Trypsin (EC3.4.21.4) from bovine pancreas type III was from Merck, Darmstadt, FRG. Concentrated stock solutions of trypsin in a 50 mM Tris/HCl buffer pH 9 were prepared and stored in a freezer. Lipase from *Penicillium simplicissimum* was a generous gift of Dr. U. Menge from GBF, Braunschweig, FRG. The enzyme showed a single band in native and SDS PAGE electrophoresis [5] and exhibited a specific activity of 142 units/mg of protein determined by titration of free fatty acids release from

Progr Colloid Polym Sci (1994) 97:227–228
© Steinkopff-Verlag 1994

triolein. Bis-(2-ethylhexyl)sulfosuccinate sodium salt (AOT) was purchased from Sigma Chemical Co., isooctane was from Fluka, Switzerland. All other chemicals were of the highest available degree of purity and doubly distilled water was used throughout this study, Microemulsions were prepared as described elsewhere [6].

The SAXS experiments were performed at the Laboratoire d'Utilisation des Rayonements Electomagnetiques (LURE), CNRS-CEA-Paris XI, Orsay, France, on the D22 diffractometer. The experimental arrangement and the theoretical treatment are described elsewhere [7].

Results and discussion

Small-angle x-ray scattering of different microemulsions containing various concentrations of either trypsin or lipase were measured. Figure 1 shows typical Porod representations [8] of such SAXS experiments, while Tables 1 and 2 regroup the experimental values of the character-

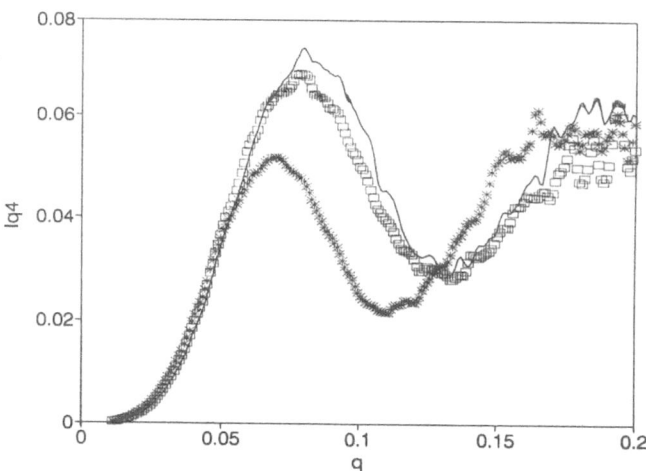

Fig. 1 Porod representation ($I(q)q^4$ vs. q) of SAXS of 0.1M AOT/isooctane microemulsions, $w_0 = 20$, containing: (*): 50 mM Tris/HCl pH 9 buffer solution, (□): 2.5×10^{-4} M Trypsin; (–): 1.1×10^{-4} M Trypsin

Table 1 Variation of the water pool radius (in Å) of the 0.1 M AOT/isooctane reverse micelles containing buffer solution and various trypsin concentrations.

	w_0	$R_c(min)^1$	$R_c(max)^1$	R_g^2	R_{sim}^3	$\sigma^3(\%)$
Buffer	20	41	38	40	35	0.20
Trypsin 1.1×10^{-4} M	20	34	34	33	31	0.23
Trypsin 2.5×10^{-4} M	20	34	34	36	33	0.23
Buffer	30	51	54	52	45	0.23
Trypsin 1.1×10^{-4} M	30	53	53	55	47	0.27
Trypsin 2.5×10^{-4} M	30	55	53	59	48	0.26

[1] from the Porod plots
[2] from the Guinier plots
[3] from the simulated plots

Table 2 Variation of the water pool radius (in Å) of the 0.1 M AOT/isooctane reverse micelles containing aqueous solution and various lipase concentrations.

	w_0	$R_c(min)^1$	$R_c(max)^1$	R_g^2	R_{sim}^3	$\sigma^3(\%)$
Water	20	36	32	38	31	0.26
Lipase 2×10^{-6} M	20	34	28	34	28	0.31
Lipase 4×10^{-6} M	20	–	28	31	25	0.32
Water	30	53	44	51	47	0.27
Lipase 3×10^{-6} M	30	51	46	54	46	0.27
Lipase 6×10^{-6} M	30	51	43	48	44	0.29

[1] from the Porod plots
[2] from the Guinier plots
[3] from the simulated plots

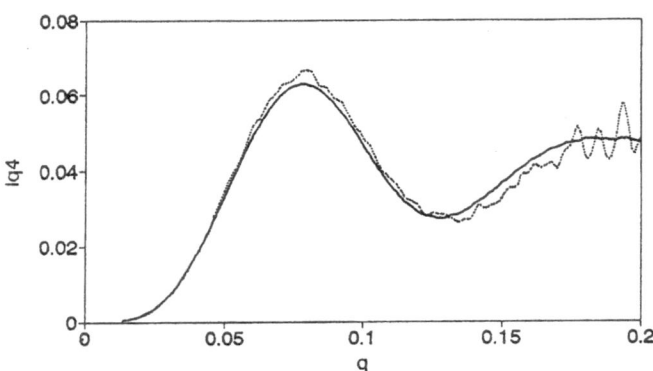

Fig. 2 Comparison of a simulated curve (–) with experimental data (..) of SAXS (Porod plots) of 0.1 M AOT/isooctane microemulsions, $w_0 = 20$, containing 2.5×10^{-4} M Trypsin

istic aqueous core radius, R_c, obtained from such plots, for various microemulsions containing different concentrations of trypsin, and lipase, respectively. The experimental data were simulated taking into account the polydispersity, σ, in the size of the reverse micelles [7] (Fig. 2). The good agreement between the experimental and simulated curves confirms that the particles are spherical entities.

In the case of trypsin, at $w_0 = 20$, the data of Table 1 show that the presence of this enzyme perturbs the reverse micellare system. The radius of the micelles decreases, the polydispersity increases, while the total inter-face does not change. These observations are consistent with previous results concerning other proteins such as cytochrome C [2]. At $w_0 = 30$ where larger micelles are formed, the perturbation is less important. In this case the quantity of water molecules surrounding the enzyme, is increased forming an aqueous core large enough to host the trypsin molecules. Similar results were reported for the same microemulsions containing other enzymes such as α-chymotrypsin [7] or ribonuclease [9].

In the case of lipase the results of Table 2 indicates that for both w_0 values (20 or 30) the size of the reverse micelles decreases, too, while the increase of the polydispersity is more important. On the contrary, the use of lipase instead of trypsin seems to increase the total interface. The low concentration of lipase imposed by its solubility in such media does not allow us to clarify this observation. Nevertheless, this indicates that the lipase molecule is located within the micellar interface, which is in good agreement with the site of localization of this lipase in the same microemulsions as determined by either activity studies [3] or fluorescence energy transfer measurements [10].

In conclusion, SAXS measurements of microemulsion containing enzymes can provide information on possible structural modifications. When trypsin is incorporated in these media the enzyme is located in the aqueous water core while the size of the reverse micelles decreases. On the other hand lipase is incorporated within the reverse micellar membrane increasing its interface.

References

1. Luisi PL, Magid L (1986) Crit Rev Biochem 20:409
2. Pileni MP (ed) (1989) Structure and reactivity in reverse micelles. Elsevier, Amsterdam
3. Stamatis H, Xenakis A, Provelegiou M, Kolisis FN (1993) Biotech Bioeng 42:102–110
4. Pileni MP, Zemb T, Petit C (1985) Chem Phys Lett 118:414
5. Sztajer H, Lunsdorf H, Erdmann H, Menge U, Schmid R (1992) Biochim Biophys Acta 1124:253–261
6. Papadimitriou V, Xenakis A, Evangelopoulos AE (1993) Colloids Surf Biointerfaces 1:295–303
7. Pitré F, Regnault C, Pileni MP (1993) Langmuir 9:2855
8. Porod G (1982) In: Glatter O, Kratky O (eds) Small Angle x-ray Scattering, Academic Press New York
9. Michel F, Pileni MP (1994) Langmuir 10:390
10. Stamatis H, Xenakis A, Kolisis FN, Malliaris A (1993) Progr Colloid Polym Sci 97:00

Progr Colloid Polym Sci (1994) 97:229–232
© Steinkopff-Verlag 1994

MICROEMULSIONS

A. Hammouda
M.P. Pileni

Synthesis of small latexes by polymerization of reverse micelles

Received: 5 October 1993
Accepted: 21 October 1993

M.P. Pileni (✉) · A. Hammouda
Université Pierre et Marie Curie
Laboratoire SRSI
B.P. 52
4 place Jussieu
75231 Paris Cedex 05, France

CEN Saclay
DRECAM/SCM
91191 Gif sur Yvette, France

Abstract The synthesis of a new polymerizable surfactant forming reverse micelles has been performed. The didecyldimethylammonium methacrylate is soluble in aromatic solvents and, by water addition, the solutions remain clear, stable over several months, and low conducting. Structural studies of these microemulsions have been performed by small-angle x-ray and dynamic quasi elastic light scattering. These microemulsions have been UV irradiated in the presence of azobisisobutyronitrile, at various water contents; nanosized latexes were obtained.

Key words Latexes – reverse micelles – polymerization

Introduction

Reverse micelles are spherical, water-in-oil droplets stabilized by surfactant molecules [1]. Previously, reverse micelles have been used to form latexes [2–5]. Two different approaches have been performed: In the first case, micelles are used as microreactors and an hydrophilic monomer, solubilized in the water pool, can be polymerized. This is the case with acrylamide, studied by Candau et al. [2–3], with formation of one chain polymer per latex and characterized by a high molecular weight. The hydrodynamic radii of such latexes are about 25 nm diameter. In the second case, a surfactant containing a double bond has been polymerized at various water contents [4–5]. The hydrodynamic radii of the particles are about 10 nm. In all these cases, the size of the latexes is larger than the size of the initial droplets. This has been attributed to coalescence between the water pool droplets inducing the formation of larger aggregates which grow progressively during the polymerization. At the end of the polymerization process, latexes containing, on average, one high molecular weight polymer are formed.

In this paper, we present the synthesis and the polymerization of a new polymerizable surfactant forming reverse micelles: didecyldimethylammonium methacrylate.

Experimental section

Materials

Didecyldimethylammonium bromide was purchased from Fluka, toluene from SDS, azobisisobutyronitrile from Serva.

Methods

The didecyldimethylammonium methacrylate is obtained from its bromide analogue by ion-exchange chromatography performed on a previously methacrylated resin (AG1 X2, Biorad). The surfactant synthesized has been characterized by proton and carbon NMR and its purity determined by titration of residual bromide ions.

The conductivity measurements were performed using platinum electrodes and a Tacussel CD 810 instrument. The small angle x-ray scattering experiments were performed at L.U.R.E. on the D22 diffractometer. The x-ray scattering intensity $I(q)$ is:

$$I(q) = P(q) \cdot S(q) ,$$

where q is the wave vector equal to $4\pi(\sin\theta)/\lambda$ (2θ is the scattering angle), $P(q)$ is the form factor, and $S(q)$ the structure factor which described the interactions. For spherical structures, in the case where the interactions can be neglected, $S(q)$ is assumed to be equal to 1. From the Porod plot ($I(q) \cdot q^4$ vs q) the characteristic radius R_c is related to the first minimum and to the first maximum of this representation, and equal to $4.5/R_c$ and $2.7/R_c$, respectively [6].

The dynamic quasi-elastic light scattering measurements were performed with an argon laser (514,5 nm), at a temperature of 25 °C, and a scattering angle of 45°. The autocorrelation functions were obtained with a 136-channel Brookhaven digital correlator. The hydrodynamic radius of the aggregates is deduced from the Stokes–Einstein equation ($R_H = kT/6\pi\eta D_0$, where k is the Boltzmann constant, T is the temperature, η the oil viscosity, and D_0 the intrinsic diffusion coefficient measured at infinite dilution).

Polymerization: Azobisisobutyronitrile (AIBN) is added to the micellar solution, keeping the water pool constant. The micellar solutions are UV irradiated.

Results and discussion

Conductivity

The conductivity of didecyldimethylammonium methacrylate ($C_{10}MA$)-toluene-water microemulsions is measured at various water contents (w = $[H_2O]/[C_{10}MA]$) until phase separation or turbidity occur. The results shown in Fig. 1 indicate that the conductivity is low and, therefore, one can assume that these microemulsions consist of a dispersion of water droplets stabilized by surfactants in the oil.

The conductivity pattern of $C_{10}MA$-toluene-water microemulsions, with a maximum at a low w value, resembles that observed for sodium bis(2-ethylhexyl) sulfosuccinate reversed micellar solutions [7]. The decrease of the conductivity beyond this value has been explained by Eicke et al. [7] in terms of a charge fluctuation model. The existence of the maximum has been demonstrated theoritically by Kallay et al. [8], using the same model, but distinguishing between the radius of the polar core where the charges are distributed and the radius of the micelle.

Scattering analysis

Structural studies of such aggregates have been performed from small-angle x-ray and dynamic quasi-elastic light scattering (S.A.X.S. and D.Q.E.L.S., respectively). Figure 2 shows a linear increase of the droplets size with the water content, with a good agreement between the data obtained

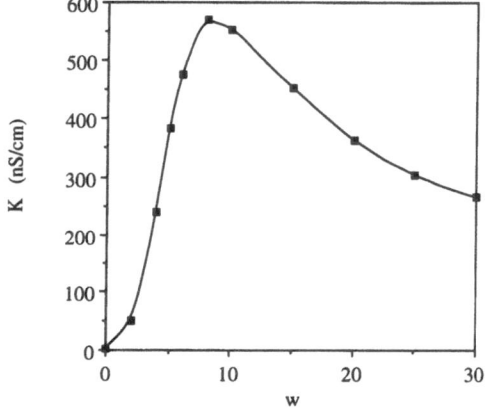

Fig. 1 Conductivity of $C_{10}MA$ water in toluene microemulsions as a function of w. $[C_{10}MA] = 0,1$ M

Fig. 2 Variation of the $C_{10}MA$ reverse micelles size in toluene as a function of the water content w. ■: D.Q.E.L.S., □: S.A.X.S

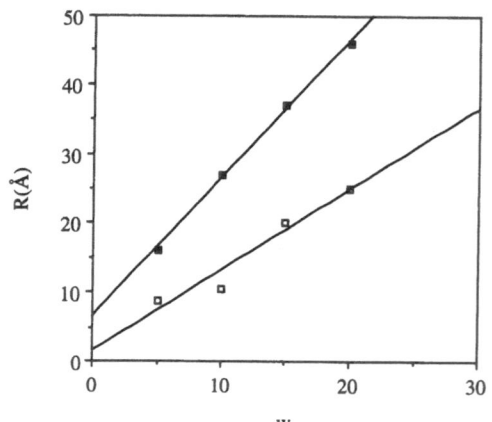

Progr Colloid Polym Sci (1994) 97:229–232
© Steinkopff-Verlag 1994

from S.A.X.S. and D.Q.E.L.S. The difference between the two radii is equal to the length of the aliphatic tails plus the thickness of the solvent layer, which varies as the micellar size increases.

Polymerization of $C_{10}MA$ reverse micelles in toluene

At a $C_{10}MA$ concentration equal to 0.05 M, microemulsions have been irradiated during 7 h, at various water contents in the presence of 1% of AIBN (in weight with respect to the monomer). Electron microscopic pictures, obtained after polymerization (Fig. 3), show the presence of small-size particles. At low w value ($w = 5$), main size of the particles is equal to 19 Å. By increasing the water content, the size increases. At w up to 20, the average size of latexes remains unchanged (28 Å), but the polydispersity decreases. This can be explained by the fact that the percentage of initiator is the same with respect to the monomer, so it decreases with respect to one micelle.

Fig. 3 Electron micrographs and histogramms of polymerized microemulsions. $[C_{10}MA] = 0,05$ M (A): w = 5, 1 mm = 10 Å; (B): w = 10, 1 mm = 14 Å; (C): w = 15, 1 mm = 14 Å; (D): w = 20, 1 mm = 14 Å

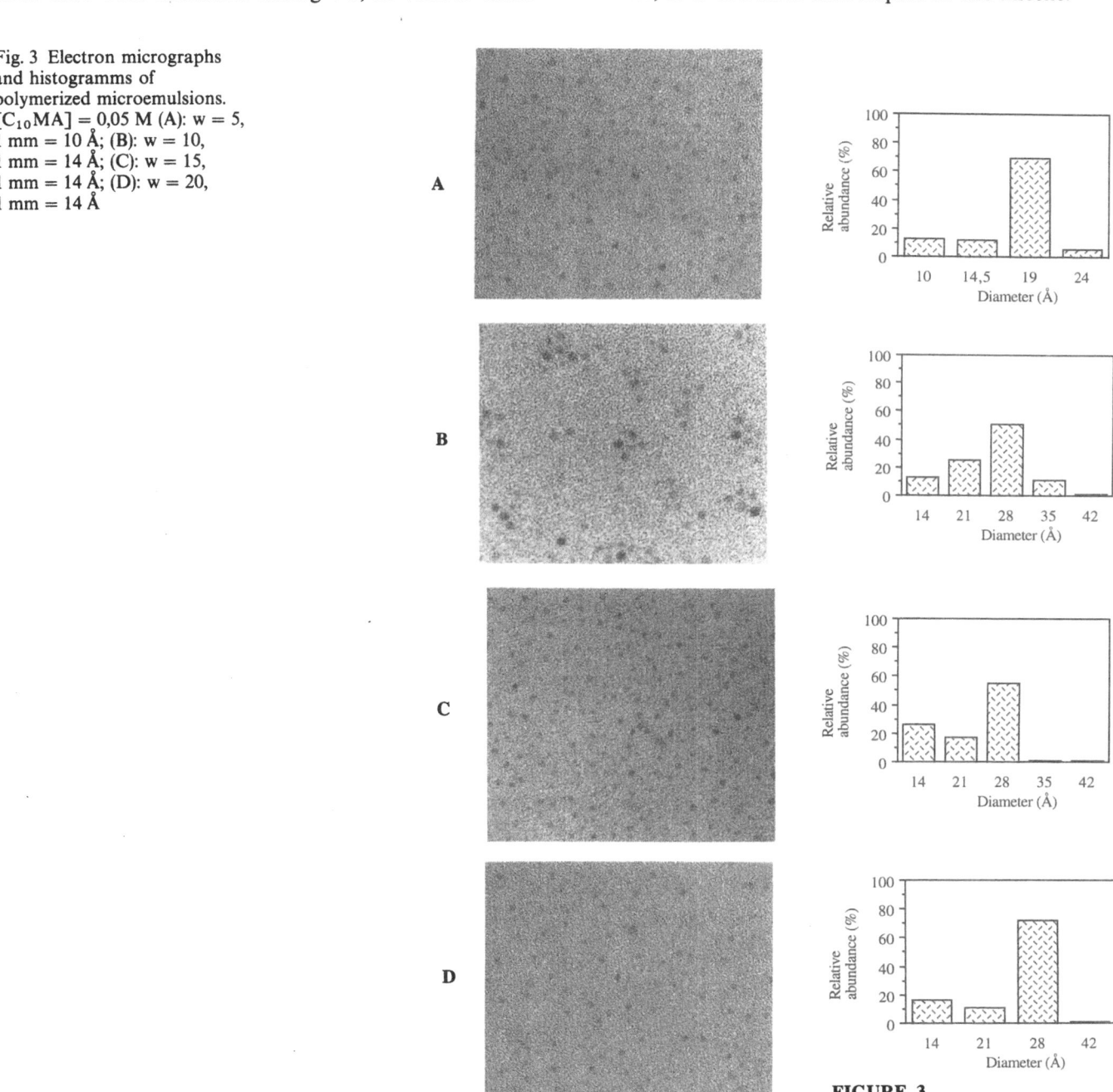

FIGURE 3

Conclusion

The didecyldimethylammonium methacrylate is a polymerizable surfactant forming reverse micelles in aromatic solvents. Conductivity measurements have shown that the system formed by this surfactant, toluene and water, consists of water droplets stabilized by surfactants and dispersed in the oil. Their size can be monitored by the water content added. Nanosized reverse latexes have been obtained by polymerization of micellar solutions containing didecyldimethylammonium methacrylate at various water content.

References

1. Pileni MP (ed) (1989) Structure and reactivity in reverse micelles; Elsevier, Amsterdam
2. Leong YS, Candau F (1982) J Phys Chem 86 (13):2269–71
3. Candau F, Carver MT (1989) Structure and reactivity in reverse micelles; Pileni MP (ed) Elsevier, Amsterdam (1989)
4. Voortmans G, Verbeeck A, Jackers C, De Schryver F (1988) Macromol 21:1977–80
5. Voortmans G, Jackers C, De Schryver F (1989) Brit Pol J 21:161–69
6. Porod G, Small Angle X-Ray Scattering; Glatter O, Kratky O, Eds.; Academic Press: New York Chapter 2 (1982)
7. Heicke HF, Borkovec M, Das-Gupta B (1989) J Phys Chem 93 (1):314–17
8. Kallay N, Chittofrati A (1989) J Phys Chem 94 (11):4755–56

Progr Colloid Polym Sci (1994) 97:233–236
© Steinkopff-Verlag 1994

MICROEMULSIONS

J. Appell
G. Porte
J.F. Berret
D.C. Roux

Some experimental evidence in favour of connections in elongated surfactant micelles

Received: 16 September 1993
Accepted: 22 October 1993

J. Appell (✉) · G. Porte
J.F. Berret · D.C. Roux
Groupe de Dynamique des Phases
    Condensées – URA no. 233 du CNRS
Université Montpellier II Science et
Techniques du Languedoc (case 26)
34095 Montpellier Cedex 05, France

Abstract In dilute aqueous solutions of surfactants a succession of phases is generally observed, namely, solutions of globular then elongated flexible micelles (worm-like micelles) followed by the phases of fluid membranes (successively a lamellar phase and the L_3 (sponge) phase). This is well accounted for by the continuous decrease of the spontaneous curvature of the surfactant film upon variation of some control parameter (temperature or concentration of added salt or of cosurfactant). However, a number of experimental facts are still open to question, e.g.: i) The solutions of worm-like micelles undergo a separation in two distinct isotropic phases upon variation of the control parameter; ii) Upon increasing the concentration of aggregates in the micellar phase, a nematic phase is found to exist in a narrow range of the control parameter. For lower and higher values of this parameter the isotropic micellar phase persists; iii) The unexpected evolution of the viscosity of the solutions of worm-like micelles. We argue that these facts can be interpreted assuming that in the course of the continuous decrease of the curvature of the amphiphilic film the cylindrical aggregates (worm-like micelles) form increasingly more connections allowing for favorable regions of smaller curvature.

Keywords Surfactant micelles
– structure – connected cylinders

Introduction

In aqueous solutions of surfactants (TA) a succession of phases is generally observed, namely, solutions of globular then elongated flexible micelles (worm-like micelles) [1–3] followed by the phases of fluid membranes [4, 5] (successively a lamellar phase and the L_3 (sponge) phase). This is well accounted for by the continuous decrease of the spontaneous curvature C_0 of the surfactant film upon variation of some control parameters (temperature, concentration of added salt or of cosurfactant). Classically, adding the constraint that one dimension of the aggregate must be approximately equal to the length of the apolar chain, this leads to the well-known sequence spherical, cylindrical and planar local structures [6] and to the explanation of the succession of phases mentioned above [7]. However, many experimental facts are still open to question. After briefly describing these facts, we will argue that they can be interpreted assuming that, during the continuous decrease of C_0, the occurrence of increasingly more connections between the cylindrical aggregates allows for increasingly more favorable regions of smaller mean curvature before the transition to a bilayer morphology.

Some experimental facts

1) The solutions of worm-like micelles undergo a separation into two distinct isotropic phases (one concentrated

micellar phase and one dilute one) upon variation of the control parameter. The corresponding critical behavior has been studied in a number of systems [8, 9]. At first sight the situation appeared analogous to that of the phase separation observed in polymer solutions (where it is driven by the evolution from good to poor solvent). This explanation is ruled out in surfactant solutions where, upon a further variation of the experimental parameter, the occurrence of a new stable phase (the lamellar phase) is observed, as illustrated in Fig. 1.

2) In similar systems (the example given here is for cetylpyridinium chloride-CPCl/hexanol/brine (0.2 M NaCl), upon increasing the concentration of surfactant, a nematic phase is found to exist in a narrow range of the control parameter which is here the ratio alcohol to surfactant. The ordering of the cylindrical micelles of L_1 upon increasing the control parameter which favors the elongation of the micelles is understandable. But very surprisingly, the nematic phase disappears after a narrow range of existence and the isotropic micellar phase is found to exist at higher values of the control parameter [10] as shown in the insert of Fig. 2.

3) The evolution of the viscosity of the micellar phase, near the nematic phase described above, is shown in Fig. 2. With the increase of the control parameter viscosity increases first tremendously then decreases [10].

Discussion

When increasing the control parameter (in the examples mentioned above it is the ratio alcohol/TA) the sponta-

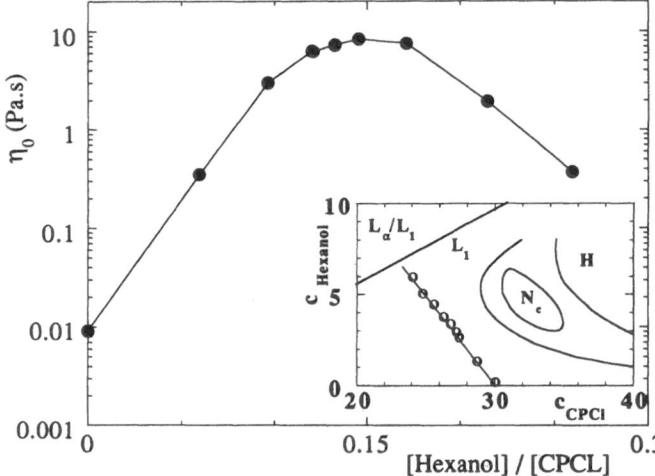

Fig. 2 Evolution of the viscosity in the micellar solution as a function of alcohol to surfactant ratio. In the insert, the phase diagram of cetylpyridinium chloride/hexanol/brine (0.2 M NaCl) is shown in the vicinity of the nematic phase (c_{CPCL} and $c_{hexanol}$ are in weight percents): note that the micellar phase (L_1) extend both below and above the nematic phase (N_c). The viscosity has been measured along the indicated line ($c_{brine} = 70\%$)

neous curvature C_0 of the amphiphilic film varies continuously, with the additional constraint for the aggregate to have at least one dimension equal to the length l_a of the apolar chain of the surfactant (no void in the center of the aggregate). It is impossible to find a succession of possible geometrical structures where the curvature varies continuously. The closest approach leads to the classical succession: a sphere of radius close to l_a, then a cylinder of radius close to l_a, then finally a bilayer of thickness close to $2l_a$. The branching of one cylinder onto another with locally a saddle-like structure (with the two principal curvatures of opposite signs) is a possible solution intermediate between the cylindrical and planar local structure as illustrated in Fig. 3. The insertion of these connections in the succession of forms could allow to parallel more closely the continuous variation of C_0.

What happens then on a larger scale? Initially at the lowest value of the control parameter the solution of quasi-spherical micelles (radius $r \sim l_a$) is the best solution when C_0 is close to $2/r$. Increasing the control parameter will decrease C_0 so that the quasi-spherical micelles are less and less "appropriate"; and gradually the micelles grow to cylindrical micelles which must be closed at both ends. In the complete description of this phenomenon, the two extremities of the cylindrical micelles are assumed to be hemispherical, the micellar solution is then described in terms of unidimensional reversible polymerization: the fusion of two micelles to give a larger one corresponds to a gain in energy (ΔE_b). ΔE_b is due to the transfer of the

Fig. 1 Schematic phase diagram in the brine-rich region of quasiternary system such as cetylpyridinium bromide (CPBr)/hexanol/brine (0.2 M NaBr) (T = 31 °C). L_1 is the micellar phase, L_α is the lamellar phase, and L_3 the sponge phase. A is the region of coexistence of two micellar solutions (one diluted and one concentrated) just below the lamellar phase region. P_c is the critical point of this phase separation. The dotted triangle is a triphasic region ($L_1 + L_1 + L_\alpha$)

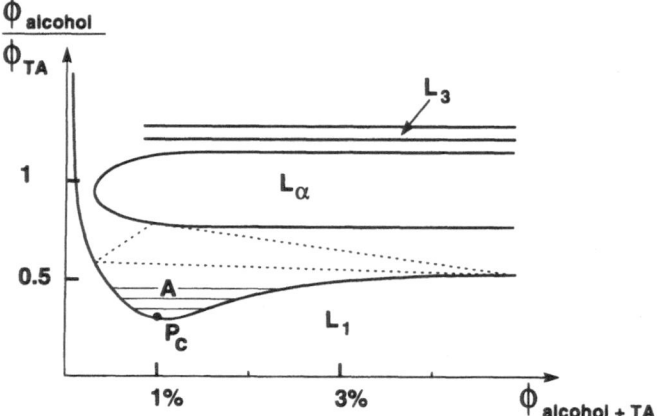

Fig. 3 Lowering of the local curvature C: a) cylinder of radius r: $C \sim 1/r$; b) a connection point between cylinders: the two principal curvatures are of opposite sign and $C < 1/t$; c) a bilayer $C = 0$

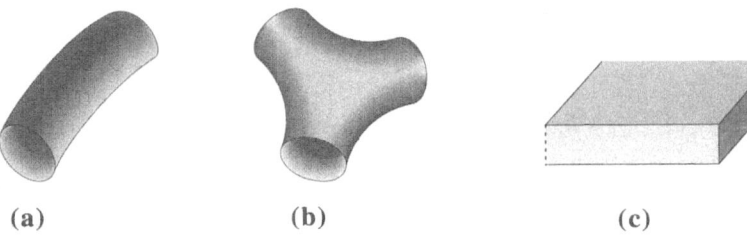

(a)   (b)   (c)

surfactant molecules in the two suppressed extremities from a spherical to a more favorable cylindrical surrounding. The micelles have a large distribution of lengths and the mean length remains finite, but can grow to very large values if ΔE_b is large. In a given system ΔE_b is expected to grow when C_0 decreases from its initial value towards that of the cylinder. However, when C_0 decreases, a second possibility to eliminate the ends of the cylinders can be thought of, namely, the branching of cylinders. As mentioned above, at a connection the curvature is lower than that of the cylinder (see Fig. 3) so that with decreasing C_0 past and beyond the curvature of the cylinder, we assume that the energetical cost of a connection (ΔE_c) decreases. Thus, ΔE_b increases and ΔE_c decreases when C_0 decreases, so that it is natural to think that the connections will gradually take over the hemispherical ends in the solution of cylindrical micelles before a further decrease of C_0 induces the transition to the fluid membrane phases.

How does this description permit the interpretation of the experimental facts described above?

Fig. 4 The small-angle neutron scattering pattern of the nematic phase after alignment under shear [10]. The SANS-data were obtained at a neutron wavelength $\lambda = 6.29$ Å on a two-dimensional detector (using 128×128 elements) located at 2 m from the sample. The contour intensities are 20, 40, 60, 80, and 120. Note the two crescent-like dots perpendicular to the director $\hat{n}$ and the absence of scattering in the $\hat{n}$ direction

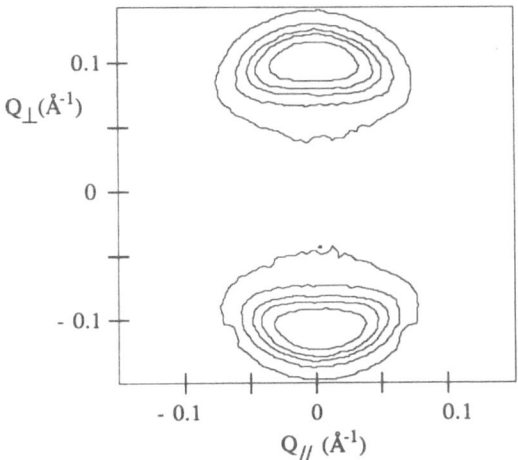

In the micellar phase L_1 when increasing the control parameter the micelles gradually grow and the entangled network of micelles begin to connect up to the point where all entanglement points have become connections: the network is saturated and upon further increase of the control parameter the network shrinks, expelling almost pure solvent in order to allow for new connections [7, 11]. This leads to the observed separation of one micellar phase in two micellar phases as described above.

The nematic phase has been studied by small-angle neutron scattering [10]. A typical scattering pattern for a sample, after its alignment under shear, is shown in Fig. 4. The observation of two peaks in the form of rather short crescents in the direction perpendicular to the direction of the director $\hat{n}$ and the absence of scattering in the direction parallel to $\hat{n}$ is a strong indication that the cylindrical micelles aligned along $\hat{n}$ in the nematic phase are very long and without defects. The nematic phase is found to exist for a narrow range of the control parameter, below and above the phase is the isotropic phase L_1.

Classically, the elastic energy per unit volume of the nematic phase is written as the sum of three terms corresponding to splay, twist, and bend (with the corresponding rigidity constants K_1, K_2, K_3) [12]:

$$\frac{F}{V} = \frac{1}{2} K_1 [\text{div}(\hat{n})]^2 + \frac{1}{2} K_2 [\hat{n} \cdot \overrightarrow{\text{curl}}(\hat{n})]^2$$

$$+ \frac{1}{2} K_3 [\hat{n} \wedge \overrightarrow{\text{curl}}(\hat{n})] .$$

We focus on the splay term. We have sketched in Fig. 5 a splay deformation acting in different situations. On a phase of "infinite" rods (Fig. 5a), splay induces a large gradient in the local concentration which is unfavorable energetically: this is reflected in a large value of K_1. This large value of K_1 limits the amplitude of the splay deformation, thus hindering the melting of the nematic order. In a phase of rods with possible free ends (Fig. 5b), splay can occur with almost no gradient of concentration. Thus K_1 is correspondingly smaller and the thermally induced splay deformation prevents the occurrence of a nematic ordering. This can explain that, at the lowest value of the control parameter (below the nematic phase), the L_1 phase

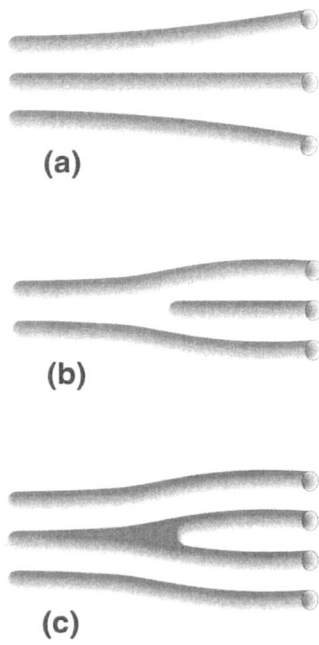

Fig. 5 Possible melting of the nematic order arising from thermal splay fluctuations. a) for "infinite" rods splay leads to a strong gradient of concentration; b) if free ends exist, splay can occur with a small gradient of concentration; c) if connections exist then again splay can occur with almost no gradient of concentration

exists with micelles of finite lengths. On a phase of rods with possible connections (Fig. 5c) in an analogous way, splay can occur with almost no gradient of concentration.

K_1 is correspondingly small and the thermally induced splay deformation allows the melting of the nematic order. Thus, when the control parameter is further increased (above the nematic phase) the nematic order disappears and a isotropic phase is recovered, presumably with connected cylindrical micelles.

The third experimental fact is the evolution of the viscosity of the micellar phase L_1 when increasing the control parameter as illustrated in Fig. 2. At the lowest values of the control parameter (alcohol/TA in the example shown) the micelles are small and the viscosity low. As the control parameter increases, the micelles grow to long flexible micelles which soon (the concentration is high) form an entangled network of long micelles (analogous to an entangled network of equilibrium polymers [13, 14]. Correspondingly, the viscosity increases steeply. Then, as explained above, connections become favorable and the entanglements are progressively replaced by connections. An entanglement is a topological constraint to the relaxation of stress while a connection is free to slide along the cylindrical micelles. Thus, the viscosity decreases as entanglements are replaced by connections [15]. A similar evolution of the viscosity has been observed in systems where the control parameter is temperature [16] or salinity [17].

In conclusion, insertion of connected cylinders in the now classical sequence of local structures for a surfactant aggregate provides an explanation for a certain number of puzzling experimental facts. However, direct evidence for the existence of these connections must still be sought.

References

1. Mazer N, Benedek G, Carey MC (1976) J Phys Chem 80:1075; Missel P, Mazer N, Benedek G, Carey M (1983) J Phys Chem 87:1264
2. Porte G, Appell J, Poggi Y (1980) J Phys Chem 84:3105; Porte G, Appell J (1981) J Phys Chem 85:2511; Appell J, Porte G, Poggi Y (1982) J Colloid Interface Sci 87:492
3. Cates ME, Candau SJ (1990) J Phys Cond Matt. 2:6869 (and refs. therein)
4. Porte G (1992) J Phys Condens Matter 4:8649–8670 (and refs. therein)
5. Roux D (1991) Physica A 172:242–257
6. Israelachvili JN, Mitchell DJ, Ninham BW (1976) J Chem Soc Faraday Trans 2, 72:1525; Israelachvili JN 1985 "Intermolecular and surface forces" (Academic Press, New York)

7. Porte G, Gomati R, El Haitamy O, Appell J, Marignan J (1986) J Phys Chem 90:5746–5751 (and refs. therein)
8. Porte G, Appell J (1985) In: Physics of Amphiphiles: Micelles, vesicles and microemulsions Degiorgio V, Corti M (eds) North Holland Pub 461–468; (1983) J Phys Lettr 44:689–695
9. Dauverchain E (1988) Thesis Université des Sciences et Techniques du Languedoc Montpellier
10. Berret JF, Roux DC, Porte G, Lindner P (1994) Europhysics Lett 25:521–526
11. Cates ME, Drye TE (1991) J Chem Phys 96:1367
12. De Gennes PG (1974) "The physics of liquid crystal" Clarendon Press Oxford

13. Petschek RG, Pfeuty P, Wheeler JC (1986) Phys Rev A 34:2391 (and refs. therein)
14. Cates ME (1987) Macromolecules 20:2289; (1988) J Phys (France) 49:1593
15. Lequeux F (1992) Europhysics Lett 19:675
16. Richtering WH, Burchard W, Jahns E, Finkelmann H (1988) J Phys Chem 92:6032–6040
17. Khatory A, Kern F, Lequeux F, Appell J, Porte G, Morie N, Ott A, Urbach W (1993) Langmuir 9:933–939

Progr Colloid Polym Sci (1994) 97:237–239
© Steinkopff-Verlag 1994

MICROEMULSIONS

G. J. M. Koper
J. Smeets

Clustering in microemulsions: aggregation of aggregates

Received: 16 September 1993
Accepted: 17 November 1993

Dr. J. Smeets (✉)
Department of Physical and
Macromolecular Chemistry
Leiden University
2300 RA Leiden, The Netherlands

Dr. G. J. M. Koper
Gorlaeus Labs.
P.O. Box 9502
2300 DA Leiden

Abstract A model is presented that describes the clustering of microemulsion droplets from a thermodynamic point of view. It is based on two assumptions: one assumption is that the microemulsion droplets aggregate linearly. The other assumption is that the chemical potential per droplet is equal for all sizes of aggregates. By means of the fundamental thermodynamic equations of self-association the distribution of aggregates can then be calculated. The predictions of this model are in agreement with the available experimental data.

Introduction

Droplet phase microemulsions can exist in the form of small pockets of water, coated with a monomolecular layer of a suitable surfactant, dispersed in an apolar solvent. Our work is usually performed with sodium di(2-ethylhexyl) sulfosuccinate (AOT) as a surfactant and 2,2,4-trimethyl-pentane (isooctane) as the solvent. Many physical observables, such as low shear viscosity, the dielectric constant, and the electro-optical birefringence, increase dramatically as a function of both the volume fraction and the temperature, see for instance [1] and references therein. A percolation transition to a conducting phase occurs at a temperature-dependent volume fraction [2]. Although the dynamic percolation phenomena are abundantly discussed, see for instance [3] for a recent account, the temperature dependence of the percolation threshold has, to our knowledge, not been addressed from a thermodynamic point of view. In this paper, we put forward a simple model that is capable of describing the "temperature-dependent clustering phenomena" as occurring in droplet phase microemulsions.

Aggregation model

The analysis given below more or less follows chapter 16 of ref. [4] specialized to the particular situation at hand. The system under consideration has a volume fraction φ of droplets of which the volume fraction single (isolated) droplets is φ_1. The volume fraction of droplets aggregated into k-clusters is φ_k. The total number of droplets is conserved, hence $\sum_{k=1}^{\infty} \varphi_k = \varphi$. All clusters are in equilibrium with one another; in particular, one has

$$nC_1 \rightleftharpoons C_n, \tag{1}$$

which states that n single droplets can cluster into one single n-cluster. When μ_k is the chemical potential of the droplets in k-clusters, the equilibrium condition implies

$$\alpha n \mu_n + m \beta \mu_m = (\alpha n + \beta m) \mu_{\alpha n + \beta m} \tag{2}$$

for all non-negative integer values of α, β, n, and m. This relation is satisfied when the chemical potential per droplet is the same for all types of clusters, i.e., $\mu_n = \mu_1$ for all $n \geq 1$.

For dilute systems the relation between the chemical potential per-droplet of k-clusters and their volume fractions is

$$\mu_k = \mu_k^\circ + \frac{k_B T}{k} \log \frac{\varphi_k}{k}, \qquad (3)$$

where k_B is Boltzmann's constant and T is temperature. Since the chemical potentials per droplet for all types of clusters are identical, we can derive a relation between the volume fraction of droplets in k-clusters and their standard chemical potentials

$$\varphi_k = k[\varphi_1 e^{(\mu_1^\circ - \mu_k^\circ)/k_B T}]^k \quad \text{for} \quad k \geq 1 . \qquad (4)$$

In order to proceed assumptions have to be made about the nature of the clusters. The first assumption is that the droplets aggregate linearly, they do not form (two-dimensional) sheets or (three-dimensional) compact structures. For low volume fractions this assumption is supported by observations with freeze fracture electron microscopy [5]. To be precise: a linear k-cluster is formed by k droplets that are connected by $k-1$ bonds. Let B be the free energy per bond relative to isolated droplets, then the standard chemical potential per droplet is given by

$$\mu_k^\circ = \mu_1^\circ + \frac{k-1}{k} B . \qquad (5)$$

Using this expression in Eq. (4) for the k-cluster volume fraction yields

$$\varphi_k = k\varphi_0 \left(\frac{\varphi_1}{\varphi_0} \right)^k , \qquad (6)$$

with $\varphi_0 \equiv \exp(B/k_B T)$. The conservation of droplets can now be used to express the volume fraction of single droplets solely as a function of both the volume fraction φ of droplets and the bond free energy B. The sum over cluster size converges because $\varphi_k \leq \varphi < 1$ and one obtains

$$\varphi = \varphi_1 \Big/ \left(1 - \frac{\varphi_1}{\varphi_0} \right)^2 . \qquad (7)$$

Solving the above equation for φ_1 gives

$$\varphi_1 = \varphi_0 \frac{1 + 2\varphi/\varphi_0 - \sqrt{1 + 4\varphi/\varphi_0}}{2\varphi/\varphi_0} \qquad (8)$$

(the positive root leads to unphysical volume fractions). The Boltzmann factor φ_0 plays the role of a "critical droplet volume fraction" much in the same way as there is a critical micelle concentration in micellar systems.

For concentrated systems Eq. (3) is no longer valid, mainly because of a reduction of the free volume available to the droplets. This effect can be accounted for by

adding a term $(1 - \varphi/\varphi_m)^p$ where φ_m is the volume fraction of random close packing of clusters and p is some power accounting for the fractal nature of the clusters. This results in a redefinition of the critical cluster volume fraction and it is the origin of the critical behavior for the physical observables as has been reported many times, see for instance [2, 3].

Application of the model

We shall now apply the model to interpret data from two different experiments on the water/AOT/isooctane microemulsion at water to surfactant ratio $[H_2O]/[AOT] = 25$, namely, the low frequency dielectric permittivity and the low shear viscosity.

Neglecting interparticle interactions the relative low frequency dielectric permittivity ε_r of a suspension of spherical droplets is given by the Clausius-Mossotti relation [6]. For higher volume fractions deviations occur that can be evaluated systematically in the following way

$$\frac{1}{\varphi} \frac{\varepsilon_r - 1}{\varepsilon_r + 2} = \alpha_p + \frac{4\pi\varphi}{V} \int_0^\infty dr\, r^2 g(r; \varphi, T) \{\alpha(r) - \alpha_p\} , \qquad (9)$$

with α_p being the electrical dipole polarizability per unit volume of a single droplet. V is the droplet volume, $g(r; \varphi, T)$ the droplet pair correlation function, and $\alpha(r)$ the directional average of the polarizability (per unit volume) of two spheres with center to center separation r. The function $\alpha(r) \to \alpha_p$ for $r \to \infty$ and the difference $\alpha(r) - \alpha_p$ is by numerical calculation found to be only nonvanishing and sharply peaked for separations r very near to contact. Therefore, the linear term in φ is almost exclusively proportional to the fraction of bound droplets. For low volume fractions the fraction of bound droplets is given by (each k-cluster has $k-1$ bonds)

$$\frac{1}{\varphi} \sum_{k=1}^\infty (k-1) \varphi_k \simeq \frac{2\varphi}{\varphi_0} , \qquad (10)$$

which leads to

$$\frac{1}{\varphi} \frac{\varepsilon_r - 1}{\varepsilon_r + 2} = \alpha_p + A\varphi e^{-B/k_B T} , \qquad (11)$$

with A being a constant. This expression has been fitted to the dielectric permittivity data and one finds approximately $B = 30.6 k_B T_a - 31.9 k_B T$ at room temperature $T_a = 298\ K$ [1].

Compared to electrical dipole interactions hydrodynamic interactions are much more complicated and also higher numbered aggregates do contribute in the hydrodynamic case so that one is forced to write for the friction moment, the hydrodynamic equivalent of the polarizability, a higher order polynomial in the droplet volume fraction [7, 8]

$$\frac{1}{\varphi}\frac{\eta_r - 1}{\eta_r + \frac{3}{2}} = \mu_p + \varphi(a_{1,0} + a_{1,1}e^{-B/k_BT})$$

$$+ a_2(T)\varphi^2 + \dots \,, \tag{12}$$

where we shall call μ_p the friction moment of a single droplet; $\mu_p = 1$ for a rigid sphere. Using this expression to fit the viscosity data gives approximately the same binding free energy as for the dielectric permittivity [1, 9].

This free energy consists of an enthalpic part and an entropic part that almost balance. In fact, they balance around 14 °C and, indeed, below this temperature aggregation disappears [9]. The origin of the binding free energy has been discussed by Van Dijk et al. [10]. They argue that at contact the surfactant tails interdigitate thereby excluding the solvent molecules. The number of surfactant molecules that are involved in this process is of the order of 30 which makes both the binding enthalpy as well as the binding entropy of the order of k_BT, which is quite reasonable. Moreover such a scheme predicts that the binding enthalpy and entropy scale with droplet size. This has also been verified by both Van Dijk et al. [10] and by Smeets et al. [9].

Acknowledgements Discussions with D. Vollmer, D. Bedeaux, J. Lucassen, and M. Borkovec are gratefully acknowledged. Part of this work has been performed under the auspices of the EC Network *Thermodynamics of Complex Systems* (contract nr. CHRX-CT92-0007).

References

1. Bedeaux D, Koper GJM, Smeets J. (1993) J Physica A 194:105–113
2. van Dijk MA, Casteleijn G, Joosten JGH, Levine YK (1986) J Chem Phys 85:626–631
3. Boned C, Peyrelasse J, Saidi Z (1993) Phys Rev E 47:468–478
4. Israelachvili J (1992) Intramolecular & Surface Forces, Academic Press Ltd., London
5. Vollmer D, private communication
6. Böttcher CJF, Bordewijk P (1987) Theory of Electric Polarization, Vol. 1 (Elsevier, Amsterdam)
7. Saitô N (1950, 1952) J Phys Soc Japan 5:4; 7:447
8. Bedeaux D (1987) J Coll Int Sci 118:80
9. Smeets J, Koper GJM, van der Ploeg JPM, Bedeaux D (1994) Langmuir 10:1387–1392
10. Van Dijk MA, Joosten JGH, Levine YK, Bedeaux D (1989) J Phys Chem 93:2506–2512

Progr Colloid Polym Sci (1994) 97:240–242
© Steinkopff-Verlag 1994

MICROEMULSIONS

P. Lianos
D. Papoutsi

TiO₂ microemulsions gels obtained by the sol-gel method using titanium isopropoxide

Received: 16 September 1993
Accepted: 30 November 1993

P. Lianos (✉) · D. Papoutsi
University of Patras
School of Engineering
26500 Patras, Greece

Abstract Titanium isopropoxide was hydrolyzed in some water-in-oil microemulsions to produce TiO₂ gels. Photophysical methods have been used to follow the evolution of the gels and to compare the effect of the different microemulsions used. Dip-coating of various substrates has also been examined. It was found that the formation of the TiO₂ gels is a factor assisting adhesion and spreading of the microemulsions on the substrates.

Key words Titania – microemulsion – gels

Introduction

The sol-gel method is a low temperature chemical process for producing inorganic glasses and ceramics. It consists of two major steps: 1) the hydrolysis of a metal alkoxide precursor and 2) the polymerization of the hydroxide produced by hydrolysis to give the corresponding oxide [1, 2]. In this work titanium isopropoxide has been used as a precursor of TiO₂. Water-in-oil microemulsions have been employed as support material to avoid Ti (OH)₄ precipitation. Microemulsion gels are then formed during polymerization.

Experimental

Titanium (IV) Isopropoxide, Ti[OCH(CH₃)₂]₄, (Aldrich); Triton X-100 (Aldrich); sodium dodecylsulfate, SDS (Fluka); n-pentanol (Fluka); cyclohexane (Fluka); methanol and ethanol (Merck); tris (2,2′-bipyridine) ruthenium dichloride hexahydrate, Ru(bpy)₃²⁺ (GFS Chemicals) and pyrene (Fluka) were of the best quality commercially available and were used as received. High purity Millipore water was used in all sample preparations.

Solutions were obtained by the following procedures: 1) w/o microemulsions in cyclohexane were prepared by mixing precalculated quantities of SDS, water and cyclohexane and finally adding n-pentanol. The weight ratio of water to surfactant was always 2.5. Several water (SDS) contents have been tried. Titanium (IV) isopropoxide was added last under vigorous stirring. Optimum conditions for making clear gels were obtained when the molar ratio of water/alkoxide/SDS was 0.4/0.2/0.01 M. 2) Reversed micelles were obtained by mixing precalculated quantities of cyclohexane, Triton X-100 and water. Titanium (IV) isopropoxide was then added under vigorous stirring. Optimum conditions were obtained for the following molar ratio: water/alkoxide/Triton: 0.4/0.2/0.2 M. 3) Water dispersions in the presence of SDS were also obtained in pure pentanol [3] without cyclohexane. Again the alkoxide was added last. The molar ratio of water/SDS was always 50 while different water/alkoxide ratios have been tried.

Dip-coating was carried out in free air at a withdrawal speed of 7 cm/min. The coated substrates were left to dry in air for several hours. Some of them were then brought to 400–450 °C at the rate of 1.5 °C/min. In that case all organic residue was pyrolyzed and the film contained only titania.

Results and discussion

When titanium alkoxide is introduced into a w/o microemulsion it dissolves in the continuous phase. By then

Progr Colloid Polym Sci (1994) 97:240–242
© Steinkopff-Verlag 1994

interacting with the water droplets it is hydrolyzed. Even though the process is complicated, a complete hydrolysis should obey the following scheme:

$$Ti[OCH(CH_3)_2]_4 + 4H_2O$$

$$\rightarrow Ti(OH)_4 + 4(CH_3)_2CHOH$$

Since, however, no $Ti(OH)_4$ precipitate results from the mixtures we actually used, we believe that the hydrolysis is incomplete and it is paralleled by polymerization according to the following scheme, which is, again, a simplification:

$$Ti(OH)_4 \rightarrow TiO_2 + 2H_2O .$$

During this process –O– bridges are developed between the Ti atoms in the oxide. Hydrolysis and polycondensation then start immediately after addition of alkoxide and they go on for a time that depends on the nature, i.e., the content, of each solution. A rough estimate of the condensation time may be obtained by visual inspection of the gelation of the solutions. We have then found that both cyclohexane-SDS-pentanol- and cyclohexane-triton-based w/o microemulsions become gels within a few hours following the alkoxide addition. On the contrary, a few days are needed for the pentanol-SDS-water dispersions. We have also utilized an analytical means to monitor the evolution of each sample. We have introduced $Ru(bpy)_3^{2+}$ into the solutions before adding alkoxide and measured the luminescence decay time τ of this substance before and several times after addition of $Ti[OCH(CH_3)_2]_4$. The results are given in Fig. 1. The original solutions were aerated, i.e., they contained oxygen gas. The luminescence

decay time of $Ru(bpy)_3^{2+}$, which is of the order of several hundred nanoseconds, is sensitive to both the oxygen content and the viscosity of the probe environment. Oxygen is actually a quencher of the probe excited state. Thus, the gelling of the solutions is followed by an increase of the probe excited-state lifetime, because the mobility of oxygen is then reduced. At the same time the motions (rotational and translational) of the probe itself is highly reduced and with it all the non-radiative processes of deexcitation are reduced. Before alkoxide addition, τ was relatively small. These low values are due to oxygen quenching. Indeed, solutions deoxygenated by the freeze-pump-thaw method gave much larger τ-values. Following alkoxide addition τ increased. The increase was very fast in the case of Triton X-100, slower in the case of SDS microemulsions, and much slower in the third sample. This result verified the conclusions obtained by visual inspection of the solutions. The gelation process is then rapid in both triton and SDS microemulsion, but it is much slower in the alcoholic solutions. Gelation is followed by a process of evaporation of the volatile components, shrinkage and formation of xerogels. Both microemulsions gave xerogels which were transparent glasses. The values of τ in xerogels decreased again. We believe that this decrease is due to quenching by excited-electron transfer to TiO_2. This process has been previously observed [4] and it is exploited for photosensitization of TiO_2 [4].

The particular behavior of the samples made with pentanol, SDS and water should be mentioned. The presence of SDS is necessary in order to avoid precipitation upon alkoxide addition. When the solution is placed into a plastic cuvette, covered with perforated aluminum foil, it takes several days before it gels. Next, it becomes opaque and it remains a soft gel practically forever. If kept in the cuvette, it never becomes a xerogel, it does not suffer from shrinkage and it does not crack.

If care is taken so that the process of evaporation is slow, then the microemulsion gels are always transparent. This is particularly true for the optical measurements. We present an example, using a fluorescent probe, pyrene. The vibronic structure of the fluorescence spectrum of pyrene is sensitive to environmental polarity [5, 6]. Thus, as shown in Table 1, the ratio of the intensity of the first over the third vibronic peak increases on going from cyclohexane

Fig. 1 Evolution of the decay time of $Ru(bpy)_3^{2+}$ during gelation: – – – – Triton-based microemulsion ——— SDS-cyclohexane-pentanol quaternary microemulsion; and .... Pentanol-SDS water dispersion.

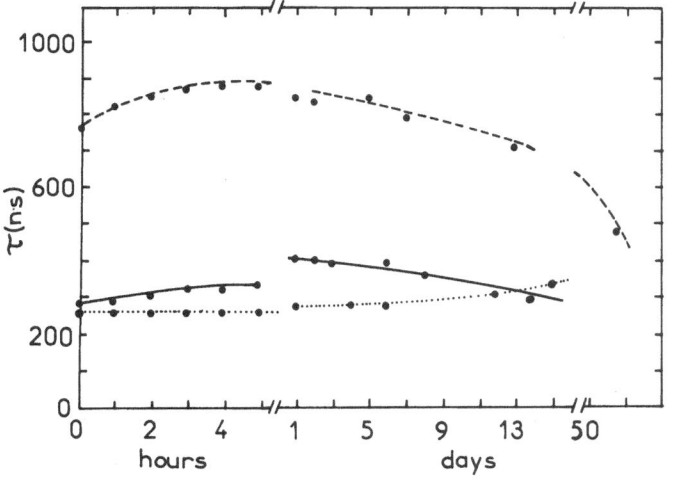

Table 1 Values of I_1/I_3 and τ pyrene monomer solubilized in various solvents.

Solvent	I_1/I_3	τ (ns)
Cyclohexane	0.56	418
Triton X-100 based microemulsion	1.00	366
Ethanol	1.26	308

to ethanol, while in the triton-based microemulsion gel it occupies an intermediate position. The increase of the intensity of the first vibronic peak of the pyrene fluorescence spectrum in polar solvents is due to the partial decrease of the symmetry of the pyrene molecule as a result of multipolar interactions. The decrease of symmetry also results in smaller decay times in polar solvents. The values of the decay time τ of pyrene are also seen in Table 1. We note that τ decreases on going from cyclohexane to ethanol while in the microemulsion it has an intermediate value.

Microscope glass slides have been coated by dip-coating using microemulsion gels at their early stages of gelation. There are some rules of thumb that one needs to follow to obtain "good" films. The faster the speed of withdrawal of the dipped substrates, the thicker the film obtained. The speed of withdrawal must be constant in order to obtain uniform films. The solution used must be macroscopically homogeneous and clear. Films are thinner and more uniform at a very early stage of gelation compared to those obtained at later stages. The extent of gelation is the most fundamental factor while the type of water dispersion used is less significant. A sufficient degree of TiO₂ polymerization is also necessary in order to obtain a good film. *If a water dispersion without alkoxide is used the material does not adhere on the substrate.* TiO₂ polymerization is thus a basis for obtaining organic films on the glass slide. This important procedure which can be used to make thin films of amphiphilic substances, is studied further in our laboratory.

The thickness of our samples was measured to be of the order of some hundreds of nanometers. In some cases the thickness can be estimated by the position of the interference bands, as seen in Fig. 2. In the same figure, we can see the characteristic absorption (< 340 nm) of TiO₂. It must be noted that TiO₂ absorption was the same before and after the pyrolysis of the organic components. This does not constitute a proof that TiO₂ is already formed in gel, since incompletely hydrolyzed alkoxide can be further hydrolyzed when exposed to atmospheric moisture. However, it is important in this respect that the film does not suffer from abrupt changes, when exposed to air, other than the evaporation of the volatile components.

X-ray measurements on films heated to $< 400\,°C$ revealed a rutile structure.

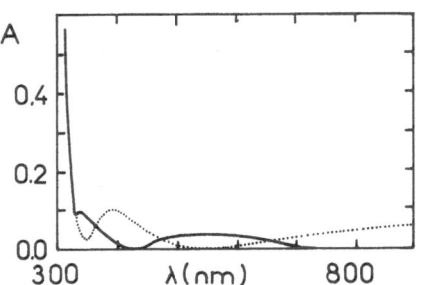

Fig. 2 Typical absorption spectrum of a film obtained before (————) and after (. . . .) heating up to 400°C.

Some films made on glass slides were analyzed with a scanning electron microscope after they have been heated to 400–500 °C. The coverage of the substrate was incomplete with rather interesting geometrical formations. It should be noted however that the obtained geometry of the particles adhering on the substrates was not reproducible and it was a question of art rather than exact science. Nevertheless, it is our feeling that it depended mainly in the extent of gelation and secondarily on the type of original solution used.

Conclusion

When Ti[OCH(CH₃)₂]₄ is dissolved in solutions containing fine water dispersions it is hydrolyzed and polymerized resulting to the gelling of the solutions. Gelation is obtained in a few hours in the case of cyclohexane-triton- and cyclohexane-SDS-based w/o microemulsions while in the case of pentanol-SDS water dispersions gelation was very slow. The evolution of gelation is monitored by observing the luminescence decay characteristics of Ru(bpy)₃²⁺. The microenvironment of the gels is also probed by observing pyrene spectra and fluorescence lifetimes. At the early stages of gelation substrates can be coated by dipping them in the solutions. The obtained films contain TiO₂. Films heated to 400–500 °C are made of rutile which partially covers the substrate.

Acknowledgements We acknowledge financial help from the program ΓΓΕΤ ΠΕΝΕΔ 91 ΕΔ 815.

References

1. Segal D (1989) Chemical Synthesis of Advanced Ceramic Materials, Cambridge University Press, New York
2. Hench LL, West JK (1990) Chem Rev 90:33–72
3. Friberg SE, Yang CC, Sjoblom J (1992) Langmuir 8:372–376
4. Duonghong D, Borgarello E, Gratzel M (1981) J Am Chem Soc 103:4685–4690
5. Nakajima A (1971) Bull Chem Soc Jpn 44:3272–3277
6. Lianos P, Georgiou S (1979) Photochem Photobiol 30:355–362

Progr Colloid Polym Sci (1994) 97:243–246
© Steinkopff-Verlag 1994

D. Papoutsi
W. Brown
P. Lianos

Effect of polyethylene glycol of varying chain length on cyclohexane-pentanol-sodium dodecylsulfate water-in-oil microemulsions

Received: 16 September 1993
Accepted: 30 November 1993

P. Lianos (✉) · D. Papoutsi
University of Patras
School of Engineering
Physics Section
26500 Patras, Greece

W. Brown
Department of Physical Chemistry
University of Uppsala
Box 532
75121 Uppsala, Sweden

Abstract Polyethylene glycol of different molecular weights ranging from a few hundred to a few hundred thousand was solubilized in water-in-oil microemulsions made of cyclohexane, pentanol, sodium dodecylsulfate, and water. The effect of the presence of polymer was studied by conductivity, fluorescence probing, and light scattering. We have found that the small size polymer affects only the water-oil interface while the large size polymer induces a percolation structure by passing through an initial growth of droplet size.

Key words Polyethylene – glycol – microemulsions

Introduction

Water-soluble polymers interact with micelles and microemulsions and affect their structure. The present paper is a continuation of two previous works [1, 2] where we examine the effect of a simple and well-known water soluble polymer, polyethylene glycol, on a quaternary water-in-oil microemulsion made of sodium dodecylsulfate, 1-pentanol, water and cyclohexane. In the first paper [1] it was found that even a relatively small quantity of the polymer produces large changes in the microemulsion both at low and high water contents. When cyclohexane was replaced by toluene [2] the effect of the polymer was qualitatively the same. In the present work, we study the effect of the size of the polymer chain and the polymer concentration on the above microemulsion. The water content was maintained such that the microemulsion is above the electrical percolation threshold [3]. We have used similar techniques as those in the previous two papers [1, 2], namely, conductivity, fluorescence probing, and dynamic light scattering.

Experimental

Cyclohexane (Fluka, UV spectroscopy), 1-pentanol (Fluka, puriss), sodium dodecylsulfate, SDS (BDH, for biochemistry), tris (2,2'-bipyridine) ruthenium dichloride hexahydrate, Ru $(bpy)_3^{2+}$ (GFS Chemicals), potassium hexacyanoferrate (III), $Fe(CN)_6^{3-}$ (Merck), polyethylene glycol, PEG, of different molecular weights (Fluka) were used as received.

The composition of the microemulsion (wt%) was 7.1 SDS, 17.8 water, 16.7 1-pentanol and 58.4 cyclohexane. This composition was chosen from the pseudoternary phase diagram close to the demixion line. Solutions were prepared and measurements were made as in previous publications [1, 2]. All measurements were performed at 25 °C.

Results and discussion

Table 1 shows conductivity data. We note that the conductivity of the microemulsion (above the electrical per-

Table 1. Values of Electrical Conductivity of Polymer Containing Microemulsion vs Polymer Content for Various Polymer Molecular Weights.

PEG content wt%	MW 200	600	Conductivity, μS cm^{-1} 1000	2000	6000	35 000	300 000
0	243	243	243	243	243	243	243
0.5	–	–	–	71	–	–	–
0.73	307	315	166	–	90	54	–
1	–	–	–	60	–	–	–
1.1	406	372	185	–	74	37	23
1.47	421	436	198	–	76	38	–
1.5	–	–	–	182	–	–	–
2.	–	–	–	255	–	–	–
2.2	536	529	288	–	102	73	61
4	–	–	–	917	–	–	–
6	–	–	–	1097	–	–	–

colation threshold) changes in the presence of polymer and the effect depends strongly on the chain length. Up to MW 1000, the conductivity steadily increases with PEG concentration. Above MW 1000 it initially decreases showing a minimum at about 1.5% wt and then increases with PEG concentration.

Table 2 shows data obtained by analysis of the luminescence decay profiles of 2×10^{-5} M Ru(bpy)$_3^{2+}$ in the presence of 10^{-3} M Fe(CN)$_6^{3-}$. The latter is a quencher of Ru (bpy)$_3^{2+}$ luminescence, frequently used with anionic surfactant assemblies [4]. Both luminophore and quencher are dissolved in the dispersed phase. The analysis was carried out by using a "Percolation Model" based on random walk interactions in fractal domains [1, 2]. The luminescence decay curves were fitted by the following equation

$$I(t) = I_0 \exp(-k_0 t) \exp(-C_1 t^f + C_2 t^{2f}), \quad (1)$$

where C_1, C_2, f are constants. f is always smaller than unity. k_0 is the decay rate in the absence of quencher. $\tau_0 = 1/k_0$ was 635 ns. The first-order time-dependent decay rate was obtained through the following equation

$$K(t) = fC_1 t^{f-1} - 2fC_2 t^{2f-1}. \quad (2)$$

In Table 2, data are summarized as a function of polymer content for PEG 35 000. Column 2 shows the effect on the value of the non-integer exponent f which is a measure of the restriction imposed by the environment. Free diffusion in a non-viscous solvent corresponds to f values close to unity. If diffusion is restricted and the quenching reaction is localized, then f exhibits low values. According to these data, f decreases with increasing PEG concentration when the latter is sufficiently low. With further increase in [PEG], f increases again. Column 3 shows the values of K_1, i.e., the decay rate at the first recorded time-channel. K_1 thus corresponds to the quenching reaction probability immediately after excitation of the luminophore. K_1 varies with increasing [PEG] and passes through a maximum.

Table 2 Data Obtained by the Analysis of the Luminescence Decay Profiles of 2×10^{-5} M Ru (bpy)$_3^{2+}$ in the Presence of 10^{-3} M Fe (CN)$_6^{3-}$ at Various Polymer Contents, for PEG 35 000.

PEG content wt%	f	K_1 10^6 S^{-1}	K_{av} 10^6 s^{-1}	K_L 10^6 s^{-1}
0	0.70	6.6	2.2	1.6
1.1	0.64	6.7	1.5	0.72
2.2	0.50	9.8	1.3	0.70
4.4	0.50	8.6	1.3	0.94
6.6	0.64	4.1	1.2	1.04

Column 4 gives K_{AV}, i.e., the average value of the quenching rate over 500 recorded time-channels. K_{AV} decreases monotonically with increasing [PEG]. Column 5 shows K_L, i.e., the quenching rate at the last recorded time-channel and corresponds to the reaction rate at long times. This rate is thus mainly associated with reactant diffusion through the reaction domain, which extends over all microemulsion droplets. K_L should thus also be associated with the electrical conductivity in the water-in-oil microemulsion. As with the electrical conductivity, the K_L-values pass through a minimum when [PEG] increases. Similar data (not shown), as in Table 2, obtained with MW 200 in the place of MW 35 000, did not prove any substantial variation of the above parameters in the presence of polymer.

Table 3 shows dynamic light-scattering data for different chain-length polyethylene glycols. Addition of short chain-length PEG produces no effect on the hydrodynamic radius R_H. Long chain-length PEG however shows two components. One of these is identified with the microemulsion droplets corresponding to the system without polymer added and the other is a species of large apparent size. The contribution of each species depends on the PEG concentration and not on the polymer molecular weight, as long as it is greater than a minimum value lying between 11 000 and 35 000. It is possible that the R_H-value

Table 3 Values of the Diffusion Coefficient D and the Apparent Hydrodynamic Radius R_H at Various Polymer Molecular Weights. Polymer Concentration: 2.2 wt%.

Molecular weight	D, $10^{-11} m^2 s^{-1}$	R_H^a, Å	Normalized contributions (%)
no polymer	2.92	75	
200	2.88	76	
400	3.49	63	
11250	1.91	115	
35000	3.06	72	19.6
	0.76	289	80.3
100000	2.81	78	24.2
	0.62	355	73.2
300000	2.52	87	20.1
	0.60	367	79.5

$^a R_H = kT/6\Pi\eta D$; $\eta = 0.992$ cP.

found for MW 11250 also corresponds to two species, but these are not resolvable.

In our previous works [1,2], we have shown that PEG 6000 has an end-to-end distance of 77 Å. The identity of this value with the size of the microemulsion droplets suggests that when the polymer is equal or larger than the apparent droplet diameter, some droplets are constrained to grow in size.

The above observations can be explained in the following manner: polymer chains of low molecular weight (i.e., < 1000) are much shorter than the diameter of the microemulsion droplets. These can dissolve in the dispersed phase without influencing the drop geometry or their size. However, they have a substantial effect on conductivity, which increases strongly. It is possible that PEG of low MW is solubilized close to the water-oil interface, modifying it and facilitating exchange of ions by droplet fusion. This may explain the observed increase in electrical conductivity in the presence of PEG 200.

The situation changes, however, when the polymer chain becomes sufficiently large since the polymer then affects the microemulsion structure. This is demonstrated in several ways. Above MW 1000 the electrical conductivity initially decreases as polymer is added to the free solution, it passes through a minimum and then increases again. We have previously [1, 2] interpreted a decrease in electrical conductivity by a model in which the droplets grow in the presence of the polymer while their number (and thus the number of their encounters) decreases. An increase in conductivity at high polymer concentrations may be explained by the formation of a percolation cluster throughout the solution. Polymer chains form the skeleton of the cluster while some water, surfactant, and cosurfactant molecules are structured in a pattern associated with the supramolecular polymer configuration. The remainder

form microemulsion droplets. Such a scheme is in agreement with the data of Table 3, where for molecular weights above MW 35000 two components are clearly distinguished.

The luminescence data are quantatively in agreement with this model. Table 2 shows that the f-values pass through a minimum as the PEG concentration increases. The variation in f demonstrates the transition from an electrically percolating structure to a more restricted structure and then finally into a new percolating structure (probably involving a percolation cluster). It should be noted that according to the so-called Alexander–Orbach conjecture [5], $f \approx 0.67$ at the percolation threshold. This theoretical value is consistent with the above interpretation. At low PEG concentrations, the microemulsion droplets grow in size and decrease in number; thus, for a given quencher concentration, the probability K_1 of an encounter between an excited luminophore and a quencher, immediately after excitation, increases. When, however, the percolation cluster is formed the reactants are again dispersed. K_1 thus decreases again on further polymer addition, as seen in Table 2. K_L then follows the variation of the above structure and changes in the same manner as the conductivity. Finally, for the same quencher concentration, K_{AV} decreases monotonically with increasing [PEG]. This finding suggests that the overall reaction efficiency decreases in the presence of the polymer and it is justified by steady-state luminescence intensity measurements (not shown).

Conclusion

When polyethylene glycol is added to the studied microemulsions, effects are produced which depend on the size of the polymer chain. For molecular weights below MW 1000, the polymer serves to modify the water-oil interface with an increase in the electrical conductivity. With large chains comparable to the droplet size or larger, the structure of the microemulsion is altered. At relatively small polymer concentrations the droplets become larger and decrease in number. At higher polymer concentration, a percolating structure is possibly formed with the polymer forming the backbone.

References

1. Lianos P, Modes S, Staikos G, Brown W (1992) Langmuir 8:1054–1059
2. Papoutsi D, Lianos P, Brown W (1993) Langmuir 9:663–668
3. Lagues M, Ober R, Taupin C (1978) J Phys Lett 39:L487–L491
4. Lianos P, Zana R, Lang J, Cazabat AM (1986) In: Mittal KL, Bothorel P (eds) Surfactants in Solution Vol. 6. Plenum Press, New York, pp 1365–1372
5. Alexander S, Orbach R (1982) J Phys Lett 43:L625–L631

Progr Colloid Polym Sci (1994) 97:247–252
© Steinkopff-Verlag 1994

MICROEMULSIONS

Z. Saidi
C. Boned
P. Xans
J. Peyrelasse

Conductivity of ternary microemulsions The pressure-percolation effect

Received: 16 September 1993
Accepted: 5 December 1993

Prof. C. Boned (✉) · Z. Saidi · P. Xans
Laboratoire Haute Pression
Centre Universitaire de Recherche
Scientifique
Avenue de l'Université
64000 Pau, France

J. Peyrelasse
Laboratoire de Physique des Matériaux
Industriels
Centre Universitaire de Recherche
Scientifique
Avenue de l'Université
64000 Pau, France

Abstract The electrical conductivity σ of aqueous ternary water/AOT/undecane micro-emulsions (AOT: sodium bis 2-ethylhexyl sulfosuccinate) was studied at constant temperature T versus pressure (up to 200 bar) and the volume fraction ϕ of dispersed matter (water + AOT). The phase diagram was also determined. The results presented in this paper are given in terms of curves plotted as $\sigma(\phi)$ at a given P and T and $\sigma(P)$ at constant ϕ and T. The curves were analyzed within the framework of percolation theory. The threshold ϕ_c decreases as P increases (at constant T), which corresponds to an increase in interactions. When ϕ and T are kept constant, a threshold pressure P_c, which increases as ϕ decreases, is introduced. The scaling exponents are almost the same as those determined at atmospheric pressure. This reflects the fact that the class of universality of this system is unchanged regardless of pressure.

Key words Microemulsion – percolation – high pressure – electrical conductivity

Introduction

The phenomenon of percolation presented by microemulsions has been known for some time. It was first identified as a result of study of the electrical conductivity σ of a water/cyclohexane/pentanol/sodium dodecylsulfate microemulsion [1]. Subsequently, a considerable volume of research has been carried out on water/AOT/oil ternary systems (AOT: sodium bis 2-ethylhexyl sulfosuccinate). The reader will find a large number of references in a recent review article [2]. This work indicates that when a microemulsion presents the phenomenon of percolation (for it should be recalled that not all microemulsions do so), it seems to obey the dynamic model of percolation of spheres in a continuous medium, one of the characteristics of which is the fact that the value of the percolation exponent below the threshold is -1.2 (whereas in the static case the figure is -0.7). References [3–7] give a good example of this for electrical conductivity and dielectric relaxation in the particular case of water/AOT/oil ternary systems. It has just been shown [8] that the description is generally valid and reflects the fact that these microemulsions belong to the same class of universality. All this research has been carried out at atmospheric pressure and, in order to confirm the generality of the description, it seemed interesting to study these systems at higher pressures. A few studies in the literature are devoted to the influence of pressure on microemulsions. It does not appear [9, 10] that pressure has a measurable effect on droplet radius in the monophasic domain. One study [11] deserves to be mentioned in this context, on pressure effects on the phase behavior of a propylene/water/surfactant mixture. The conductivities of the nominally propylene continuous upper phases in the systems examined are high enough to suggest electrical percolation. However, the phenomenon of percolation proper was not specifically studied as a function of pressure.

Experimental technique

Measurements were made of the electrical conductivity σ versus pressure P and volume fraction ϕ in dispersed matter (water + AOT) at a given temperature T. Because of the influence of pressure on the density of components, ϕ varies slightly with pressure for a given system. The volume fraction at $P = 1$ bar is indicated by ϕ_0 and at pressure P by ϕ_P. The microemulsions studied were water/AOT/undecane type systems which had been studied in previous work [12] at $P = 1$ bar and $T = 20\,°C$. We assumed $n = $ [water]/[AOT] as the value of the molar ratio, and the conductivity measurements were carried out at $20\,°C$. The substances used were distilled water, AOT (Sigma, purity > 99%) and undecane (Fluka purity > 99%). The density of AOT was assumed to be independent of pressure. The density of water and undecane were measured using a DMA 45 Anton Paar KG densitometer to which was added a supplementary DMA 512 cell which allows pressure measurements to be taken up to 400 bar. Use of this cell with double reference calibration has been described in a specialized review [13]. For water at $T = 20\,°C$ and $P = 1$ bar, $\rho = 0.9982$ g/cm^3, and at 150 bar $\rho = 1.0047$ g/cm^3. For undecane at the same temperature $\rho = 0.7399$ g/cm^3 at 1 bar and $\rho = 0.7503$ g/cm^3 at 150 bar. It follows that the microemulsion $\phi_0 = 0.300$ corresponds to $\phi_P = 0.303$ at $P = 150$ bar.

Conductivity is determined using a semi-automatic Wayne-Kerr B 331 precision bridge operating at a frequency of 1591.5 Hz. The measuring cell supplied by Bioblock Scientific (reference C93032) comprises a plane condenser with parallel electrodes. The electrodes are platinum and are completely immersed in the liquid under investigation. The vessel containing the sample is a spiral bellows made of Gaflon, type G00661 supplied by Plastic Omnium, equipped with a stainless steel bridge. The complete apparatus – measuring cell and bellows – is placed in an airtight chamber filled with compression oil and surrounded at its surface by a tube through which circulates water, thermostatically regulated to within 0.1 °C by a Julabo Paratherm regulator. Pressure is transmitted to the sample by means of the bellows using an apparatus including, in particular, an oil compression pump linked to one end of the cylindrical airtight chamber. The sample volume required to fill the chamber is approximately 10 ml. This system was tested at 1 to 1500 bar ($\Delta P = 5$ bar) and for temperature between 10° and 60 °C.

Finally, we also examined the behavior of the phase diagram with a view to locating precisely the domain of existence of the monophasic zone. In order to do this, we used a Rop brand full-vision sapphire cell, the operating performance of which will be found in the literature [14], with which pressure measurements can be carried out up to 500 bar.

Experimental results

1) Determination of the monophasic domain: first we examined the behavior of the phase diagram under pressure for the samples studied, so as to identify the domain of existence of the monophasic zone. Using the sapphire cell the sample was subjected to isothermal pressure variation and the liquid-liquid type transition is clearly identifiable since the fluid becomes opaque and scatters light significantly. The system then gradually separates into two phases. For example with $n = 30$ and $\phi_0 = 0.20$ (at 20 °C) the boundary P–T pairs which mark the passage from single phase microemulsion to two-phases system are: $P = 161$ bar at $T = 20.2\,°C$, $P = 147$ bar at $T = 22.8\,°C$, $P = 121$ bar at $T = 24.8\,°C$ and $P = 24$ bar at $T = 29.7\,°C$. Figure 1 represents these results. It will be noted that the water/AOT/undecane system has a domain which shrinks with increasing pressure, an observation which has already been made elsewhere [15] for similar water/AOT/octane systems. This is also true for other values of ϕ_0.

2) Study of electrical conductivity: figure 2 represents the curve of variations of σ versus P for the $n = 30$, $\phi_0 = 0.15$ at 20 °C system. Figure 3 shows the associated $\log_{10} \sigma = f(P)$ curve. It will be seen that the interval of variations extends over approximately 1.5 decades. Figure 4

Fig. 1 Realm of existence of the monophasic zone ($n = 30$; $\phi_0 = 0.20$; at 20 °C)

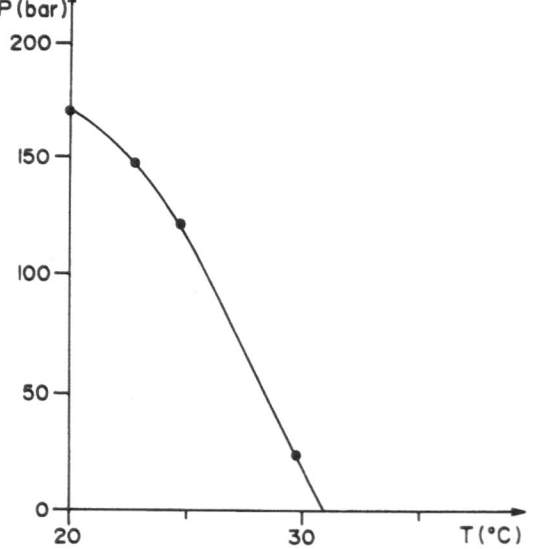

Progr Colloid Polym Sci (1994) 97:247–252
© Steinkopff-Verlag 1994

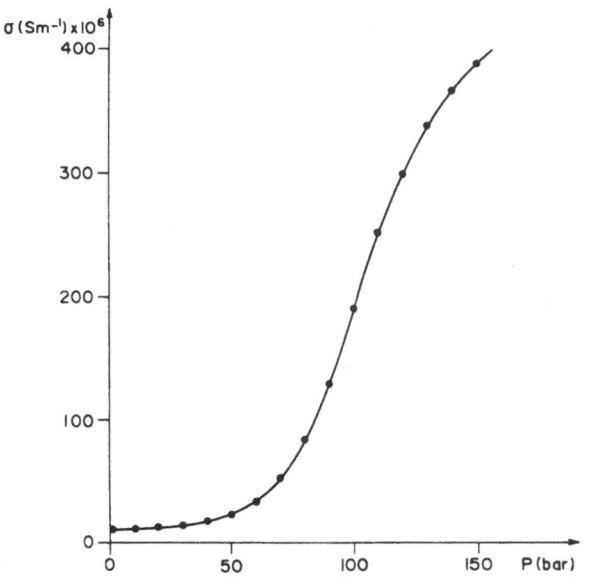

Fig. 2 Variations of σ versus P ($n = 30$; $\phi_0 = 0.15$; $T = 20\,°C$)

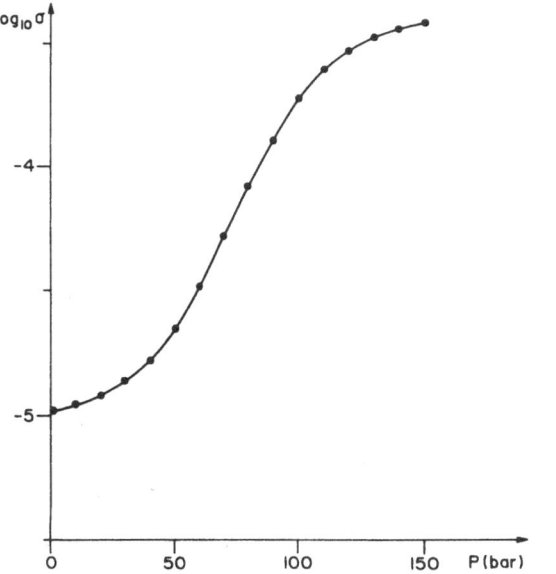

Fig. 3 Variations of $\log_{10} \sigma$ versus P ($n = 30$; $\phi_0 = 0.15$; $T = 20\,°C$)

represents the curves of variations of $\log_{10} \sigma$ versus ϕ_P for pressures 1, 50, 100 and 150 bar. The sigmoid shape of the curves and the large variation interval (approximately 5 decades) will be observed.

Discussion of results

1) Variations at constant pressure P: analysis of the numerous previous studies carried out on percolation indi-

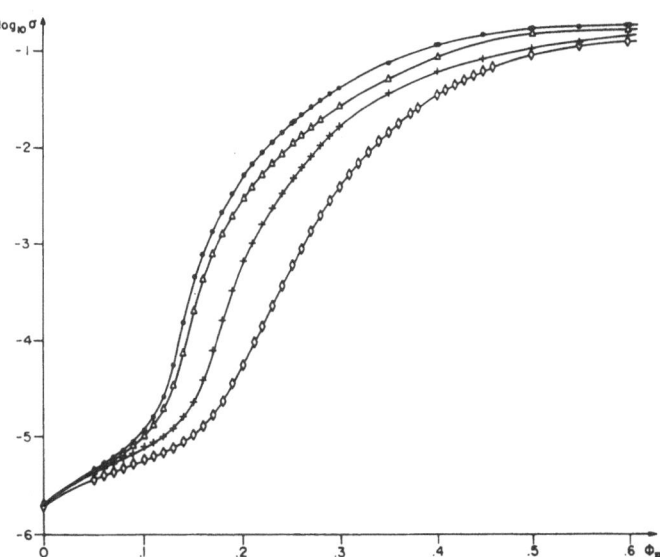

Fig. 4 Variations of $\log_{10} \sigma$ versus ϕ_P ($n = 30$; $T = 20\,°C$) for different values of P. $\diamondsuit$:1 bar; $+$:50 bar; $\triangle$:100 bar; $\bullet$:150 bar

cate that the following relationships can be adopted as asymptotic laws of behavior. It is generally assumed that the system is made up of two components (1) and (2) and that ϕ_P is the volume fraction of component 1 at pressure P,

$$\begin{cases} \phi_P > \phi_c + \delta & \sigma = C_1 \sigma_1 (\phi_P - \phi_c)^\mu \\ \phi_P < \phi_c - \delta' & \sigma = C_2 \sigma_2 (\phi_c - \phi_P)^{-s} \end{cases} \quad (1)$$

in which $\Delta = \delta + \delta'$ is the size of the transition interval (the "cross-over region") which is of the order [16] of $(\sigma_2/\sigma_1)^{(1/(\mu+S))}$. These relationships indicate that $1/\sigma \cdot d\sigma/d\phi_P$ tends to infinity at the percolation threshold ϕ_c. In fact, experimentally, there is a continuous transition within the transition interval Δ close to ϕ_c and $1/\sigma \cdot d\sigma/d\phi_P$ moves through a maximum at the threshold. A recent investigation [8] into dynamic percolation of spheres in a continuum, which is the case with microemulsions, indicates that $\mu = 2 \pm 0.25$ and $s = 1.2 \pm 0.2$; these values clearly express the dynamic aspect of the phenomenon [17, 18]. However this investigation only relates to atmospheric pressure.

The $\log_{10} \sigma = f(\phi_P)$ curves at a given pressure P have a sigmoid shape which merits analysis. To achieve this, by the least squares method, the coefficients of a 4th degree polynomial are adjusted on to ϕ_P in the most sharply rising part of the curve; the value of ϕ_P for which the second derivative of the polynomial is cancelled out is then determined. This value is that of the point of inflection, in other words, at the maximum for $1/\sigma \cdot d\sigma/d\phi_P$ and corresponds to the position of the percolation threshold ϕ_c. Figure 5 represents variations of $1/\sigma \cdot d\sigma/d\phi_P$ versus ϕ_P for

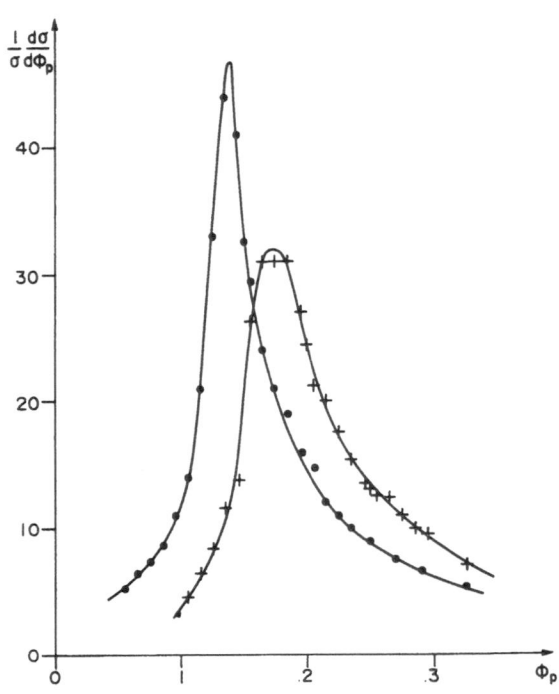

Fig. 5 Variations of $\dfrac{1}{\sigma}\cdot\dfrac{d\sigma}{d\phi_P}$ versus ϕ_P for different values of P (● + :numerical evaluation ●: $P = 150$ bar + : $P = 50$ bar)

Table 1 Values of ϕ_c, μ and s versus pressure ($n = 30$, $T = 20°C$)

P (bar)	ϕ_c	μ	s
1	0.225	2.25	1.36
50	0.176	2.16	1.05
100	0.146	2.14	1.15
150	0.141	2.11	1.15
		Mean value: 2.16	Mean value: 1.18

$P = 50$ and 150 bar ($n = 30$ and $T = 20\,°C$) at which the points correspond to the numerical derivative.

Based on knowledge of ϕ_c the theoretical expressions (Eqs. (1)) are adjusted by the least squares method so as to determine the exponents μ and s. The values are indicated in Table 1. The mean values are $\mu = 2.16$ and $s = 1.18$. Figures 6a and 7a represent the experimental and theoretical $\log_{10} \sigma$ curves for $P = 50$ bar and $P = 150$ bar. We have also represented in Figs. 6b and 7b variations of $(\sigma)^{1/\mu}$ and $(\sigma)^{-1/S}$ versus ϕ_P by normalizing at 1 for $\phi_P = 0$ and $\phi_P = 1$. Finally, curves 6c and 7c represent variations of $\log_{10} \sigma$ versus $\log_{10} |\phi_P - \phi_c|$. For all three types of representation an excellent fit is observed between experimental values and theoretical curves (Eqs. (1)). It can be seen that the representations 6a and 6c and 7a and 7c clearly show the existence of the transition interval Δ when

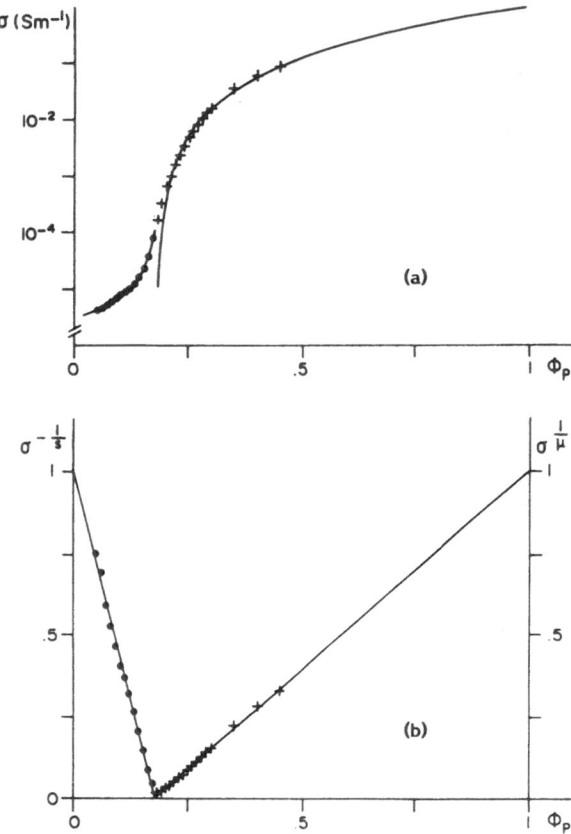

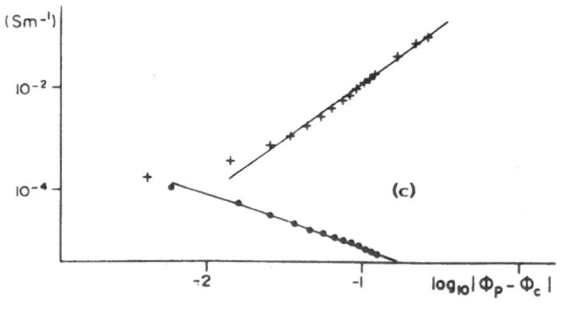

Fig. 6 ($n = 30$; $T = 20°C$; $P = 50$ bar) ●, + :experimental data; — :theoretical curves. a) Variations of $\log_{10} \sigma$ versus ϕ_P. b) Variations of $(\sigma)^{1/\mu}$ and $(\sigma)^{-1/S}$ versus ϕ_P (normalized to 1 for $\phi_P = 0$ and $\phi_P = 1$). c) Variations of $\log_{10} \sigma$ versus $\log_{10} |\phi_P - \phi_c|$

$\phi_P \to \phi_c$. This is not observed in Figs. 6b and 7b because $(\sigma)^{1/\mu}$ and $(\sigma)^{-1/S}$ tend towards zero when $\phi_P \to \phi_c$. Finally, Fig. 8 represents variations of ϕ_c versus P ($n = 30$, $T = 20\,°C$). A clear decrease in ϕ_c is observed with increasing pressure. This result should be linked to the very sharp decrease in ϕ_c as a function of temperature T observed for identical [12] or similar systems [7] at $P = 1$ bar. This decrease reflects an increase in interactions.

2) Variations for variable pressure P: Figure 8 shows that ϕ_c varies with pressure P, whereas μ and s can be con-

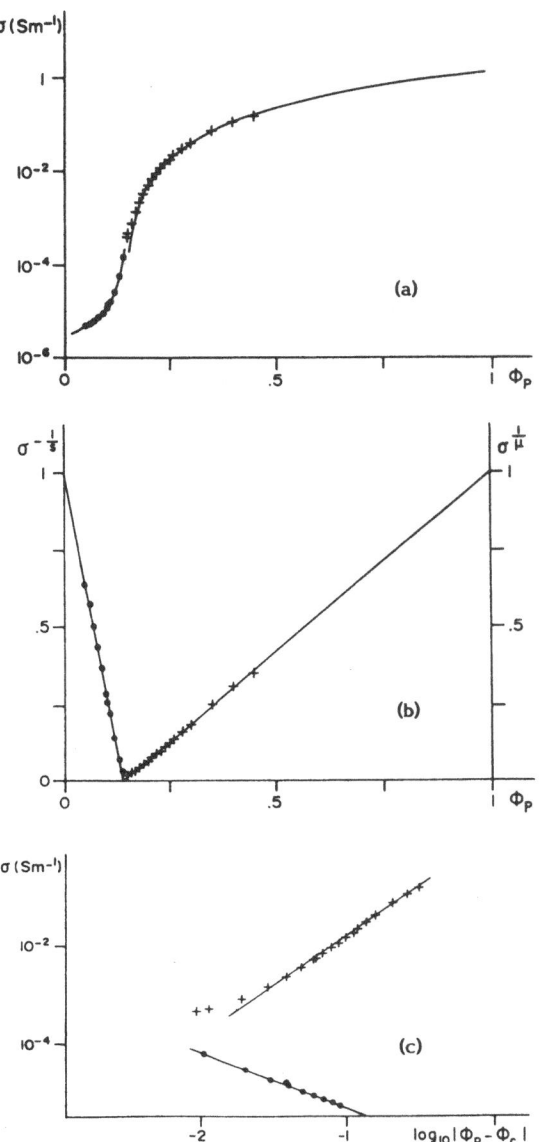

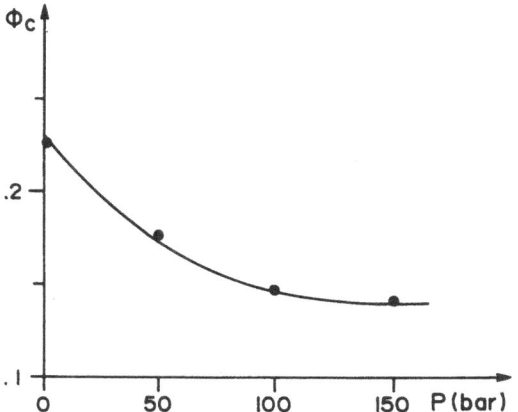

Fig. 8 Variations of the percolation threshold ϕ_c versus P ($n = 30$; $T = 20\,°C$)

Fig. 7 ($n = 30$; $T = 20\,°C$; $P = 150$ bar) ● + :experimental data; — :theoretical curves. a) Variations of $\log_{10} \sigma$ versus ϕ_P. b) Variations of $(\sigma)^{1/\mu}$ and $(\sigma)^{-1/S}$ versus ϕ_P (normalized to 1 for $\phi_P = 0$ and $\phi_P = 1$). c) Variations of $\log_{10} \sigma$ versus $\log_{10} |\phi_P - \phi_c|$

sidered as practically constant. In Eqs. (1) the quantities $\sigma_1, \sigma_2, C_1, C_2$ are a priori functions of pressure. However, if at variable pressure, P, $\phi_c (P)$ remains close to the value ϕ_P, one can write, developing to the first order, $\phi_c (P) = \phi_P + K(P - P_c)$ in which $K = (d\phi_c/dP)_{P_c}$ and $\phi_c (P_c) = \phi_P$, which defines the percolation pressure P_c. If $K < 0$, which corresponds to the case in Fig. 8, then $\phi_P > \phi_c$ if $P > P_c$. It follows that:

$$\begin{cases} \sigma = C_1(P)\,\sigma_1(P)\,[K(P_c - P)]^\mu & \text{if} \quad P > P_c + \delta_P \\ \sigma = C_2(P)\,\sigma_2(P)\,[K(P - P_c)]^{-S} & \text{if} \quad P < P_c - \delta'_P \end{cases} \quad (2)$$

These relationships are not valid in the immediate vicinity of P_c (in other words of ϕ_c) where there is continuous variation over a narrow interval of pressure around the percolation threshold P_c. The cross-over regime is $\Delta P = \delta_P + \delta'_P$. Thus, there are scaling laws for variations of σ with P, with the same exponents μ and s as for variations of σ with ϕ_P. It has already been shown [2, 19] that one also has the same exponents at constant ϕ_P and P with varying temperature T. It should be emphasized that analysis of variations with respect to pressure P with Eqs. (2) is only simple if K is independent of P_c and if σ_1, σ_2, C_1, C_2 are also independent of P, which is not generally the case. Moreover, there is also a small variation of volume fraction with pressure. The result of this is that in fact, for a given sample, the P_c value at the threshold for this sample is itself a function of the pressure, because one has $\phi_P (P)$. This variation will however only exert an influence when the pressure is such that the sample is close to the percolation threshold. It is only if $\sigma_1, \sigma_2, C_1, C_2$ are independent of P that $\log_{10} \sigma = f(\log_{10} |P - P_c|)$ corresponds to two straight lines with slopes μ and $- s$. If this is not the case, then it is difficult or even impossible to determine experimentally the values of μ and s with this experimental pathway (variable P) and generally, achieving a satisfactory analysis of the experimental curves is complicated. Let us recall at this juncture that the same is also true for the experimental pathway: variable temperature T, which can yield curious results that can only be analyzed with caution [19, 20].

Conclusion

In an earlier article [21], we stressed the importance of the experimental pathway. In this work we were able to vary the pressure while maintaining the other parameters constant. Equations (1) show that the distance $[\phi_P(P) - \phi_c(P)]$ is essential. As the threshold varies at each point of the pathway it is possible to move through a percolation point which defines the percolation pressure P_c. Equations (2) show that the scaling exponents remained unchanged. As the prefactors (and also ϕ_P) of the asymptotic equations depend on P, correct analysis of experimental $\sigma(P)$ curves is difficult.

Analysis on the $P = C^{te}$ and variable ϕ_P pathway is much easier. It is this analysis which shows that the dynamic description of the percolation phenomenon applies to the systems studied. The scaling exponents are not dependent on P (≈ 2 above the threshold and ≈ -1.2 below the threshold) but, however, the threshold does depend on P. These results correspond to the universality of the description of the percolation phenomenon for these systems.

References

1. Lagues M, Ober R, Taupin C (1978) J Phys Lett (Paris) 39:L 487–L491
2. Boned C, Peyrelasse J (1991) J Surf Sci Techn 71:1–31
3. Van Dijk MA (1985) Phys Rev Lett 55:1003–1005
4. Bhattacharya S, Stokes JP, Kim MW, Huang JS (1985) Phys Rev Lett 55:1884–1887
5. Peyrelasse J, Moha-Ouchane M, Boned C (1988) Phys Rev A 38:904–917
6. Cametti C, Codastefano D, Tartaglia P, Chen SH, Rouch J (1989) Phys Rev A 40:1962–1966
7. Cametti C, Codastefano D, Tartaglia P, Chen SH, Rouch J (1992) Phys Rev A 45:R 5358–R 5361
8. Boned C, Peyrelasse J, Saidi Z (1993) Phys Rev E 47:468–478
9. Eastoe J, Young WJ, Robinson BH, Steytler DC (1990) J Chem Soc Faraday Trans 86:2883–2889
10. Kaler EW, Billman JF, Fulton JL, Smith RD (1991) J Phys Chem 95:458–462
11. Beckman EJ, Smith RD (1991) J Phys Chem 95:3253–3257
12. Moha-Ouchane M, Peyrelasse J, Boned C (1987) Phys Rev A 35:3027–3032
13. Lagourette B, Boned C, Saint-Guirons H, Xans P, Zhou H (1992) Meas Sci Technol 3:699–703
14. Daridon JL, Saint-Guirons H, Lagourette B, Xans P (1992) High Pressure Research 9:309–312
15. Eastoe J, Robinson BH, Steytler DC (1990) J Chem Soc Faraday Trans 86:511–517
16. Efros AL, Shklowskii BL (1976) Phys Status Solidi B 76:475–485
17. Lagues M (1979) J Phys Lett (Paris) 40:L 331–L 333
18. Grest GS, Webman J, Safran SA, Bug ALR (1986) Phys Rev A 33:2842–2845
19. Saidi Z, Boned C, Peyrelasse J (1992) Progress in Colloid and Polymer Science 89:156–159
20. Mathew C, Saidi Z, Peyrelasse J, Boned C (1991) Phys Rev A, 43:877–882
21. Peyrelasse J, Boned C (1990) Phys Rev A 41:918–953

Progr Colloid Polym Sci (1994) 97:253–255
© Steinkopff-Verlag 1994

MICROEMULSIONS

H. Stamatis
A. Xenakis
F.N. Kolisis
A. Malliaris

Lipase localization in W/O microemulsions studied by fluorescence energy transfer

Received: 6 October 1993
Accepted: 15 December 1993

H. Stamatis · A. Xenakis (✉)
The National Hellenic Research Foundation
Institute of Biological Research
& Biotechnology
48, Vas. Constantinou Ave
11635 Athens, Greece

F.N. Kolisis
National Technical University
Dept. of Chemical Engineering
Athens, Greece

A. Malliaris
N.R.C. "Democritos"
Aghia Paraskevi, Greece

Abstract The localization of *Penicillium simplicissimum* lipase in AOT/isooctane microemulsions has been investigated by fluorescence energy transfer. This method is based on the nonradiative transfer from the tryptophan residues of the enzyme which act as the donor, to cis-parinaric acid which act as the acceptor molecule. The energy transfer efficiency which depends on the distance separating this pair, was examined as a function of the molar ratio of donor to acceptor for different water contents of the microemulsion system.

Key words Microemulsions – reverse micelles – lipase – fluorescence energy transfer

Introduction

Water in oil microemulsion systems have been employed for the solubilization of enzymes which can retain their catalytic activity [1]. The use of enzymes in these systems allows the catalysis of reactions in a direction opposite to that observed in aqueous solutions [2]. It has been reported that lipases can catalyze the esterification of aliphatic alcohols with fatty acids in AOT/isooctane/water microemulsions [3]. The observed lipase selectivity in these systems appeared to be related to the localization of the enzyme molecule within the micellar microstructure [4, 5]. Specifically, enzyme molecules can be localized either in the aqueous core of the reverse micelles or in the region of the micellar membrane. In the present study, in order to confirm the site of the enzyme residence in the reverse micellar system, we have used the fluorescence energy transfer technique.

Experimental

Lipase from *Penicillium simplicissimum* was a generous gift of Dr. U. Menge from GBF, FRG. The enzyme showed a single band in native and SDS PAGE electrophoresis and exhibited a specific activity of 142 units/mg of protein determined by titration of free fatty acids release from triolein. The enzyme molecule contains seven tryptophanyl residues [6]. Bis-(2-ethylhexyl)sulfosuccinate sodium salt (AOT) and 99% oleic acid were purchased from Sigma Chemical Co. Isooctane was purchased from Merck, FRG. 99% cis-parinaric acid (9,11,13,15-*cis,trans,trans,cis*-octadeca-tetranoic acid-*cis*-PnA) was obtained from Molecular Probes, USA. A stock solution of *cis*-PnA in an 100 mM AOT in isooctane can be stored under argon at − 20 °C in the presence of 0.1 mg/l BHT (2,6-di-tert-butyl-4-methylphenol) as antioxidant [7]. These precautions ensure negligible polyene decomposition.

Microemulsions were prepared as described elsewhere [4, 5]. The final w_0 value ($w_0 = [H_2O]/[AOT]$) was adjusted by the addition of the required amount of an 25 mM acetate buffer, pH 5.0, to an isooctane solution containing 100 mM AOT. The final lipase concentration in the microemulsion was $2.6\ 10^{-7}$ M.

The fluorescence energy transfer technique employed here is based on the nonradiative transfer of the excited state energy from the fluorescent amino acid residue of lipase tryptophans which act as the donor molecule to cis-PnA which acts as the acceptor molecule [8].

Energy transfer was examined by measuring the fluorescence of lipase with various concentrations of cis-PnA. Aliquots of this conjugated polyene fatty acid were added from a stock solution to give a final concentration of 0 to $2.8\ 10^{-5}$ M. This procedure was carried out in a series of microemulsions with different w_0 values. The energy transfer efficiency (T) was calculated from the reduction of the emission of the donor according to the relationship [9]

$$T = \left(1 - \frac{F}{F_0}\right) \times 100 , \qquad (1)$$

where F_0 refers to the unquenched fluorescence intensity of the donor and F refers to the fluorescence intensity of the donor in the presence of acceptor.

The fluorescence emission spectra were monitored using a Perkin-Elmer 650-40 fluorometer at 25 °C. The excitation wavelength for cis-PnA and lipase in AOT/isooctane microemulsion were 325 nm and 280 nm, respectively. Absorption spectra of lipase and cis-PnA in AOT/isooctane microemulsions were recorded by a thermostatted Hitachi U-2000 double-beam spectrophotometer.

Results and discussion

In principle, energy transfer between the pair cis-PnA/tryptophan residues of lipase, is possible as proved from the considerable overlap occurring between the absorption spectrum of cis-PnA and the fluorescence spectrum of lipase (Fig. 1). The possibility of such energy transfer in our microemulsions was examined by measuring the fluorescence of lipase in the presence of various cis-PnA concentrations present at constant w_0. The results shown in Fig. 2 indicate that as the concentration of cis-PnA increases the fluorescence of the lipase decreases. This has been attributed to Förster type energy transfer as already published for similar systems in homogeneous media [8].

It is known that in this type of reverse micelles lipase resides in the dispersed phase [5] while the cis-PnA being highly hydrophobic prefers the continuous oil phase. In

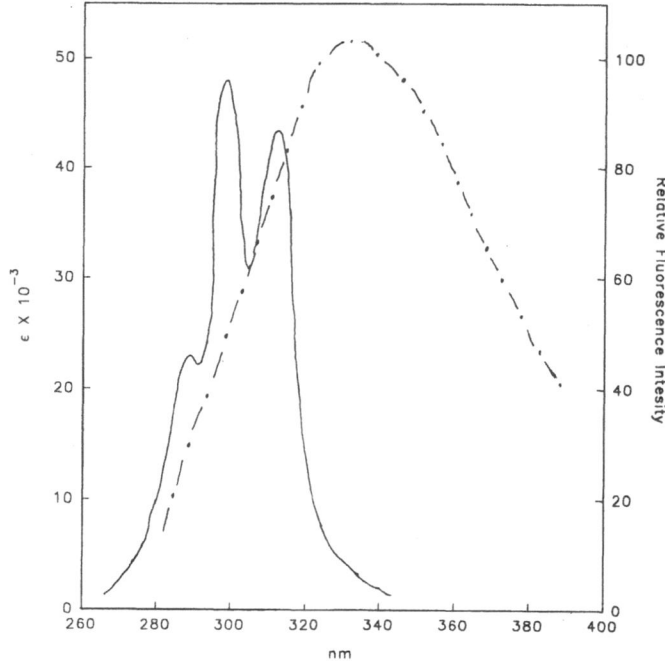

Fig. 1 Absorption spectrum of cis-PnA and fluorescence spectrum of lipase in a 0.1 M AOT/isooctane, $w_0 = 6$, microemulsion. Concentrations of lipase and cis-PnA, 2.6×10^{-7} M and 2.8×10^{-5} M, respectively. $\lambda_{ex} = 280$ nm. T = 25 °C

Fig. 2 Florescence spectra of lipase in the presence of various cis-PnA concentrations, (0, 1.4×10^{-6} M, 2.8×10^{-6} M, 6.9×10^{-6} M, 1.4×10^{-5} M, 2.8×10^{-5} M) in 0.1 M AOT/isooctane, $w_0 = 6$, microemulsions. Experimental conditions as in Fig.1

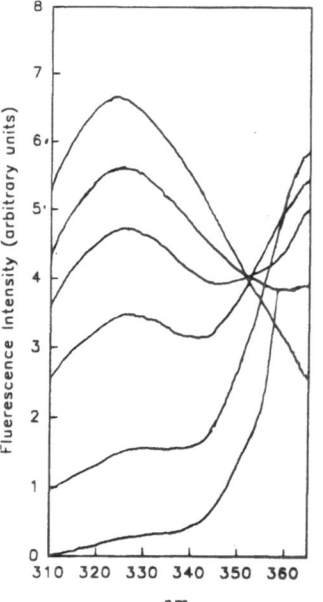

Progr Colloid Polym Sci (1994) 97:253–255
© Steinkopff-Verlag 1994

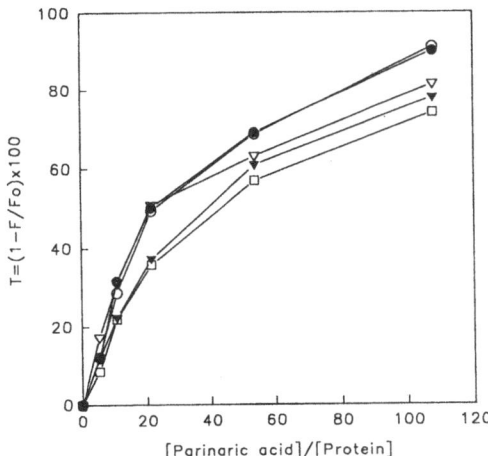

Fig. 3 Energy transfer expressed as $T = (1 - F/F_0)100$ vs. the ratio [cis-PnA]/[lipase], in various 0.1 M AOT/isooctane microemulsions with different molar ratios $w_0 = [H_2O]/[AOT]$. (○): $w_0 = 6$; (●): $w_0 = 9$; (▽): $w_0 = 15$; (▼): $w_0 = 20$; (□): $w_0 = 30$. Concentration of lipase 2.6×10^{-7} M. $\lambda^{ex} = 280$ nm. T = 25 °C

order to determine the precise site of solubilization of the enzyme within the dispersed phase we have measured the fluorescence intensity of lipase at various cis-PnA concentrations and increasing w_0 values, i.e., increasing droplet size. The variation of the energy transfer efficiency T, as a function of the ratio of the concentration of cis-PnA to protein for various w_0 values is shown in Fig. 3. It is seen that T is high for low w_0 values and slightly decreases as larger micelles are formed. This minute decrease of T suggests that the enzyme resides on the average near the interface rather than in the center of the reverse micelle. This is suggested by the fact that if the enzyme was near the center of the droplet then the cis-PnA would have been located at a distance of ca. 50 Å away, which is the average diameter of the droplet, at least at the high w_0 range [10]. On the other hand, it is known that at these distances Förster type energy transfer is practically zero.

These results agree with our previous findings [4, 5] concerning indirect evidence that the solubilization site of this type of lipases is the interfacial region of the reverse micelles.

References

1. Luisi PL, Magid L (1986) CRC Crit Rev Biochem 20:409–474
2. Kolisis FN, Valis TP, Xenakis A (1990) An New York Acad Sci 613:674–680
3. Stamatis H, Xenakis A, Kolisis FN, Sztajer H, Menge U (1992) In: Tramper H et al (eds) Fundamentals of Biocatalysis in Non-Conventional Media, Progress in biotechnology, Elsevier, Amsterdam. Vol 8, pp 733–738
4. Stamatis H, Xenakis A, Provelegiou M, Kolisis FN (1993) Biotech Bioeng 42:103–110
5. Xenakis A, Stamatis H, Kolisis FN, Malliaris A (1993) Progr Colloid Polym Sci 93:373–376
6. Sztajer H, Lunsdorf H, Erdmann H, Menge U, Schmid R (1992) Biochim Biophys Acta 1124:253–261
7. Sklar LA, Hudson BS, Petersen M, Diamond J (1977) Biochemistry 16:813–818
8. Sklar LA, Hudson BS, Simoni RD (1977) Biochemistry 16:5100–5108
9. Stryer L (1978) Ann Rev Biochem 47:819–846
10. Zulauf M, Eicke HF (1979) J Phys Chem 83:480–486

Progr Colloid Polym Sci (1994) 97:256–261
© Steinkopff-Verlag 1994

BIO-COLLOIDS

M.N. Jones
M. Kaszuba

Molecular interactions and the targeting of vesicles to biosurfaces

Received: 1 October 1993
Accepted: 15 January 1994

Dr. M.N. Jones (✉) · M. Kaszuba
Department of Biological Sciences
Stopford Building
University of Manchester
Manchester M13 9PT, United Kingdom

Abstract Phospholipid vesicles have been prepared from mixtures of dipalmitolyphosphatidylcholine (DPPC) and phosphatidylinositol (PI) covering a range of mole % PI. A theoretical approach to vesicle adsorption as a function of PI content of the vesicles has been developed based on a lattice model for interactions between the bacterium glycocalyx and the inositol head groups in the vesicle surface. The model leads to optimum levels of PI for adsorption in the glycocalyx and vesicle surface and cross hydrogen bonding interactions between the inositol head groups and glycocalyx monosaccharides.

Key words: Phospholipid vesicles – proteovesicles – lectins – bacteria – adsorption biofilms

Introduction

The interactions between vesicles and biological cells have important applications in the field of drug delivery and it is becoming increasingly evident that vesicle delivery of toxic drugs such as amphotericin B, effective against systemic fungal infections [1, 2] and anti-cancer drugs such as adriamycin [2] have clinical advantages over the use of the free drug. In the field of personal hygiene, particularly oral hygiene, vesicles also have potential for the delivery of bactericides such as Triclosan [3].

The value of vesicular systems for transport and delivery of therapeutic agents resides in our ability to modify in a controllable way the nature of the vesicular surface in relation to the target cell to which it is directed. The targeting of vesicles to either mammalian or bacterial cells can be brought about by either the use of antibodies raised to cell surface antigens [4] or sugar-binding proteins (lectins) which bind to the polysaccharides in the cellular glycocalyx [5–13]. The interactions between such chemically engineered vesicles and their target sites present a challenging problem in colloid and interface science in that specific biological interactions are superimposed on non-specific intermolecular physical interactions. In this paper, we consider the interactions between phospholipid proteovesicles with surface-bound lectins and "naked" phospholipid vesicles, incorporating what has been found to be a site-directing phospholipid (phosphatidylinositol) with a range of oral and skin-associated bacteria.

Materials and Methods

Materials

Succinyl concanavalin A (sConA) product No. L3885, wheat germ agglutinin (WGA) from *Triticum vulgaris* product No. L9640, L-α-dipalmitoylphosphatidylcholine (DPPC) product No. P0763 and L-α-dipalmitoylphosphatidylethanolamine (DPPE) product No. P0890 and L-α-dipalmitoylphosphatidyglycerol (DPPG) product No. P9789 were from Sigma Chemical Company, Poole, Dorset, U.K. Phosphatidylinositol (PI) (from wheat germ) was from Lipid Products, South Nutfield, U.K. m-Maleimidobenzoyl-N-hydroxysuccinimide ester (MBS) was from Pierce and Warriner, Chester, U.K. and N-succinimidyl-S-acetylthioacetate (SATA) was from Calbiochem, Cambridge, U.K. [³H]-DPPC (specific activity 55 Ci/m mol) was from Amersham International, Amersham, U.K. Bacteriological agar No. 1 (code L11), brain heart infusion (BHl) (code CM 255), yeast extract powder (code L21) and phosphate buffered saline (PBS) tablets (code BR14a) were from Oxoid Ltd, Basingstoke, Hants, U.K. All other reagents were of analytical grade and aqueous solutions were made up with double distilled water.

Vesicle preparation and characterisation

Vesicles and proteovesicles were prepared using a range of methods including sonication (SUV) and reverse phase evaporation (REV) as previously described [8–13] and also by the vesicle extrusion technique (VETs) [14]. The coupling of lectins (s con A and WGA) to vesicles was achieved by incorporating the MBS derivative of DPPE (DPPEMBS) during vesicle preparation followed by conjugation with the SATA derivative of the protein. Routinely, DPPC (27 mg), PI (3 mg) and DPPEMBS (0–6 mg) together with either 5μ Ci [³H] DPPC or 1μCi [³H] DPPC were dissolved in dry chloroform-methanol (4:1 by volume) in a 50-ml round-bottomed flask. The solvent was removed by rotary evaporation at 60 °C to leave a uniform lipid layer to which was added 3 ml of PBS at 60 °C. The suspension was vigorously mixed on a vortex mixer to form multilamellar liposomes (MLVs) followed by extrusion 10 times at 60 °C through two stacked polycarbonate filters under a pressure of 200 psi. The resulting VETs were conjugated to the SATA derivative of the lectin (routinely, 2 ml, protein concentration, 1.5 mg/ml), previously deacetylated with hydroxylamine by incubating at 4 °C overnight. The unreacted lectin was separated from the proteoliposomes by gel filtration on a Sepharose 4B column (30 cm × 2 cm).

The proteoliposome fractions were analysed for protein content by a Lowry assay using the corresponding lectin calibration curve [15]. The lipid content was determined by scintillation counting for [³H]DPPC. During the course of the preparation of the proteoliposomes, size measurements were made by photon correlation spectroscopy on the MLVs, unconjugated VETs and conjugated VETs using a Malvern autosizer model RR146. The scattering data were fitted to an equivalent normal weight distribution $W(d_i)$ to give the weight-average diameter ($\bar{d}_w$). This data together with the molar ratio of protein to lipid were used to compute the weight-average number of protein molecules per proteoliposome ($\bar{P}_w$) from the relations

$$\bar{P}_w = \frac{\sum P_i W_i}{\sum W_i} = \frac{\sum P_i W_i(d_i)}{\sum W(d_i)}, \tag{1}$$

where P_i and W_i are the number of protein molecules per proteoliposome and weight of proteoliposome of species is respectively as previously described [9].

Growth of bacteria

Bacteria were obtained from the University of Manchester collection. They were used to inoculate agar plates prepared from 3.7 g of BHI in 100 ml of double distilled water to which was added 1.5 g of bacteriological agar. The plates were inoculated by streaking and the inverted streaked plates were incubated at 37 °C for 18 h. The resulting colonies were used to inoculate 10 ml aliquots of nutrient broth prepared by mixing 3.7 g BHI and 0.3 g yeast extract powder in 100 ml of double distilled water. The 10 ml aliquots in capped bottles were incubated at 37 °C for 18 h after which the bacterial suspensions were centrifuged at 200 rpm for 15 min, the supernatant discarded and the pellet resuspended in sterile PBS. The centrifugation and resuspension was repeated a further three times and the bacterial concentration adjusted to give an absorbance of 0.5 at 550 nm.

Proteovesicle and vesicle adsorption by bacterial biofilms (targeting assay)

Assays were carried out in wells of microtitre plates (Dynatech M129B). Aliquots (200 μl absorbance 0.5) of bacterial suspension were incubated overnight at room temperature to form an adsorbed biofilm. After adsorption the bacterial suspension was removed and the biofilm washed twice with sterile PBS. Vacant potential binding sites on the microtitre wells were blocked by incubating the wells with 300 μl of 0.02% w/v casein in PBS for 1 h. After removal of the casein solution, wells were exposed to the test vesicle or proteovesicle suspension for 2 h or less as required. After incubation with the vesicle/proteovesicle suspensions, the wells were washed three times with PBS and the biofilm dispersed by addition of 200 μl of 1% w/v sodium n-dodecylsulphate, followed by sonication and a 1-h incubation. Aliquots of the dispersed biofilm (180 μl) were taken for scintillation counting. Control wells containing only bacteria, only PBS and only liposomes were used to assess background levels of activity.

The results of the targeting assays are expressed in terms of the percentage apparent monolayer coverage (% amc) given by

$$\% \, \text{amc} = \frac{N_{\text{obs}}}{L_{\text{a}}} \times 100, \tag{2}$$

where N_{obs} is the observed number of moles of lipid adsorbed to the biofilm and L_a the number of moles of lipid which would be adsorbed if the biofilm was covered with a close-packed layer of liposomes. L_a was calculated form the equation.

$$L_a = \frac{A_{bf}}{\Pi(d_w/2)^2}\bar{N}_w,\tag{3}$$

where $\bar{d}_w$ is the weight average diameter of the liposomes having a weight average number of lipid molecules per liposome of $\bar{N}_w$ and A_{bf} is the geometric area of the biofilm. $\bar{N}_w$ was calculated from $\bar{d}_w$ assuming an area per lipid molecule in the liposomal bilayer (taken as 0.50 nm^2) and a bilayer thickness (taken as 7.5 nm) as previously described [9]. The area of the biofilm was taken as 2.202×10^{-4} m^2 which was measured in a previous study for the surface of microtitre plate wells exposed to $200\ \mu l$ of solution [16].

Results

Table 1 summarises the results of the targeting of proteovesicles with surface-bound lectins to several species of oral and skin-associated bacteria. The data covers ranges of proteovesicle size ($\bar{d}_w$), concentration and surface-bound lectin level ($\bar{P}_w$). The effectiveness of lectin targeting is expressed as the lectin targeting enhancement (LTE) defined as the ratio of % apparent monolayer coverage (% amc) of lectin-bearing vesicles to lectin-free vesicles of the same lipid composition and concentration. The lectin s con A which has a specificity for D-mannose and D-glucose residues is effective for targeting *Streptococcus mutans* and *sanguis* and *Coryneform hofmanni*, whereas WGA which has specificity for *N*-acetylneuraminic acid and *N*-acetylglucosamine is effective for targeting to *Staphylococcus epidermidis*. Neither of these lectins could be used for targeting to *Proteus vulgaris*.

Studies with lectin-free vesicles incorporating the hydroxy-containing phospholipid phosphatidylinositol (PI)

upto approximately 15 mole % PI revealed that there was an optimum level of PI for targeting to the bacteria biofilms. Figure 1 shows data for targeting to *S. epidermidis* biofilms where the optimum PI level is 11.4 mole %. Table 2 summarises the optima for a number of bacteria-vesicle systems. *Proteus vulgaris* biofilms are unusual in that high targeting levels occur at two specific mole % of PI. These results demonstrate that the adsorption of vesicles to bacterial biofilms is influenced by relatively small changes in the surface composition of the vesicles.

Discussion

The adsorption of vesicles to bacterial biofilms can be brought about either by exploiting biochemical specificity such as carbohydrate-mediated interactions with lectins on the vesicle surface or by the use of selected phospholipids such as PI. In the case of lectin-mediated adsorption it would be expected that the LTE would be

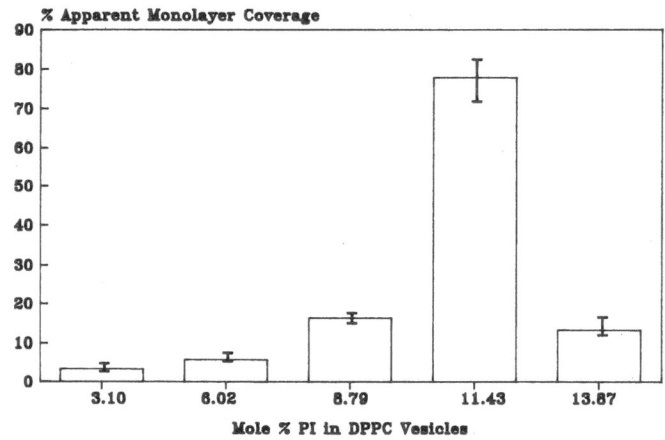

Fig. 1 Dependence of adsorption of DPPC/PI vesicles to *Staphylococcus epidermidis* biofilms on mole % PI content. The vesicles were incubated with the biofilm for 2 hours. The vesicle lipid concentration was 1.34 ± 0.06 mM, $\bar{d}_w = 78.8 \pm 6.7$ nm

Table 1 Targeting of proteovesicles to bacteria

Bacterium-proteovesicles	Site-directing lectin	Ranges $\bar{d}_w$ nm	c mM	$\bar{P}_w$	LTE
Streptococcus mutans strain D282 (DPPC/PI/DPPEMBS) SUVs*	s ConA	54–87	0.5–7.0	46–357	7–47
Streptococcus sanguis (DPPC/PI/DPPEMBS) REVs*	s ConA	106–240	0.25–1.13	54–3188	8–20
Streptococcus sanguis (DPPC/PI/DPPEMBS) VETs	s ConA	74–78	3.9–5.1	77–135	4.5–5.3
Staphylococcus epidermidis (DPPC/PI/DPPEMBS) VETs	WGA	84–108	0.23–4.9	11–38	2–16
Proteus vulgaris (DPPC/PI/DPPEMBS) VETs	WGA and s ConA	72–87	4.8–5.3	3.3–8.5	≤ 1
Coryneform hofmanni (DPPC/PI/DPPEMBS) VETs	s conA	78–86	4.9–6.0	5.4–6.6	2

* From ref. [3].

Table 2 Targeting of PI-contaiing vesicles to bacteria

Bacterium-vesicle system	Optimum mole% PI	[Lipid]mM	$\bar{d}_w$(nm)	Adsorption % Apparent monolayer coverage*
Streptococcus mutans strain NCTC 10449 (DPPC/PI) VETs	8.8	0.14 ± 0.01	109 ± 5	42 ± 10
Streptococcus mutans strain D282 (DPPC/PI) VETs	8.8	0.14 ± 0.01	105 ± 9	73 ± 10
Streptococcus sanguis strain CR2b (DPPC/PI) REVs	17.1	0.30	191 ± 56	26 ± 0.44
Staphylococcus epidermidis strain NCTC 11047 (DPPC/PI) VETs	11.4	1.34 ± 0.06	79.8 ± 6.7	75 ± 6.5
Staphylococcus epidermidis strain NCTC 11047 (DPPG/PI) VETs	6.8	1.34 ± 0.08	86.6 ± 3.0	106 ± 4.9
Proteus vulgaris (DPPC/PI) VETs	8.8 and 13.9	1.36 ± 0.07	85.0 ± 9.6	75 ± 4.7 and 85 ± 5.3

* Adsorption is measured as % monolayer coverage calculated from the projected area of the vesicles $((\pi d_w/2)^2)$ and the "geometric" area of the * biofilm. Since this does not allow for the surface roughness of the film it is only an "apparent" monolayer coverage and figures greater than 100% do not necessarily imply multilayer formation

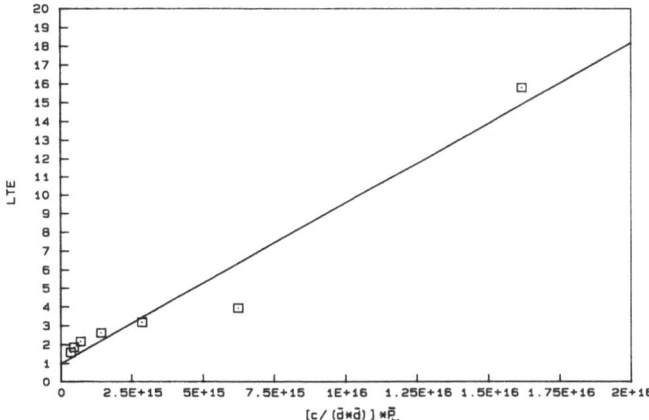

Fig. 2 The dependence of lectin targeting enhancement (LTE) on adsorption of wheat germ agglutinin-bearing vesicles (lipid composition DPCC/PI/DPPE) to *Staphylococcus epidermidis* biofilms on the parameter $(c/\bar{d}_w^2)\,\bar{P}_w$ (see text)

related to the number of lectin molecules on the vesicle surface ($\bar{P}_w$) and the number of vesicles per unit volume. The latter will depend directly on the lipid concentration (c) and inversely on the amount of lipid per vesicles, ie. on surface area of the vesicle $(4\pi\,(\bar{d}_w/2)^2)$. Thus adsorption and hence LTE should be proportional to $(c/\bar{d}_w^2)\,\bar{P}_w$. Figure 2 shows such a plot for the adsorption of WGA-bearing vesicles to *Staphylococcus epidermidis* biofilms which supports the above postulates.

The PI-mediated adsorption of vesicles to bacteria might be expected to be interpretable in terms of the classical forces between a surface and colloidal particles together with a contribution from the hydrogen bonding interactions between the hydroxy head groups of PI and the glycocalyx of the bacteria. PI is a negatively charged

lipid and the surfaces of bacteria are in general negatively charged at neutral pH. Qualitatively an optimum vesicle composition (mole % PI) suggests opposing contributions to the interactions between vesicles and bacteria. No optimum compositions for adsorption are found with DPPC-phosphatidylserine vesicles (DPPS is also a negatively charged phospholipid) so that the polyhydroxy nature of the PI head groups would appear to be significant. These observations suggest that perhaps the existence of optimum PI levels arise from a balance of hydroxy (H-bonding) attractive interactions and repulsive electrostatic effects. At low mole % PI the H-bonding contributions predominate whereas at high mole % PI electrostatic repulsion between the negative bacterium surface and the negative vesicle surface would reduce adsorption. To test this hypothesis we have developed a theoretical expression for H-bonding interactions based on a three-dimensional (3D) lattice model of the bacterium surface and a two-dimensional (2D) lattice for the vesicle surface in which the lattice sites are taken as hexose units. The theory is an adaptation of a previous treatment of H-bonding in cellular cohesion [17] and leads to the following expression for the potential energy of interaction (V_H, Jm^{-2}) at separation (2d) between the bacterium surface (excluding the glycocalyx) and the vesicle bilayer surface.

$$V_H = \{(n_s + n_s')^2/d - 2(n_s^2/l + n_s'^2/l')\}\,2V_s E_H \exp(E_H/kT),$$
(4)

where n_s and n_s' are the number of segments of volume V_s and l and l' are the thickness of the surface lattice of the bacterium (the glycocalyx) and the inositol head group on the vesicle surface respectively, E_H is the H-bonding energy, k the Boltzmann constant and T the absolute temperature. In applying Eq. (4) the thickness of the inositol head group was taken as the cube root of the volume of a lattice

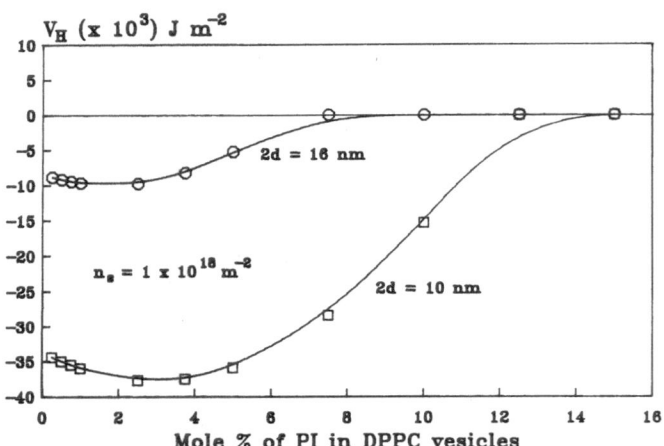

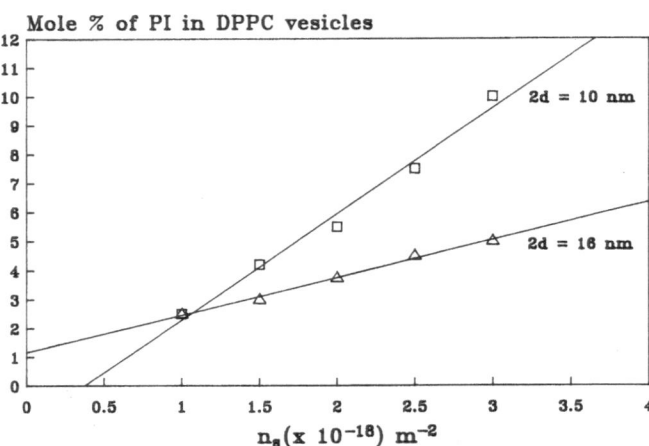

Fig. 3 Theoretical curves for the energy of interaction between bacterium and vesicle surface based on Eq. (4) as a function of the PI content of the vesicles for two interparticle separations. The minima in V_H suggest optimum PI levels for adsorption

Fig. 4 Dependence of optimum PI level for adsorption (ie minima in V_H vs. mole % PI curves (Fig. 3)) on the surface concentration (n_s) of monosaccharide residues in the bacterium glycocalyx for two interparticle separations

site ($V_s = 180 \times 10^{-30} \, \text{m}^3$ [17] hence $V_s^{1/3} = l' = 5.6 \times 10^{-10} \, \text{m}$) and the thickness of the glycocalyx was taken as $2 \times 10^{-8} \, \text{m}$ from electron microscopic studies on *Staphylococcus epidermidis* and *Streptococcus salivarius* [18]. Estimates of the number of glucose residues associated with the glycerol teichoic acid in the surface of *Staphylococci* gave a value of the order of $10^{18} \, \text{m}^{-2}$ for n_s [19]. The H-bonding energy was taken as $10 \, \text{kJ mol}^{-1}$. This is lower than the energy of a single H-bond ($25 \, \text{kJ mol}^{-1}$). In the formation of an inositol-hexose interaction, inositol–H_2O and hexose–H_2O H-bonds must be broken so that E_H is a difference in H-bond energies and would be expected to be lower than the energy of formation of a single H-bond. From Eq. (4), V_H versus bacterium–vesicle separation (2d) was calculated for vesicles with a range (0–15 mole%) of phosphatidylinositol (PI) content. Figure 3 shows the values of V_H as a function of mole % PI at two separations (10 mm and 16 mm). Interestingly, the curves show minima at particular PI levels which at least qualitatively would correspond to an optimum PI level for interaction between bacterium and vesicle. The existence of minima and their position with respect to PI level are critically dependent on the values of n_s (ie. the glucose density in the glycocalyx). No minima are found outside the n_s range 10^{17}–$10^{19} \, \text{m}^{-2}$. The minima arises from the balance of "cross" interactions (inositol-hexose) as given by the first term in Eq. (4) and the "self" interactions (hexose-hexose and inositol-inositol) as given by the second term. Figure 4 shows the relationship between the optimum mole % PI (ie the PI level at minima) and n_s for two bacterium-vesicle separations.

These theoretical studies suggest that phosphatidylinositol-mediated adsorption of vesicles to bacterial biofilms and in particular the existence of optimum levels of PI is related to the surface concentration of hydroxy moieties (sugar residues) in the glycocalyx of the bacterium. The extent of adsorption will depend on the magnitude of V_H/kT, this however is not easy to estimate under the conditions of our experiments in that the contact area between vesicle and bacterium biofilm is not precisely known nor is the separation between the bacterium cell wall and the vesicle surface. There are also shear forces involved on separation of the vesicle suspension from the biofilm during the experiments. The theoretical minimum in Fig. 3 at 2d = 16 nm is approximately $- 0.1 \, \text{Jm}^{-2}$, assuming a contact area between bacterium and vesicle ($\bar{d}_w = 100$ nm) of ca. 1/100th of the vesicle area gives V_H/kT a value of approximately 760 which would lead to extremely strong adsorption. V_H however decreases very rapidly with separation (2d) so that if the vesicle penetrated less into the bacterium glycocalyx the interactions would be very much weaker. Although other surface forces will be involved in the interaction (double layer repulsion and dispersion force attraction) these would not lead to optimum PI levels for adsorption although the double layer repulsion would contribute to decreasing adsorption on increasing the PI level.

Acknowledgements We thank the SERC for financial support for M.K. and Miss Julie Wassel for experimental assistance.

Progr Colloid Polym Sci (1994) 97:256–261
© Steinkopff-Verlag 1994

References

1. Chopra R, Fielding A, Goldstone AH (1992) Leukemia and Lymphoma 7:73–77
2. Gray A, Morgan J (1991) Blood Reviews 5:258–272
3. Jones MN, Francis SE, Hutchinson FJ, Handley PS, Lyle IG (1993) Biochim Biophys Acta 1147:251–261
4. Wright S, Huang L (1989) Adv Drug Deliv Rev 3:343–389
5. Carpenter-Green S, Huang L (1983) Anal Biochem 135:151–155
6. Liautard JP, Vidal M, Philipot JR (1985) Cell Biol Int Rep 9:1123–1137
7. Margolis LB, Bogdanov AA, Jr Gordeva LV, Torchilin VP (1988) in Liposomes as Drug Carriers (Gergoriadis G ed.) Chapter 52, pp 727, John Wiley and Sons, New York
8. Hutchison FJ, Jones MN (1988) FEBS Lett. 234:493–496
9. Hutchinson FJ, Francis SE, Lyle IG, Jones MN (1989) Biochim Biophys Acta 978:17–24
10. Hutchinson FJ, Francis SE, Lyle IG, Jones MN (1989) Biochim Biophys Acta 987:212–216
11. Francis SE, Jones MN (1990) Biochem Soc Trans 18:876–877
12. Francis SE, Lyle IG, Jones MN (1990) Biochim Biophys Acta 1062:117–122
13. Francis SE, Hutchinson FJ, Lyle IG, Jones MN (1992) Colloids Surfaces 62:177–184
14. Mayer LD, Hope MJ, Cullis PR (1986) Biochim Biophys Acta 858:161–168
15. Lowry OH, Rosebrough NJ, Farr AL, Randall RJ (1951) J Biol Chem 193:265–275
16. Chapman V, Fletcher SM, Jones MN (1990) J Immunol meth, 131:91–98
17. Jones MN (1976) FEBS Letters 62:21–24
18. Handley PS, Hargreaves J, Harty DWS (1988) J General Microbiology 134:3165–3172
19. White PJ personal communications

Progr Colloid Polym Sci (1994) 97:262–266
© Steinkopff-Verlag 1994

BIO-COLLOIDS

S.P.F.M. Roefs
C.G. de Kruif

Heat-induced denaturation and aggregation of β-lactoglobulin

Received: 4 October 1993
Accepted: 12 January 1994

S.P.F.M. Roefs (✉) · Dr. C. G. de Kruif
Netherlands Institute for Dairy Research
(NIZO)
Kernhemseweg 2
6718 ZB Ede, The Netherlands

Abstract The heat-induced denaturation and aggregation of β-lactoglobulin in water can be quantitatively modeled by analogy with a radical polymerization reaction. The model contains an initiation, a propagation and a termination step and it predicts that linear polymeric aggregates of β-lg monomers will be formed. The β-lg monomers are assumed to be linked together via intermolecular disulphide bonds, which are formed via thiol group/disulphide bond exchange reactions.

The decrease in native β-lg is predicted to follow order 3/2, which is in full agreement with experimental results. The size of the protein polymer particles is predicted to be proportional to the square root of the initial β-lg concentration. The increase in scattered light intensity at the beginning of heating, which is proportional to the product of concentration increase and size of the protein polymer particles, should be proportional to the initial β-lg concentration squared, which is indeed found.

Key words β-lactoglobulin – denaturation – aggregation – light scattering

Introduction

Cow's milk contains about 3.4% w of protein. In cheese manufacture the major part of the milk protein is coagulated and transformed into the cheese curd. the residual fluid, which is called whey, still contains around 0.5–0.6% w of protein, the whey proteins, of which β-Lactoglobulin (β-lg) is the main protein. β-Lg is a globular protein ($M_w = 18400$ Dalton) containing two intramolecular disulphide bonds and one –SH group; it's isoelectric pH is 5.2. At ambient temperature and pH = 5.3–8.0 it is found as a non-covalently linked dimer.

Like many other proteins, β-lg is sensitive to heating at temperatures exceeding 60 °C, where intra- and intermolecular changes and reactions occur which are generally denoted as denaturation and aggregation. In dairy and non-dairy products containing β-lg, these reactions may lead to (undesired) colloidal instability during pasteurization and sterilization treatments. A proper control of the denaturation and aggregation reactions will prevent instability, but will also make a desired colloidal structure possible. Knowledge of the precise mechanisms of denaturation and aggregation is still inadequate, and there is no quantitative model, nor is there a clear physical picture on which to base predictions. Predictions are of great interest for industrial processing and for finding new applications of β-lg as an (food) ingredient.

Upon heating, the protein undergoes conformational changes which are thought to result in an unfolding of the molecule. This process is usually described as denaturation. The denatured molecules are unstable and will aggregate. Therefore experimentally, denaturation, which is in principle a reversible reaction, mostly appears to be irreversible [1, 2] and a distinction between denaturation and aggregation is hard to make. Native molecules are rather inert as long as solvent conditions are non-critical and aggregation seems not to occur before denaturation has taken place.

Several phenomena have invariably been observed in the denaturation and aggregation of β-lg. The appearance of heated β-lg dispersions can vary from transparent to opalescent and milk-white turbid, and not only dispersions but also firm gels may be formed [3]. Electron microscopy studies have confirmed that turbid systems have a particulate microstructure, whereas transparent systems have a so-called fine stranded microstructure [3], as may be expected on the basis of light-scattering theory. The final appearance strongly depends on the medium conditions such as ionic strength, type of ions (especially Ca^{2+}) and pH. Transparent dispersions and gels are found at low ionic strength and a pH distinct from the iso-electric pH (e.g., above 6.5). In a rheological characterization transparent gels exhibit elastic properties [2, 3] quite similar to those of polymer gels. In a recent study on ovalbumin [4] similar features were observed and in transparent ovalbumin dispersions the formation of large linear aggregates (like a string of beads) was suggested [5].

Intermolecular –S–S– bonds are involved in the aggregation of β-lg and even –SH/–S–S– exchange reactions are reported to occur [6, 7]. In the latter reaction a free reactive–SH group reacts with an –S–S– bond to form a new –S–S– bond with one of the two sulphur atoms, while the other forms a new free –SH group. There is no doubt that intermolecular –S–S– bridges play a role in the aggregation of β-lg; however, their relevance with respect to all the other interactions involved in the aggregation process is usually considered to be minor.

The concentration decrease of β-lg in heated milk and milk salt solutions [1, 8] can always be described by a one-component reaction. Only the reaction order is reported to vary from 1 to 2 or even higher than 2. Often a value of 1.5 is found. Reaction orders of 1, 1.5 or 2 point to a relatively simple overall reaction for the denaturation and aggregation of β-lg.

Here, we present a model for the aggregation of β-lg dissolved in a low ionic strength solvent at neutral pH. It incorporates the aforementioned characteristics and can explain many of the phenomena not understood hitherto. For brevity and clarity we will give here only an outline of the theory and present some results. The derivations of all equations will be published elsewhere.

Theory

The basic assumption in the model is that β-lg aggregates like ethene monomers in an ordinary polymerization reaction; the β-lg molecules act as polymeric monomers and

linear strands of aggregated monomers are formed. The protein polymerization reaction is treated in a similar way to an ordinary radical addition polymerization reaction [9, 10], in which –SH groups play the role of the radicals. The total reaction scheme contains an initiation, a propagation and a termination step. β-Lg acts both as initiating and as propagating monomer.

The initiation step consists of a number of reversible reactions followed by a (pseudo) irreversible reaction, which is the real *initiation reaction*. In the reversible reactions the dimer will be split into two monomers. The initiation reaction is a first order reaction, in which the free –SH group of native β-lg (B) is transformed in such a way that it becomes reactive (indicated by B^*):

$$B_2 \leftrightarrows \ldots \ldots \leftrightarrows B \xrightarrow{k_1} B^* \tag{1}$$

The reactive –SH group of B^* reacts via an –SH/–S–S– exchange reaction with one of the two *intra*molecular –S–S– bonds of a non-denatured monomer (B) to form a reactive dimer B_2^*; an *inter*molecular –S–S– bond is formed and a new reactive free –SH group is now available on the initially non-denatured molecule. The reactive dimer in turn can react in the same way with a non-denatured monomer and this step (*propagation* reaction) can be repeated many times:

$$B + B_i^* \xrightarrow{k_2} B_{i+1}^* \quad i \geq 1 \tag{2}$$

Considering the conformation of β-lg [11] and the conformational changes occurring during the propagation reaction it is supposed that only one of the two *intra*-molecular –S–S– bonds and only one –SH group per monomer is reactive. Consequently, linear aggregates will be formed. The "polymerization" process stops when one reactive intermediate multimer B_i^* reacts with another intermediate B_j^* (*termination* reaction) forming a polymer without a reactive –SH group:

$$B_i^* + B_j^* \xrightarrow{k_3} B_{i+j} \quad i, j \geq 1 \tag{3}$$

During the polymerization process the concentration of reactive intermediates will reach a steady state situation according to the Bodenstein principle [9]. Using this steady-state principle the above presented reaction scheme can be worked out by analogy with addition radical polymerization reactions [9]. If the rate of initiation is small and much smaller than the rate of propagation, and taking

$$k_2 \left(\frac{k_1}{2k_3} \right)^{1/2} = k', \tag{4}$$

the reaction rate $d[B]/dt$ is quantified by:

$$\frac{d[B]}{dt} = -k'[B]^{3/2}, \tag{5}$$

where $[B]$ (g/l) is the concentration of native β-lg and k'

$((g/l)^{-1/2}s^{-1})$ is the reaction rate constant. The decrease in concentration of native β-lg during heating is thus described by a simple reaction of order 3/2 and if the initial β-lg concentration is given by $[B]_0$, the initial concentration decrease will be proportional to $[B]_0^{3/2}$.

From the reaction scheme it then follows that protein polymer particles will be formed with a distribution in size and a weight averaged size, $M_{p,w}$, given by:

$$M_{p,w} = 3k_2 \left(\frac{1}{2k_1 k_3}\right)^{1/2} [B]^{1/2} M_{monomer} \,. \qquad (6)$$

As the reaction proceeds the β-lg concentration decreases and the average size of the polymer particles formed will decrease. The initial average particle size will be proportional to $[B]_0^{1/2}$.

The scattered light intensity of a β-lg protein polymer particle dispersion can be expressed in the Rayleigh-ratio, $R(Q)$ [12]:

$$R(Q) \cong K^* C_p M_p P(Q) S(Q, C_p) \,, \qquad (7)$$

in which $C_p(= [B] - [B]_0$ is the (polymer) particle concentration (g/l), M_p the molar mass of the (polymer) particles (g/mole) and K^* a constant, which depends on apparatus and difference in refractive index between particles and solvent. $P(Q)$ is the particle form factor, whereas $S(Q, C_p)$ is the structure factor of the dispersion. The scattered intensity of the protein polymer particles was much larger than of the native β-lg and the increase in intensity during aggregation was not affected by the decreasing concentration of native β-lg. $P(Q)$ was assumed to be constant and close to unity for the polymeric particles formed in the range of β-lg concentrations (10–100 g/l) studied. At the beginning of heating ($t = 0$), where polymer particle concentration is zero and $S(Q,C_p) = 1$, the increase in time of $R(Q)$ is found by combining Eqs. (5, 6 and 7). It is given by:

$$\frac{dR(Q)}{dt} \simeq K^* \frac{3k_2^2}{2k_3} [B]_0^2 M_{monomer} \qquad (8)$$

So, the slope of $R(Q)$ versus heating time extrapolated to $t = 0$ is proportional to $[B]_0^2$.

The rate constants k_i of the propagation and termination reactions depend on the encounter frequency of reactants and on the fraction of encounters that lead to a reaction. The encounter frequency depends on the diffusional (translational and rotational) motion of the reactants, which is related to the visosity of the medium. When the concentration and/or the size of the polymer particles formed increases, viscosity will increase and, more importantly, the rotational and diffusional motion of especially long reactive intermediates is reduced; the termination rate constant k_3 will decrease much more strongly than the

propagation constant k_2, and the overall reaction rate constant k' will increase as the reaction proceeds. In polymerization chemistry this auto-acceleration phenomenon is called the *Trommsdorf* [9] effect.

Experimental

We used a purified β-lg sample (mixture of β-lg A and β-lg B), which was prepared at NIZO from whey, basically following the procedure of Maubois [13]. The sample contained 91% w β-lg, a few percent α-la, less than 1% w salt (less than 0.1% w Ca) and about 3% w water. Dynamic light-scattering measurements gave a particle diameter of 10.4 nm independent of temperature and scattering angle. β-Lg was dissolved in double distilled water.

For determination of concentration decrease, a series of test tubes with protein concentrations in the range of 2–95 g/l was heated at 65 °C for different times. The tubes were cooled in ice water, the pH was adjusted to pH = 4.7 $\pm$ .3 and the precipitates of denatured/aggregated protein were separated by centrifugation for 30 min at 20 000 g. The native β-lg concentration present in the supernatant was determined by HP-GPC [1].

We measured the size of the particles formed during heating by applying a dynamic light-scattering set-up (90° configuration, wave vector, Q, 0.0186 nm^{-1}). Both the time-averaged scattering intensity $R(Q)$ and the effective Stokes Einstein particle diameter were evaluated at the heating temperatures, 61.5°, 65° and 68.5 °C, as a function of heating time. Samples (β-lg concentration 14–91 g/l) were double filtered (0.1 μm non-protein adsorbing filters) before use. The particle diameter was an apparent diameter and it was only interpreted qualitatively, since the intensity auto-correlation function was fitted with a so-called cumulant fit. This is quantitatively not valid for non-spherical particles, where the correlation function is a multi-exponential decay curve because of separate rotational and translational diffusion contributions. In addition, we did not take into account the effect of particle concentration on the apparent diameter [14].

Results and discussion

In Fig. 1, we plotted the relative concentration of native β-lg in the supernatant as a function of heating time. The order (see Eq. (5) of the overall denaturation/aggregation reaction was derived from the initial concentration decrease. The logarithm of the slope of the concentration curves at $t = 0$ plotted as a function of the logarithm of the β-lg concentration at $t = 0$ (Fig. 2) resulted in a straight

Progr Colloid Polym Sci (1994) 97:262–266
© Steinkopff-Verlag 1994

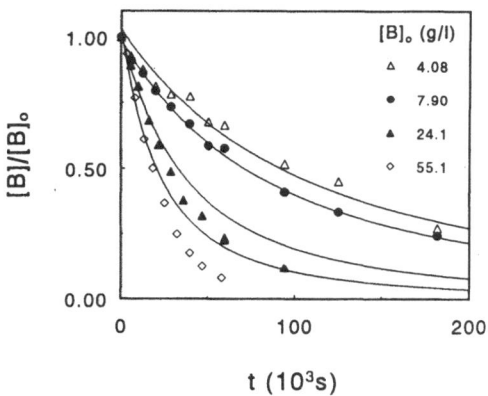

Fig. 1 Decrease of native β-lg concentration as a function of time at 65 °C. Four different initial concentrations of β-lg are reported: $\triangle$, 4.08 g/l; $\bullet$, 7.90 g/l; $\blacktriangle$, 24.1 g/l; $\diamond$, 55.1 g/l. The drawn lines represent 3/2-order reaction kinetics

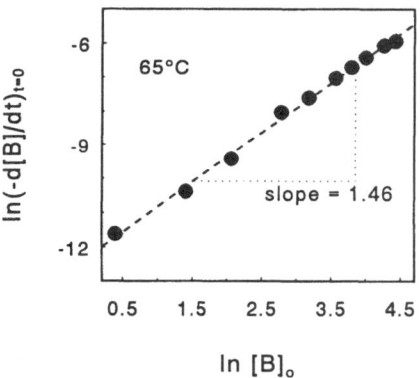

Fig. 2 The logarithm of the initial reaction rate against the logarithm of the initial β-lg concentration. The slope of the curve is 3/2 in accordance with the model prediction

effect is even stronger (see Fig. 1) since not only more, but also longer (Eq. (6)) protein polymer particles are formed. As a result, the particle size distribution tends to diverge.

In Figs. 3 and 4 the apparent particle diameter and scattering intensity are shown of samples heated and measured at 61.5 °C. The measured apparent diameter of heated β-lg solutions was in good agreement with the model at all three temperatures. Particle diameter rapidly increases from the beginning of the aggregation reactions due to the formation of protein polymer particles and after a relatively short time a more or less constant value is reached, which corresponds to the steady state in the aggregation reactions. The time to establish the steady state is, however, much shorter than the time to reach

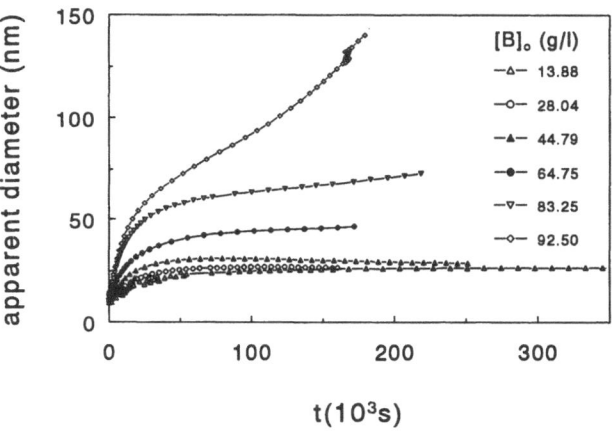

Fig. 3 Apparent particle size as measured with dynamic light scattering. Temperature was 61.5 °C, pH = 6.7–7.0 and at various β-lg concentrations.

Fig. 4 Measured scattering intensity as a function of reaction time at 61.5 °C. The initial slope of each curve is plotted in Fig. 5

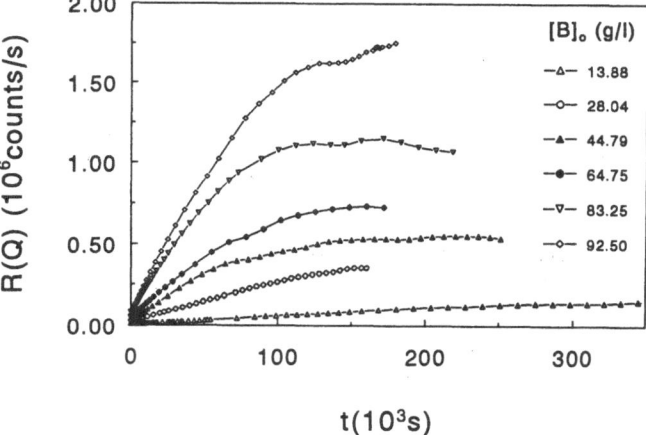

line with a slope (1.46) in good agreement with the value of 1.5 as predicted by the model (Eq. (5)). From 11 different initial concentrations and with a reaction of order 1.5 (see drawn lines in Fig. 1) we find an average $k_{1.5}$- value of $5.1 (\pm 1.0)*10^{-6}$ (l/g)$^{0.5}$s^{-1}. At low $[B]_0$ concentrations the protein polymer particles which are formed do not influence the reaction rate and the concentration decrease follows the order 1.5 reaction until all β-lg is transformed (see Fig. 1). At $[B]_0$ larger than 15 g/l viscosity increases significantly as the reaction proceeds and the protein particle intermediates formed decrease in diffusivity; the "overall" reaction rate increases and the β-lg concentration decreases more strongly than the theoretical curve based on the initial reaction rate (compare symbols and drawn lines in Fig. 1) as a result of the *Trommsdorf* effect [9]. At still higher β-lg concentrations (60–90 g/l) this

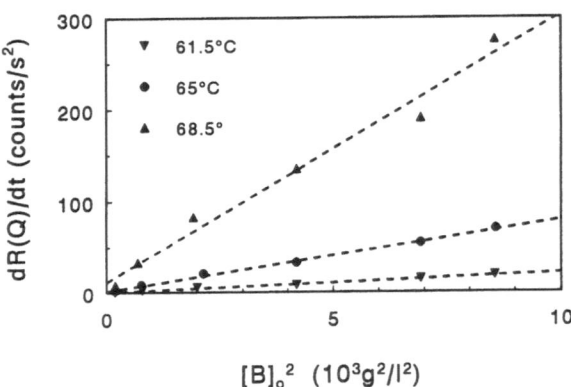

Fig. 5 The initial change of the scattered intensity as a function of initial β-lg concentration at temperatures of 61.5°, 65°, and 68.5 °C

a constant diameter, since a minimum number of protein polymer particles has to be formed before the measured diameter entirely depends on the size of the polymer particles. As predicted by Eq. (6), the constant value found for the apparent diameter increases with the initial β-lg concentration and it slightly decreases as the reaction proceeds for the lowest initial β-lg concentrations. At the highest concentrations the Trommsdorf effect becomes increasingly important and, as the reaction proceeds, protein polymer particles of increasing size are formed. This effect dominates the measured overall particles size, and the rapid initial increase in particle size is followed by a gradual growth (see Fig. 3).

The averaged scattering intensity, $R(Q)$, strongly increased with initial β-lg concentration, not only because of a higher rate of particle formation, but also because of the increasing size of the polymeric particles (see Fig. 4, 61.5 °C). Since the structure factor $S(Q, C_p)$ decreases with increasing particle concentration, the scattered intensity (see Eq. (7)) passed through a maximum after some reaction time for each initial β-lg concentration. The initial increase of the scattered intensity, i.e., the slope of $R(Q)$ at $t = 0$, was evaluated quantitatively according to Eq. 8 for a series of concentrations at three heating temperatures (61.5°, 65°, and 68.5 °C, see Fig. 5) and, as predicted by the model, it scales quantitatively well with $[B]_0^2$ for the three temperatures investigated (Fig. 5).

It can be concluded that the heat-induced denaturation and aggregation of β-lg in water can be described quantitatively by an addition polymerization reaction. The β-lg monomers aggregate into linear polymeric particles with intermolecular disulphide bonds, which are formed via $-SH/-S-S-$ exchange reactions. The initial concentration decrease of native monomers follows a reaction of order 3/2, while the size of the protein polymer particles is proportional to the square root of the initial β-lg concentration.

Acknowledgement We thank Caroline van der Horst and her group for preparing the β-lactoglobulin sample, and Eric Driessen and Jan Klok for carrying out most of the experiments. Dr. Peter van Mil is thanked for his comments and fruitful discussions. Ivon Munter is thanked for typing the manuscript. Part of this research was supported by the Ministry of Economic Affairs through the program "IOP-Industriële Eiwitten" and by the dairy companies Coberco and FRIESLAND Frico Domo.

References

1. Wit JN de (1990) Thermal stability and functionality of whey proteins, Journal of Dairy Science 73:3602–3612
2. Paulsson M (1990) Thermal denaturation and gelation of whey proteins and their adsorption at the air/water interface. Ph.D thesis. University of Lund, Sweden
3. Stading M, Langton M, Hermansson A-M (1990) Inhomogeneous fine stranded β-lactoglobulin gels. Food Hydrocolloids 6:455–470
4. Koseki T, Kitabatake N, Doi E (1989) Irreversible thermal denaturation and formation of linear aggregates of ovalbumin. Food Hydrocoll 3:123–134
5. Nemoto N, Koike A, Osaki K, Koseki T, Doi E (1993) Dynamic light scattering of aqueous solutions of linear aggregates induced by thermal denaturation of ovalbumin. Biopolymers 33:551–559
6. Shimada K, Cheftel JC (1989) Sulfhydryl group/disulfide bond interchange reactions during heat-induced gelation of whey protein isolate. J Agr Food Chem 37:161–168
7. Sawyer WH (1967) Heat denaturation of bovine β-lactoglobulins and relevance of disulphide aggregation. J Dairy Sci 51:323–329
8. Dannenberg F, Kessler HG (1988) Application of reaction kinetics to the denaturation of whey protein in heated milk, Milchwissenschaft 43, p. 3–7
9. Hiemenz PC (1984) Polymer Chemistry. The Basic Concepts. Marcel Dekker, Inc, New York
10. Flory PJ (1953) Principles of Polymer Chemistry. Cornell University Press, Ithaca, New York.
11. Papiz MZ, Sawyer L, Eliopoulos EE, North ACT, Findlay JBC, Sivaprasadarao R, Jones TA, Newcomer ME, Kraulis PJ (1986) The structure of β-lactoglobulin and its similarity to plasma retinol binding protein. Nature 324:383–385
12. Hulst HC van der. Light Scattering by Small Particles. Dover Publications, Inc. New York
13. Maubois JL (1979) Industrial fractionation of main whey proteins. Bulletin of the IDF 212:154–159
14. Kruif CG de (1992) Casein micelles: diffusivity as function of renneting time. Langmuir 8:2932–2937

Progr Colloid Polym Sci (1994) 97:267-270
© Steinkopff-Verlag 1994

BIO-COLLOIDS

S. Egelhaaf
M. Müller
P. Schurtenberger

Spontaneous vesiculation: mixed lecithin-bile salt solutions as a biologically relevant model system

Received: 16 September 1993
accepted: 14 January 1994

PD Dr. P. Schurtenberger (✉)
Institut für Polymere
ETH-Zentum
Universitätsstrasse 6
8092 Zürich, Switzerland

S. Egelhaaf · M. Müller
Labor für Elektronenmikroskopie 1
ETH - Zentrum
Schmelzbergstrasse 7
8092 Zürich, Switzerland

Abstract Spontaneous formation of vesicles which occurs upon dilution of lecithin-bile salt mixed micellar solutions was studied using static (SLS) and dynamic (DLS) light scattering. Special attention was paid to the properties of the vesicles formed, and also to the mixed micellar precursors, since the structure of these mixed micelles is still controversial. We show how a self-consistent interpretation of the SLS and DLS data can be achieved using concepts from colloid and polymer physics and incorporating the effects of polydispersity and interparticle interactions. The micelle to vesicle transition is interpreted within the framework of current theoretical models for the spontaneous formation of vesicles.

Key words Spontaneous vesiculation – mixed micelles – light scattering

Introduction

Considerable attention has recently been devoted to a theoretical and experimental characterization of the spontaneous formation of vesicles. Several theoretical models have been developed in order to characterize the formation and equilibrium properties of vesicles. A "microscopic" model for the spontaneous vesiculation has been described by Israelachvili and coworkers [1, 2]. An "optimal" surface area per head group of the surfactant molecules is calculated by minimizing the total interaction free energy per surfactant molecule in the aggregate. This area, the volume and the length of the surfactant molecule form packing constraints, which together with entropic considerations determine the shape of the aggregates and, in the case of vesicles, the critical packing radius and thus the size of the vesicles. Safran and coworkers recently presented a more "macroscopic" theory [3] which considers the curvature elastic free energy of the aggregates formed in surfactant solutions. By decreasing the so-called spontaneous curvature of the surfactant monolayer, a transition from spheri-

cal micelles to cylindrical micelles, bilayers and then inverse spherical micelles is predicted. The theory predicts the formation of vesicles only for surfactant mixtures. Interactions between the two species lead to asymmetry in the composition and the spontaneous curvature of the inner and outer layers, which are equal but of opposite sign and energetically stabilize the vesicles.

Whereas in recent experiments [4, 5] spontaneous formation of single-walled, equilibrium vesicles was reported for aqueous mixtures of single-tailed cationic and anionic surfactants, we used in our experiments mixtures of lecithin and bile salt in buffer. They serve not only as model systems, but also have a great relevance in biology and physiology as well as in pharmaceutical applications. The solutions were prepared by diluting a mixed micellar stock solution with buffer (for details see [6]). Due to the much higher monomer solubility of the bile salt, the lecithin-to-bile-salt ratio in the aggregates is changed upon dilution. This causes a transition from mixed micelles to vesicles if the solution is diluted beyond the micellar phase limit [7]. The thus prepared solutions were investigated by static (SLS) and dynamic (DLS) light-scattering experi-

268

S. Egelhaaf et al.
Spontaneous vesiculation in lecithin-bile salt solutions

ments under different scattering angles. We then aimed for a self-consistent description of these measurements using concepts from colloid and polymer physics.

Results

The experimental details are given elsewhere [6]. Figure 1 summarizes the concentration dependence of the mean hydrodynamic radius, R_h, extrapolated to zero scattering angle, of a lecithin-bile salt solution with a lecithin-to-bile salt molar ration of 0.9. With increasing dilution, the hydrodynamic radius increases in the mixed micellar region, reaches a maximum near the phase limit and decreases in the vesicular region. The polydispersity data (not shown) determined by the method of cumulants from the DLS measurements indicate a quite high polydispersity of the micellar aggregates, whereas the vesicles which form spontaneously at concentrations below the phase limit are fairly monodisperse.

The radius of gyration, R_g, and the molecular weight, M_{app}, (data not shown) as determined in SLS experiments exhibit a similar dependence on dilution, indicating an increase of the aggregate size in the mixed micellar region and a decrease in the vesicular region.

Mixed micelles

We also carefully investigated the mixed micellar precursors of the vesicles, since the structure of these mixed

Fig. 1 Mean hydrodynamic radius R_h extrapolated to zero scattering angle of a lecithin-bile salt stock solution (L/BS = 0.9, $C_{tot} = 50$ mg/mL, $T = 25°C$) as a function of dilution. In addition the phase boundary and the models used to describe the aggregates in the micellar and vesicular region, respectively, are shown

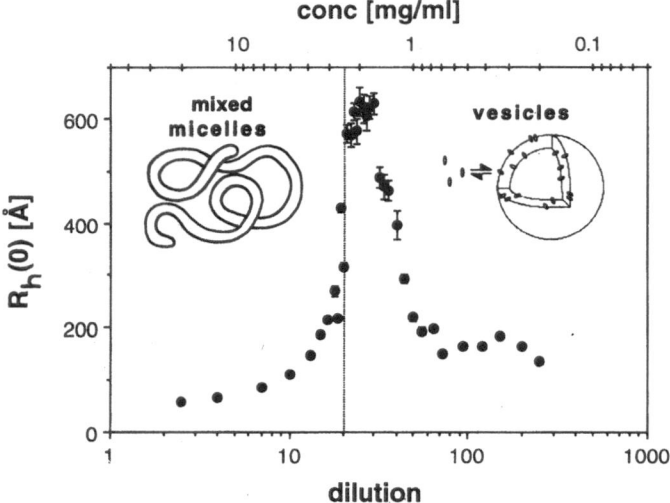

micelles is still controversial. Mazer et al. [8] proposed a mixed disc model based on light-scattering measurements. This model was widely accepted until Hjelm et al. [9,10] performed small-angle neutron scattering (SANS) experiments which were in clear disagreement with the disc model, but could be explained by a cylindrical structure of the mixed micelles. We therefore tested whether the static and dynamic light scattering could be analyzed in a self-consistent way based on the assumption of locally cylindrical mixed micelles.

If one takes into account the flexibility of the cylinder, the mixed micelles can be described by the wormlike chain model known from the theory of polymer solutions. In addition, we also accounted for the polydispersity of the size distribution in our analysis. The parameters of the model were either taken from the literature (diameter of the cylinder [10]), or determined from a simultaneous iterative fit of the angular dependence of the normalized scattered intensity (i.e., Rayleigh ratio), $R(Q)$, and the hydrodynamic radius, $R_h(Q)$, (contour length L, persistence length l_p and polydispersity):

$$R(Q) = R(O)\langle P(Q)\rangle$$

$$\text{where } \langle P(Q)\rangle = \frac{\int N(L)M(L)^2 P(Q,L,l_p)dL}{\int N(L)M(L)^2 dL} \quad (1)$$

$$R_h(Q) = \frac{\int N(L)M(L)^2 P(Q,L,l_p)dL}{\int N(L)M(L)^2 P(Q,L,l_p)\dfrac{F_{int}(R_h,Q)}{R_h(L,l_p)}dL}, \quad (2)$$

where F_{int} accounts for the contribution from internal modes [11].

The formfactors $P(Q,L,l_p)$ used in this fit were calculated from the Debye formula [12], and the theoretical expression for the hydrodynamic radius of wormlike chains was taken from Yamakawa and Fuji [13]. These fits resulted in a self-consistent quantitative description of the angular dependence of the SLS and DLS measurements. An example for the data and the corresponding fit as obtained from a sample near the phase limit is given in Fig. 2.

Towards the phase limit the polydispersity as determined by the method of cumulants increases, and a detailed analysis of the intensity autocorrelation functions using an inverse Laplace transform program indicates that an additional second peak with a very weak angular dependence develops. This can, without additional free parameters, quantitatively be explained by the internal modes of the flexible micelles and is not a result of a drastic increase in the micellar polydispersity or due to the coexistence of micelles and vesicles.

In a next step, we can test the applied structural model on an absolute scale by comparing the measured apparent

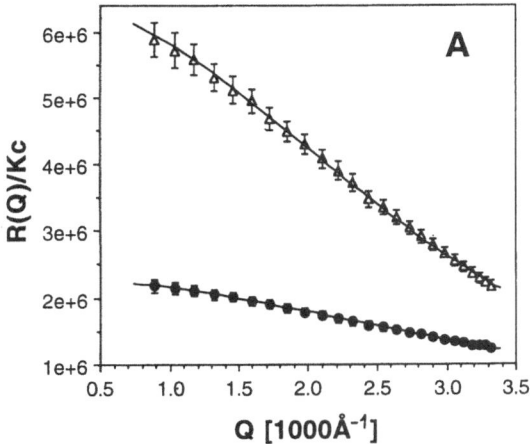

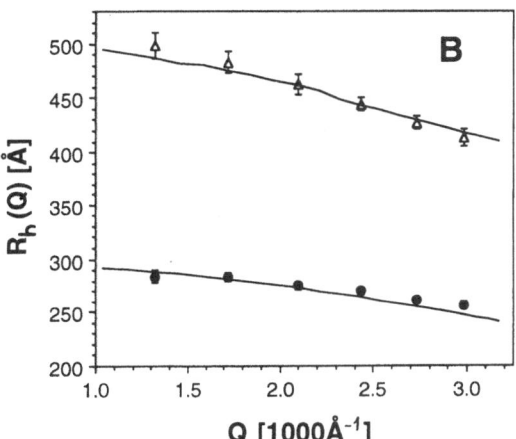

Fig. 2 Reduced Rayleigh ratio $R(Q)/Kc$, where $K = 4\pi^2 n^2 (dn/dc)^2/(N_A \lambda_0^4)$, and hydrodynamic radius $R_h(Q)$ (B) of a micellar (●, dilution 1:20.25) and a vesicular (Δ, dilution 1:21) solution as functions of the scattering vector Q. The curves are calculated on the basis of the wormlike chain and of the shell model, respectively (see text for details)

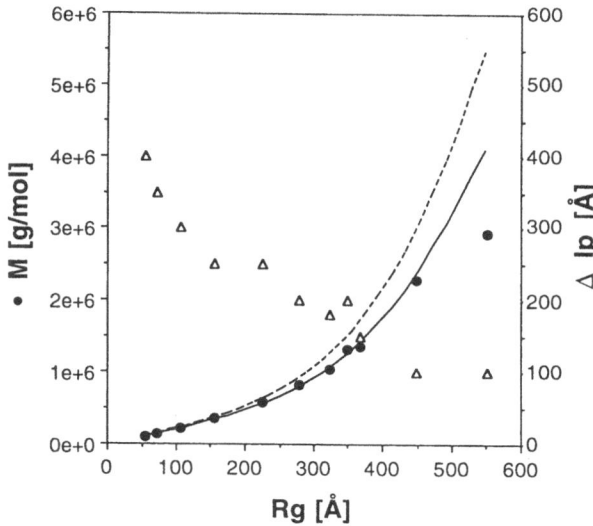

Fig. 3 Apparent molecular weight M_{app}(●) and persistence length l_p (Δ) of the mixed micellar solutions as a function of the radius of gyration R_g. The molecular weight M_0 and the apparent molecular weight M_{app}, which takes into account the interactions between the micelles, calculated on the basis of the wormlike chain model are shown as a broken and a full line, respectively

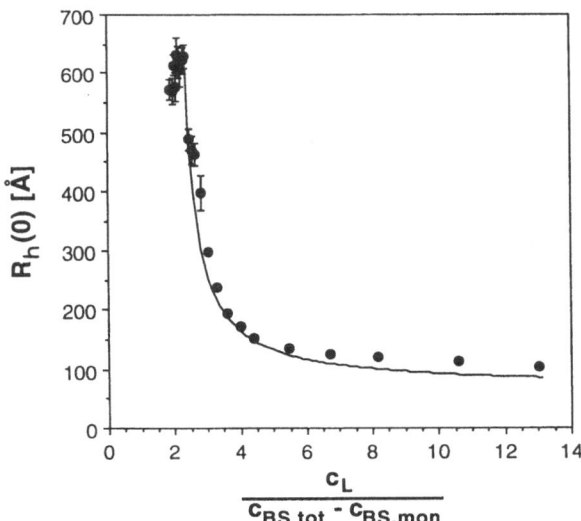

Fig. 4 Hydrodynamic radius R_h of the vesicles as a function of the composition of the bilayer $c_L/(c_{BS,tot} - c_{BS,mon})$ for the dilution series. The curve is calculated on the basis of the theoretical model by Israelachvili et al. [2, 7] for two-component vesicles

molecular weight of the mixed micelles as a function of concentration with the values of the apparent molecular weights calculated using the parameters determined by the fit to the Q-dependence of $R(Q)$ and $R_h(Q)$ described above (Fig. 3). In the calculation of the apparent molecular weight, intermicellar interactions were taken into account on the level of a virial expansion for excluded volume interactions between semi-flexible polymer coils [12]:

$$M_{app} = M_0(1 - 2A_2 M_0 c)$$

where $A_2 = 4\pi^{3/2} N_A \left(\frac{R_g^3}{M_0^2}\right)\psi(z)$ (3)

Vesicles

The vesicles are modeled as shells. The size and polydispersity of these shells were determined by simultaneously fitting the SLS and DLS data analogous to the procedure

270

S. Egelhaaf et al.
Spontaneous vesiculation in lecithin-bile salt solutions

applied to the data from the micellar region. This also yields a self-consistent and quantitative description of the angular dependence of $R(Q)$ and $R_h(Q)$ in the vesicular region assuming a single phase of fairly monodisperse vesicles (Fig. 2). The size of these vesicles decreases (Fig. 4) and the polydispersity increases slightly on dilution, which is caused by the decreases of the charge of the vesicles and thus a greater tendency to aggregate or fuse. The determined dependence of the radius on the composition of the bilayer, which is calculated by considering the amount of bile salt present as monomers, is in agreement with the critical radius predicted by the theoretical model of Israelachvili and coworkers [2, 7] as shown in Fig. 4.

Values for the geometrical parameters of the surfactants in the packing equation and the partition equilibrium constant for the bile salt were taken from the literature [7].

Conclusions

We were able to show that the data from static and dynamic light-scattering measurements at concentrations above the mixed micellar phase limit can self-consistently be interpreted using a structural model of wormlike mixed micelles. The molecular weight and overall size of these micelles increases and the persistence length decreases upon dilution (Fig. 3). This is caused by the decrease of the bile-salt-to-lecithin ratio in the aggregates due to dilution, which appears to lower the spontaneous curvature of the mixed micelles and forces them to avoid endcaps and therefore grow, and decreases the relative charge density of the micelles and hence allows for a greater flexibility. At concentrations below the micellar phase limit, mixed lecithin bile salt vesicles form spontaneously. Their size depends strongly upon the bilayer composition, and the partition equilibrium for the bile salts between the bilayer and the aqueous phase results in a monotonic decrease of the vesicle size with increasing dilution [7]. Within the experimental resolution of our study (dilution/size distribution), no indication for the coexistence region of mixed micelles and vesicles could be found for the very low lecithin and bile salt concentrations used in these experiments.

References

1. Israelachvili JN, Mitchell DJ, Ninham BW (1976) J Chem Soc Faraday Trans II 72:1525–1568
2. Israelachvili JN, Mitchell DJ, Ninham BW (1977) Biochem Biophys Acta 470:185–201
3. Safran SA, Pincus PA, Andelman D, MacKintosh FC (1991) Phys Rev A 43:1071–1078
4. Kaler EW, Murthy AK, Rodriguez BE, Zasadzinski JAN (1989) Science 245:1371–1374
5. Ambühl M, Bangerter F, Luisi PL, Skrabal P, Watzke HJ (1993) Langmuir 9:36–38
6. Egelhaaf SU, Schurtenberger P, J Phys Chem 98:8560–8573.
7. Schurtenberger P, Mazer N, Känzig W (1985) J Phys Chem 89:1042–1049
8. Mazer NA, Benedek GB, Carey MC (1980) Biochemistry 19:601–615
9. Hjelm RP, Thiyagarajan P, Alkan H (1988) J Appl Cryst 21:858–863
10. Hjelm RP, Thiyagarajan P, Alkan-Onyuksel H (1992) J Phys Chem 96:8653–8661
11. Brown W, Nicolai T (1993) In: Brown W (ed) Dynamic light scattering. Clarendon Press, Oxford, pp 272–318
12. Yamakawa H (1971) Modern Theory of Polymer Solutions. Harper & Row, New York
13. Yamakawa H, Fujii M (1973) Macromolecules 6:407–415

Progr Colloid Polym Sci (1994) 97:271–274
© Steinkopff-Verlag 1994

BIO-COLLOIDS

D.S. Horne
J. Leaver
D.V. Brooksbank

Electrostatic interactions in adsorbed β-casein layers

Received: 16 September 1993
Accepted: 17 March 1994

Dr. D.S. Horne (✉) J. Leaver
D.V. Brooksbank
Hannah Research Institute
Ayr KA6 5HL, Scotland, United Kingdom

Abstract Dynamic light scattering has been employed to study the influence of electrostatic interactions on the thickness and structure of β-casein layers adsorbed onto polystyrene latex particles. The influence of protein charge has been investigated by varying the background ionic strength or by including $CaCl_2$, the calcium ion being a specific binding agent of the protein in solution. Both moderators cause the protein layer to contract, the monovalent salt following an $I^{1/2}$ dependence, the divalent ion a more rapid change, indicating a specific binding effect. These observations are consistent with the previously proposed loop-and-train model for the adsorbed molecule, the extent of the highly charged loop being controlled by electrostatic repulsion effects.

Key words β-casein – dynamic light scattering – electrostatic interactions – adsorbed layer thickness – polystyrene latex

Introduction

Proteins are frequently employed to stabilize emulsions, forming adsorbed films at the oil/water interface. To progress a theoretical understanding of the stability of these emulsion systems, knowledge of the structure and thickness of the adsorbed protein layers is an essential prerequisite. As charged polyelectrolytes, the response of the protein layers to variations in the ionic strength of the aqueous phase is an area requiring further study. Using dynamic light scattering, we have been monitoring changes in the layer thickness of the milk protein, β-casein, adsorbed onto polystyrene latex particles as model emulsion droplets.

β-casein forms about 40% of the casein proteins of milk. It is therefore a major component of the sodium caseinate widely used in the food industry to stabilize emulsions. Unlike globular proteins such as β-lactoglobulin, β-casein has been found to form relatively thick adsorbed layers [1–3]. No information has yet been published on how their structure responds to changes in the ionic environment of the particles. Here, results are presented and contrasted for two situations, the introduction of calcium, known to bind specifically to β-casein, and the addition of sodium chloride anticipated to be a simple moderator of ionic strength.

Materials and methods

Polystyrene latex of nominal diameter 91 nm was purchased from Sigma Chemical Co. β-Casein was purified from bulk milk using preparative methods described by Leaver and Law [4].

Changes in the hydrodynamic radius of the polystyrene latex in the presence of β-casein in imidazole buffer (20 mM; pH7.0) with added $CaCl_2$ or NaCl as required were measured as detailed by Dalgleish [1] with the exception of a Malvern 7032 correlator being employed for signal analysis. The amount of protein adsorbed on the latex was determined by centrifuging the suspensions to pellet the latex and assaying the concentration of protein remaining in the supernatant [5].

Results

The increase in the recorded hydrodynamic radius (ΔR) of the latex particles in the presence of β-casein in imidazole buffer (20 mm; pH 7.0) is shown in Fig. 1. As the protein content of the solution is increased, there was a smooth increase in radius to a plateau value of 15.9 ± 0.8 nm, representing the thickness of the protein monolayer at saturation coverage. The value here is close to that found in earlier studies [1–3].

These measurements were carried out in a low ionic strength buffer to maximise the possible effects of electrostatic interaction between segments of the protein themselves or between individual segments and the negatively charged latex surface. The experiments were then repeated either in the presence of 25 mM NaCl or 5.33mM CaCl$_2$, both added subsequent to loading the latex with protein. The thickness of the adsorbed layers in the presence of these salts are also shown in Fig. 1 as a function of the applied protein concentration i.e. the amount of protein added per unit area of available latex surface.

At low levels of applied protein (< 1.5 mg m^{-2}), the addition of NaCl produced large increases in latex particle diameter, of the order of several hundred nm. This we attribute to bridging flocculation. At higher levels of bulk protein, this aggregation was no longer observed but instead the lower layer thicknesses plotted in Fig. 1 were recorded, levelling to a constant value of 9.7 ± 1.0 nm.

When 5.33 mM CaCl$_2$ was added to the coated latex dispersion, aggregation of the latex was observed in the absence of the β-casein. The addition of Ca^{2+} also induced bridging flocculation at low levels of applied protein but above 2 mg m^{-2} the results plotted in Fig. 1 were obtained. Though the apparent particle radius increased monotonically with applied protein the hydrodynamic layer thickness were much thinner in the case of added Ca^{2+} than for added NaCl, with a plateau value of 7.3 ± 0.2 nm being obtained. Since the total ionic strength of the bathing buffer system is greater in the case of the NaCl, this effect of added Ca^{2+} on the layer thickness must include some other specific contribution additional to the non-specific influence of ionic strength.

Assays on the protein contents of the supernatants of these mixtures demonstrated that changes in layer thickness were not due to losses in the amounts of protein adsorbed to the latex particles. Surface coverages were unchanged within experimental error.

Further experiments were carried out varying the amounts of Ca^{2+} or NaCl added to polystyrene latex particles pre-coated with sufficient protein to achieve monolayer saturation coverage. Changes in the thickness of these pre-adsorbed layers are shown in Fig. 2a as a function of the final NaCl concentration and in Fig. 2b as a function of the final Ca^{2+} concentration. The latter is given for two levels of applied protein, the lower just below the point of saturation coverage, the higher well into the plateau region. Initially as salt level is increased, the layer thickness decreases, the response to added Ca^{2+} being much greater than for NaCl. The layer is thinned to a minimum value, both Ca^{2+} and NaCl giving similar values though at very different ionic concentrations.

Fig. 1 Adsorbed layer thickness as a function of applied β-casein (■) measured in 20 mM imidazole/HCl buffer, pH7.0,20 °C. Points for NaCl (◇) and CaCl$_2$ (▲) measured following addition of 25mM NaCl or 5.33mM CaCl$_2$ to buffered suspensions where protein has been pre-adsorbed at plotted applied concentration

Fig. 2(A) Influence of sodium chloride on the hydrodynamic thickness of a pre-adsorbed layer of β-casein (applied protein concentration 9.2mg m^{-2})

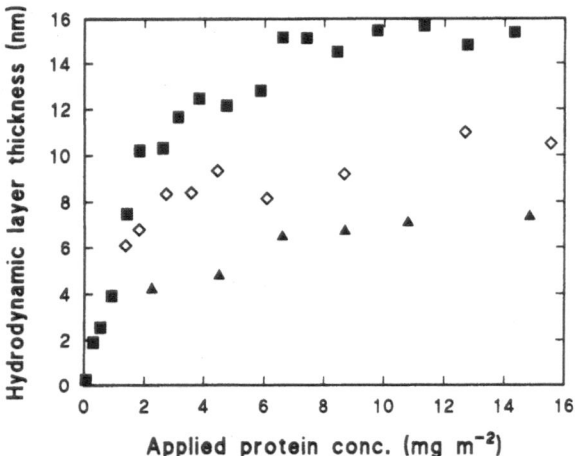

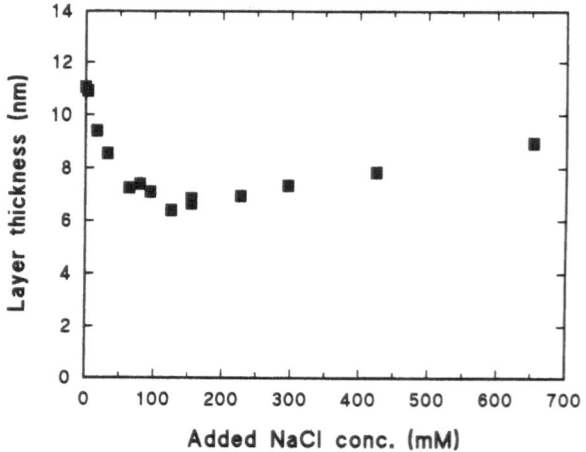

Progr Colloid Polym Sci (1994) 97:271–274
© Steinkopff-Verlag 1994

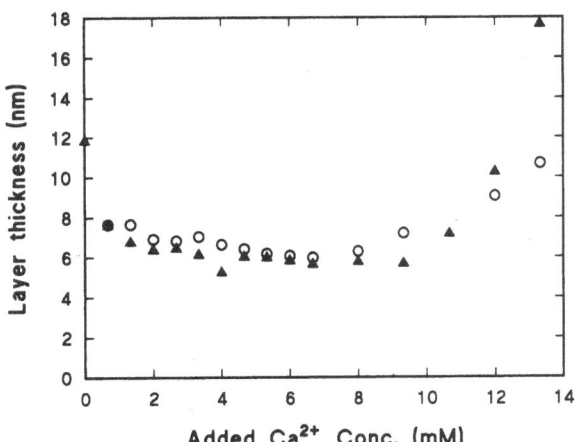

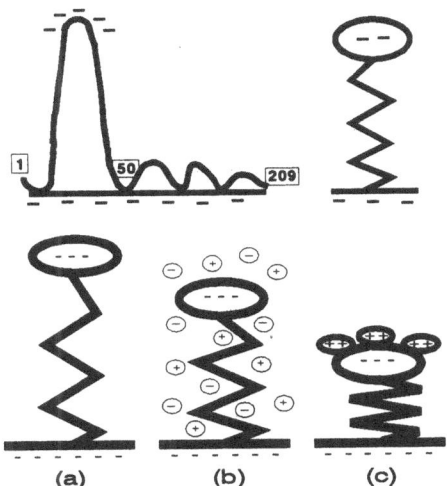

Fig. 2(B) Influence of calcium addition on the thickness of pre-adsorbed layers of β-casein. Layer thickness is plotted as a function of Ca^{2+} concentration for applied β-casein concentrations of $6\,mg\,m^{-2}$ (▲) and $15\,mg\,m^{-2}$ (○)

Fig. 3 Representation of the loop-and-chain configuration postulated for conformation of adsorbed β-casein and its blop-and-spring equivalent. The negatively charged blob or loop is repelled by the negatively charged surface and is retained by the spring. Diagram (a) represents the low ionic strength buffer situation, (b) the high ionic strength contraction and (c) the effect of the specific binding of Ca^{2+} ions

Thereafter both salt systems induce aggregation, the NaCl system at several hundred millimolar, the $CaCl_2$ at less than 10 millimolar, though the response here is dependent on the surface loading with β-casein (Fig. 2b), the latex being more readily destabilized at the lower applied protein level.

Discussion

The β-casein molecule is distinctly amphipathic. The N-terminal is highly charged whilst the remainder of the molecules has no net charge and a high content of hydrophobic residues. Evidence has accumulated for a model which represents the adsorbed β-casein molecule as having this train of hydrophobic residues lying along the interface and the N-terminal peptide stretching into the aqueous phase as a loop or tail as depicted in the cartoon (Fig. 3). This structure was initially inferred from observed changes in the hydrodynamic radius and analysis of the peptides released on treatment of emulsion droplets with the enzyme, trypsin [2, 6]. Direct evidence indicating such a structure has also been derived from neutron reflectivity studies of β-casein adsorbed at air/water and oil/water interfaces [7]. The results obtained in this series of experiments add more detail to the loop-and-train picture. The net negative charged carried by the N-terminal is approximately -10. Effectively the loop and thereby the hydrodynamic extent of the protein layer is the result of its repulsion from the negatively charged surface of the polystyrene latex.

Increasing the ionic strength would be expected to diminish the effectiveness of the electrostatic repulsion,

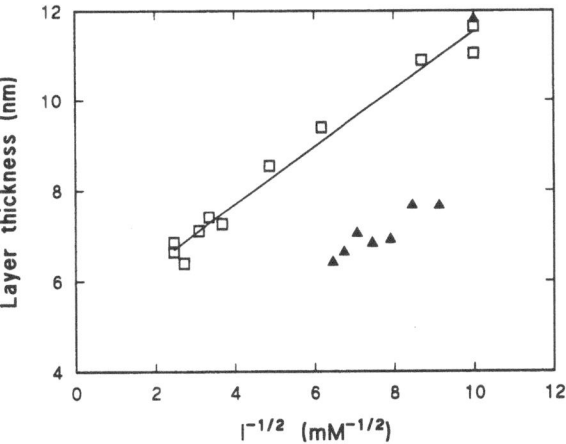

Fig. 4 Layer thickness as a function of the reciprocal of the square root of suspension ionic strength. Symbol □ indicates measurements where ionic strength was varied by addition of NaCl, ▲, by addition of $CaCl_2$

allowing the loop to relax back to the surface and thin the protein layer. The linear plot of layer thickness in NaCl solutions versus the inverse square root of the ionic strength (Fig. 4) is consistent with such a mechanism, since it suggests the product of the Debye–Huckel parameter, $\kappa (= 3.288\sqrt{I}$ in units of nm^{-1}) and the layer thickness to be constant.

When $CaCl_2$ is used to modify the ionic strength, the measured layer thickness does not follow the identical

274
D.S. Horne et al.
Interaction in adsorbed protein layers

pattern of behaviour (Fig. 4). Instead there is a rapid drop in thickness before an approximately linear trend parallel to the NaCl is observed. Ca^{2+} is known to bind strongly to α-casein in solution [8], most probably to the phosphoserine residues located near the N-terminus of the molecule in the putative loop. In this case it is suggested that such binding partially neutralizes the loop charge, allowing the layer to relax back to the latex surface as depicted in Fig. 3. The slow further contraction could then be an effect of increasing ionic strength or the approach to saturation of the calcium binding. The early precipitation encountered on Ca^{2+} addition may also be the result of an isoelectric precipitation rather than a straightforward salting-out effect but further studies of the kinetics are required to confirm this. It is, however, clear that the conformation adopted by the adsorbed casein molecule is a function of the ionic composition of the aqueous phase and that the bound molecule responds to such changes in a predictable fashion.

References

1. Dalgleish DG (1990) Colloids Surf 46:141–155
2. Dalgleish DG, Leaver J (1991) J Colloid Interface Sci 141:228–294
3. Mackie AR, Mingins J, North AN (1991) J Chem Soc Faraday Trans 87: 3043–3049
4. Leaver J, Law AJR (1992) J Dairy Res 59:557–561
5. Bradford MM (1976) Anal Biochem 72:249–254
6. Leaver J, Dalgleish DG (1990) Biochem Biophys Acta 1041:217–222
7. Dickinson E, Horne DS, Phipps JS, Richardson RM (1993) Langmuir 9:242–248
8. Parker TG, Dalgleish DG (1981) J. Dairy Res 48:71–76

Progr Colloid Polym Sci (1994) 97:275–280
© Steinkopff-Verlag 1994

BIO-COLLOIDS

C. Durrer
J.M. Irache
D. Duchêne
G. Ponchel

Study of the interactions between nanoparticles and intestinal mucosa

Received: 16 September 1993
Accepted: 9 June 1994

Dr. G. Ponchel (✉)
C. Durrer · J.M. Irache · D. Duchêne
Centre d'Etudes Pharmaceutique
URA CNRS 1218
5, rue Jean-Baptiste Clément
92296 Chatenay-Malabry, France

Abstract Nanoparticles are colloidal polymeric drug carriers of great promise for the peroral delivery of new drugs such as peptides, which are highly inactivated by the digestive juices. In this work, the influence of the surface charge and particle size of latexes and the pH of the suspension medium on the mucoadhesion were studied by adsorption experiments.

Key words Mucoadhesion – adsorption – nanoparticle – latex

Introduction

In the pharmaceutical field, bioadhesion has the objective of improving the therapeutic efficiency of drugs by increasing their residence time at their site of optimal activity or resorption [1]. If the site of attachment is a mucosa of the organism, the term *mucoadhesion* may be used. On the one hand, macroscopic bioadhesive systems such as Polycarbophil coated pellets [2] were not found to resist gastro-intestinal clearance *in vivo* for a prolonged time period. On the other hand, some encouraging results have been obtained from the peroral administration of suspensions of poly(alkylcyanoacrylate) nanoparticles to mice *in vivo* [3, 4]. From a theoretical standpoint, nanoparticulate systems are characterized by their huge specific surface that implies the possibility of an expansive contact with the mucosa, and by their small particle diameter that implies a very small hydrodynamic force of detachment for adsorbed particles [5]. Therefore, colloidal drug carriers, such as nanoparticles, are of great promise for peroral delivery of new drugs such as peptides, which are attacked by and highly inactivated by the digestive juices. Mucoadhesion would not just a) prolong the gastro-intestinal residence time, but would also b) protect the drug from inactivation, c) increase the local drug concentration, and d) allow the drug to diffuse directly from its carrier into the mucosa. For better understanding of the mucoadhesive behavior of these colloidal systems, some physico-chemical approaches have been made. *Ex vivo* desorption experiments of polymeric colloids from mucosa strips [6–8] have shown that the amount of nanoparticles resisting a flux of physiological saline (0.9% NaCl) was increased when the zeta potential of the particles was close to zero or positive. In this paper, the influence of the surface properties and particle size of model colloids and the pH of the suspension medium on the mucoadhesion were studied by adsorption experiments.

Materials and methods

Poly(styrene) latexes were chosen as a model colloidal system. Surfactant-free carboxylate and amino latexes

Table 1 Characteristics of Latexes and Mucin

Latex	Size (μm)	S (groups/nm^2)	zeta (mV) pH 4.5	zeta (mV) pH 6	zeta (mV) pH 7.4
PCM-200	0.210	9.7	− 18.3	− 34.9	− 34.8
CML-350	0.340	5.3	n.d	n.d	n.d
PAM-500	0.470	23.1	− 2.5	− 12.3	− 25.8
PAM-750	0.750	18.7	− 9.0	− 16.7	− 29.1
PCM-750	0.790	18.0	− 21.2	− 33.6	− 37.2
PAM-1000	1.000	11.2	− 11.7	− 27.2	− 30.7
PCM-1000	1.0650	11.1	− 23.2	− 34.3	− 34.0
PCM-2000	2.010	25.9	− 16.9	− 29.8	− 29.8
Mucin			− 12.4	− 12.5	− 24.6

were supplied as Polybead Carboxylate Microspheres (PCM) and Polybead Amino Microspheres (PAM), respectively, by Polysciences, Eppelheim, FRG. A carboxylate latex (CML-350) with adsorbed sodium dodecyl sulphate from a manufacturing process was furnished by Polymer Laboratories, Church Stretton, Shropshire, UK. The number in the latex designation indicates the nominal diameter (nm). Surfactant-free latex CML-350 was prepared by mixed-bed ion-exchange cleaning [9]. The surface groups were then measured by potentiometric titration with 0.1 N NaOH (carboxylate latexes) or 0.1 N HCL (amino latexes). The surface densities (S) of the titrated functional groups (groups/nm^2) are listed in Table 1. The particle sizes (TEM diameter) of the latexes were given by the supplier. The latex concentrations were determined by turbidimetry [10]. The zeta potentials of the latexes and of the suspended mucin were measured in physiological buffer solutions (Table 2) on a Malvern Zetasizer 4 (Malvern Instruments, Orsay, France). The latex characteristics and the zeta potentials of mucin are listed in Table 1.

As a model mucosa, the fresh small intestine of sacrificed male Wistar rats (IFFA CREDO, L'Arbresle, France) was excised, rinsed with physiological saline, and cut into segments of 5 cm length. Each segment was then opened lengthwise along the mesentery with scissors and spread on a glass slide previously covered with an aluminum film. A plate of aluminum with a slit 40 mm in length and 5 mm in width in the centre was then fixed on the mucosa samples thus prepared [11, 12]. For the present study, the slit was placed between Peyer's patches to avoid segmental differences, since Peyer's patches were reported to internalize particulate matter [13].

Suspensions of the latexes (4 g/L) were made in physiological buffer solutions (154 mM) at different pH. The composition of the buffer solutions is shown in Table 2. The latex suspensions were put in contact with the mucosa samples for 30 min, until equilibrium was reached [11]. The latexes were then sucked off and the mucosa samples were rinsed with 5 mL of the corresponding buffer solution to eliminate non-attached particles. All experiments were

Table 2 Composition of Physiological Buffer Solutions

Salt	pH 4.5 (g/L)	pH 6 (g/L)	pH 7.4 (g/L)
NaCl	4.9	9.0	8.0
KH$_2$PO$_4$	3.0	-	0.2
NaH$_2$PO$_4$.2H$_2$O	5.1	-	-
Na$_2$HPO$_5$.12H$_2$O	-	-	2.9
KCl	-	-	0.2

conducted at room temperature. The adsorbed amount was measured either by Fourier transform infrared spectroscopy, combined with the attenuated total reflection technique (FTIR-ATR), or by turbidimetry.

For the turbidimetric measurements, the mucous layer including the adsorbed particles was scraped off the membrane with a micro spatula and dispersed in 10 mL of a solution with 1% sodium hydroxide (NaOH) and 2% sodium dodecyl sulphate (SDS). The samples were treated for 2 h in an ultrasonication bath and left overnight at room temperature until the mucus was completely dissolved. The turbidity of the latex was then measured [11].

For the IR measurements, the dried mucosa samples, including the backing film of aluminum, were separated from the glass slides and spread, without any further sample preparation, on to the zinc selenide (ZnSe) crystal of the flat sampling plate of the ATR accessory (Specac Limited, Orpington, Kent, UK). As shown previously [12], the good contact between the sample and the crystal required for this technique was obtained using a special adjustable pressure device of the ATR accessory. The samples were run on a Mattson 5000 FT-IR spectrometer (Unicam, Argenteuil, France). The sampling conditions were set for 75 scans at 2-cm^{-1} resolution.

Results

The surface densities (S) of titrated groups were in agreement with the zeta potentials (Table 1). A higher surface

Progr Colloid Polym Sci (1994) 97:275–280
© Steinkopff-Verlag 1994

Table 3 Latex Adsorption on Apparent Surface of Rat Intestinal Mucosa

Latex	pH 4.5 (g/m²)	pH 6 (g/m²)	pH 7.4 (g/m²)	pH 6 (particles/μm²)
PCM-200	n.d.	0.660 ± 0.154	n.d.	129.6
CML-350	n.d.	0.558 ± 0.057	n.d.	25.8
PAM-500	0.754 ± 0.065	0.979 ± 0.124	0.460 ± 0.112	17.2
PAM-750	0.813 ± 0.147	0.880 ± 0.058	0.384 ± 0.092	3.8
PCM-750	0.611 ± 0.047	0.924 ± 0.062	0.226 ± 0.091	3.4
PAM-1000	0.817 ± 0.098	0.954 ± 0.078	0.265 ± 0.091	1.7
PCM-1000	0.670 ± 0.082	0.779 ± 0.038	0.215 ± 0.105	1.2
PCM-2000	0.947 ± 0.112	0.930 ± 0.130	0.263 ± 0.100	0.2

density of the amino groups yielded a less negative zeta potential, whereas a higher surface density of the carboxyl groups gave a slightly more negative zeta potential, especially at pH 7.4 where the carboxyl groups were ionized. The zeta potential of the latexes and the dispersed mucus were less negative at lower pH. For the amino latexes, the zeta potential approached zero at 4.5.

The adsorption values were expressed as mass of polymer per apparent surface of mucosa (g/m²), and are listed in Table 3. At pH 7.4, all the tested latexes were significantly less adsorbed (Anova- and Fisher test, $p = 0.05$)

than at pH 6 and 4.5 (Fig. 1). At pH 4.5, the latexes were less adsorbed than at pH 6, but the difference was not significant, except for PCM-750 and PAM-500.

At pH 7.4, the adsorption of the amino latexes (PAM) was increased when the zeta potential of the particles was less negative (Fig. 2). PAM-500 was significantly more adsorbed than PAM-1000 and the carboxylate latexes (PCM), which were all less adsorbed than the amino latexes.

At pH 6, adsorption increased with particle size, as shown by the linear regression (slope = 0.18, R > 0.84) of the carboxylate latexes (Fig. 3). PCM-200 and CML-350 were significantly less adsorbed than the larger latexes. The higher adsorption of PCM-200 and PCM-750, compared respectively with CML-350 and PCM-1000, corresponded well to the higher surface density (S) of the carboxyl groups. Generally, the amino latexes were more adsorbed than the carboxylate latexes of the same particle size. The difference was significant for PAM-1000 and PAM-500.

Figure 4 shows that at pH 4.5, adsorption increased with particle size, especially for the carboxylate

Fig. 1 Dependence of Latex Adsorption on pH

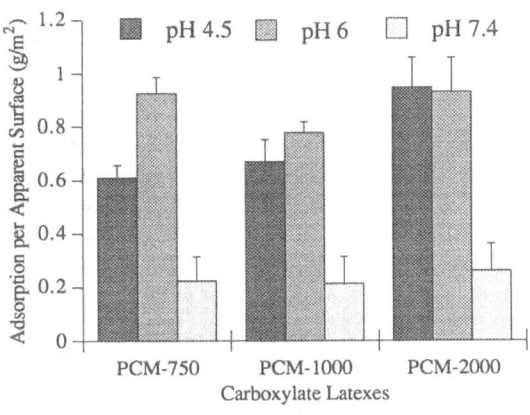

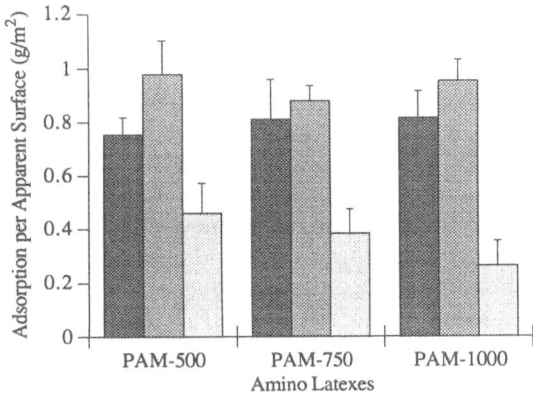

Fig. 2 Latex Adsorption at pH 7.4 against Zeta Potential

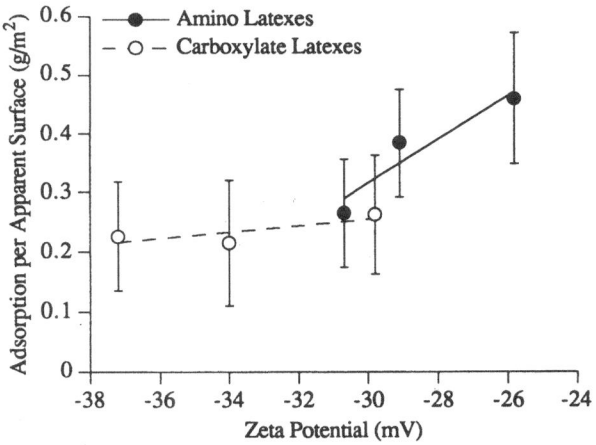

278
C. Durrer et al.
Interactions between nanoparticles and intestinal mucosa

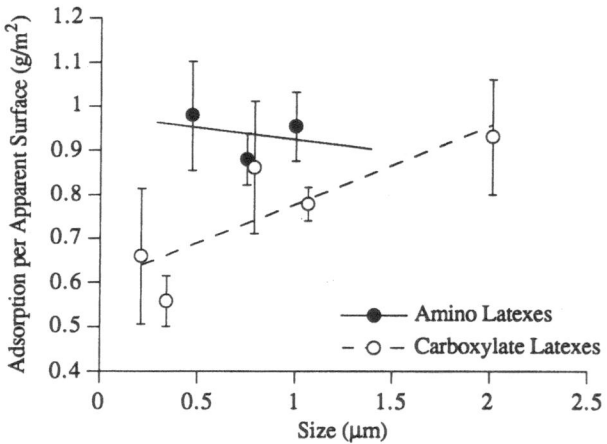

Fig. 3 Latex Adsorption at pH 6 against Particle Size

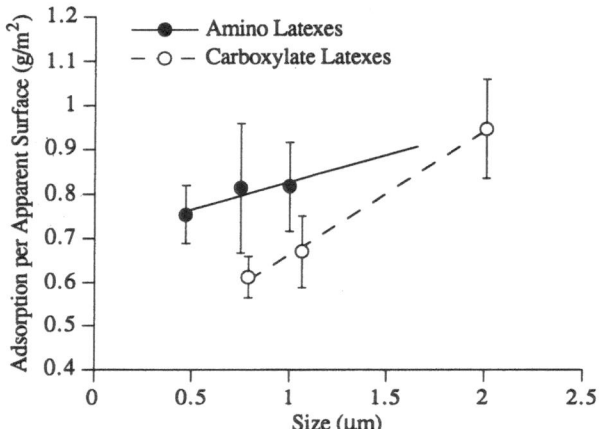

Fig. 4 Latex Adsorption at pH 4.5 against particle Size

(slope = 0.28, $R > 0.99$) but also for the amino latexes (slope = 0.12, $R > 0.90$). PCM-750, PCM-1000, and PAM-500 were significantly less adsorbed than PCM-2000. Contrary to pH 6, adsorption was not increased with higher surface density of the functional groups. The amino latexes were still more adsorbed than the carboxylate latexes of the same particle size, and the difference was significant for PAM-500 and PAM-750.

Discussion

At pH 7.4, mucus and latex particles were more negatively charged than at lower pH, as shown by their zeta potentials (Table 1). The mucous gel layer consists of a highly hydrated network of glycoprotein molecules. The terminal acidic substituents of the oligosaccharides in the

glycosylated parts are responsible for the polyanionic character of the mucin glycoproteins [14]. The pK_a of mucus was reported to be around 2.6 [13]. For the amino latexes, a dependence of the adsorption on their electrokinetic potential was observed (Fig. 2). Thus, it was concluded that adsorption at pH 7.4 was controlled by long-range interactions as known from the DLVO theory [15]: the repulsion forces and the energetic barrier of adsorption were higher when the surface density (S) of the amino groups was smaller and the zeta potential therefore more negative. No difference in adsorption was found between the carboxylate latexes. The presence of the carboxyl groups increased the negative charge of the poly(styrene) latexes slightly. Therefore, the repulsion forces were stronger for all the carboxylate latexes than for the amino latexes.

Since the mucous gel layer and the latexes were less negatively charged at pH 6 than at 7.4 (Table 1), adsorption increased. The higher adsorption of the amino latexes compared with the carboxylate latexes may be explained by the less negative zeta potentials of the amino latexes, and consequently by the smaller repulsion forces. The differences in adsorption between the carboxylate latexes could not be explained by the zeta potential, since their zeta potentials were all similar. As shown in the results, the adsorption of carboxylate latexes increased with higher surface density (S) of the functional groups. For PAM-500, it was not clear if the adsorption was very high only due to its small zeta potential or also because of its high surface density of the amino groups. However, a higher surface density of the functional groups increases the hydrophily of the polymer surface [13]. The results were in agreement with other work, where the surface tension of polymer films and the wetting behavior of mucin were studied, and adsorption increased with the presence of hydrophilic sites at the polymer surface [16, 17]. Furthermore, functional groups, such as carboxyl and amino groups, were described to promote adhesion to different adsorbents [18]. According to the dependence of adsorption on the surface density of the functional groups at the polymer surface, and according to the high proportion of hydrophilic groups in the mucus i.e., carbohydrates $> 60\%$ dry weight glycoprotein [19], it is likely that chemical bonds, for example hydrogen bonds [18], act as driving short-range forces in the mucoadhesion of functionalized latexes. However, the presence of hydrophobic interactions, as proposed for protein adsorption [20], cannot be excluded.

Theoretically, the repulsion forces are smaller at pH 4.5 than at pH 6, since the acidic groups of the mucus and the carboxyl groups of the latexes are less dissociated at lower pH, and the amino groups are protonated and therefore positively charged to some extent. Van der Waals attraction may also be involved in the adsorption of the amino latexes at pH 4.5 [15, 20]. This could explain the differences between the amino and the carboxylate latexes (Fig. 4), but not the differences between pH 4.5 and 6. Generally, the latexes were less adsorbed at pH 4.5 than at pH 6, except the largest latex. PCM-2000, which did not change (Fig. 1).

The model of adsorption on a smooth surface could hardly give a satisfying explanation of these results. The influence of the structure of the adsorbent, i.e., the mucous layer, was therefore considered. As mentioned above, the mucus is formed by a polyanionic network [14]. The glycoproteins constitute 5% or less of the mucus, and the rest is mainly water [19]. Adsorption isotherms [21], which were performed under the same experimental conditions as this study, have shown that the mucous gel layer behaved as a porous adsorbent for latexes up to 1 μm in particle size, whereas a Langmuirian isotherm shape was observed for a latex with 2 μm in particle size. At a latex bulk concentration of 4 g/L, as used in this work, the results were situated in the plateau of the adsorption isotherms for all latexes, except for PCM-2000.

At pH 7.4, the mucus gel is widely expanded due to the intramolecular repulsion forces between the ionized groups of the mucous glycoproteins [22]. The adsorption was controlled by electrostatic repulsion, as seen above, and no obvious dependence of the adsorption on the particle size was observed. At lower pH, the repulsion forces were less important than at pH 7.4, and the latex particles not only could reach the mucosa, but also could penetrate into the mucous network. At pH 4.5, the mucous glycoproteins were less dissociated and the intramolecular repulsion forces of the glycoproteins were reduced. Consequently, the network was less expanded and the pores were smaller. Therefore, latex diffusion into the mucous gel became more difficult. Furthermore, since the electrostatic barrier of adsorption was reduced, penetrating particles might have been attached earlier and might have blocked the pores for the following particles. On the one hand, for a diffusion controlled system, one would expect higher adsorption for smaller particles, because diffusion of particles is inversely proportional to the square of the particle size. On the other hand, interparticulate repulsion between adsorbed particles, increases with decreasing particle size, due to the higher specific surface. Taking into account the much higher number of small particles which was necessary to adsorb the same mass of latex as with larger particles (Table 3), the higher number of interactions between adsorbed particles was evident. We propose that adsorption was controlled by the balance of these two opposite phenomena. At lower PH, when diffusion was reduced, due to the changed mucous structure, interactions between the particles became more obvious. This was in agreement with the increased size dependence at pH 4.5

compared with pH 6, as shown by the slopes of the linear regression lines (Figs. 3 and 4). In other words, the mucus gel layer tended to behave more and more like a smooth adsorbent at lower pH, and interparticulate repulsion was responsible for the size dependent change in packing of the adsorbed particles., At pH 6, diffusion tended to dominate interparticulate repulsion, leading to inverse slopes of the size dependence of the adsorption, as observed for the amino latexes (Fig. 3). For PCM-2000, changes in the porosity of the mucus gel did not affect adsorption, and the mucosa behaved like a smooth surface. In fact, the adsorbed amount of PCM-2000 corresponded quite well to a monolayer of adsorbed particles on the apparent surface of mucosa.

Generally, it was found in this study that adsorption was higher for the amino than for the corresponding carboxylate latexes, since the repulsion forces between mucus and particles and between the particles were smaller. This was in agreement with earlier studies, where the adsorption of latexes to the mucosa under flux conditions [6], or the desorption of latexes [7, 8] were studied.

Conclusion

The adsorption experiments gave evidence of the influence of the surface charge and particle size on the latex adsorption on mucosa. At pH 7.4, all latexes were two to three times less adsorbed than at pH 6 and 4.5, due to repulsion forces of the negatively-charged mucous layer. The adsorption barrier was smaller for the less negatively-charged amino latexes. At pH 6, the adsorption was not just increased by the diminished repulsion forces, but also by the hydrophilic character of the latex surface. Chemical bonds, such as hydrogen bonds, were therefore proposed as driving adhesive forces. At pH 4.5, where the mucous gel probably lost a part of its expanded porous character, the dependence of adsorption on particle size became obvious. The higher specific surface of the smaller latexes may involve not just reduced adsorption due to more repulsion forces between the adsorbed particles, but also stronger adhesion of the particles. In experimental work, especially on a biological adsorbent, the situation will always be very complex, since several factors interact at the same time. However, considering the intestinal surface of over 10 m^2 in humans, these studies proved that mucoadhesion of nanoparticles is a hopeful strategy to improve drug therapy.

Acknowledgements We thank Mr. Sauvaire, Malvern, Orsay, France, for his help and hospitality during the zeta potential measurements.

280

C. Durrer et al.
Interactions between nanoparticles and intestinal mucosa

References

1. Duchêne D, Ponchel G (1989) STP Pharma 5:830–838
2. Khosla R, Davis SS (1987) J Pharm Pharmacol 39:47–49
3. Lenaerts V, Couvreur P, Grislain L, Maincent P (1990) In: Lenaerts V, Gurny R (eds) Bioadhesive drug delivery systems. CRC Press, Boca Raton, Florida, pp 93–104
4. Kreuter J, Müller U, Munz K (1989) Int J Pharm 55:39–45
5. Visser J (1970) J Colloid Interf Sci 34:26–31
6. Teng CLC, Ho NFH (1987) J Control Rel 6:133–149
7. Lenaerts V, Pimienta C, Juhasz J, Cadieux C (1989) Proceed Intern Symp Control Rel Bioact Mater 16:199–200
8. Pimienta C, Lenaerts V, Cadieux C, Raymond P, Juhasz J, Simard MA, Jolicoeur C (1990) Pharm Res 7:49–53
9. Van den Hul HJ, Vanderhof JW (1972) J Electroanal Chem 37:161–182
10. Irache JM, Durrer C, Ponchel G, Duchêne D (1993) Int J Pharm 90:R9-R12
11. Durrer C, Irache JM, Puisieux F, Duchêne D, Ponchel G (1994) Pharm Res: 11:674–679
12. Durrer C, Ponchel G, Duchêne D (1992) 6th Congr Int Technol Pharm 2:66–73
13. Gupta PK, Leung SHS, Robinson JR (1990) In Lenaerts V, Gurny R (eds) Bio-adhesive drug delivery systems. CRC Press, Boca Raton, Florida, pp 65–92
14. Carlstedt I, Sheehan JK, Corfield AP, Gallagher JT (1985) Essays Biochem 20:40–76
15. Rutter PR, Vincent B (1980) In Berkeley RCW, Lynch JM, Melling J, Rutter PR, Vincent B (eds) Microbial Adhesion to Surfaces. Society of Chemical Industry, Ellis Horwood, Chichester, England, pp 79–92
16. Proust JE, Baszkin A, Boissonnade MM (1983) J Colloid Interf Sci 94:421–429
17. Baszkin A, Proust JE, Monsenego P, Boissonnnade MM (1990) Biorheology 27:503–514
18. Wu S (1982) Polymer interface and adhesion. Marcel Dekker, New York
19. Creeth JM (1978) Br Med Bull 34:17–24
20. Norde W (1986) Adv Colloid Interf Sci 25:267–340
21. Durrer C, Irache S.M, Puisieux F. Duchêne D, Ponchel G (1994) Pharm Res 11:680–683.
22. Hesselink FT (1977) J Colloid Interf Sci 60:448–466

Progr Colloid Polym Sci (1994) 97:281–284
© Steinkopff-Verlag 1994

BIO-COLLOIDS

G.A. van Aken
M.T.E. Merks

Dynamic surface properties of milk proteins

Received: 16 September 1993
Accepted: 24 June 1994

Dr. G.A. van Aken (✉) · M.T.E. Merks
Department of Biophysical Chemistry
NIZO (Netherlands Institute for
Dairy Research)
P.O. Box 20
6710 BA Ede, The Netherlands

Abstract Surface-dilational properties of adsorbed layers of the pure milk proteins β-casein, β-lactoglobulin-A, and bovine serum albumin were studied by monitoring the surface pressure after a step-wise alteration of the area of the surface. The surface pressure instantaneously responds to the applied area variation, by an amount that is dependent on the particular protein and the initial surface pressure. For β-lactoglobulin and bovine serum albumin at surface pressures above $10\ \mathrm{mN \cdot m^{-1}}$ it was found that the instantaneous change of the surface pressure is followed by a decay towards a smaller change. The decay curves are well-fitted by bi-exponential functions. An important feature is that all the processes appeared reversible at time scales longer than the relaxation times. The surface-dilational behavior of the proteins was quantified by the values of the surface-dilational modulus at short and long time-scales, and the measured decay times.

Key words Protein adsorption – dynamic surface tension – surface rheology – surface dilational modulus – milk proteins

Introduction

Proteins play an important role in the formation and stability of emulsions and foams in various applications. In order to improve our understanding in this matter, we study the relationship between the behavior of proteins at interfaces and their emulsifying and foaming capacity. The subject is very complex because it involves many factors such as the process of formation of foam bubbles and emulsion droplets, the adsorption of proteins on the freshly formed interfaces, the sizes of the droplets and foam-bubbles, and the stabilization towards several breakdown mechanisms (e.g., disproportionation and coalescence).

A requisite for the understanding of all these factors is a detailed knowledge of the interfacial behaviour of proteins.

An important aspect of an adsorbed layer is the manner in which the surface pressure responds to imposed area changes. This can be quantified by the surface-dilational modulus, which is defined as [1–4]

$$\varepsilon = \frac{d\gamma}{d\ln A} = -\frac{d\Pi}{d\ln A}, \tag{1}$$

where A is the area of the adsorbed layer, γ is the surface tension, Π is the surface pressure defined by $\Pi = \gamma_0 - \gamma$, where γ_0 is the surface tension of the solvent.

As it turns out for many proteins, the adsorbed layer is viscoelastic. If the area is altered stepwise, an instantaneous (elastic) change of the interfacial tension is observed, followed by a (viscous) decay of this change. This relaxation behavior is due to conformational changes that take place within the adsorbed layer. It is customary to describe this viscoelastic behavior by the frequency-dependent surface-dilational modulus ε, defined by $\varepsilon = |\varepsilon| \cdot e^{i\theta} = \varepsilon'(\omega) + i\varepsilon''(\omega)$ in Eq. (1), where ω is the frequency and θ is the loss-angle.

The surface-dilational modulus is usually measured by imposing sinusoidal area variations with various frequencies ω to the adsorbed layer and recording the resulting variations of the surface pressure. A much faster alternative is to apply a Fourier transformation to the curve of the time-dependent response of the surface pressure to a stepwise area variation [2–4], which yields a wide frequency spectrum for ε from a single curve. The two methods were shown to yield similar results for an adsorbed layer of bovine serum albumin (BSA) [4]. In this paper we report on the surface-dilational behavior for the pure proteins β-casein, BSA and β-lactoglobulin, as studied by step-wise variations of the area.

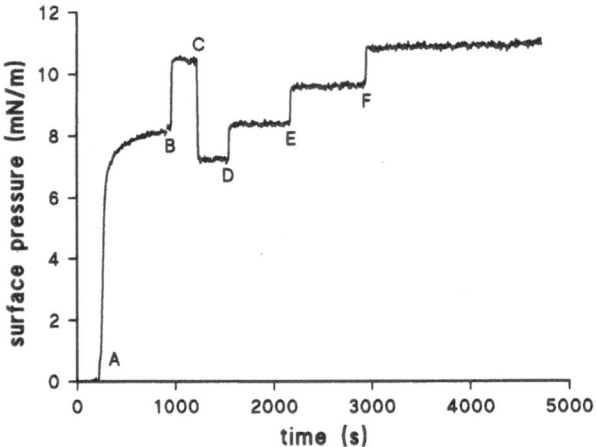

Fig. 1 Surface pressure of an adsorbed layer of β-casein. At A the protein is introduced to the surface (ca. $1\,\text{mg}\cdot\text{m}^{-2}$). Initial area = $6.901\,\text{cm}^2$. Subsequent area variations are (in cm^2): B: $6.901 \rightarrow 6.011$; C: $6.011 \rightarrow 7.346$; D: $7.346 \rightarrow 6.901$ E: $6.901 \rightarrow 6.456$; F: $6.456 \rightarrow 6.011$

Materials and methods

β-casein was isolated and purified in our laboratory by the method described in ref. [5]. BSA was of "essentially fatty acid free" quality purchased from Sigma-Aldrich N.V./S.A. (Brussels, Belgium). β-lactoglobulin-A was isolated by a procedure similar to the method of Maubois [6]. A small amount of dry protein was transferred to the surface of a buffered solution (0.020 M imidazole in double-distilled water, adjusted to pH 7.0 with a hydrochloric acid solution) in a small-sized Langmuir trough (dimensions ca. 1×6 cm) specially designed for this purpose. The area was varied stepwise by changing the short (ca. 1 cm) dimension of the trough by means of a micrometer screw, and the resulting change of the surface tension was monitored by measuring the force acting on a roughened platinum Wilhelmy plate sticking through the surface.

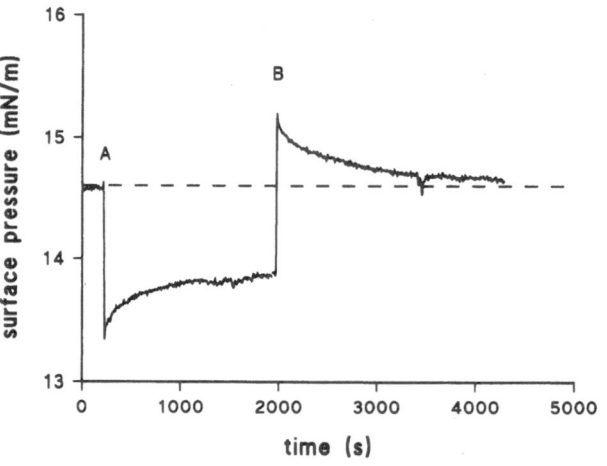

Fig. 2 Surface pressure of an adsorbed layer of β-lactoglobulin (ca. $1.5\,\text{mg}\cdot\text{m}^{-2}$). The dotted line shows that the expansion/compression cycle is reversible. Subsequent area variations are (in cm^2): A: $6.901 \rightarrow 7.346$; B: $7.346 \rightarrow 6.901$

Results

β-casein

From Fig. 1 it can be seen that at surface pressures between 7 and $11\,\text{mN}\cdot\text{m}^{-1}$ the surface pressure immediately responds to the area variations. This shows that relaxation effects are absent, which is indicative of fully elastic behavior for time scales accessible with our equipment (> 1 s). The compression/expansion cycles appear to be reversible. From these measurements we calculate $\varepsilon = 16.3 \pm 0.4\,\text{mN}\cdot\text{m}^{-1}$ for β-casein under these circumstances.

β-lactoglobulin-A

Figure 2 shows the behavior of the adsorbed layer at a surface pressure of about $15\,\text{mN}\cdot\text{m}^{-1}$. Relaxation of the surface pressure is observed, revealing visco-elasticity of the surface. Reversibility is indicated by the fact that the surface pressure eventually resumes its original value after the expansion/compression cycle is completed. The similarity of the shapes of the two decay curves suggests that the decay times for expansion and compression are approximately equal. The decay curves are well-fitted by

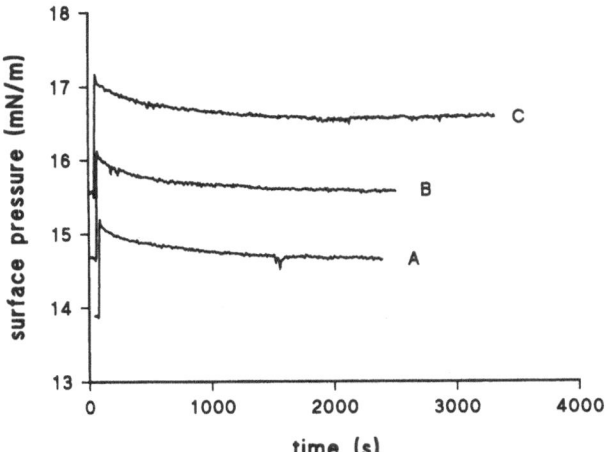

Fig. 3 Surface pressure of an adsorbed layer of β-lactoglobulin (same layer as in Fig. 2). Subsequent area variations are (in cm^2): A: $7.346 \to 6.901$; B: $6.901 \to 6.456$; C: $6.456 \to 6.011$

Table 1 Values for the surface-dilational moduli ε_0 and ε_∞ for β-lactoglobulin-A at a few values of the surface pressure Π.

Π ($mN \cdot m^{-1}$)	ε_0 ($mN \cdot m^{-1}$)	ε_∞ ($mN \cdot m^{-1}$)
14.5	21.0	11.7
15.5	21.6	13.2
16.5	22.6	13.5

bi-exponential functions

$$\Pi = P_1 \exp\left(-\frac{t}{\tau_1}\right) + P_2 \exp\left(-\frac{t}{\tau_2}\right) + \Pi_\infty \,, \qquad (2)$$

and the decay times τ_1 and τ_2 were found to be approximately 90 s and 1000 s, respectively, and $|P_1| \approx 0.14 \, mN \cdot m^{-1}$ and $|P_2| \approx 0.40 \, mN \cdot m^{-1}$. Figure 3 shows decay curves recorded at different surface pressures. Analysis of these curves showed that the decay times τ_1 and τ_2 are almost independent of the surface pressure.

As an alternative to plotting the real and imaginary parts of $\varepsilon(\omega)$ calculated by Fourier transformation of the relaxation curves, Table 1 gives the values $\varepsilon_0 = -(\Pi_0 - \Pi_i)/\Delta\ln A$ and $\varepsilon_\infty = -(\Pi_\infty - \Pi_i)/\Delta\ln A$, where Π_i is the equilibrium surface pressure before the step-wise area variation and Π_∞ is the surface pressure at infinite time thereafter. Π_0 is the value for the surface pressure at the moment of the area variation, which is found by extrapolating the biexponential fit according to $\Pi_0 = P_1 + P_2 + \Pi_\infty$. We note that Π_0 may have no physical meaning at time scales below the experimental limit of 1 s. The values of ε_0 and ε_∞ correspond to respectively the short-time scale (high frequency) and long-time scale (low frequency) limits of the surface-dilational modulus.

BSA

At surface pressures below $10 \, mN \cdot m^{-1}$ no relaxation effects were observed (Fig. 4), indicating that the adsorbed layer is elastic at measurable time scales (> 1 s). Also for this protein the compression/expansion cycle appears to be reversible. Table 2 shows ε-values calculated from similar experiments (not shown). From Table 2 we see that ε increases with increasing Π.

At surface pressures above $10 \, mM \cdot m^{-1}$ a relaxation behavior gradually appears (Fig. 5). At these pressures we also found reversibility of the compression/expansion cycle after relaxation (not shown). From the similarity of the curves of Fig. 5 we see that the time constants are again independent of the surface pressure. The decay curves were well-fitted by Eq. (2), and yielded decay times of approximately 50 and 300 s. P_1 was found to increase from ca. 0.4 to ca. 0.8 and P_2 from ca. 0.2 to ca. 0.5 for Π_∞-values increasing from 13.4 to 17.0 $mN \cdot m^{-1}$. Table 3 gives the values for ε_0 and ε_∞.

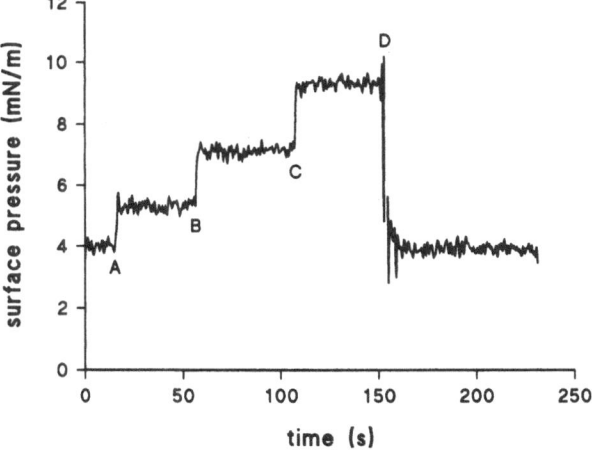

Fig. 4 Surface pressure of an adsorbed layer of BSA (ca. 1 $mg \cdot m^{-2}$) at low surface pressures. The area variations are (in cm^2): A: $7.346 \to 6.901$; B: $6.901 \to 6.456$; C: $6.456 \to 6.011$ D: $6.011 \to 7.346$

Table 2 Values for the surface-dilational modulus ε for BSA at low surface pressures Π.

Π ($mN \cdot m^{-1}$)	ε ($mN \cdot m^{-1}$)
3.0	16.1
3.5	18.1
4.1	20.3
4.8	21.1
5.5	25.6
6.4	27.2
7.3	27.1
8.3	28.2
9.4	28.1

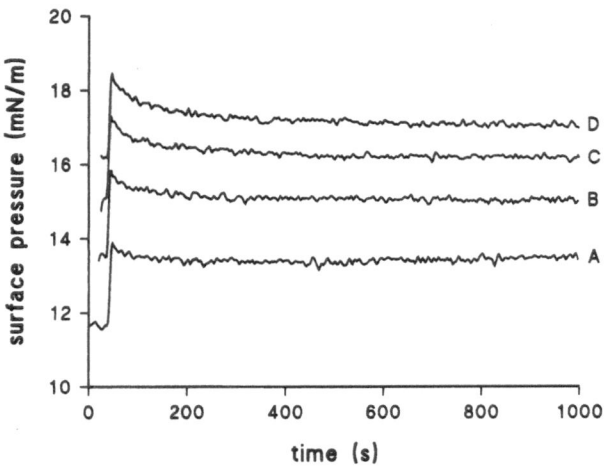

Fig. 5 Surface pressure of an adsorbed layer of BSA (ca. $2 \, \text{mg} \cdot \text{m}^{-2}$) at high surface pressures. The area variations are (in cm^2): A: $7.791 \rightarrow 7.346$; B: $7.346 \rightarrow 6.901$; C: $6.901 \rightarrow 6.456$; D: $6.456 \rightarrow 6.011$

Table 3 Values for the surface-dilational moduli ε_0 and ε_∞ for BSA at high surface pressure Π.

$\Pi \, (\text{mN} \cdot \text{m}^{-1})$	$\varepsilon_0 \, (\text{mN} \cdot \text{m}^{-1})$	$\varepsilon_\infty \, (\text{mN} \cdot \text{m}^{-1})$
13.5	36.4	26.5
15.5	35.6	23.4
16.5	33.9	17.7
17.5	30.5	12.2

Discussion and conclusion

For all pure proteins in this study, we found that the final value of the surface pressure after relaxation is a function of the area of the surface layer. This shows that the relaxation processes are reversible, and also that no significant desorption of protein from the surface layer occurred under the given conditions. Because exchange of protein with the bulk solution is absent, the dilational behavior of adsorbed layers of the pure proteins can be studied independently. The results of these measurements can be used in the modeling of other aspects of the interfacial behavior.

Acknowledgements β-casein was kindly supplied by B.W. van Markwijk and β-lactoglobulin-A was kindly supplied by D.G. Schmidt.

References

1. Lucassen-Reijnders EH (1981) In: Lucassen-Reijnders EH (ed) Anionic Surfactants. Marcel Dekker, New York, pp 173–216
2. Loglio G, Tesei U, Cini R (1979) J Colloid Interface Sci 71:316–320
3. Miller R, Loglio G, Tesei U, Schano KH (1991) Adv. Colloid Interface Sci 37:73–96
4. Van Aken GA, Merks MTE (1993) In Dickinson E, Walstra P (eds) Food Colloids and Polymers. The Royal Society of Chemistry, Cambridge, pp 402–406
5. Payens TAJ, Heremans K (1969) Biopolymers 8:335–345
6. Maubois JL (1979) Bulletin of the IDF 212:154–159

Progr Colloid Polym Sci (1994) 97:285–292
© Steinkopff-Verlag 1994

BIO-COLLOIDS

F. Aliotta
M. E. Fontanella
G. La Manna
V. Turco-Liveri

Dynamic properties of lecithin reverse micelles: an investigation of the sol-gel transition

Received: 16 September 1993
Accepted: 24 June 1994

F. Aliotta (✉) · M. E. Fontanella
Istituto di Tecniche Spettroscopiche
del C.N.R.
98166 Messina, Italy

G. La Manna · V. Turco-Liveri
Dipartimento di Chimica-Fisica
Universita
90123 Palermo, Italy

Abstract We present some results from the spectroscopic investigation of the hydrodynamic triplet in lecithin/cyclohexane/water reverse micelles. The investigation at high lecithin volume fraction clearly shows the dependence of the gel-formation process on the water content and the temperature. The observed phenomena are interpreted as originated by the competition between interfacial exchange processes and collisional effects whose balance is strongly temperature dependent. The investigation at low volume fraction, at fixed water content, allows us to determine the influence of the micelle population on the establishment of the entangled network. The data are compared with other results from literature and from rheological measurements.

Key words Giant micelles – gels – Brillouin scattering

General considerations

Very recently, a growing interest has been devoted to the study of the static and dynamic properties of lecithin-based gels [1–7]. Soybean lecithin, in fact, appears to act as a surfactant when the lecithin/organic solvent/water system is taken into account. In particular, the whole body of the experimental results seems to indicate that the sol-gel transition is driven by the one-dimensional growth of giant cylindrical reverse micelles, induced by the addition of small quantities of water. The gel formation takes place above a critical water content w_0 (w_0 = number of water molecules per lecithin molecule) whose value depends on the solvent under consideration, while appears quite independent on the lecithin volume fraction ϕ. Above a cross-over value ϕ^*, the entanglement of micelles becomes highly favorable, with the subsequent formation of a transient network very similar to that observed in semidilute polymer solutions.

This later consideration, together with the observation that micelles are not static entities but transient objects whose dynamic equilibrium depends on both temperature and concentration, makes a good model system for understanding the properties of living polymers.

On this basis, a good starting point to model the dynamics of stress relaxation can be found in the theoretical study of Cates [8] on entangled networks of living polymers. In that work, a simple model for the reaction kinetics is assumed in which i) a chain can break everywhere along the chemical sequence with fixed probability per unit length, and ii) the rate at which two chains can combine is proportional to the product of their concentrations. The dimensional chain distribution $N(l, \phi)$ is then exponential with mean $\bar{L}(\phi)$

$$N(L, \phi) \sim \exp\left[-\frac{L}{\bar{L}(\phi)} \right], \qquad (1)$$

where $\bar{L}(\phi)$ is taken large enough that $\alpha = L_e/\bar{L}(\phi) \leq 1$, being L_e the entanglement length. It is further assumed that the stress relaxation is driven by the reptation mechanisms. In such a way, the two relevant time scales are τ_{rep} (the reptation time for a chain of length $\bar{L}$) and $\tau_{break} = \tau_{rep}\zeta$ (the mean time for such a chain breaks in two pieces). It is found that for $\zeta \geq 1$ the main stress relaxation process is simple reptation, while for $\alpha \leq \zeta \leq 1$ a new intermediate time scale $\tau = \tau_{rep}\zeta^{1/2} = (\tau_{rep}\tau_{break})^{1/2}$ is predicted, induced by the breaking of the chain in a position close enough to a given tube segment that reptative relaxation of that segment can occur, before the new chain end is lost by recombination. Crossovers are interpreted in terms of similar mechanisms in which breathing motions of the chains and local Rouse-like [9] motions are involved.

The aim is the investigation, in the lecithin/cyclohexane/water system, of the dependence on w, T and ϕ of the involved relaxation processes. In particular, we will show how the results from a Brillouin scattering experiment, namely, the dependence of the hypersonic velocity v_h and of the normalized absorption α/f^2 on the above parameters, can be interpreted in terms of a micelle size distribution described by Eq. (1). Furthermore, it will be shown how the sol-gel transition is driven by increasing the system polydispersivity with ϕ, which results in an increase of $\bar{L}(\phi)$. Finally, an estimation of the micelle sizes and the entanglement length will be obtained.

Experimental procedures and results

High purity (97%) soybean lecithin was obtained starting from a commercial product (Sigma), following the purification procedure described elsewhere [1, 4]. The purified lecithin was then dissolved, under continuous stirring at 20 °C, in cyclohexane, reagent grade quality (Baker Chemical), to obtain a 197 mM solution (corresponding to a lecithin volume fraction $\phi = 0.142$). Water was then added to obtain samples with different water/lecithin ratios ($w = 0, 3, 5, 8$). Using the same procedure, samples were prepared at different concentrations ($\phi = 0, 0.002, 0.005, 0.009, 0.019, 0.034, 0.066, 0.102$) at fixed w value ($w = w_0 = 10$).

Brillouin-scattering measurements were performed in a VV polarization geometry at different scattering angles ($35° \leq \theta \leq 135°$). The measurements at fixed ϕ, as a function of w, were performed in the temperature range $10°C \leq T \leq 40°C$, while the investigations at fixed $w = w_0$, as a function of ϕ, were performed at the constant temperature of 25 °C. In Fig. 1 we report, as an example, the experimental results for the $w = 10$ samples, at three different ϕ values ($\theta = 90°$, $T = 25°C$).

The experimental spectra were analyzed by fitting them to the usual expression [10]:

$$I_{vv}(\omega) = \frac{A_R \Gamma_R}{\omega^2 + \Gamma_R^2} + \left[\frac{A_B \Gamma_B}{[\omega - (\omega_B^2 - \Gamma_B^2)^{\frac{1}{2}}]^2 + \Gamma_B^2} \right.$$

$$+ \left. \frac{A_B \Gamma_B}{[\omega + (\omega_B^2 - \Gamma_B^2)^{\frac{1}{2}}]^2 + \Gamma_B^2} \right] + \frac{\Gamma_B}{\omega_B^2 + \Gamma_B^2}$$

$$\times \left[\frac{\omega - (\omega_B^2 - \Gamma_B^2)^{\frac{1}{4}}}{[\omega - (\omega_B^2 - \Gamma_B^2)^{\frac{1}{4}}]^2 + \Gamma_B^2} \right.$$

$$+ \left. \frac{\omega + (\omega_B^2 - \Gamma_B^2)^{\frac{1}{4}}}{[\omega + (\omega_B^2 - \Gamma_B^2)^{\frac{1}{4}}]^2 + \Gamma_B^2} \right], \qquad (2)$$

where the first term describes the quasi-elastic, resolution enlarged, central line, the next the symmetrical Brillouin contributions, and the last the asymmetric contributions. The fitting procedure furnishes the frequency shift ω_B and the HWHM Γ_B of the Brillouin lines. The continuous lines in Fig. 1 represent the results of such a procedure.

Fig. 1 Brillouin spectra for the lecithin/cyclohexane/water system at fixed water contents and at different lecithin volume fractions ϕ ($\theta = 90°$, T = 25 °C). Continuous lines represent the fitting results by Eq. (2)

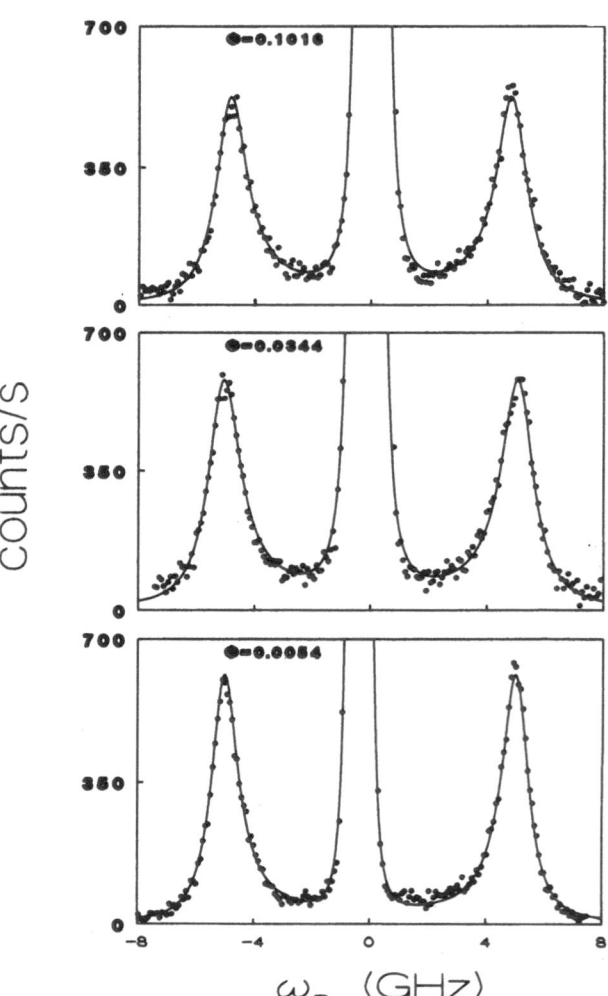

Progr Colloid Polym Sci (1994) 97:285–292
© Steinkopff-Verlag 1994

From the fitting results, we can calculate the values of the acoustic parameters, namely the hypersonic velocity v_h and the normalized absorption α/f^2, according to the expressions:

$$v_h = \frac{\omega_B}{k}$$

$$\frac{\alpha}{f^2} = \frac{2\pi\Gamma_B}{v_h\omega_B^2} \ .$$

Complementary measurements of refractive index, density, and viscosity were performed by means of an Abbé refractometer, a flux densimeter and an Ubbelohde viscometer. The systems were tested using viscometers with different shear rates ($100 \div 500$ s^{-1}). Since the measured viscosity is constant under such a shear rate, and taking into account the experimental results from ultrasonic absorption that given indication for a relaxation process centered at higher frequency (~ 5 MHz) [4, 6], we can assume that the resulting data represent $\eta(0)$.

The values of the refractive index n, density ρ and shear vicosity η_S are reported in Fig. 2 for the two sets of data.

Discussion

w, T dependence

An inspection of the viscosity data in Fig. 2 shows that, while a single T-dependence is observed for the sample at $w = 0$, two distinct regimes are detected for the samples containing water. The stronger temperature dependence at the lower temperatures is a clear evidence of the establishment of some more extended structures, while the shift of the transition temperature towards higher values, when w increases, represents the first evidence for a network formation strongly dependent on a number of internal parameters (the mean size of the micelles, their population, the degree of polydispersity) whose equilibrium changes with the external parameter T.

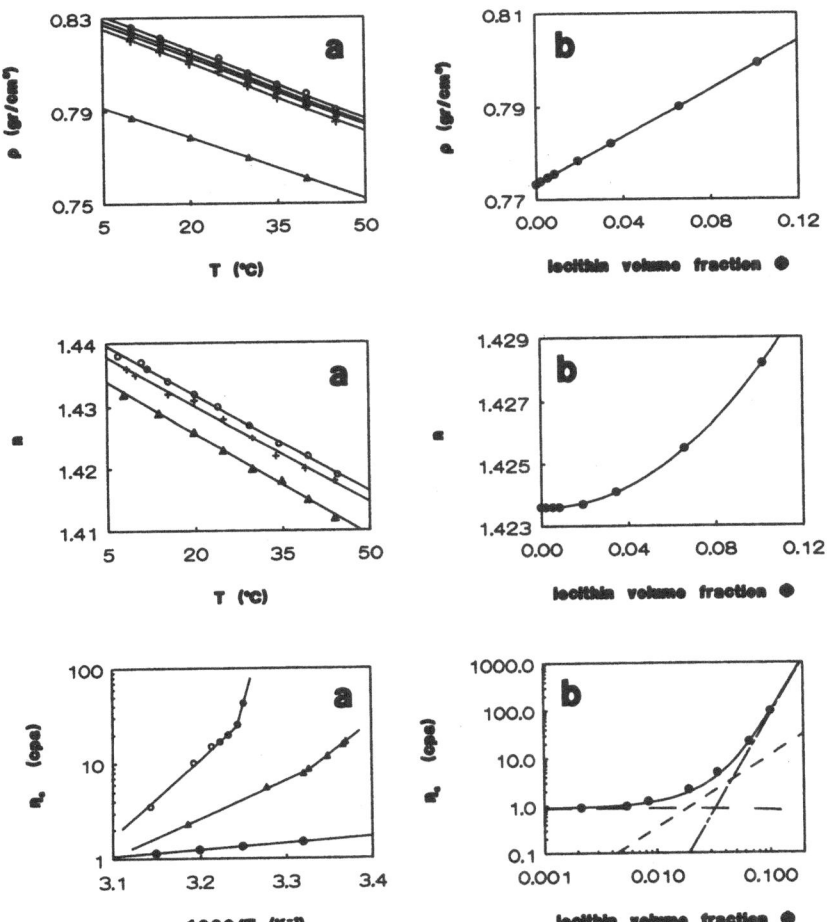

Fig. 2 Temperature (a) and ϕ (b) dependence of the density ρ, refractive index n and viscosity η_S, for the lecithin/ciclohexane/water system. (a) data at fixed lecithin volume fraction, $\phi = 0.142$, and at different water contents: crosses $w = 0$, triangles $w = 3$, full circles $w = 5$, circles $w = 8$. Full triangles represent data for pure cyclohexane (b) data at fixed water content ($w = w_0 = 10$) as a function of ϕ. Continuous line in the η_S plot: fitting result (see text)

In fact, the addition of water results in an increase of the mean micellar size while the overall micelle population has to decrease, due to the fact that we are working at fixed ϕ values. An opposite role is played by the temperature. When the temperature increases, the micelle size distribution given by Eq. (1) narrows and the mean length $\bar{L}$ lowers.

Furthermore, due to the fact that micelles are not charged, they can easily collide, merge and then break. Percolation occurs at a critical lecithin volume fraction ϕ^*, whose value depends on the position of the system in the (w, T) space. Such a mechanism, obviously, will give contributions to the excess sound absorption of the system with respect to the continuous phase. Other contributions to the acoustic losses can be found in some exchange processes at the interface. The balance between the above-mentioned processes will be a function of the position of the system in the (w, T) space. The possibility of such effects was reported in the literature for a number of micellar solutions close to a percolation threshold [11].

A suggestion for a dynamics driven by the competition between two effects is given in Fig. 3 where the normalized hypersonic absorption is reported as a function of frequency at different temperatures for the systems at $w = 0$ (a), $w = 5$ (b) and $w = 8$ (c). It is evident that, for the two extreme systems, α/f^2 is nearly temperature independent, while the intermediate one moves from a situation very similar to that of the more connected system, at the lower temperatures, toward values very close to those of the $w = 0$ case, when temperature increases.

A more direct indication for a continuous evolution in the structure of our system as a function of the water content comes from the behavior of the normalized absorption as a function of w (see Fig. 4, where the data are reported at different frequencies). In particular, in the lower temperature range (10–20 °C) α/f^2 shows a diverging behavior as w increases. Such a result could be interpreted in terms of a growing up of the mean micelle size with w that induces the building-up of the extended structure of the gel network. When temperature increases, the breaking and reforming mechanisms of the micelles become increasingly faster, the kinetic equilibrium between these two processes shifts toward lower $\bar{L}$ values (see Eq. (1), and no further extended structure is allowed for our system. Now, the main energy-loss channels are the interfacial exchange processes and the α/f^2 dependence on w becomes smoother. It is to be stressed that both viscosity and hypersonic data indicate $T \simeq 25$ °C as the limiting temperature above which the gel network is destroyed.

Summing up, our results seem to indicate that it is the equilibrium between collisional effects and interface exchange processes that determines the behavior of our system: when temperature decreases the mean micellar size increases, and the same polydispersity does, so allowing

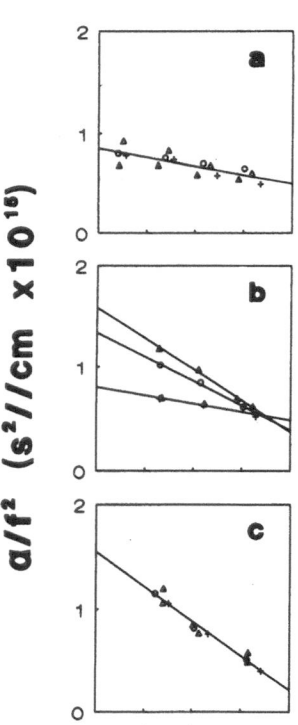

Fig. 3 Normalized hypersonic absorption for the lecithin/cyclohexane/water system at different temperatures as a function of frequency. Data are reported for systems at different water contents: $w = 0$ (a); $w = 3$ (b); $w = 8$ (c). Symbols (crosses) $T = 10$ °C, (triangles) $T = 20$ °C, (circles) $T = 30$ °C, (full triangles) $T = 40$ °C. Continuous lines are guides for eye

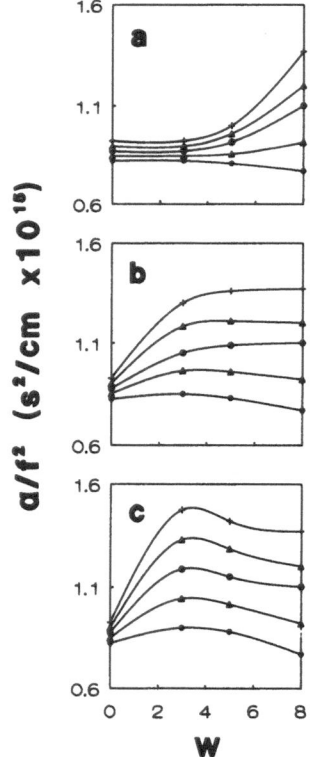

Fig. 4 Normalized hypersonic absorption data extrapolated at different frequencies, reported as a function of the water content w. Data are reported for different temperatures: 10 °C $\leq T \leq 20$ °C (a); $T = 30$ °C (b); $T = 40$ °C (c). Symbols: (crosses) 3 GHz, (triangles) 4 GHz, (circles) 5 GHz, (full triangles) 6 GHz, (full circles) 7 GHz

Progr Colloid Polym Sci (1994) 97:285–292
© Steinkopff-Verlag 1994

the building up of the gel network. On the contrary, at higher temperature the micelle size distribution narrows, hindering the formation of any extended structure.

ϕ dependence

In order to analyze the dependence of the system properties on the lecithin volume fraction, we tried to extract the micellar contribution from the density, refractive index and viscosity data, according to the expressions:

$$\rho_{mic} = \frac{\rho - (1 - \phi) \cdot \rho_{cyc}}{\phi}; \quad n_{mic} = \frac{n - (1 - \phi) \cdot n_{cyc}}{\phi};$$

$$\eta_{mic} = \frac{\eta - (1 - \phi) \cdot \eta_{cyc}}{\phi},$$

where the indexes mic and cyc refer to the micellar and cyclohexane contributions respectively.

The micellar density ρ_{mic} turned out to be quite independent of the lecithin volume fraction ($\rho_{mic} = 1.03$), indicating that lecithin concentration influences only the size of the micelles and not their structure.

The micellar refractive index looks linearly dependent on ϕ, indicating that some aggregative phenomena are taking place, reflecting in the dielectric constant of the system (see Fig. 5).

The viscosity data η_S, reported in Fig. 2b, seem to be in good agreement with the results of Ott. and coworkers [2] on the self diffusion coefficient of the cylindrical micelles. The viscosity data are almost constant until the treshold concentration value, ϕ^*, is reached. After this, the percolation effects start and the viscosity assumes a diverging behavior with ϕ. We tried to fit our data with

a simple Stokes-like law at the lower temperatures, while a scaling law ϕ^x was assumed for the higher concentrations, where x is a parameter to determine. The fitting procedure furnished a value $x = 3$, in agreement with the Cates model [8], when some breathing motions are assumed to take place, overimposed to the breaking-reptation mechanism. The result of the fitting is reported as a continuous line in Fig. 2b, while the dashed lines represent the above described components.

In an analogous way, the micelle contributions to the hypersonic velocity and normalized absorption were extracted. The results are reported in Fig. 6 as a function of ϕ and at different frequencies. A strong concentration dependence is shown by the data. The velocity, in fact, slows-down rapidly when ϕ increases, while the absorption increases.

In order to explain the observed behaviors, a simple model was adopted. In this approach, we assumed that the observed dependence of the acoustic parameters on ϕ is originated by the changes in the micelle size distribution. Such an idea was suggested by the observation that the wave-length of the measured hypersound is of the order of

Fig. 6 Micellar contributions to the hypersonic phase velocity and to the normalized absorption as a function of the lecithin volume fraction, at $T = 25\,°C$. Symbols: (full triangles) 2 GHz, (circles) 3 GHz, (triangles) 4 GHz, (reversed triangles) 5 GHz, (rhombuses) 6 GHz, (squares) 7 GHz. Continuous lines represent the fitting results with eq. 8

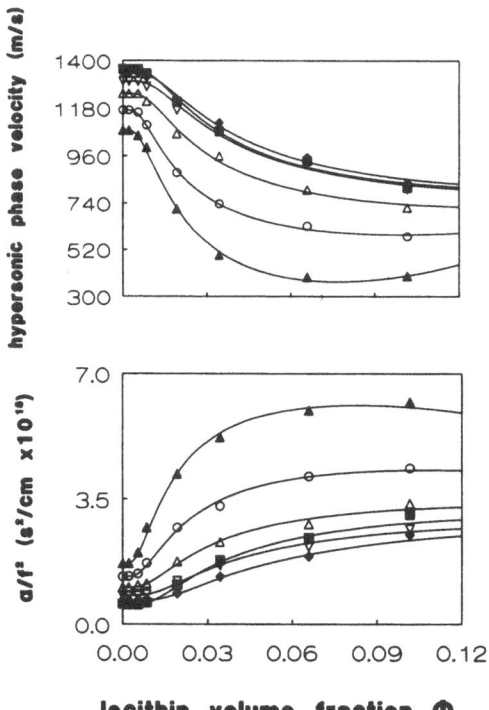

Fig. 5 Micellar refractive index vs lecithin concentration at $T = 25\,°C$

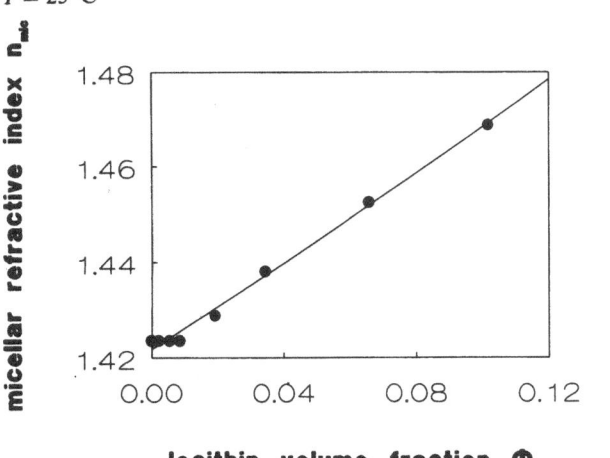

some thousands of $\mathring{A}$, the same order of magnitude of the micellar contour length determined from light scattering data [3]. From a mechanical point of view, each micelle can be looked at as a string of length L. Such an object behaves like an oscillator with eigen-frequency

$$\omega_0 = \sqrt{\frac{k}{m}} = \sqrt{\frac{K}{\rho_{\mathrm{mic}}\sigma L}} , \qquad (2)$$

where σ is the constant area of the micelle section. The last term in Eq. (2) comes from the experimental observaton of the concentration independence of the micellar density.

An obvious result is that when a wave of frequency ω interacts with our oscillator, the phase will undergo a delay, whose amplitude will be dependent on the quantity $\omega - \omega_0$. In other words, the phase velocity of the acoustic wave will be delayed according to the expression:

$$v_{\mathrm{hyp}}(\Omega) \sim \frac{1}{1 + \dfrac{\Omega^2}{\Gamma^2}} , \qquad (3)$$

where Γ is a parameter taking into account the anharmonicity of the oscillator and $\Omega = \omega - \omega_0$. Of course, the phase loss will be reflected in an energy loss,

$$\alpha(\Omega) \sim \frac{1}{1 + \dfrac{\Omega^2}{\Gamma^2}} . \qquad (4)$$

In the case under examination, hypersonic wave at fixed frequency traveling across the system will interact with oscillators whose mean length $\bar{L}$ will be dependent on the lecithin volume fraction. As a consequence, also the mean resonance frequency and the mean width of the resonance curve will be ϕ dependent. After these considerations, and introducing the quantity,

$$\Omega(\phi) = \omega - \langle\omega_0(\phi)\rangle ,$$

Eq. (3) can be rewritten as

$$v_{\mathrm{hyp}}(\Omega(\phi)) \sim \frac{1}{1 + \dfrac{[\Omega(\phi)]^2}{[\Gamma(\phi)]^2}} , \qquad (5)$$

and the same applies for Eq. (4).

Such an equation could be successfully applied only to monodisperse systems. In that case, in fact, the oscillators will be coherent (all the strings will oscillate at the same frequency and with the same phase).

In our system, on the contrary, at a fixed ϕ the micelles are dispersed in size according to Eq. (1). Taking Eq. (2) into account and transforming to the (ω_0, ϕ) space, we can write

$$N(\omega_0, \phi) \sim \exp\left[-\frac{\langle\omega_0(\phi)\rangle^2}{\omega_0^2} \right] .$$

As a consequence, the number of oscillators that will follow in phase a vibration at frequency Ω will be

$$N(\Omega, \phi) \sim \exp\left[-\frac{\langle\omega_0(\phi)\rangle^2}{\Omega^2} \right] . \qquad (6)$$

We can look at Eq. (6) as an expression of the degree of coherence of the oscillators. In fact, it gives the number density of oscillators that are moving in phase after a time $\tau = 1/\Omega$ from the beginning of the motion.

After these considerations, we rewrite Eq. (5) as

$$v_{\mathrm{hyp}}(\Omega, \phi) \sim \exp\left[-\frac{\langle\omega_0(\phi)\rangle^2}{(\Omega(\phi))^2} \right] \cdot \frac{1}{1 + \dfrac{[\Omega(\phi)]^2}{[\Gamma(\phi)]^2}}$$

$$+ \left[1 - \exp\left(-\frac{\langle\omega_0(\phi)\rangle^2}{(\Omega(\phi))^2} \right) \right] . \qquad (7)$$

Equation (7) differs from the case of the monodisperse system for the factor $N(\Omega, \phi)$ that weights coherent term, while the summed complementary factor decreases the maximum absorbance of the system. If $N(\Omega, \phi) = 0$, there will not be any interference between the oscillators and hence the measured phase velocity will be just the average of the contribution from each oscillator, while for $N(\Omega, \phi) = 1$, we come back to the hypothesis of a monodisperse system.

Transforming from the space (Ω, ϕ) to the space (k, ϕ), where k is the wave-vector of the hypersound, related with the wave-length λ_{hyp} according to

$$k = \frac{2\pi}{\lambda_{\mathrm{hyp}}} ,$$

we obtain

$$\bar{v}_{\mathrm{hyp}}(\lambda_{\mathrm{hyp}}, \phi) \sim \exp\left[\frac{1}{\bar{L}(\phi) \cdot 4\sin^2\left(2\pi\dfrac{\bar{L}(\phi)}{\lambda_{\mathrm{hyp}}} \right)} \right]$$

$$\times \left[\frac{1}{1 + \left(\dfrac{\sin\left(2\pi\dfrac{\bar{L}(\phi)}{\lambda_{\mathrm{hyp}}} \right)}{\Delta L(\phi)} \right)^2} - 1 \right] + 1 , \qquad (8)$$

and, taking Eq. (4) into account, it is possible to obtain the analogous expression for the normalized absorption. In fitting the experimental data with Eq. (8), a scaling law $\bar{L} \sim \phi^{1/2}$ was assumed for the mean micelle size, in agreement with the Cates model [8]. The results of such a fitting are represented as continuous lines in Fig. 6.

The above described procedure, furnishes only the fractionary part of the $\bar{L}(\phi)/\lambda_{\mathrm{hyp}}$ ratio. In order to obtain our goal, namely the determination of the $\bar{L}$ values at the different lecithin volume fractions, we have to found the set of integers n, for which the quantities

$$\frac{\bar{L}(\phi)}{\lambda_{hyp}} + n \cdot \frac{\lambda_{hyp}}{2}$$

become the same, independent of the λ_{hyp} values. The result is presented in Fig. 7.

Looking at the viscosity data, the crossover from the dilute to the semidilute regime can be observed at a concentration $\phi^* \simeq 0.018$, very near to the value $\phi^* \simeq 0.014$ determined by Schurtenberger and coworkers [3]. At that concentration, the mean micelle length has to be coincident with the entanglement length L_e. So, by assuming that L_e behaves like ϕ^{-1}, the values of the entanglement lengths can be deduced. The result is presented in Fig. 7. It is to be stressed that our result for $\bar{L}$ at $\phi = 0.0036$ ($\bar{L} \simeq 900$ Å) appears underestimated when compared with the determination of Schurtenberger and coworkers [3] at the same concentration ($\bar{L} \geq 4500$ Å). In any case, when the Schurtenberger value is assumed, if one tries to extrapolate the result at $\phi = 0.12$, a mean micelle length of $\simeq 30000$ Å is deduced, and this is in disagreement with the experimental observation of an optically clear sample.

Conclusions

Summing up, the Brillouin-scattering results presented in this paper are able to furnish a good idea of the processes giving rise to the gel formation in our sample. In agreement with the previous results, it appears that the sol-gel transition is taking place just for topologic reasons: when the mean micelle size reaches the value of the entanglement length, the extended network is established and the reptative motion dominates the observed relaxation processes. As a consequence, any dependence of the dynamics of our system on T, w or ϕ has to be connected with the dependence of the mean micelle size, and of the micelle size distribution, on the above parameters. In particular, an increase of w (in the range $0 \leq w \leq w_0$) induces an increasing of the maximum allowed micelle length or, in other words, increases the polydispersity of the system. As a consequence, the sol-gel transition takes place at lower $1/T$ and ϕ values.

ϕ and $1/T$ play the same role: both of them shift in the same way the kinetic equilibrium of the breaking and reforming mechanisms of the micelles. A lowering of their values reflects in a narrowing of the micelle size distribution, while an increase induces an higher polydispersity. In any case, the sol-gel transition observed in such kind of systems turns out to be a consequence of the variation of the polydispersity of the system.

The hypersonic probe, coupling directly with the characteristic lengths of the system, appears able to follow the sizes distribution crossing the transition from the dilute to the semidilute regime.

The deduced dependence of Eq. (1) on ϕ is shown in Fig. 8.

Fig. 7 ϕ dependence of the mean micelle size, $\bar{L}$, and of the entanglement length

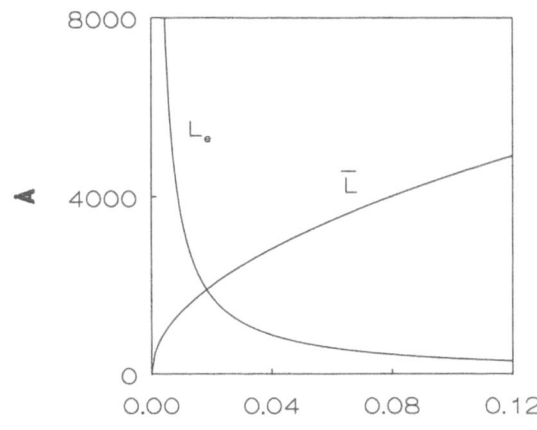

lecithin volume fraction Φ

Fig. 8 Dependence on ϕ of the micelle size distribution

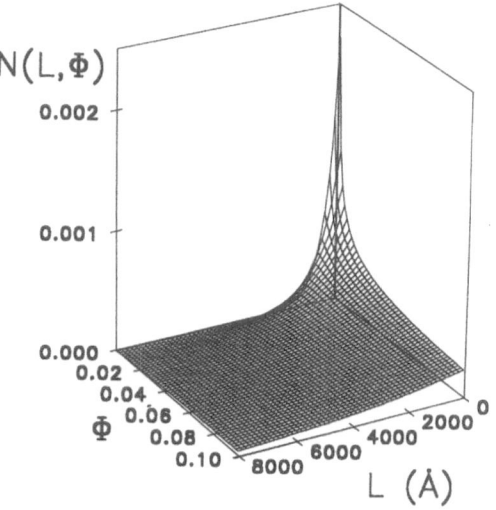

292

F. Aliotta et al.
Dynamic properties of lecithin reverse micelles

References

1. Schurtenberger P, Scartazzini R, Luisi PI, (1989) Rheol Acta 28:372
2. Ott A, Urbach W, Langevin D, Schurtenberger P, Scartazzini R, Luisi PL (1990) J Phys Condens Matter 2:5907
3. Schurtenberger P, Magid LJ, King SM, Lindner P (1991) J Phys Chem 95:4173
4. Aliotta F, Fontanella ME, Magazú S, Vasi C, Crupi V, Maisano G, Majolino D (1992) Mol Cyrst Liq Cryst 212:255
5. Aliotta F, Fontanella ME, Magazú S, Maisano G, Majolino D, Migliardo P (1992) Prog Colloid Polym Sci, 89:253
6. Aliotta F, Fontanella ME, Galli G, Lanza M, Miliardo P, Salvato G (1993) J Phys Chem, 97:733
7. Aliotta F, Fontanella ME, Squadrito G, Migliardo P, La Manna G, Turco-Liveri V (1993) J Phys Chem 97:6541
8. Cates ME (1986) Macromolecules, 21:2289
9. Rouse PE (1953) J Chem Phys 21:1272
10. See, e.g., Boon JP, Yip S (1980) Molecular hydrodynamics, McGraw-Hill, New York
11. Zana R, Lang J, Sorba O, Cazabat AM, Langevin D (1982) J Phys Lett, L829:43

Progr Colloid Polym Sci (1994) 97:293–297
© Steinkopff-Verlag 1994

BIO-COLLOIDS

Stressing phospholipid membranes using mechanical effects of light

M.I. Angelova
B. Pouligny
G. Martinot-Lagarde
G. Gréhan
G. Gouesbet

Received: 16 September 1993
Accepted: 17 March 1994

B. Pouligny (✉)
G. Martinot-Lagarde
Centre de recherche Paul Pascal
av. Schweitzer
33600 Pessac, France

G. Gréhan · G. Gouesbet
Laboratoire d'énergétique des systèmes et
procédés
INSA de Rouen
BP 8
76131 Mont-Saint-Aignan, France

M.I. Angelova
Central Laboratory of Biophysics
Bulgarian Academy of Sciences
1113 Sofia, Bulgaria

Abstract We describe a few applications of mechanical effects of light to study bilayer membranes. We show that lipid vesicles can be distorted by direct optical coupling. Individual charged latex microspheres are optically manipulated and brought in contact with vesicles. We systematically observe adhesion of membranes on the polystyrene surface and sometimes a total inclusion of the bead inside the vesicle.

Key words Giant lipid vesicles
– membrane electrostriction
– radiation pressure – latex
microspheres

Introduction

We describe a few novel experiments involving giant phospholipid vesicles (phospholipid bilayer membranes), focused laser beams, and latex micro-particles. As we show in the following, mechanical effects of light on matter can be used to tweeze and distort such membranes, or to hold and move solid particles in contact with membranes. Such manipulations allow us to study sphere-membrane adhesion, particle endocytosis, Brownian motion and interactions between solid particles bound to membranes.

Experimental

In our experiments, giant vesicles a few tens of microns in size, are grown in pure water, either spontaneously or by electroformation [1]. We used L-α phosphatidylcholine from frozen egg yolks (EPC). The sample cell has a $1 \times 10 \text{ mm}^2$ cross-section and is held horizontal inside an optical levitation trap [2]. The heart of this set-up is made of two vertical counter propagating (up and down) laser beams, which are focused inside the sample cell (see Fig. 1). The source is a c.w. argon ion laser (Spectra-Physics 2025). For a detailed description of the set-up, see ref [3].

Different tuning conditions allow us to produce the different geometries shown in Fig. 2. In Figs. 2a and b the beams are tightly focused and have beam-waists (ω_0) of about 1 μm. In a both beams coincide, while in b they are only coaxial, with a finite "positive" separation ($2H = 40 \mu$m) between the waists. Such a configuration is known to provide a stable optical trap for small particles [4]. By "small", we mean a radius smaller than about 3 μm, which is the beam radius in the horizontal symmetry plane in Fig. 2b. In other words, the region between the

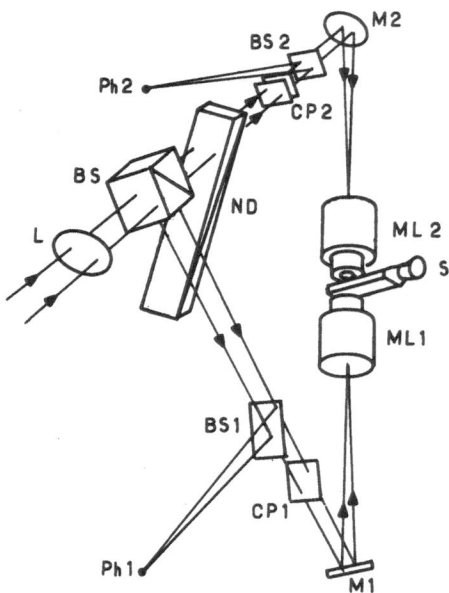

Fig. 1 Scheme of the optical trap. The sample cell (S) is located between two microscope lenses (ML1, ML2). Light from an argon ion laser is split in two beams by BS, which are then folded by two mirrors (M1, M2), and focused inside the sample by ML1 and ML2. M1 and M2 are highly reflecting for green light, but transparent for red light. A whole microscope, not shown in this figure, is set up around the optical trap for observation of the sample in red light. For details, see ref. [3].

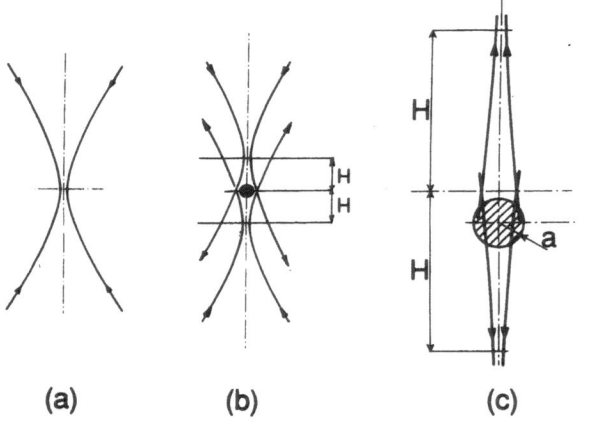

Fig. 2 Different optical trapping geometries (see text). In (b) and (c), a "small" (2 μm in diameter) and a "large" (16 μm in diameter) trapped particle is shown for illustration

two beam-waists acts as a small electromagnetic box for small particles in there. In Fig. 2c the set-up is tuned to produce a "negative" separation between the beams. $\omega_0 = 2.7 \mu m$ and $H = 100 \mu m$ are the characteristics of this geometry, which is stable for "large" (radius $> 6 \mu m$) spheres [4]. In all these geometries, the maximum avail-

able power per beam is about 0.1 Watt (c.w). Corresponding maximum electric fields (in Fig. 2a and b) are of the order of 4×10^6 V.m^{-1}.

We used "large" latex particles (diameter about 15μm) and "small" particles (diameter about 2 μm). These were considerably diluted in pure water, so that there was never more than one single particle in the volume spanned by the two beams. The particles were trapped very far from the phospholipid vesicles to avoid their surface being contaminated by lipid molecules prior to adhesion on a membrane [5]. To check this point, we made a few extra experiments with small amounts of a fluorescent marker (rhodamine-DHPE, Molecular Probes, catalog reference L-1392) mixed with EPC (marker concentration about 1:200 mol/mol). Latex spheres which were far outside the lipid-rich region did not show any fluorescence.

Membrane capture and distorsion

Focused beams as in Fig. 2a can be coupled directly to a membrane. In Fig. 3a, a flaccid vesicle is viewed from above and the white spot is the horizontal cross-section of the beams in their common beam-waist plane. When this spot is moved to overlap the vesicle contour, the membrane is seen to "hook" on the beam, provided the power is high enough (about 0.1 Watt per beam). Then, by slowly moving the whole cell (speed about 0.5 μm.s^{-1}), one can observe a considerable distorsion of the vesicle contour (Figs. 3b–d).

Assuming that the coupling mechanism between light and the membrane is mainly electrostrictive [6–8], we estimate the coupling energy involved in this process to be of the order of 100 $k_B T$ and the corresponding maximum force acting on the membrane to be of the order of 0.4×10^{-12} N.

This force is opposed by membrane elasticity and restriction of shape fluctuations (entropic forces [9]). In the future, the possibility of tweezing a membrane with laser beams could be used for directly estimating the membrane curvature elasticity. However, this would need more elaborate experiments (essentially two optical traps acting on an isolated vesicle) and a specific model to be worked out.

Adhesion of solid particles

"Small" or "large" latex particles can be very easily levitated and trapped with moderate beam powers (milliwatts) and brought in contact with vesicles. Here, the forces acting on the particles are due to radiation pressure [8]. Using the so-called "Generalized Lorenz-Mie theory" [10], we calculated these forces as a function of beam

Progr Colloid Polym Sci (1994) 97:293–297
© Steinkopff-Verlag 1994

Fig. 3 Capture and deformation of a vesicle membrane by tightly focused laser beams (configuration of Fig. 2a). Bar length = 20 μm

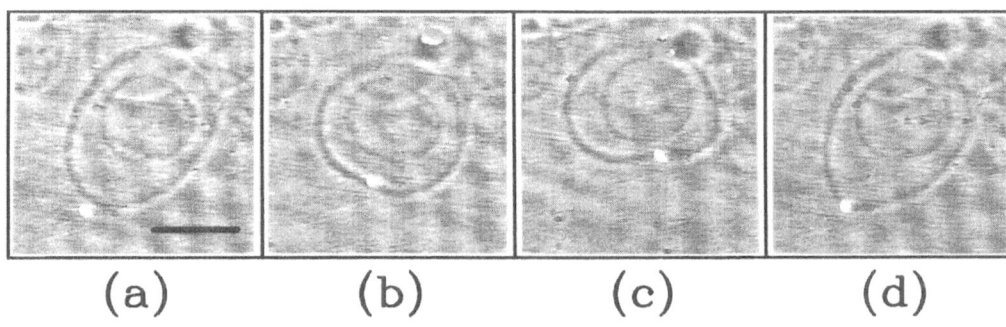

(a) (b) (c) (d)

Fig. 4 In (a), a "small" polystyrene latex sphere (2 μm in diameter) on the right of the vesicle contour is held by the optical trap at about 7 μm from the membrane. In (b), the sphere is brought in contact with the membrane. The optical trap is then switched off. The sphere stays bound to the vesicle surface and undergoes Brownian motion (c). Bar length = 20 μm

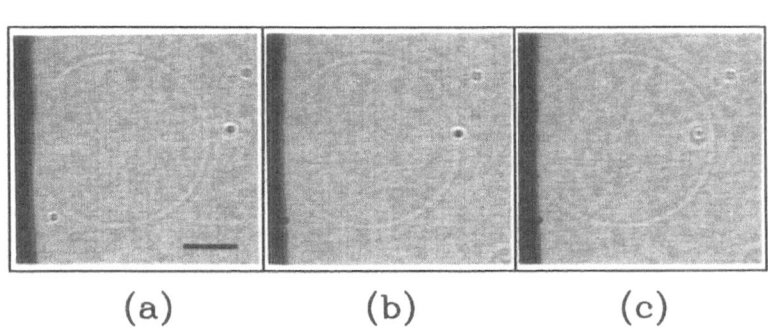

(a) (b) (c)

position and found maximum trapping energies about 4×10^{-16} J and 10^{-18} J for "large" and "small" spheres, respectively.

Our systematic observation with both types of spheres is that when contact occurs, the solid particle jumps out of the optical trap and adheres to the membrane surface. In almost all situations, this adhesion is found to be irreversible, in that it is impossible to detach an adhered sphere by means of only the laser beams.

However, the two sorts of particles may behave differently after adhesion, depending on the vesicle flaccidity. We made many experiments, some of them with vesicles that showed no sizable contour shape fluctuations, some others with fairly flaccid vesicles, corresponding to relative excess surface areas up to a few percent. We observed that the small spheres always stay on the membrane surface whatever the flaccidity of the vesicle (Fig. 4). Within the resolution of our observation system, it is however impossible to decide if the particle is located *on* or *across* the membrane.

The ways in which "large" spheres behave are illustrated in Fig. 5. The vesicle shown in Fig. 5a is apparently spherical prior to collision with solid particle. After collision (Fig. 5b) the sphere is adhered to the membrane *outside* the vesicle volume. Apparently, only a fraction (about 20%) of the solid sphere outer area is wet by the membrane. In this situation one may make a rough esti-

mate of the adhesion energy density [7]. We find: $3 \times 10^{-6} < \gamma < 3 \times 10^{-5}$ J.m^{-2}.

The vesicle shown in Fig. 5c has a large excess area, as evidenced by the definitely non-circular contour shape. In this case, the sphere, when jumping out of the optical trap, was seen to literally "tunnel" through the membrane contour. This tunneling is very fast and is well beyond the time resolution limit of the video system we used for observation (40 ms). Figure 5d shows the final position of the sphere in the interior of the vesicle contour. The fact that the sphere is inside the vesicle can be proved by moving it with the laser beams. In this case, the sphere can be easily detached from the vesicle surface and moved in all directions inside. These observations suggest that the initial latex sphere was bagged by the membrane available from the large excess area and that the whole process is a "physical endocytosis".

Brownian motion and interaction between solid particles on membranes

Fifteen μm spheres in bulk water are nearly non-Brownian, while 2 μm spheres are. Interestingly, 2 μm particles adhered on a membrane still definitely undergo Brownian motion. This motion can be observed very easily under an optical microscope, and this can be considered as

Fig. 5 Adhesion of "large" polystyrene latex spheres (about 15 μm) on vesicles (see text). Bar length = 20 μm

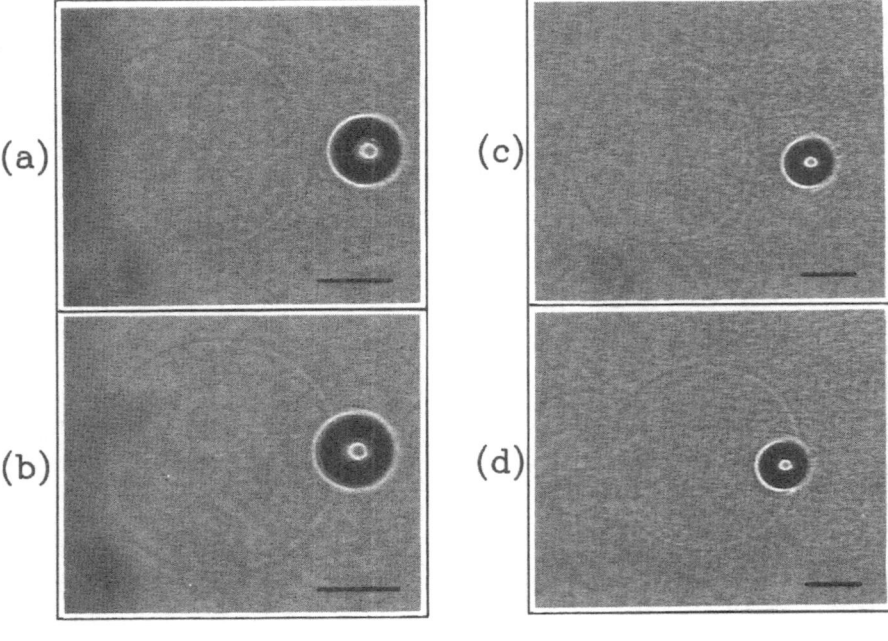

Fig. 6 Transient aggregation of 2 μm particles bound to a membrane. The three photographs shown are top views. In (a) most of the particles are below the vesicle "equator". In (b) three particles are aggregated near the equator. In (c) an aggregate is formed near the top of the vesicle, i.e., in conditions where gravity cannot be responsible for aggregation. In this last view, the image was focused on the particles, which makes the vesicle contour hardly visible. Bar length = 20 μm

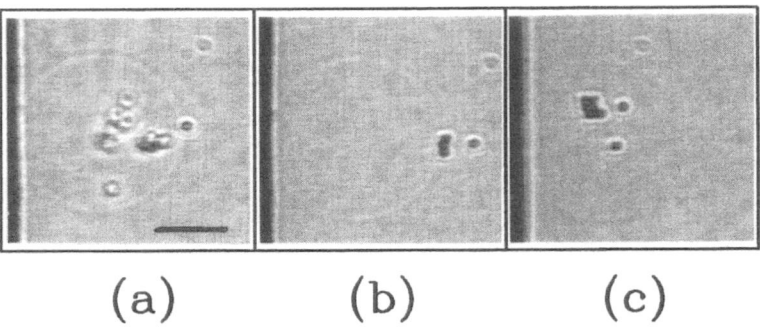

(a)   (b)   (c)

a direct illustration of the liquid character of EPC bilayers at room temperature [11].

Several small spheres can be adhered on the surface of the same vesicle. In this case, we systematically observed an *attraction* between the particles which leads to the build-up of aggregates, as shown in Fig. 6. However, these structures are easily broken and reassembled under thermal agitation, which means that the interaction energies involved in the process may be of the order of $k_B T$. By comparison, the same particles sedimented on the glass bottom of the sample cell undergo Brownian motion but have no tendency to aggregate (at least within a few hours).

Recently, Goulian et al. [12] proposed a theory of the interaction between inclusions in membranes mediated by membrane thermal shape fluctuations. They found an attraction energy proportional to $k_B T.R^{-4}$, where R is the distance between two inclusions. This might well explain our observations.

A more quantitative data analysis is in progress, including a characterization of the single particle Brownian motion and an analysis of the correlation between the positions of two adhered particles.

Acknowledgements This work is a part of the *International Scientific Cooperation Program* (PICS 107) between the Centre de recherche Paul-Pascal, CNRS, France, the Institute of Solid State Physics and the Central Laboratory of Biophysics, Bulgarian Academy of Sciences. We acknowledge support of ULTIMATECH program of CNRS.

We thank I. Bivas, P. Méléard, M. Mitov, N. Toulmé and P. Richetti for helpful discussions. and P. Bothorel for his interest in this work.

One of us (M.I.A.) is gratefully indebted to CNRS and to the Commission of European Communities (ref. ERB-CIPÅ-CT-92-0174, project 2441) for supporting her stay in France.

Progr Colloid Polym Sci (1994) 97:293–297
© Steinkopff-Verlag 1994

References

1. Angelova MI, Dimitrov DS (1986) Liposome electroformation, Faraday Disc Chem Soc 81: 303–311; disc.: 345–349; Angelova MI, Soleau S, Méléard P, Faucon JF, Bothorel P (1992) AC field controlled formation of giant fluctuating vesicles and bending elasticity measurements. Springer Proc in Physics 66: 178–182
2. Buican TN, Neagley DL, Morrison WL, Upham BW (1989) Optical trapping, cell manipulation and robotics. New technologies in cytometry. SPIE 1063: 190–197
3. Angelova MI, Pouligny B (1993) Trapping and levitation of a dielectric sphere with off-centred Gaussian beams: I Experimental Pure Appl Opt 2:261–276
4. Roosen G, Imbert C (1976) Optical levitation by means of two horizontal laser beams: a theoretical and experimental study Phys Lett 59 A:6–8; Roosen G (1977) A theoretical and experimental study of the stable equilibrium positions of spheres levitated by two horizontal laser beams. Opt Commun 21: 189–194
5. Carmona-Riabeiro AM, Herrington TM (1993) Phospholipid adsorption onto polystyrene microspheres. J Coll Interface Sci 156:19–23
6. Palmer AJ (1980) Nonlinear optics in aerosols. Opt Lett 5:54–55
7. Angelova MI, Martinot-Lagarde G, Pouligny B (1994) Interaction of lipid vesicles with latex spheres. Preprint submitted to Eur Biophys J
8. Ashkin A, Dziedzic JM, Bjorkholm JE, Chu. S (1986) Observation of a single beam gradient force optical trap for dielectric particles. Opt Lett 11:288–290
9. Evans E, Rawicz W (1990) Entropy driven tension and bending elasticity in condensed fluid membranes. Phys Rev Lett 64:2094–2097
10. Gouesbet G, Maheu B, Gréhan G (1988) Light scattering from a sphere arbitrarily located in a Gaussian beam, using a Bromwich formulation. J Opt Soc Am A5: 1427–1443; Gouesbet G, Gréhan G, Maheu B (1990) Localized interpretation to compute all the coefficients g_{nm} in the generalized Lorenz-Mie theory. J Opt Soc Am A7:998–1007
11. Quinn PJ (1984) Phases of membrane polar lipids in aqueous systems. Natural Products Reports: 513–531
12. Goulian M, Bruinsma R, Pincus P (1993) Long-range forces in heterogeneous fluid membrane. Europhys Lett 22: 145–150; erratum 23:155

Appendix

Our most recent observations [7] show that the latex particles are completely coated by the membrane material, whatever may be the initial vesicle tension. We recently proposed a model to explain full encapsulation of "hydrophilic" particles (Pouligny B, Martinot-Lagarde G, Angelova MI (1994) Encapsulation of solid microspheres by bilayers, VIIIth ECIS Conference, Montpellier, France, September 26–30). According to this model, a lower boundary of the membrane-polystyrene adhesion energy can be estimated. We find $\gamma \geq 0.04$ J.m^{-2}, a value much larger than the one based on the "partial wetting" hypothesis.

Progr Colloid Polym Sci (1994) 97:298–301
© Steinkopff-Verlag 1994

O. Regev
A. Khan

Vesicle - lamellar transition events in DDAB-water solution

Received: 1 October 1993
Accepted: 14 January 1994

O. Regev[†] (✉) · A Khan
Division of Physical Chemistry 1
Chemical Centre, Box 124
University of Lund
22100 Lund, Sweden

[†]Present address: Department of
Chemical Engineering, Ben-Gurion
University of the Negev
Box 653
84105 Beer-Sheva, Israel

Abstract Double-tailed ionic surfactants form lamellar liquid crystalline phases with water. Here, we have used cryo-transmission electron microscopy (cryo-TEM) technique to study the transition to a lamellar phase at dilute aqueous solutions of the cationic surfactant didodecyldimethylammonium (DDA) with different counterions: acetate (DDAAc), hydroxide (DDAOH), bromide (DDABr), sulphate (DDAS), as well as the anionic surfactant sodium bis(2-ethylhexyl) sulfosuccinate (AOT). Vesicles are observed in all the double-tailed surfactants systems at low concentrations, with their structure being concentration dependent.

Key word Vesicles – didodecyldimethylammonium-bromide – sodium di-(2-ehtylhexyl) sulfosuccinate – cryo-TEM – microstructure

Introduction

Double-tailed cationic and anionic surfactants are known to form lamellar liquid crystals with water as the first liquid crystalline phase [1–4]. Depending on counterions, some of the surfactants are practically insoluble in water and form lamellar dispersions [4], others are soluble in water, forming isotropic micellar solution phases prior to the formation of the single lamellar phase. From a limited study [2], it is revealed that most of these systems form vesicles at high dilution as the first aggregate structure in the surfactant self-assembly processes. However, the microstructural transformations that takes place in these systems (vesicles – – → lamellar dispersion – – → single lamellar phase for insoluble surfactants, and vesicles – – → micelles – – → single lamellar phase for soluble surfactants) are poorly understood.

The lamellar structure of the DDAB-water system has been studied extensively. Rich phase behavior has been reported at low surfactant concentration (0.15–3.0 wt% DDAB), which was attributed to the delicate interplay between electrostatic repulsion and Helfrich undulation forces. A phase transition from a dispersion of lamellar crystallites in L_1 phase to a pure lamellar phase with increasing concentration above the chain melting temperature (18 °C) has been monitored by SAXS, SANS and light-scattering techniques [5]. The lamellar dispersion contains spherulites or lamellar scattering droplets [6] which show a characteristic length, ξ, obtained by scattering techniques. Two different characteristic spacings found in the lamellar dispersion region were attributed to the bilayer periodicity, and a long range ordering of superstructures.

Cryo-TEM appears to us to be a powerful technique that can directly image microstructures formed in very dilute aqueous solution of surfactants. Moreover, we have recently observed that the self-diffusion of the solvent (water) in vesicular solution can be related to some physico-chemical properties of vesicles.

In this work, we focus on the structural changes induced by increasing the concentration at a low surfactant concentration of DDAB (< 3 wt%) and AOT (< 1.3 wt%).

Progr Colloid Polym Sci (1994) 97:298–301
© Steinkopff-Verlag 1994

Experimental Section

Sample preparation

Purified DDAB (Tokyo Kasei Kogyo Co.) and Millipore water were used to prepare the samples. These were gently mixed by shaking for several days and then allowed to equilibrate at room temperature for 1 day. The sample homogeneity was periodically checked between cross polaroid sheets

Cryo-transmission electron microscopy

Direct visualisation of the vitrified specimen of the DDAB-water by transmission electron microscopy (TEM) was carried out by a technique described by Bellare et al. [7]. The specimen was prepared by depositing a 5 μl drop of the solution on a TEM grid, coated by a holey carbon film [8] in a controlled environment vitrification chamber at room temperature where relative humidity is kept at about 100% to prevent drying of the sample. A thin (20–200 nm) film of the solution suspended over the holes of the grid was formed by blotting the grid to remove excess fluid. The specimen was vitrified by plunging it into liquid ethane at its freezing point. The vitrified specimens were transferred under liquid nitrogen to a JEOL 2000FX microscope equipped with a cold stage (Model 626, Gatan, Inc., Warrendale, PA), and examined under an acceleration voltage of 100 kV. The working temperature was kept below $-168\,°C$ and the images were recorded on a SO-163 Kodak film.

Results and discussion

DDAB system

DDAB is sparingly soluble in water (≈ 0.1 DDAB wt%) giving a non-viscous, bluish color solution. On increasing the surfactant concentration to 2 wt%, the solution becomes viscous, and the sample is flow-birefringent. Between 2–3 surfactant wt% a flow birefringent is observed between crossed polaroid sheets, and above 4 wt% the system yields single lamellar D phase.

We have investigated the low concentration region of the DDAB-water system by direct imaging with the cryo-TEM technique. For samples with surfactant concentration of about 0.5 wt% or below, single-wall (bilayer) vesicles are recorded (Fig. 1). With increase of surfactant concentration to 3 wt%, single-wall vesicles (SWV) are found to be in equilibrium with predominantly double

wall vesicle (DWV) (Fig 2) as well as with a few tubules (Fig. 3). The spacing between the lamellae in all vesicles and tubules is kept constant and equal to 25 nm (Figs 2a, 3a)

We observed an intermediate state between single- and double-wall vesicles. The SWV deforms (Fig. 2b) and transforms into a DWV by "deflating", i.e., expelling the water within the SWV, resulting in an uncompleted DWV or a "deflated vesicle," which is clearly observed in Fig. 2c from top and side view and in Fig. 3b from side view. The opening in Fig. 2c – side view is projected as a hole in the top view (indicated by an arrow).
which is followed by a complete closure of the outer bilayer to form DWV (Fig. 2e).

It is important to note that a cryo-TEM micrograph is a photograph of dynamic processes that take place in surfactant solution (Fig. 2). Therefore, one micrograph can give a global picture which reflects different events occurring during the progression of a process.

The fact that the vesicle is spherical and contains two bilayers can be shown by mass contrast differences in the micrographs as demonstrated in Fig. 2e; the inner volume of the two bilayer vesicle is surrounded by two bilayers and therefore appears darkest in the figure, where the inter-lamellae zone appears brighter. The amorphous water outside the vesicles is the brightest.

AOT system

AOT has a narrow solubility range in water (1.3 wt%), which is followed by a two-phase region, lamellar + water, prior to the formation of a single lamellar phase at about 10 wt% surfactant. However, the system does not form micelles of the type shown by a large number of water soluble surfactant, e.g., SDS. Micrographs of samples in the AOT-water system for the concentration between 1–3 wt% show a similar concentration-dependent sequence of aggregation as in the DDAB-water system (not shown): single-walled vesicles ---→ double-walled vesicle ---→ tubules ---→ lamellar liquid crystal

On the other hand, water soluble double-tailed quaternary ammonium surfactants, e.g., DDAAc, DDAOH form micellar solution prior to the formation of the lamellar phase. For these systems, concentraion-dependent aggregate transitions recorded by cryo-TEM and ^{1}H NMR line width measurements are as follows [2]: Vesicles ---→ small micelles ---→ large micelles ---→ lamellar liquid crystal

Cryo-TEM observations in the lamellar phase (> 3 wt% DDAB or > 10 wt% AOT) did not show the expected D phase. Probably the reasons are that the lamellae are arranged parallel to the surface of the sample, and therefore posses a very low phase and mass-thickness

Fig.1 TEM micrograph of single wall vesicles in DDAB-water at 0.5 wt% surfactant system, temperature 298 K (bar = 50 nm)

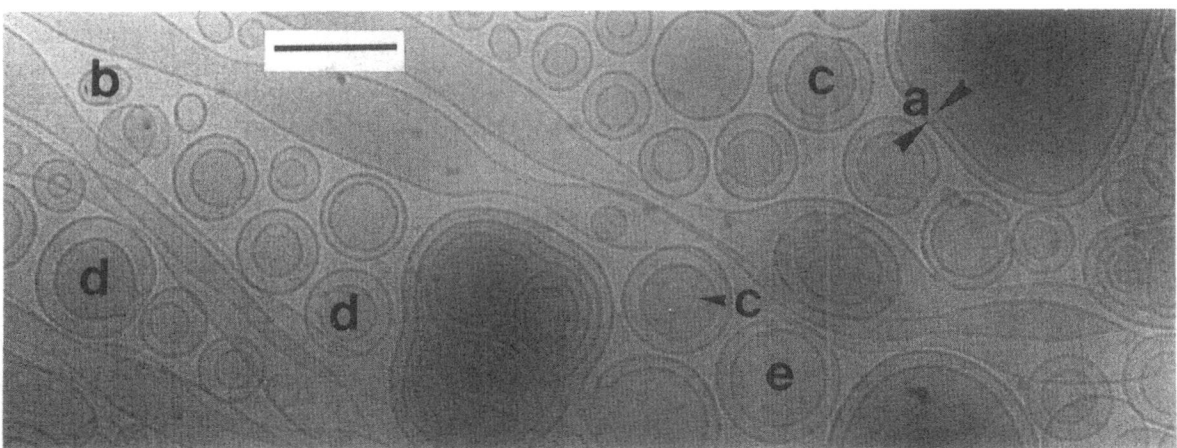

Fig. 2 TEM micrograph of multi-wall vesicles in DDAB-water at 3 wt% surfactant system, temperature 298 K: a) 25 nm inter-lamellae spacing, b) deformation of a single-wall vesicle, c) deflated single-wall vesicle, d) fusion of the outer wall, e) complete double-wall vesicle (bar = 200 nm)

Fig. 3 TEM micrograph of long tubular structures in DDAB-water at 3 wt% surfactant system, temperature 298 K. a) 25 nm inter-lamellae spacing b) deflated single-wall vesicle (bar = 200 nm)

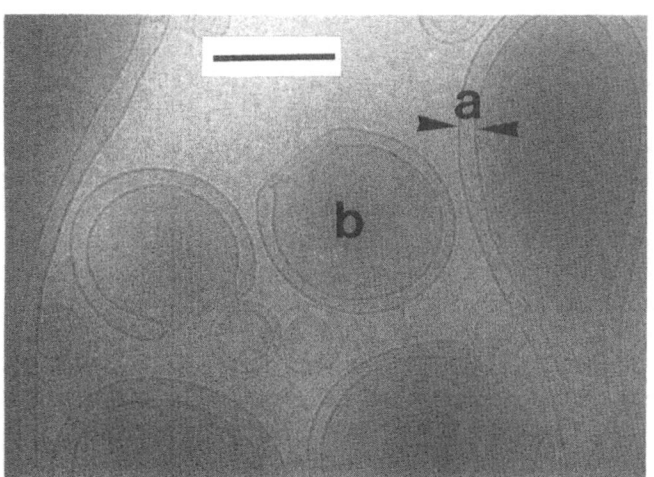

contrast which makes it impossible to observe their structure.

Comparison with other works

It is interesting to compare our findings to small-angle x-ray and neutron-scattering results obtained by Fontell et al. [4] and Dubois et al. [5]. Both authors reported a very broad peak in SAXS measurements below 3 wt% surfactant, where the location of the peak changes with surfactant concentration. The wide polydispersity observed by cryo-TEM in the lamellar dispersion explains the reported broad x-ray peak. At 10 wt% surfactant (pure D phase) a peak corresponding to a 24 nm spacing was found by small-angle x-ray scattering measurements [5]. This spacing length fits the interlamellar distance found in the double bilayer vesicles shown in Figs 2 and 3. The smaller spacing value obtained from cryo-TEM measurements

Progr Colloid Polym Sci (1994) 97:298–301
© Steinkopff-Verlag 1994

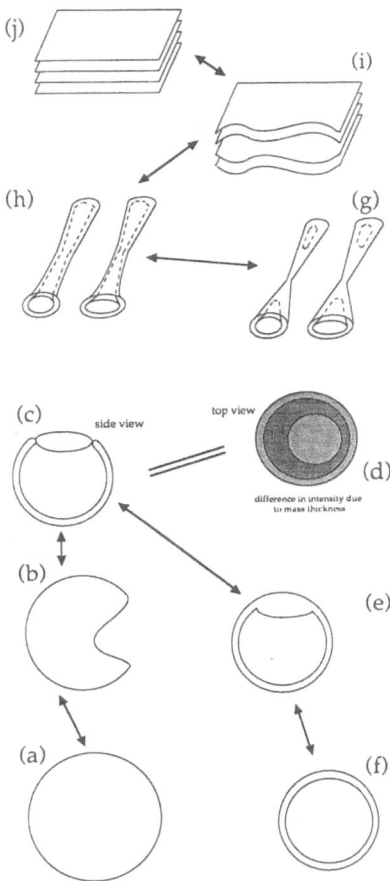

Fig. 4 Experimental model for the transition from single-wall vesicle to double-wall vesicle (for details see text)

300–400 nm which corresponds to the tubules' diameter (Fig. 3).

Inflated-deflated vesicle transition due to temperature changes has already been reported by Berndl et al. [10] for lecithin vesicles, and termed discocyte-stomatocyte transition

Experimental model

Our results suggest a simple mechanism of vesicles-to-lamellar transition which is schematically presented in Fig. 4. SWV (4a) is deformed (4b) to an open DWV (4c). This process resembles deflation of a football. It is important to remember that the micrographs obtained from the electron microscope show two-dimensional projection, but describe three-dimensional objects. Therefore, the open DWV in 4c presents a side view which could also be observed from the top (Fig. 4d). Both structures, 4c and 4d, correspond to the experimental results shown in Fig. 2c. Since the energy differences between the different vesicular structures prevailing in the solution are very small [11], one should expect reversible processes in the system. The open DWV could extend or fuse to form tubules (4f, 4g) where the latter collapse finally to the lamellar phase (4h, i)

Conclusion

Micrographs of the DDAB and AOT systems in the dilute region show that upon increasing the surfactant concentration a single-wall vesicle transforms, through a "deflated vesicle" state, to a double-wall vesicle. Long double-wall tubules are believed to be precursors to the lamellar phase.

Acknowledgements Dr. Marc Leaver is acknowledged for valuable discussion. The project is partially financed by the Swedish Natural Science Research Council. The stay of O.R. was made possible by a grant from the Swedish Institute

may be due to concentration effects induced by the thinning of the film during sample preparation and results in concentration gradient along the sample [9]. Furthermore, light-scattering curves of DDAB-water system in [lamellar] phase [5] show a second characteristic length of

References

1. Ekwall P (1975) In: Brown G H (Ed) Advances in Liquid Crystals, Vol 1
2. Regev P, Kang C, Khan A (1994) J Phys Chem 98:6619
3. Kang C, Khan A (1993) J Colloid Interface Sci. 156:218
4. Fortell K, Ceglie A, Lindman B, Ninham BW (1986) Acta Chem Scand Ser A 40:247
5. Dubois M, Zemb T (1991) Langmuir 7:1352.
6. Van De Pas CJ (1992) Colloid and Surfaces 68:127
7. Bellare JR, Davis HT, Scriven LE, Talmon Y (1988) J Electr Microsc Tech 10:87
8. Vinson PK (1987) In: Baird GW. (Ed) the 45the Annual Meeting of the Electron Microscopy Society of America, San Francisco Press, Inc, pp 644.
9. Vinson PK, Bellare JR, Davis HT, Miller WG, Scriven LE (1991) J Colloid Interface Science 142:74
10. Berndl K, Käs J, Lipowsky R, Sackmann E, Seifert U (1990) Europhysics Letters 13:659
11. Talmon Y (1986) Colloids and Surfaces 19:237

Progr Colloid Polym Sci (1994) 97:302–306
© Steinkopff-Verlag 1994

BIO-COLLOIDS

U. Gehlert
D. Vollhardt

The phase behavior of an ether lipid monolayer compared with an ester lipid monolayer

Received: 16 September 1993
Accepted: 12 January 1994

Dr. D. Vollhardt · U. Gehlert
Max-Planck-Institut für Kolloid-
und Grenzflächenforschung
Rudower Chaussee 5
12489 Berlin

Abstract The main features of the morphological structure changes during monolayer compression of 1-O-hexadecyl-rac-glycerol are studied by Brewster angle microscopy (BAM). Images of the monolayer related to the corresponding points of the π-A-isotherms are presented and discussed. The results are compared with those obtained for 1-monopalmitoyl-rac-glycerol having only small distinctions in the chemical structure.

Key words Glycerol ester – glycerol ether – monolayer – Brewster angle microscopy – π-A isotherm – phase behavior

Introduction

There has been a continuous interest in monolayers recently enlarged by new techniques which provide new insights into the morphological structure of monolayer phases. The recent development of the BAM [1–8] enables the visualization of morphological features of monolayers. This features can also be seen by fluorescence microscopy with polarized excitation [9, 10], but the addition of fluorescent probes is necessary. Images of coexisting phases and domain structures in monolayers can be recorded simultaneously with the surface pressure π as a function of the area/molecule A.

At present, it is largely unknown whether and how far small changes of the molecule structure of the amphiphile influence the morphological structure and the phase behavior of the monolayers if the surface pressure (π) –area (A) isotherms are very similar. In looking at the molecular structure, Hauser et al. [11] emphasized the important role of carbonyl groups comparing surface potential data resulting from glycerol esters and glycerol ethers. They discussed the biological significance of the replacement of the ether linkage by an ester linkage in Dialkyl glycerol phospholipids.

In an early work, Knight [12] compared the π-A isotherms of 1-O-hexadecyl-sn-glycerol (chimyl alcohol) and α-monoglyceride monolayers and concluded that the monolayer properties are very similar.

The primary objective of the present work is to compare the morphological properties of 1-O-hexadecyl-rac-glycerol and 1-monopalmitoyl-rac-glycerol monolayers during compression. As discussed, these compounds have only small distinctions in the chemical structure, namely, a long hydrocarbon chain linked by an ether or an ester group to a glycerol group. We would like to demonstrate the important role of the carbonyl group for the morphological structure and the phase behavior although the π-A isotherms of 1-O-hexadecyl-rac-glycerol and 1-monopalmitoyl-rac-glycerol are similar at the same temperature.

Experimental section

A Langmuir film balance from LAUDA is used to obtain surface pressure π as a function of the area/molecule A. The film balance is combined with a Brewster Angle Microscope for simultaneous recording of the monolayer structure during compression.

Progr Colloid Polym Sci (1994) 97:302–306
© Steinkopff-Verlag 1994

The physical principle of BAM and experimental details were described previously [2]. Pictures are recorded and stored with a video system. The spatial resolution of the Brewster angle microscope is ca 4 μm. Because the visual angle of the camera is about 53°, the images are compressed in the horizontal direction.

After spreading and evaporating the spreading liquid, the monolayer is compressed at a slow rate of 9.3×10^{-3} nm²/molecule minute.

The trough is placed in a thermostatted box, so that the temperature is maintained at $23.1 \pm 0.2\,°C$.

Material 1-O-hexadecyl-rac-glycerol and 1-monopalmitoyl-rac-glycerol were obtained from Sigma with a purity of approximately 99 mol. %. The lipids were spread from a 9:1 mixture of heptane/ethanol solution. The subphase water was twice distilled with the second distillation from an alkaline permanganate solution.

Results and discussion

The chemical structure of 1-O-hexadecyl-rac-glycerol is very similar to those of 1-monopalmitoyl-rac-glycerol. As demonstrated by the formulae, the ether linkage in the 1-O-hexadecyl-rac-glycerol molecule (I) is replaced by an ester linkage in 1-monopalmitoyl-rac-glycerol (II).

$$H_2C-O-\overset{\overset{H}{|}}{\underset{\underset{H}{|}}{C}}-(CH_2)_{14}-CH_3$$
$$HO-\overset{|}{\underset{|}{C}}-H$$
$$CH_2OH$$

I

$$H_2C-O-\overset{\overset{O}{||}}{C}-(CH_2)_{14}-CH_3$$
$$HO-\overset{|}{\underset{|}{C}}-H$$
$$CH_2OH$$

II

A comparison of the π-A compression isotherms of both substances at 23.1 °C is shown in Fig. 1. As expected, both isotherms resemble one another so that also similar surface energetical properties of the monolayers should be concluded.

It is often assumed that monolayers of substances with similar π-A isotherms have also similar two-dimensional phase behavior. However, the following BAM study demonstrates impressively that this assumption fails in the case of 1-O-hexadecyl-rac-glycerol and 1-monopalmitoyl-rac-glycerol.

The images shown in Fig. 2 are taken with the Brewster Angle Microscope and characterize the monolayer of 1-O-hexadecyl-rac-glycerol in the points a–g designated on the π-A isotherm. The condensed phase domains surrounded by a homogeneous fluid phase of low density are first observed after the sharp break in the π-A isotherm at the beginning of the plateau region (Fig. 2a). In Fig. 2b,

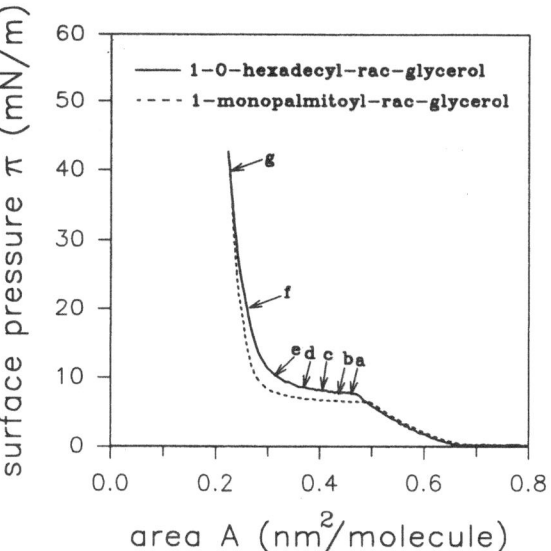

Fig. 1 Surface pressure–area per molecule (π-A) isotherms for 1-monopalmitoyl-rac-glycerol and 1-O-hexadecyl-rac-glycerol, measured parallel to the recording of the Brewster angle microscope at 23.1 °C

the domains have grown in area at the expense of the fluid phase. There, it can be seen that all domains are divided into segments reflecting differently meeting at a point on the edge of a domain. The substructure results from different directions of the molecule chains in the several segments. The boundary between domains and surrounding fluid phase is realized by a straight line for each segment edge.

Continuing the compression to a certain molecular area, the domain shape becomes unstable and three-armed structures are formed (Fig. 2c). Each arm is divided by a sharp boundary at which the molecular orientation is changed. Watching the video display during compression with a rotating analyser, one can see a small continuous change of reflectivity within both regions of the domain arms.

With further compression (Fig. 2d), the domains impinge on each other due to a decrease in area/molecule.

It is interesting to note that, at the end of the plateau upon further compression, a new monolayer phase which

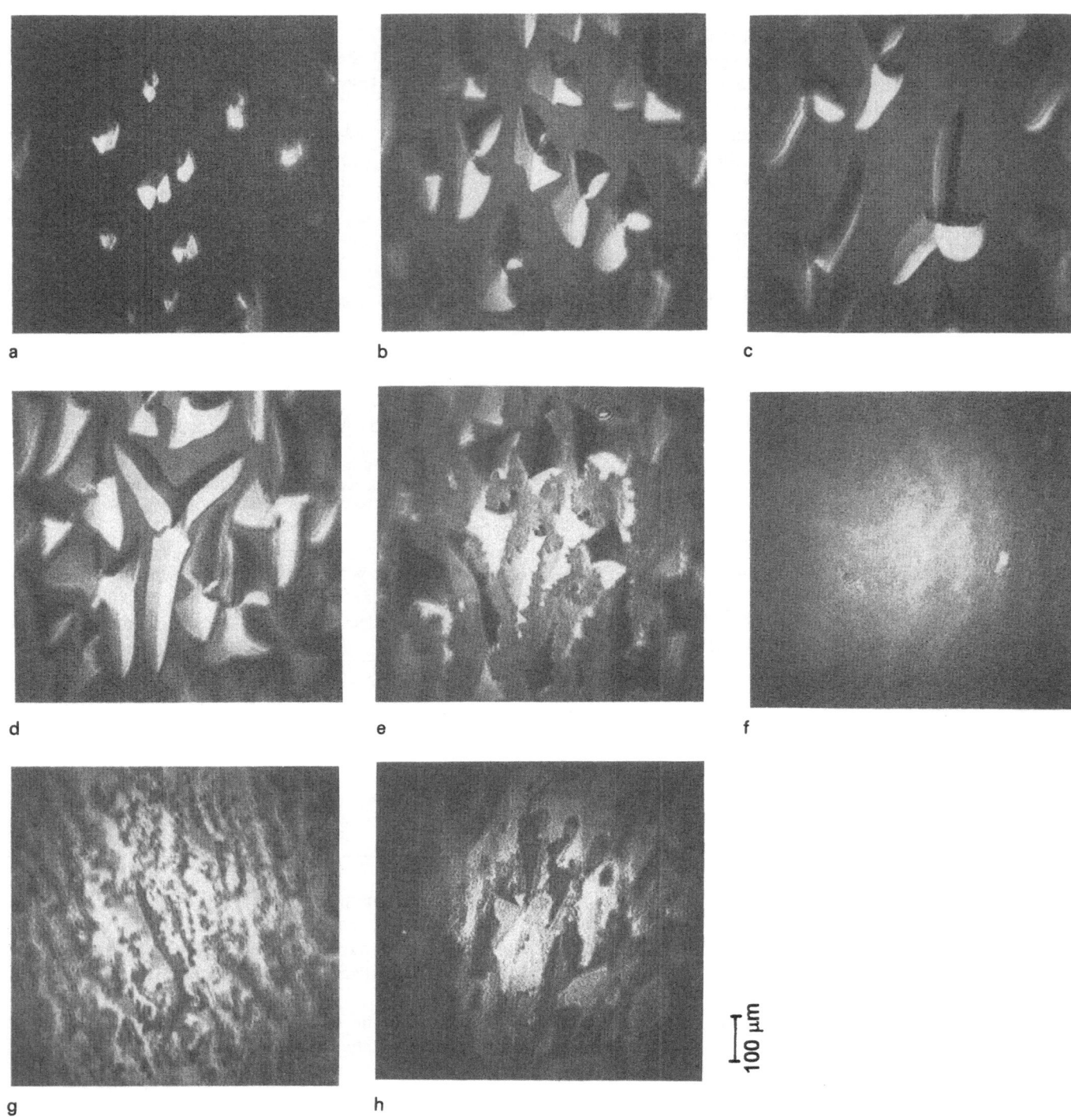

Fig. 2. Brewster angle microscopy of the 1-O-hexadecyl-rac-glycerol monolayer at 23.1 °C a) –g): corresponding to the points designated in the π - A isotherm of (Fig. 1) h): $\pi = 40$ mN/m 30 min after compression stop

reflects homogeneously is formed (Fig. 2e), starting at that point of the domains at which the different reflecting regions contact. At 10 mN/m (Fig. 2e) the monolayer image shows remarkable zig-zag lines as boundaries between the condensed phases. At the end of the phase transition ($\pi \sim 20$ mN/m), the monolayer is homogeneously reflecting. No structure is visible on the scale of the resolution (Fig. 2f). On compression, at a surface pressure exceeding

23 mN/m some contrast evolves again, occurring in diffuse patterns. Finally, at 40 mN/m the monolayer displays regions with long-range orientational order (Figure 2g). If compression is stopped now, adjacent regions of different molecular orientation are observed confined by straight or zigzag boundaries against the remaining homogeneous phase. Figure 2h shows the monolayer at 40 mN/m after 30 min.

Now, the morphological monolayer structures of 1-monopalmitoyl-rac-glycerol are considered. They resemble those of 1-monostearoyl-rac-glycerol described in detail in our previous work [8]. Therefore, the discussion is confined to some main features necessary for the comparison with the 1-O-hexadecyl-rac-glycerol monolayer. Note that the plateau of the isotherm and the formation of condensed domains is shifted to higher surface pressures with the decrease of the alkyl chain length at the same temperature.

On compression, the first formation of condensed phase domains of 1-monopalmitoyl-rac-glycerol begins immediately after the sharp break at the plateau corresponding to the phase behavior of 1-O-hexadecyl-rac-glycerol. However, there are essential differences in the morphological structures of domains at the same experimental conditions of a low compression rate. For example, the geometrical shape of the domains formed in the plateau region of the isotherm is nearly a disk (Fig. 3a) and quite different to the corresponding morphological structures of 1-O-hexadecyl-rac-glycerol (Figs 2a–d). The shape of the domains is assumed to be dependent on the line tension and the long-range electrostatic repulsion [13]. The anisotropic line tension forces the condensed phase into the shape observed.

The substructure of the domains of both amphiphiles is rather different. Almost all domains of 1-monopalmitoyl-rac-glycerol are subdivided into seven segments of different molecular orientation meeting at the domain center. In some domains the point at which the boundaries intersect is situated at the edge of a domain. In this case the number of segments is smaller than seven.

The inner structure of the domains can be regarded as multiple twin. The sharp boundaries between regions of different molecular orientation are thought to be realized by dense lattice rows [14].

The monolayers of the amphiphiles considered show essential differences in the behavior of the condensed phase. At the plateau of the π-A isotherm the condensed phase domains of 1-monopalmitoyl-rac-glycerol increase in number and size. Accompanied by an increase in the surface pressure, at further compression the condensed phase domains start to contact and deform each other (Fig. 3b), filling up the gaps in the fluid phase of low density up to the surface pressure of the weak kink in the π-A isotherm ($\pi = 25$ mM/m). Above a surface pressure of 25 mN/m, the domains are compressed so heavily that they prefer the hexagonal shape without any visible gaps (Fig. 3c). The contrast of the segments with different molecular orientation can be observed up to the point of irreversible collapse.

A comparison with the ester lipid monolayer (Fig. 3) shows clearly that the phase behavior of the 1-O-hexadecyl-rac-glycerol monolayer is more complex. Several phase transitions can be observed during monolayer compression of 1-O-hexadecyl-rac-glycerol. The origin of the remarkable zigzag lines at high surface pressures (Fig. 2e) is not yet clear; possibly the molecular transition occurs preferentially along favored lattice rows.

The vanishing of the contrast at 20 mN/m can be explained either by the vertical orientation of the hydrocarbon chains or the long-range order of the tilt

Fig. 3 Brewster angle microscopy of the 1-monopalmitoyl-rac-glycerol monolayer at 23.1 °C a): $\pi = 6.7$ mN/m, b): $\pi = 13.5$ mN/m, c): $\pi = 35$ mN/m

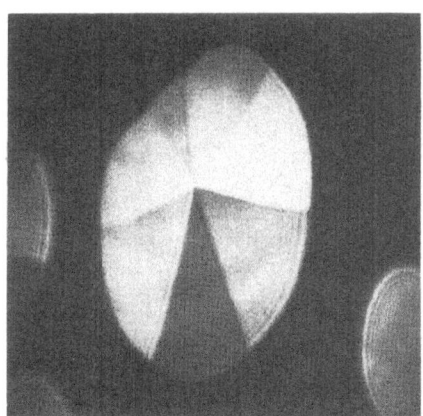
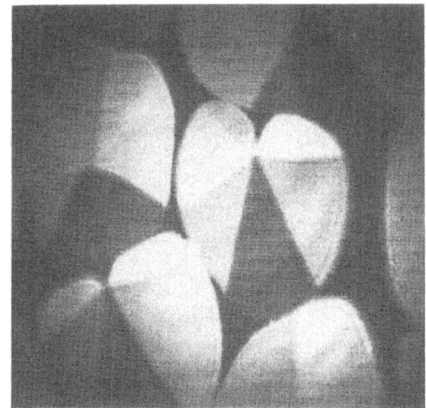
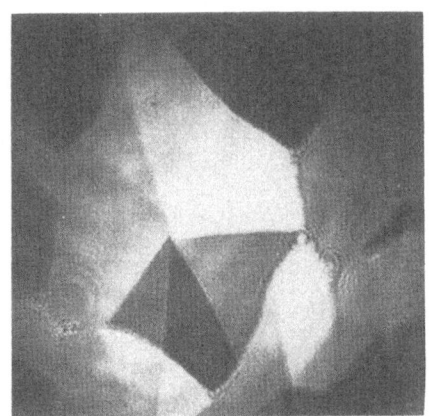

a   b   c

azimuth becomes short range so that the regions of uniform chain orientation are smaller than the spatial resolution of the microscope.

Summing up, this work provides evidence that the replacement of the ether linkage by the ester linkage has a significant effect on the monolayer structure, in particular, on the morphology of condensed domains and the phase behavior. It seems likely that the more complex phase behavior of the 1-O-hexadecyl-rac-glycerol is caused by the greater configurational freedom for the ether linkage compared to the ester group. The C=O-dipoles in the ester group might cause a fixed configuration of chain and head group.

Conclusions

New insight in the two-dimensional phase behavior of 1-O-hexadecyl-rac-glycerol and 1-monopalmitoyl-rac-glycerol visualised by BAM are reported.

A comparison of the monolayers of both amphiphiles at slow compression yields the following conclusions:

1) The replacement of the ether linkage by the ester linkage has no significant effect upon the π-A isotherms.
2) For similar π-A isomers, pronounced differences in the morphological structure and the phase behavior are observed by BAM.

Both monolayers have one common feature: The formation and the growth of condensed phase domains starts at the main transition point of the plateau of the π-A isotherms. However, in size and substructure, the domains of both compounds are quite different.

At compression of 1-O-hexadecyl-rac-glycerol monolayers three conspicuous phase transitions can be visualized. First, at the plateau region of the π-A isotherm the main transition occurs with the formation of condensed phase domains from the fluid phase. With compression the transition to a homogeneously reflecting monolayer state is observed and is finished at approximately 20 mN/m. A further increase of the surface pressure leads to an appearance of a state with a long-range orientational order again.

The phase behavior of 1-monopalmitoyl-rac-glycerol is characterized by the growth of the condensed phase domains at the plateau of the π-A isotherm. The steep surface pressure increase of the isotherm is coupled with the deformation of the circular shaped monolayer disks, which prefer a hexagonal shape at higher surface pressure.

Acknowledgement The authors are indebted to the Koordinierungs- und Aufbau- Initative, KAI e.V., the Deutsche Forschungsgemeinschaft, DFG and the Fond der Chemischen Industrie for financial support.

References

1. Henon S, Meunier J (1991) Rev Sci Instrum 62:936–939
2. Hönig D, Möbius D (1991) J Phys Chem 95:4590–4592
3. Hönig D, Overbeck GA, Möbius D (1992) Adv Mater 4:419–424
4. Henon S, Meunier J (1992) Thin Solid Films 210/211:121–123
5. Overbeck GA, Hönig D, Möbius D (1993) Langmuir 9:555–560
6. Hönig D, Möbius D (1992) Thin Solid Films 210/211:64–68
7. Siegel S, Hönig D, Vollhardt D, Möbius D (1992) J Phys Chem 96:8157–8160
8. Vollhardt D, Gehlert U, Siegel S, (1993) Coll & Surf 76:187–195
9. Moy VT, Keller DJ, McConnell HM (1988) J Phys Chem, 92:5233–5238
10. Qiu X, Ruiz-Garcia J, Stine KJ, Knobler CM (1991) Phys Rev Lett 67:703–706
11. Paltauf F, Hauser H, Phillips MC (1971) Biochim Biophys Acta 249:539–547
12. Knight BC (1930) J Biochem J 24:257–261
13. McConnell HM (1991) Annu Rev Phys Chem 42:171–95
14. Weidemann G, Gehlert U, Vollhardt D (to be published)

Progr Colloid Polym Sci (1994) 97:307–310
© Steinkopff-Verlag 1994

BIO-COLLOIDS

E. Bottari
M. R. Festa

Sodium salts of bile acids in aqueous micellar solutions

Received: 1 October 1993
Accepted: 17 March 1994

E. Bottari (✉)
Dipartimento di Chimica
Università "La Sapienza"
P. le Aldo Moro 5
00185 Roma, Italy

M. R. Festa
Dip. Scienze An. Veg. Ambiente
Universitá del Molise
Via Cavour 50
86100 Campobasso, Italy

Abstract Aqueous micellar solutions of deoxycholate (DC), glycodeoxycholate (GDC), and taurodeoxycholate (TDC) were studied to find the predominant species and their formation constants.

Key words Micellar solutions
– deoxycholate – glycodeoxycholate
– taurodeoxycholate

Introduction

Several studies have been carried out on bile sodium salts and, in particular, on aqueous micellar solutions of deoxycholate (DC), glycodeoxycholate (GDC), and taurodeoxycholate (TDC). Data obtained from different experimental approaches (NMR, circular dichroism, x-ray analysis, SAXS, EXAFS) were explained by assuming a helical structure for these compounds [1–8].

Our study on aqueous micellar solutions of sodium deoxycholate and its conjugated salts with glycine and taurine was performed to find the predominant species in such solutions and their formation constants in a wide concentration range of the reagents. The study was also carried out to deduce the influence of different parameters on the formed species.

This work collects the results obtained on the behavior of deoxycholate (DC), and glycodeoxycholate (GDC) micellar solutions published previously [9–12]. A comparison between them is presented together with the behavior of taurodeoxycholate towards barium ions.

Method of Investigation

All the measurements were performed at 25 °C and in a constant ionic medium. The method of constant ionic medium proposed by Biedermann and Sillén [13] was adopted to minimize the variations of activity coefficients so that it could be possible to substitute activities with concentrations in all calculations.

By assuming Na^+, H^+, and the anions of the bile acids (BS) to be independent reagents, the following equilibrium can be formulated without preliminary hypotheses:

$$q\,Na^+ + p\,H^+ + r\,BS \Leftrightarrow Na_q H_p L_r;$$

$$\beta_{q,p,r} = [Na_q H_p L_r][Na^+]^{-q}[H^+]^{-p}[BS]^{-r},$$

where $q \geq 1$, $p \geq \; \leq 0$ and $r \geq 1$.

In solutions of high complexity, it was necessary to measure more parameters at equilibrium to correctly explain the system. For this purpose, electromotive force (e.m.f.) measurements were carried out at 25 °C by means of the following cells:

$$(-) \text{ R.E./Solution S/G.E. } (+) \qquad (1)$$

$$(-) \text{ R.E/Solution S/Na E. } (+), \qquad (2)$$

where R.E. is a reference electrode, Ag, AgCl/XM N(CH$_3$)$_4$Cl saturated with AgCl and X is the concentration of the ionic medium, G.E. and Na E. are two glass electrodes. They give free hydrogen ion and sodium ion concentrations, respectively. Solution S is prepared in constant ionic media 0.100, 0.200, 0.300, 0.400, 0.500, 0.600 and 0.750 M N(CH$_3$)$_4$Cl. From preliminary data it could be concluded that it was necessary to know also the free concentration of the bile salts (BS). To measure the free concentration of DC, GDC, and TDC a suitable electrode was prepared.

In the case of DC and GDC, the solid lead(II) deoxycholate and lead(II) glycodeoxycholate was prepared. Their sight solubility was determined in the absence and in the presence of sodium ions by polarographic measurements. Furthermore, the e.m.f of the following cell was measured:

$$(-) \text{ Pb(Hg)/Solution S/R.E. } (+), \qquad (3)$$

where Pb(Hg) was an amalgam electrode, alternatively a membrane electrode.

In the case of TDC, because of the high solubility of lead(II)-TDC, barium(II)-TDC was precipitated and analyzed. Its solubility was determined polarographically in the same concentrations of N(CH$_3$)$_4$Cl, used as ionic medium.

Results and discussion

The results obtained can be divided into two parts.

In the first part, solubility and species formed in aqueous solutions in the absence of sodium ions have been determined, and in the second part experimental data have been obtained in the presence of sodium ions and they have been explained by assuming the presence of aggregates of different complexity.

Tables 1 and 2 collect the solubility products of Pb(DC)$_2$ and Pb(GDC)$_2$, respectively, in N(CH$_3$)$_4$Cl as an ionic medium at seven different concentrations and at 25 °C. In the same tables the species formed between lead(II) and DC and GDC are reported, as well.

The main conclusion of the comparison of the data collected in Tables 1 and 2 is that Pb(DC)$_2$ is less soluble than Pb(GDC)$_2$ and that GDC forms with lead(II) species with ratio 1:2 and 1:3, while the latter forms only species with a ratio 1:3.

The constants of association are indicated by γ, because they have been obtained in the presence of solid

Table 1 Values of $\log K_s$ and $\log \gamma_3$ for lead(II) - deoxycholate at 25 °C

Ionic medium N(CH$_3$)$_4$Cl (M)	$-\log K_s$	$\log \gamma_3$
0.100	11.95	8.25
0.200	11.60	8.2
0.300	11.55	8.0
0.400	11.45	7.8
0.500	11.40	7.5
0.600	11.15	7.5
0.750	10.9	7.4

Table 2 Values of $\log K_s$, $\log \gamma_2$ and $\log \gamma_3$ for lead(II) - glycodeoxycholate at 25 °C

Ionic medium N(CH$_3$)$_4$Cl (M)	$-\log K_s$	$\log \gamma_2$	$\log \gamma_3$
0.100	11.50	6.36	8.04
0.200	11.00	6.10	7.65
0.300	10.91	6.08	7.62
0.400	10.65	6.06	7.59
0.500	10.58	6.00	7.50
0.600	10.45	5.86	7.29
0.750	10.35	5.8	7.2

Pb(DC)$_2$ and Pb(GDC)$_2$, respectively and thus it is difficult to have information on the q value. The results obtained on the solubility of Ba(TDC)$_2$ are presented in another paper [14] and also in Table 3.

An inspection of the three tables shows that Ba(TDC)$_2$ is more soluble than Pb(DC)$_2$ and Pb(GDC)$_2$ and species with ratio 1:2 and 1:3 are formed in N(CH$_3$)$_4$Cl at 0.1, 0.2 and 0.3M. At higher values of concentration of the ionic medium only species with ratio 1:2 are present in an appreciable concentration. The second part of the results is relative to the aggregation between sodium, hydrogen ions and deoxycholate and glycodeoxycholate, respectively.

Results on the behavior of taurodeoxycholate are not yet ready for elaboration and comparison with the other bile salts. Data obtained for both DC and GDC in aqueous micellar solutions were explained by assuming species of the type Na$_q$H$_p$(DC)$_r$ and Na$_{q'}$H$_{p'}$(GDC)$_{r'}$. The values of q, p and r and q', p' and r' were different at different concentrations of the ionic medium. The proposed values of q, p, r and q', p' and r' and the constants relative to the aggregation of the species found are collected in [10] and [12]. In the same papers the distribution curves for the proposed species as a function of pH and concentration of ionic medium are reported.

In this paper, we present a different point of view relative to the buffer properties of micellar solutions of

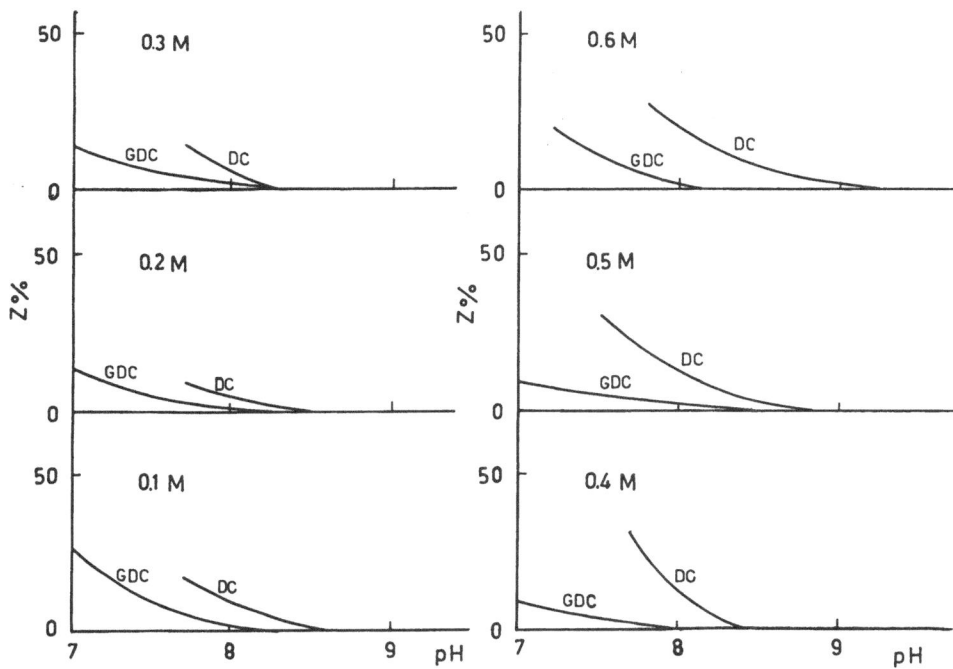

Fig. 1 Percentage of protonated species for sodium deoxycholate (DC) and sodium glycodeoxycholate (GDC)

Table 3 Values of log K_s, log γ_2 and log γ_3 for barium(II) taurodeoxycholate at 25 °C

Ionic medium $N(CH_3)_4Cl$ (M)	$-\log K_s$	$\log \gamma_2$	$\log \gamma_3$
0.100	7.92	4.54	6.91
0.200	7.25	4.26	6.49
0.300	6.85	3.94	5.97
0.400	6.75	4.46	—
0.500	6.42	4.40	—
0.600	6.29	3.44	—
0.750	6.1	3.4	—

NaDC and NaGDC. In the range $7 \leq pH \leq 9$, approximately, the formation of protonated species takes place and solutions of NaDC and NaGDC present buffer properties.

In particular, in the case of sodium deoxycholate protonated species are present in the range $7.6 \leq pH \leq 9$, while at $pH \leq 7.4$ precipitation of HDC begins.

Glycodeoxycholate forms protonated species at lower pH and its solutions are buffered in the range $7 \leq pH \leq 8.5$. The comparison between the behavior of DC and GDC is evident from Fig. 1. The percentage of protonated species (Z%) for both compounds is plotted as a function of pH for the different concentrations of ionic medium. The percentage is higher for DC than that of GDC.

The knowledge of the solubility product of Ba(TDC)$_2$ and the constants of Table 3 allows one to measure the increase of solubility of Ba(TDC)$_2$ in the presence of sodium ions in order to have information on the presence of aggregates in aqueous micellar solutions of NaTDC. From preliminary experiments it seems that TDC behaves differently from DC and GDC towards protonation.

The different behavior of DC, GDC, and TDC can be considered parallel to the different values of pk_a of HDC, HGDC, and HTDC.

The buffer properties of the three compounds follows the pk_a values. The species HTDC is not present in appreciable quantity at $pH \geq 3$, HGDC has $pk_a = 5,0$ and HDC has $pk_a = 5,3$. Solutions of TDC are still under study and results relative to the aggregation are in progress.

Acknowledgement This work was supported by the National Research Council of Italy (CNR), Progetto finalizzato "Chimica Fine II" and by Ministero dell'Università e della Ricerca Scientifica e technologica (MURST) of Italy.

References

1. Conte G, Di Biasi R, Giglio E, Porretta A, Pavel NV (1984) J Phys Chem 88:5720–5724

2. Campanelli AR, Ferro D, Giglio E, Imperatori P, Piacente V (1983) Thermochim Acta 67:223–232

3. D'Alagni M, Forcellese ML, Giglio E (1985) Colloid Polym Sci 263:160–163

4. Esposito G, Zanobi A, Giglio E, Pavel NV, Campbell ID (1987) J Phys Chem 91:83–89

5. Giglio E, Loreti S, Pavel NV (1988) J Phys Chem 92:2858–2862

6. Campanelli AR, Candeloro De Sanctis S, Giglio E, Scaramuzza L (1987) J Lipid Res 28:483–489

7. D'Alagni M, Giglio E, Petriconi S (1987) Colloid Polym Sci 265:517–521

8. Chiessi E, D'Alagni M, Esposito G, Giglio E (1991) J Incl Phen 10:453–469

9. Bottari E, Festa MR, Jasionowska R (1988) Ann Chim 78:261–271

10. Bottari E, Festa MR, Jasionowska R (1989) J Incl Phen 7:443–454

11. Bottari E, Festa MR (1990) Ann Chim 80:217–229

12. Bottari E, Festa MR (1993) Mh Chemie 426:1119–1132

13. Biedermann G. Sillén LG (1953) Ark Kemi 5:425–440

14. Bottari E, Festa MR (1994) Analyst 119:469–472

Progr Colloid Polym Sci (1994) 97:311–315
© Steinkopff-Verlag 1994

BIO-COLLOIDS

G. Caminati
G. Gabrielli
R. Ricceri

Effect of valinomycin on P.E.T. partners in L-B mimetic membranes

Received: 16 September 1993
Accepted: 17 March 1994

Dr. G. Caminati (✉)
G. Gabrielli · R. Ricceri
Dipartimento di chmica
Via G. Capponi, 9
50123 Firenze, Italy

Abstract Spreading phospholipid monolayers at the water-air interface were used as model membrane systems in order to investigate the effect of a large ionophore molecule on the interactions between two photo-induced electron transfer. partners. We therefore studied monolayers containing the covalently bound donor (pyrene) diluted in a matrix phospholipid where valinomycin was also embedded. We studied the interactions of a water soluble acceptor (methylviologen) with the monolayer either in the presence or in the absence of the ionophore. The investigations were carried out measuring surface pressure and surface potential–area isotherm as a function of valinomycin content as well as acceptor concentration in the subphase.

Key words Spreading monolayers – acceptor interactions at interfaces – surface potential – valinomycin

Introduction

The transfer of energy or charge across a membrane or at a membrane interface is one of the processes most commonly encountered in biological systems. In particular, the mechanism and conditions for the production of charge separation and consequent vectorial electron transfer are the subject of active research, not only to understand the biological process, but also to exploit solar energy conversion by mimicking natural photosynthesis [1, 2]. One of the key questions is the accomplishment of a spatial organization of the reaction centers in lipid membranes in order to provide high efficiency of light-to-chemical energy conversion. Monolayers at the water-air interface and L-B films provide a simple way to define and control the molecular organization of the membrane components [3]. Recent papers on lipid vesicles suggested that the presence of a ionophore such as valinomycin may influence the rate of photo-induced electron transfer (PET) across the model membrane [4]. Previous studies on Langmuir–Blodgett films and vesicles of pyrene-labeled dipalmitoylphosphatidilcholine diluted in a dipalmitoylphosphatidic acid matrix showed that the presence of the ionophore may induce phase segregation in the bilayer systems [5]. The aim of the present work is to understand and control this phenomenon at a molecular level by studying the simplest model of biomembranes, that is to say, spreading a monolayer at the water-air interface. In this way, we can investigate the distribution of the molecules in the monolayer and their reciprocal interactions [6]. We therefore studied spreading monolayers at the water-air interface formed by the same constituents of the previously studies Langmuir–Blodgett films and vesicles systems [5]. The distribution and the interactions of the phospholipid molecules in the monolayer were determined measuring the surface pressure and surface potential–area isotherms as a function of the relative concentration of the monolayer components. Secondly, we studied the process of adsorption of a water soluble acceptor (methylviologen) at the monolayer-water interface and how this process was affected by the presence of valinomycin in the monolayer.

Experimental

Materials

L-α-dipalmitoylphosphatidic acid (DPPA and 1-pal-mitoyl-2-(1-pyrenedecanoyl)-sn-glycero-3-phosphocholine (PyDPPC) were supplied by Aldrich. Valinomycin (VAL) was purchased from Sigma. Chloroform (Merck) was used as spreading solvent. Methylviologen dichloride (MV^{2+}) was supplied by Sigma. Water was twice distilled and further purified with a Milli-Q system (Millipore).

Methods

Surface pressure (π)-area isotherms were recorded using a Lauda Filmwaage with continuous compression at a compression rate of 5 mm/min. Surface potential was recorded using radioactive ^{241}Am electrodes as a function of monolayer compression. Langmuir–Blodgett films were prepared using a KSV 5000 instrument; all layers were transferred at 4 mm/min for both the up- and the down-stroke. The layers were transferred keeping the surface pressure constant at 40 mN/m with a compression rate of 4 mm/min.

Results and discussion

PyDPPC/DPPA monolayers on water subphase

Surface pressure-area isotherms were recorded for several monolayer compositions. In particular, we studied mixtures of PyDPPC/DPPA with a molar ratio ranging from 1:99 to 1:999; typical results are shown in Fig. 1. Previous studies for the same mixed monolayer with higher PyDPPC content [7] showed that the monolayer could be considered as a homogeneous mixture at a molar fraction equal or lower than 0.09. In the present study, we found that even at very low PyDPPC content, the isotherms differ from the one obtained for the pure DPPA. The isotherms of the PyDPPC/DPPA system, although much more condensed than for pure PyDPPC (not reported here), show a shape which is definitely different from the pure DPPA, with phases which are considerably more expanded, especially in the low surface pressure regimes. This means that the PyDPPC molecules do not merely insert in the monolayer, but change the packing and the interactions between the chains and between the head-groups. The effect of PyDPPC can be investigated in more detail by computing the reciprocal of the surface compressional moduli defined as:

$$C_S^{-1} = -A\,(\partial\pi/\partial A) \qquad (1)$$

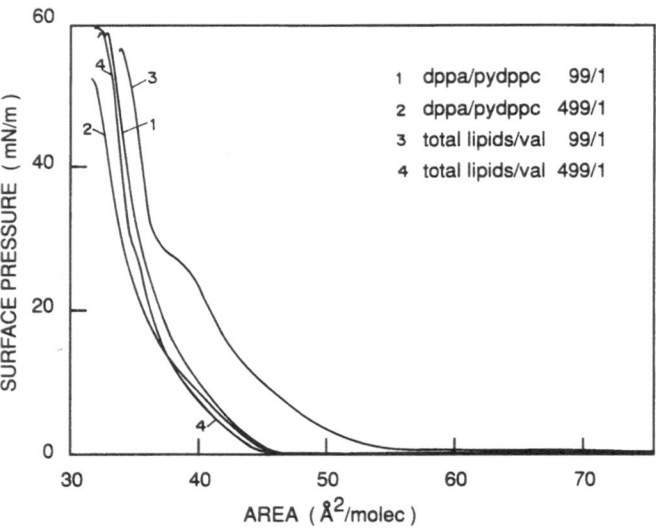

Fig. 1 Surface pressure–area isotherms for mixed monolayers on a water subphase at $T = 20\,°C$

from the experimental π-A isotherms [8]. From the values of C_S^{-1} reported in Table 1, we observe that PyDPPC induces a lowering of surface compressibility, especially in the high surface pressure regimes; this corresponds to less rigid phases even if molecular areas for the mixture are lower than for pure DPPA [7]. Further evidence on the influence of pyrene-labeled phospholipid was found from the measurements of surface potential ΔV-area isotherms. In Table 1, we report the values of ΔV in the condensed phase ($\pi = 40$) for the various mixtures, the values are 100 mV lower than for the pure DPPA, reflecting the decrease in charge density of the polar heads region of the monolayer. All the above results illustrate the increase in flexibility of the monolayer packing due to the breaking of hydrogen bonds between phosphatidic acid groups and to the presence of the bulky pyrene group at one end of the aliphatic chain.

In analogy with previous studies on vesicles [5], we added valinomycin in different concentrations at the monolayer containing PyDPPC/DPPA 1:499; two typical isotherms are reported in Fig. 1, curves (3) and (4). In the presence of valinomycin a discontinuity appears around 28 mN/m, the inflection becomes a plateau when the content of valinomycin is higher than 1:99. This value of surface pressure corresponds to the collapse of valinomycin [9] and suggests immiscibility of the components in the monolayer. However, when the valinomycin molar fraction in the mixed monolayer is as low as 0.002, no inflection can be directly monitored and there is only a negligible effect on the limiting areas A_0, as well as on the surface compressional moduli C_S^{-1} (see Table 1) which increases only slightly in the high surface pressure region.

Progr Colloid Polym Sci (1994) 97:311–315
© Steinkopff-Verlag 1994

Table 1 Surface compressional moduli C_S^{-1} and ΔV for the systems PyDPPC/DPPA and PyDPPC/DPPA/VAL on water.

MONOLAYER COMPOSITION	C_S^{-1} (mN/m) $\pi = 5$ mN/m	$\pi = 40$ mN/m	ΔV_{max}(mV)	MONOLAYER	C_S^{-1} (mN/m) $\pi = 5$ mN/m	$\pi = 40$ mN/m	ΔV_{max}(mV)
PyDPPC	35	50	270	PyDPPC/DPPA	70	370	280
PyDPPC/DPPA 1::99	80	440	250	LIPID/VAL 999:1	80	475	270
PyDPPC/DPPA 1::499	70	370	280	LIPID/VAL 499:1	80	515	230
PyDPPC/DPPA 1:999	70	230	280	LIPID/VAL 99:1	65	480	145
DPPA	140	900	365				

Mixed monolayers on methylviologen subphase

Methylviologen solution was used as monolayer subphase to study the interactions of this well-known acceptor [10] and the mixed monolayer containing the donor. Different concentrations of methylviologen were used in the range 10^{-7} M to 10^{-4} M and some typical isotherms are reported in Fig. 2. The molar ratio between the labeled and unlabeled phospholipid was kept constant at 1:499 and experiments were performed either with or without valinomycin in the monolayer.

In the case of mixed monolayers without VAL, the presence of methylviologen induces a variation in the position and in the shape of π-A isotherms. Firstly, all the isotherms with MV^{2+} show a discontinuity marking the passage to a more condensed phase, the position of this discontinuity shifts to higher surface pressures with increasing concentration of methylviologen. Secondly, the isotherms shift to larger areas with increasing content of MV^{2+} in the subphase. Furthermore, the value of C_S^{-1} shows that MV^{2+} interactions with the monolayer induce a different response in the molecular packing depending on the molecular density. C_S^{-1} values at 5 mN/m were found to be of the order of 80 mN/m in the whole MV^{2+} concentration range and these figures are typical of liquid expanded phases. On the contrary, the compressibility of the monolayer in the condensed region (500 mN/m) is higher than the corresponding value on the water subphase (370 mN/m) suggesting a more rigid distribution of the molecules in the monolayer. The change in molecular areas at constant surface pressure is also larger in the expanded than in the condensed phase and increases with increasing concentration of methylviologen. In Table 2, we summarize the change in the limiting areas, A_0, upon addition of MV^{2+} together with C_S^{-1} values in the condensed phase. The increase in the limiting area with methylviologen concentration is small and reaches at maximum 6 Å^2/lipid molec.; considering that the methylviologen

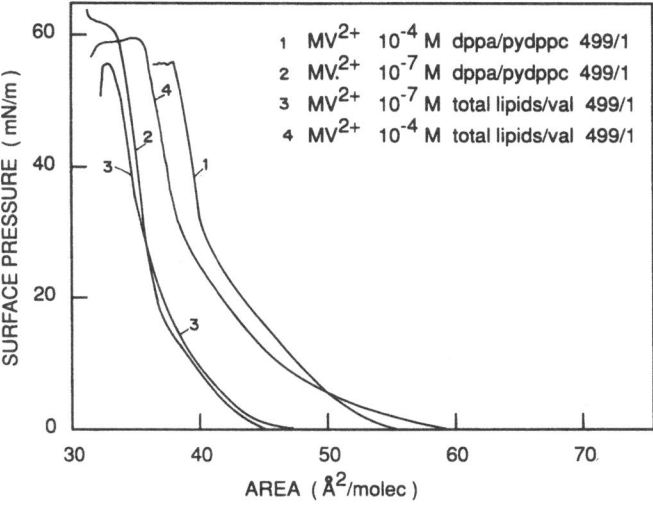

Fig. 2 Surface pressure–area isotherms for mixed monolayers on a methylviologen solution subphase at $T = 20$ °C

molecule occupies an area of 13.3×3.5 Å^2 [11] we can exclude MV^{2+} penetration in monolayer and consider MV^{2+} as adsorbed underneath the polar head groups. MV^{2+} electrostatic interactions with the partly ionized DPPA polar head groups may also explain the increase in the rigidity of the monolayers shown by the increase in C_S^{-1} values (see Table 2) if we consider that one MV^{2+} molecule may interact with two neighboring DPPA groups, forcing them to keep an optimal distance to maximize interaction with MV^{2+} positive charges. This would, in turn, result in a lower compressibility of the monolayer compared to the water subphase.

Information on the adsorption of MV^{2+} at the monolayer-water interface can also be extracted from the data of surface potential as a function of methylviologen concentration. Again, two distinct behaviors may be observed for the PyDPPC/DPPA mixture depending on the phase of

Table 2 Changes of limiting areas ΔA and surfaces compressional moduli C_S^{-1} at $\pi = 40$ mN/m

PyDPPC/DPPA 1:499			PyDPPC/DPPA/VAL 1:498:1		
[MV^{2+}]	ΔA	C_S^{-1}	[MV^{2+}]	ΔA	C_S^{-1}
0	–	370	0	–	515
0.1	1.5	600	0.1	2	450
1	4.5	495	1	5	480
10	3.5	400	10	6	470
100	5.6	525	100	3.5	400

the monolayer at molecular areas lower than 50 Å^2/molec. and hence low surface charge density; ΔV increases with the concentration of methylviologen in the subphase. This same trend was also previously observed for pure DPPA monolayers in the whole surface pressure range [12]. We find a reversal of this tendency at higher surface charge densities in the mixed system: compressing the monolayer in the condensed state we observe a decrease of surface potential with increasing MV^{2+} concentration.

The same experiments were repeated with VAL embedded in the monolayer at concentrations low enough to avoid phase segregation in the monolayer. π and ΔV isotherms were recorded on subphases containing MV^{2+} at various concentrations and the results are reported in Fig. 2, curve (3) and (4). The isotherms are shifted towards larger areas with increasing acceptor concentration but, in this case, the shift in the molecular areas at constant surface pressure, induced by the presence of methylviologen ions in the subphase, is slightly lower. Methylviologen insertion in the monolayer can thus be excluded also in the PyDPPC/DPPA/VAL system. The migration of MV^{2+} at the monolayer-water interface is confirmed by the change

in surface potential with MV^{2+} concentration (see Fig. 3) although the values obtained for C_S^{-1}, both at high or low surface pressure, are the same as in the system on water subphase (80 mN/m at $\pi = 5$ mN/m and ~ 500 mN/m at $\pi = 40$ mN/m) as can be seen from Table 2. When VAL is present in the monolayer, adsorption of methylviologen at the monolayer interface does not alter significantly the monolayer compressibility. This is in agreement with previous NMR studies [13] on phospholipid bilayers which have shown that VAL reduces the mobility and flexibility of the alkyl chains of the phospholipids and therefore the interactions of the polar head groups with MV do not alter the packing and the mechanical properties of the monolayer.

To rationalize more clearly the difference in methylviologen adsorption with and without valinomycin, we report in Fig. 3 the behavior of the difference of ΔV in the condensed phase and before starting the compression. For the PyDPPC/DDDPPA monolayer, $\Delta(\Delta V)$ decreases with concentration and level off at 10 µM indicating adsorption saturation at this concentration. The behavior in the case of PyDPPC/DPPA/VAL monolayers is similar but saturation is reached at higher MV^{2+} concentrations. This may be explained by the reduced charge density in the latter system: valinomycin molecule occupies a large surface area, i.e., 280 Å^2 [9], corresponding to at least 7 DPPA molecules and, therefore, even small amounts of VAL may affect the number of charges per unit area.

Langmuir–Blodgett films

Monolayers of the mixture DPPA/PyDPPC were transferred on a quartz substrate previously covered with four

Fig. 3 Difference of surface potential in the condensed phase and before starting compression as a function of methylviologen concentration

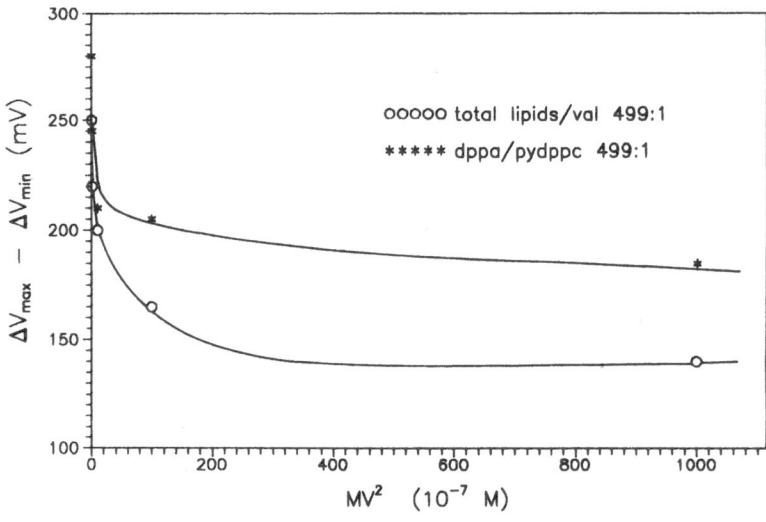

Progr Colloid Polym Sci (1994) 97:311–315
© Steinkopff-Verlag 1994

layers of DPPA to prevent disturbing interactions with the quartz surface. The transfer of the mixed monolayers was performed in the liquid condensed region ($\pi = 40$ mN/m). The system was transferred homogeneously with an optimal transfer ratio (transfer ratio = 1.01–1.013) in the whole covered surface. On the contrary, Langmuir–Blodgett film deposition was not as satisfactory when the multilayers were prepared with the mixture PyDPPC/DPPA/VAL. In fact, good transfer ratios were obtained only for a low VAL mole fraction and only in the first layer (1.05), whereas for the following layers the transfer ratio was always lower than 0.82, this was probably due to the disruptive effect of valinomycin in the monolayer already discussed.

Conclusions

From the above results, we may possibly draw the following conclusions:

1) analysis of surface pressure and surface potential isotherms are a powerful tool to investigate the interactions between different components in microheterogeneous systems and to establish their relative distribution at molecular level, a task which is often neglected in studies in confined systems such as vesicles, bilayers, etc.

2) Even small amounts of pyrene-labeled phosphatidylcholine may induce a change in the properties of the monolayer, and consequently of the bilayer, even if the distribution of the probe molecules may be considered statistically homogeneous. The perturbing effect of PyDPPC is due not only to the presence of the aromatic pyrene but also to the bulky polar head group which prevents hydrogen bonding between DPPA polar groups. Furthermore, we observed that the addition of valinomycin induced phase segregation in the mixed phospholipid monolayer, unless extremely low concentrations of valinomycin are used.

3) Methylviologen, which is very often used in PET reactions in microheterogeneous systems, was found to adsorb at monolayer-water interface due to electrostatic interactions with the fraction of ionized phosphate groups of DPPA, the presence of valinomycin in the monolayer shifts adsorption saturation to higher MV concentration.

The above findings may be useful for all the problems connected to reactions in confined bidimensional systems (in particular for PET) where the knowledge of the distribution of the reactions partner and the properties of their environment are of paramount importance to understand and exploit the process.

Acknowledgements Thanks are due to the Italian CNR (Consiglio Nazionale delle Ricerche) and to MURST (Ministero dell' Università e della Ricerca Scientifica e Tecnologica) for financial support.

References

1. Fendler J (1985) J Phys Chem 89:2730–2740
2. Robinson NJ, Cole-Hamilton DJ (1991) Chem Soc Rev 20:49–44; Lymar SV, Parmen VN, Zamareev KI (1991) Top in Current Chem (1991) 159:3–165
3. Möbius D (1981) Acc Chem Res 14:63; Caminati G, Ahuja R, Möbius D (1992) Thin Solid Film 210/211:335–337
4. Rong S, Brown RK, Tollin G (1989) 49:107–119
5. Ricceri R, Thesis (1993) Florence
6. Gabrielli G (1991) Advances in Coll Interface Sci 34:31–72
7. Caminati G, Ahuja R, Möbius D (1994) Thin Solid Film 243:651–55
8. Davies JT, Rideal EK (1963) Interfacial Phenomena, Academic Press, New York
9. Gabrielli G, Puggelli M, Prelazzi G (1991) 84:232
10. Kuhn AT, Bird CL (1981) Chem Soc Rev 10:49–82; Caminati G, Tomalia D, Turro NJ (1991) Progr Coll Interface Sci 84:219–22
11. Cotton TM, Kim J-K, Uphaus RA, (1990) Microchemical J 42:44–71
12. Caminati G, Ahuja R, Möbius D (1992) Progr Coll Interface Sci 89:218–22
13. Finer EG, Hauser H, Chapman D (1969) Chem Phys Lipids 3:386–392

Progr Colloid Polym Sci (1994) 97:316–320
© Steinkopff-Verlag 1994

BIO-COLLOIDS

R.K. Heenan
S.J. White
T. Cosgrove
A. Zarbaksh
A.M. Howe
T.D. Blake

SANS studies of the interaction of SDS micelles with gelatin, and the effect of added salt

Recieved: 15 October 1993
Accepted: 28 January 1994

Dr. R.K. Heenan (✉)
Building R3
Rutherford Appleton Laboratory
Chilton
Didcot, OX11 OQY, United Kingdom

S.J. White · T. Cosgrove · A. Zarbakhsh
University of Bristol
Chemistry Department
Bristol , United Kingdom

A.M. Howe · T.D. Blake
Kodak Limited
Research Division
Harrow , United Kingdom

Abstract The structural interaction between the anionic surfactant sodium dodecyl sulphate (SDS) and the bipolymer gelatin has been studied using small-angle neutron scattering. In particular, neutron scattering length contrast variation has been used to highlight the structures of the components of the mixture both separately and together. Some aspects of the structure of gelatin gels are discussed in the light of other publications. The effect of adding excess salt has been examined. Micelles of SDS are strongly coupled to gelatin; intermicellar interactions dominate the structure at low ionic strength whilst at high ionic strength the polymer network dictates the intermicellar separations.

Key words Small-angle neutron scattering – surfactant-polymer interactions – gelatin – SDS sodium dodecyl sulphate

Introduction

The formation of complexes between the anionic surfactant sodium dodecyl sulphate (SDS) and the biopolymer gelatin has been studied extensively by surface tension [1, 2], and viscosity measurements [3, 4]. Adsorption of SDS micelles onto gelatin has been envisaged in these studies as occurring in a stepwise fashion. Native collagen consists of triple helices covalently bonded into rod-like structures. The covalent bonds are broken on processing, in this case with alkali, to form the polydisperse fragments of protein known as gelatin. Aqueous solutions of greater than about 2% gelatin are a gel at room temperature, which become fluid above 37 °C, when spectroscopic evidence points to a rod-to-coil transition [5]. At neutral pH the gelatin chain is amphoteric, its positively charged residues, lysine and arginine (≈ 7.5% of residues) [6] providing the possibility of binding with SDS whilst others (e.g., glutamic and aspartic acid ≈ 12%) will be negative. Further residues are strongly hydrophobic (leucine, isoleucine, methionine and valine, ≈ 6%), with essentially neutral, hydrocarbon side chains that have been suggested by NMR [7] to be involved in hydrophobic bonding to SDS. The bulk of the chain, consisting of glycine, proline and hydroxyproline (≈ 58%), is weakly hydrophobic and has the regular periodicity which promotes triple helix formation.

Pezron et al. [8] have studied the sol state of gelatin in the presence of 0.1 M NaCl by use of light, small-angle neutron, and x-ray scattering. Their results indicate that in the dilute regime the chains appear isolated, with a persistence length or stiffness of the chain of 20 Å and a radius of gyration R_G of ≈ 350 Å. In the semi-dilute regime the small-angle scattering could be described by a model in which two different length scales were identified.

Surface tension measurements on gelatin-SDS solutions [1, 2] show two transition points which depend

Progr Colloid Polym Sci (1994) 97:316–320
© Steinkopff-Verlag 1994

on the gelatin concentration, one below and one above the normal critical micelle concentration CMC (7 mM = 0.2%) of pure SDS. The first has been attributed to induced micellisation brought about by a condensation of the surfactant onto the gelatin backbone. The second transition corresponds to a bulk micellisation. The accompanying large increase in macroscopic viscosity need not necessarily be directly related to structural changes. Indeed relaxation times determined from rheological data [4] do not vary over a wide range of viscosities and concentrations. Similarly, in other polymer/surfactant systems, binding isotherms may not provide direct evidence of structural changes as recent results have shown that virtually identical binding isotherms can give very different viscosities [9].

Small-angle neutron scattering (SANS) is a technique which provides direct structural information, with the possibility of using "contrast variation" by selective deuteration to see separately the structures of individual components. Data for a related series of samples containing 2% w/v SDS and 5% w/v gelatin with and without an excess of salt, are discussed below.

Experimental

Solutions were prepared using D_2O (Fluorochem), distilled H_2O, deuterated SDS (d-SDS 98% pure) (MSD), protonated SDS (h-SDS 99% pure) (BDH). The purity of the SDS samples was checked by surface tension measurements. The gelatin used was deionised photographic gelatin which had been alkali processed (Type IV, supplied by Kodak Limited) and has a nominal molecular weight of 107 000 and a density of 1.4 g/ml. By use of d-SDS and

water solvent contrast variation with D_2O/H_2O mixtures it is possible to observe scattering from either gelatin or SDS alone, or both together. Scattering length densities assumed for each component were (in units of 10^{10} cm^{-2}) H_2O − 0.56, D_2O 6.39, Gelatin 2.11, d-SDS 6.73, h-SDS 0.39. It was expected, and experimentally verified, that d-SDS micelles would be contrast matched, i.e. invisible to SANS, in pure D_2O whilst gelatin was matched at around 40% D_2O : 60% H_2O. The gelatin and SDS were dissolved in water at 50 °C for 1 h and then cooled to 25 °C. (With the exception of the 0.5 M salt data in Fig. 1 which were at 45 °C, though the scattering at 25 °C in other series of runs showed minimal changes, as reported by Pezron et al. [10] for aqueous gelatin solution).

SANS experiments were carried out at the Rutherford Appleton Laboratory, Didcot, Oxfordshire, using the LOQ instrument. Time of flight methods with neutron wavelengths of $\lambda = 2.2 - 10$ Å, and a sample-detector distance of 4.4 m gave a Q range of 0.007 to 0.22 Å^{-1}. Data are combined after allowing for the wavelength dependence of the incident spectrum, the sample transmission and detector efficiencies, to give the absolute scattering cross-section $\partial \Sigma(Q)/\partial \Omega$. The magnitude of the scattering vector is $|Q| = 4\pi \sin\theta/\lambda$, where 2θ is the scattering angle.

Small-angle scattering data

SDS and gelatin in the absence of salt

The scattering from SDS micelles alone in Fig. 1 (●●) is typical of the scatter from interacting charged particles, where the form factor P(Q) for scatter from the individual particles is formed into a peak at low Q by a structure

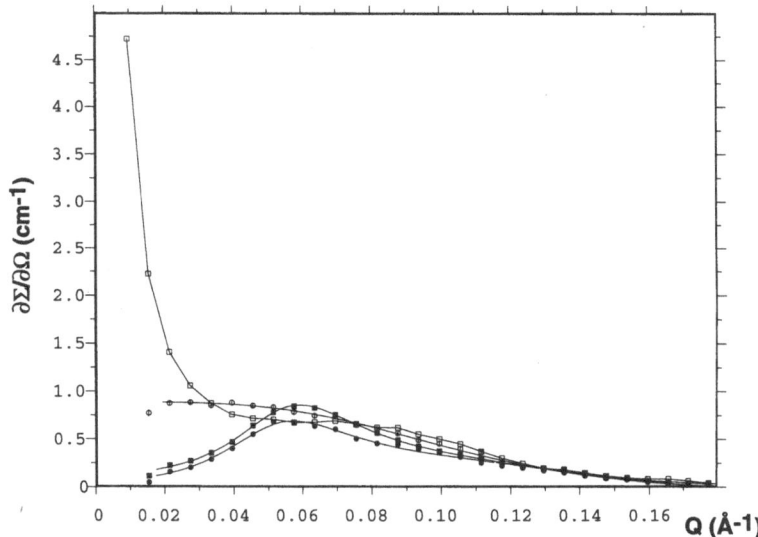

Fig. 1 SANS from aqueous sodium dodecyl sulphate (SDS) micelles with and without gelatin or salt. Scattering only visible from SDS micelles, water contrast matched to gelatin (40% D_2O): 2% d-SDS ●● (with fit), 2% d-SDS + 5% gelatin ■ ■ (with fit), 2% d-SDS + 0.5 M NaCl ○ ○(with fit), 2% d-SDS + 5% gelatin + 0.5 M NaCl □ □. Lines are guides to the eye, except where noted as a fit. Note that the SDS scatter here should be 1.69 times less than for the h-SDS in D_2O of Figs. 2 and 3

factor for interparticle correlations $S(Q)$. Here, the SDS micelle has a slightly elliptical shape and the data may be fitted by the methods of Hayter and Penfold [11, 12, 13] to give an aggregation number of around 65 and axial ratio of 1.2, and with around 21 of the SDS molecules not having an associated Na^+ counterion, to give a net negative charge. The model assumes an elliptical core plus shell structure in which the core contains say 90% of the surfactant hydrocarbon tails. The axial ratio is adjusted so as to keep the minimum core radius less than the fully stretched length of the hydrocarbon chain (ca 16.7 Å). This avoids a "hole" at the centre of the micelle, apparent in earlier spherical models, and allows a smooth transition to rod-like structures seen with higher concentrations of SDS or salt. The 4-5 Å thick "shell", which makes a significant contribution to the scattering, contains the remainder of the surfactant tails, the surfactant head groups and their associated hydrated counter ions (which are constrained to vary according to the charge on the micelle). Further parameters adjusted in each fit are an inverse Debye screening length, and an effective volume fraction for the charged sphere $S(Q)$. Best fits must reproduce the absolute neutron scattering intensity, and the known volume of SDS. Some parameter correlation is present, especially between the inverse Debye length (which it is not easy to calculate in the presence of gelatin) and the micellar charge. There is also some dependence on the assumed parameters such as the molar volumes, bulk densities and degrees of hydration of the several components.

On adding 5% gelatin the scattering from the SDS micelles (Fig. 1, ■ ■) changes little and can be fitted with parameters very similar to before. A small extra signal seems to appear at high Q, which if included in the fitting might suggest a change in micelle size, but the absolute

scaling is then unrealistic. It is possible that on short distance scales the scattering length density at the surfactant head group/gelatin/water interface is not as expected. It is quite likely that the neutron scattering length density of the gelatin amino acids residues which interact with the SDS will not be the same as the average for the whole gelatin molecule.

SANS signals from gelatin alone have a quite different shape (Fig. 2, △ △). These have been considered in detail by Pezron et al. [8], where here we are in the semi-dilute region, above the overlap concentration c*. In their model, they treat the scattering as a sum of two terms, one originating from the screening length or average mesh size [ξ] and one from larger scale structural inhomogeneities, the origins of which are not clear. For the first term, De Gennes [14] gives, for $Q < \xi^{-1}$

$$I_\xi(Q) = I_\xi(0)/[1 + Q^2\xi^2] . \qquad (1)$$

For the second regime the model of Debye and Bueche was proposed to characterise the extent of the spatial inhomogeneities over a length scale ζ.

$$I_\zeta(Q) = I_\zeta(0)/[1 + Q^2\zeta^2]^2 . \qquad (2)$$

Here we find, typically, a screening length ξ of around 35–40 Å, and a correlation length ζ of around 70 Å, consistent with ref. [8], though their ζ of 135 Å is probably a better estimate due to their reaching lower Q values. Perzon et al. worked at a temperature of 50 °C, though our own results and their [10] agree in showing that over the Q range studied here temperature has a minimal effect, despite gelation occurring on cooling to 25 °C. The length of ageing at 25 °C does however determine the amount of

Fig. 2 Scattering only visible from gelating (solvent D_2O). 5% gelatin △ △, 2% d-SDS + 5% gelatin ■ ■, 5% gelatin + 0.1 M NaCl ▽ ▽, 2% d-SDS + 5% gelatin + 0.1 M NaCl □ □. (Other details as in Fig. 1)

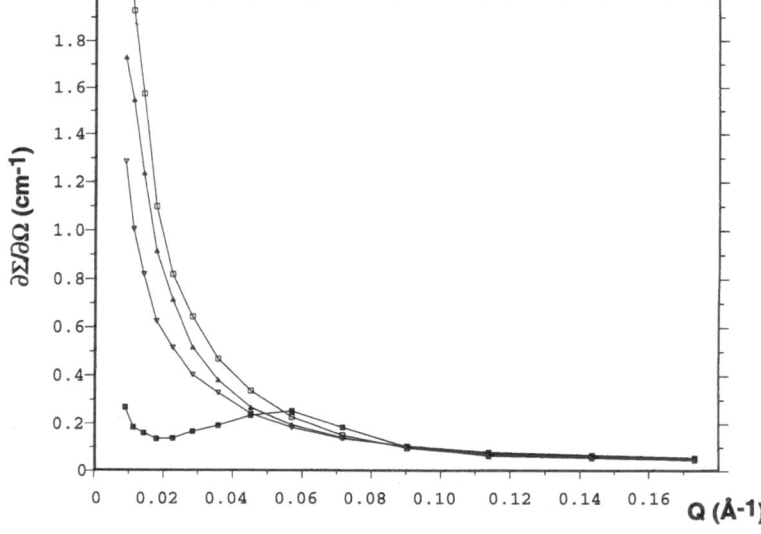

triple helix gelatin seen by optical rotation spectroscopy [5].

When SDS and gelatin are mixed however the SANS profile from gelatin itself (Fig. 2, ■ ■) adopts a shape similar to that imposed by the SDS micelle structure factor. The gel network is locally rearranged to reflect the spacing of SDS micelles.

The cross correlation between the separate gelatin and SDS structure may be deduced from Fig. 3, where both gelatin and SDS are visible (Fig. 3, ■ ■). Note however that the relative scattering signals of the two components, proportional to the square of the scattering length density of each with the D_2O solvent, are not the same. Here again the SDS intermicellar interaction dominates the structure.

The effect of excess salt

As is well known, adding salt to an SDS micellar solution reduces the Debye screening length, suppressing the strong charge-charge interaction between the micelles. The scattering is then much closer to the pure form factor $P(Q)$ for an isolated micelle as shown by Fig. 1 (○ ○). The micelles grow in size and become more elliptical, the fit in Fig. 1 being for an aggregation number of 95, charge 18, and axial ratio of 1.5.

The addition of salt to gelatin alone is illustrated in Fig. 2 (▽ ▽), where the scattering loses intensity and the length scale ξ of the short range correlation decreases. With gelatin, SDS and salt, the gelatin term (Fig. 2, □ □) gains some intensity, and length scale ξ increases.

The most dramatic change of SANS signal on adding salt to SDS/gelatin occurs when both SDS and gelatin are visible, when there is a large increase in scatter at low Q (Fig. 3, □ □). The scatter from SDS micelles alone, in excess salt, is included for comparison in Fig. 3 (○ ○), to show that above about 0.07 $Å^{-1}$, as seen already in Fig. 1, the scattering of the individual components is not greatly altered on short distance scales.

Discussion

Assuming gelatin to have a diameter of 7 Å (ref. [8] gives a cross-sectional radius at high Q of 3.2 ± 1.0 Å), the mean length of a single molecule is around 3300 Å. If these were to uniformly fill space via a diamond-like lattice with tetrahedral nodes, then at 5% w/v the lattice size would be 120 Å and the distance between nodes about 100 Å. In fact, optical spectroscopy suggests [5] that a fraction of the gelatin (say 30%, depending on ageing time) is in its triple helix form of diameter 14 Å [5], so the real mesh size might be slightly larger. The observed correlation lengths of order 35 Å (perhaps controlled by the lower molecular weight fractions, by entanglements or by the spacing of binding sites along the chains) implies that there must be less densely packed regions on longer distance scales. It is not surprising therefore to find evidence for a second, longer, distance correlation in the structure, which makes gelatin rather different from other polymers. Such a description is similar to images obtained by freeze fracture, heavy metal shadowed, electron micrographs of gelatin solutions [5].

The nearest neighbour, centre-centre spacing of SDS micelles at 2% w/v, assuming a face centred cubic lattice, is of order 140 Å, predicting the peak in the scattering, as

Fig. 3 Scattering only visible from both SDS and gelatin (solvent D_2O). 2% h-SDS + 5% gelatin ■ ■, 2% h-SDS + 0.1 M NaCl ○ ○, 2% h-SDS + 5% gelatin + 0.1 M NaCl □ □. (other details as in Fig. 1)

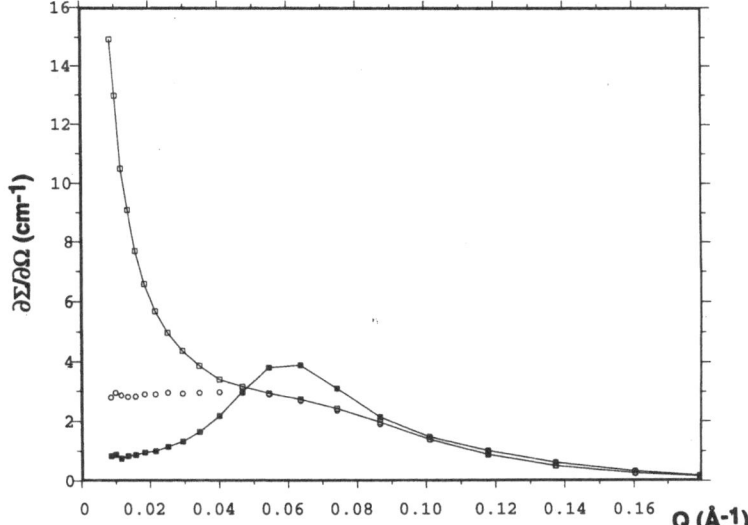

320
R.K. Heenan et al.
Interaction of SDS micelles with gelatin

observed, around 0.06 Å^{-1}. At the concentrations illustrated here there are an average of around two SDS micelles for each 3300 Å gelatin strand. Neglecting the distribution of any specific binding sites, it would take at least 150 Å of gelatin to wrap once around an SDS micelle, or at least 1200 Å to completely cover the surface of a micelle. The presence of SDS micelles, even if closely attached to the gelatin might not be expected to greatly perturb the network structure, their preferred spacings and mesh sizes being apparently compatible.

In practise, at these concentrations a *strong coupling* is seen as shown by the SANS signal from gelatin itself (Fig. 2, ■ ■), where in the absence of salt the intermicellar interactions greatly distort the gelation network. This strong coupling is further illustrated by the SDS plus gelatin signal in Fig. 3 (■ ■) which is very similar to that of SDS micelles alone. In fact, it can be reasonably well fitted by taking structural parameters from the fit to Fig. 1 (■ ■) and setting the scattering length of the roughly 5 Å thick micellar "shell" to be that of gelatin. With excess salt the coupling remains, as seen in Figs. 2 and 3 (□ □) when extra scattering occurs at low Q, indicating that the SDS micelles are now clustered, on long distance scales, in an arrangement apparently similar to the gelatin network, rather than being uniformly distributed throughout the solution.

There are not yet clear answers to questions such as how much gelatin is in contact with each SDS micelle and where the SDS micelles are located relative to the different parts of the gelatin network. The shorter correlation length or mesh size ξ is of course close to the diameter of an SDS micelle, so it is perhaps not surprising in Fig. 2 that the gelatin mirrors the SDS intermicellar structure at low salt, though at high salt the changes to the gelatin mesh are more stable. Further data are to be obtained at low gelatin

concentrations, below c^*, in order to determine whether the strongest scattering in Fig. 3 (□ □) is from micelle-micelle correlations or from micelle-gelatin cross terms, though comparison with Fig. 1 (□ □) suggests the former is more likely.

Conclusions

The results are very similar to those reported for SDS interacting with polyethylene oxide, PEO, [15] where three distinct regions of behaviour were identified dependent upon the relative concentrations of surfactant, polymer and salt. In this study, with around two micelles per gelatin molecule, the SDS intermicellar charge-charge interactions determine the gelatin network structure at low salt concentrations. In contrast, at high salt concentrations the SDS micelles adopt long range correlations similar to the gelatin mesh. In both cases there is a strong coupling of the SDS to the gelatin, rather than two co-existing structures. The detail structure generating the two different distance scales seen for the gelatin network, and its relationship to the different types of amino acids remains to be elucidated. SANS studies of its interactions with micelles such as SDS, in conjunction with other techniques, will help to provide further information on this important, but complex, bi-polymer.

Acknowledgements We thank T. Whitesides of Eastman Kodak Company for useful discussions; M. Djabourov for providing us with a pre-print of ref. [8]; D. Miller for a preprint of ref. [7] and J.B. Hayter and J. Penfold for copies of their unpublished data and models for analysis of SDS micelles. SW acknowledges funding from SERC and Kodak Limited. SERC are thanked for allocations of neutron beam time at ISIS.

Reference

1. Knox WJ, Parshall TO (1990) J Coll Int Sci 33:16
2. Dickinson E, Woskett CM (1989) In: Bee RD, Richmond P, and Mingins J (eds) Food Colloids. RSC, Cambridge, p. 74.
3. Greener J, Constable BA, Bale MD (1987) Macromolecules 20:2490
4. Howe AM, Wilkins AG, Goodwin JW (1992) J Photo Soc, 40:234
5. Djabourov M, Leblond J, Papon P (1988) J Phys France 49:319–332
6. Rose PI (1977) in: T.H. James (ed) The

theory of the photographic process, 4th ed. Eastman Kodak, Rochester.
7. Miller DD, Lenhart W, Antalek B, Williams A, Hewitt M (1994) Langmuir 10:68–71
8. Pezron I, Djabourov M, Leblond J (1991) Polymer 32:3201–3210
9. Fruhner H, Kretzschmar G (1992) Colloid Polym Sci 270:177
10. Pezron I, Herning T, Djabourov M, Leblond J (1990) In: Ross-Murphy SB & Burchard W (eds) Physical Networks.

Elsevier Applied Science, ch 18
11. Hayter JB, Penfold J (1983) Colloid Polym Sci 261:1022–1030
12. Kotharchyk M, Chen SH (1983) J Chem Phys 79:2461–2469
13. Hayter JB, Penfold J, Unpublished paper
14. de Gennes PG (1985) Scaling concepts in polymer physics 2nd ed Cornell Univeristy Press, Ithaca, NY
15. Cabane B, Duplessix R (1987) J de Physique 48:651–662

Progr Colloid Polym Sci (1994) 97:321-322
© Steinkopff-Verlag 1994

AUTHOR INDEX

Progr Colloid Polym Sci (1994) 97:323–324
© Steinkopff-Verlag 1994

SUBJECT INDEX